ARTIFACTS, REPRESENTATONS AND SOCIAL PRACTICE

BOSTON STUDIES IN THE PHILOSOPHY OF SCIENCE

VOLUME 154

MARX W. WARTOFSKY

ARTIFACTS, REPRESENTATIONS AND SOCIAL PRACTICE

Essays for Marx Wartofsky

Edited by

CAROL C. GOULD
Stevens Institute of Technology

and

ROBERT S. COHEN
Boston University

SPRINGER-SCIENCE+BUSINESS MEDIA, B.V.

Library of Congress Cataloging in Publication Data

Artifacts, representations, and social practice : essays for Marx Wartofsky / edited by Carol C. Gould and Robert S. Cohen.
p. cm. -- (Boston studies in the philosophy of science ; v. 154)
Includes bibliographical references and index.
ISBN 978-94-010-4390-8 ISBN 978-94-011-0902-4 (eBook)
DOI 10.1007/978-94-011-0902-4
1. Philosophy. 2. Aesthetics. 3. Science--Philosophy. 4. Social sciences--Philosophy. I. Wartofsky, Marx W. II. Gould, Carol C. III. Cohen, R. S. (Robert Sonné) IV. Series.
B73.A69 1994
100--dc20 93-38216

ISBN 978-94-010-4390-8

Printed on acid-free paper

TABLE OF CONTENTS

CAROL C. GOULD

PREFACE

The essays collected here in honor of Marx Wartofsky's sixty-fifth birthday are a celebration of his rich contribution to philosophy over the past four decades and a testimony to the wide influence he has had on thinkers with quite various approaches of their own. His diverse philosophical interests and main themes have ranged from constructivism and realism in the philosophy of science to practices of representation and the creation of artifacts in aesthetics; and from the development of human cognition and the historicity of modes of knowing to the construction of norms in the context of concrete social critique. Or again, in the history of philosophy, his work spans historical approaches to Hegel, Feuerbach, and Marx, as well as contemporary implications of their work; and in applied philosophy, problems of education, medicine, and new technologies. Marx's philosophical theorizing moves from the highest levels of abstraction to the most concrete concern with the everyday and with contemporary social and political reality. And perhaps most notably, it is acutely sensitive to the importance of historical development and social practice.

As a student of John Herman Randall, Jr. and Ernest Nagel at Columbia, Marx developed an exemplary background in both the history of philosophy and systematic philosophy and subsequently combined this with a wide acquaintance with analytic philosophy. He is at once aware of the requirements of system and of the need for rigorous and careful detailed argument. Interestingly, too, as a practicing artist and violinist from early on, he became attuned to the role of the aesthetic in experience and in science as well. And as a politically engaged person, he has always been keenly aware of the situatedness of philosophical thought and of the impact of both theories and actions on those who are exploited or oppressed.

What has perhaps not been sufficiently noted about Marx's thought is the degree to which he anticipated so many of the recent trends in philosophy years before it became fashionable to do so. For example, his early innovative concern with historical epistemology and with historical approaches to the philosophy of science has been widely echoed

C.C. Gould and R.S. Cohen (eds.), Artifacts, Representations and Social Practice, ix–x.

recently and his analysis of representation and artifacts in aesthetics has also been reiterated in present discussions. It is our expectation that this book will elicit further attention to Marx's creative and important contributions to a wide range of key philosophical problems and will stimulate further thought in the directions he has emphasized.

The essays collected here touch on a number of the themes that have been central in Marx's theorizing, though always from the unique standpoint of each of the distinguished contributors. The authors of these essays are friends or colleagues of Marx from the various "eras" of his work – from Boston University and from Baruch College, from colleagues in the U.S. philosophical community to friends from Western and Eastern Europe, and include some of his older and more recent students as well. Inevitably, many philosophical friends, colleagues, and students of Marx are missing here, either because of the editors' oversight or because the authors were asked to produce their contributions in some haste. We are sorry for these omissions but look forward to editing another *Festschrift* on Marx's seventy-fifth birthday, by which time he will undoubtedly have produced a range of new ideas and themes on which we may comment and which will inspire us to new sorts of thinking of our own.

ROBERT S. COHEN

MARXOLOGY

Observation is *praxis*-laden . . .

First the facts. Marx Wartofsky, the philosopher and violinist, was first a musician and an artist. He graduated from that wonderful High School of Music and Art in New York City in January 1945. After a semester at Brooklyn College, he went to Columbia University, all the way to his Ph.D. in philosophy in 1952. He came to Boston University in 1957, and after 26 years returned to New York as Distinguished Professor of Philosophy in Baruch College and the Graduate Center of the City University of New York, where he flourishes today. I met him in the fall of 1958 when we shared a graduate seminar on Hume. I thought then, and now too, that he is the ideal colleague, teacher, friend, comedian, and critic. For six years he was the chairman of the philosophy department at Boston University, a time of turmoil without and within the University, and yet we experienced a Renaissance of philosophical quality due to him as first among equals. He seems to be a natural mediator while also a firm leader, qualities which were so very valuable in his years as Secretary-Treasurer of the American Philosophical Association, Eastern Division, as a local, state, and national official of the American Association of University Professors, and as one of the main figures in the development of the Boston Colloquium for the Philosophy of Science in 1960. He was the creative and innovative figure even as a department chairman when he decided to cut through university bureaucratic budget-making by inventing a new unit of exchange value for full-time, part-time, tenure-track, non-tenure-track, teaching fellows, visiting professors, adjunct faculty . . . he called it 'philosophon'. No dean would recognize it. This was his only failure known to me.

What has he published thus far? Three books: *Conceptual Foundations of Scientific Thought* (1968, with translations published in Madrid, 1973, Budapest, 1977, and Beijing, 1984); *Feuerbach* (1977); *Models: Representation and the Scientific Understanding* (1979, with translation published in Moscow, 1988). He has published more than three score and ten philosophical papers, beginning with his fine essay on 'Diderot

C.C. Gould and R.S. Cohen (eds.), Artifacts, Representations and Social Practice, xi–xiv.

and the Development of Materialist Monism' in *Diderot Studies II* (1953). Some of his titles will show his interests and tempt the reader: "Marx on the Jewish Question – A Critical Review" (1961); "Metaphysics as Heuristic for Science" (1967); "Aesthetic Deprivation and the Drugged Scene" (1970); "From *Praxis* to *Logos*: Genetic Epistemology and Physics" (1971); "Is Science Rational? Repressive Reason and Liberating Reason" (1972); "Actions and Passions: Spinoza's Construction of a Scientific Psychology" (1973); "Technology and Art as Conflicting Models of Education" (1975); "Art as Humanizing *Praxis*" (1975); "Organs, Organisms and Disease: Human Ontology and Medical Practice" (1975); "The Mind's Eye and the Hand's Brain: Toward an Historical Epistemology of Medicine" (1976); "On Doing it for Money" (1976); "Politics, Political Philosophy and the Politics of Philosophy" (1979); "Picturing and Representing" (1979); "Art, Artworlds and Ideology" (1980); "Cameras Can't See: Representation, Photography and Human Vision" (1980); "The Critique of Impure Reason II: Sin, Science and Society" (1980); *Homo Homini Deus Est*: Feuerbach's Religious Materialism" (1982); "The Child's Construction of the World and the World's Construction of the Child: From Historical Epistemology to Historical Psychology" (1983); "Karl Marx and the Outcome of Classical Marxism, or: Is Marx's Labor Theory of Value Excess Metaphysical Baggage?" (1983); "Virtue Lost, or Understanding MacIntyre" (1984); "The Paradox of Painting: Pictorial Representation and the Dimensionality of Visual Space" (1984); "Virtues and Vices: The Social and Historical Construction of Medical Norms" (1984); "Good Science, Bad Science, or: Dr. Frankenstein's Dilemma" (1985).

Wartofsky is the consummate editor – encouraging, connecting, flexible, a pioneer in recognizing how very far philosophical inquiry and analytical rigor should go into human affairs. He has edited *The Philosophical Forum* since 1970, and we were responsible together for the first hundred volumes of the *Boston Studies in the Philosophy of Science* (1960–1987). With Carol Gould he edited *Women and Philosophy: Toward a Theory of Liberation* (1976). In 1981 he was elected chairperson of the Society of Philosophy Journal Editors.

I once wrote that Wartofsky is a philosopher's philosopher, but also philosopher for Everyman. He has thought through the problems, indeed the philosophical anatomy, of the natural sciences, the social sciences, medicine, psychology, the fine arts, politics and political economy, the conceptual development of philosophy, especially in the last three centuries. He is Jack-of-all-philosophical-trades, and master of them too.

It is fashionable these days to identify 'key words'. Wartofsky's key words I suppose are *praxis*, model, and representation. I have selected some revealing phrases and passages from his writings.

(From *Feuerbach*)
"[He] articulates himself in the very process of his critique. And we come to know him through it as well. This is why this study is devoted to the struggling, emerging, Feuerbach – not to the mature, complete one". (p. 11)

" . . . an 'adequate materialism' is a touchstone for criticism rather than a theory *as such* . . . a goal, not an achievement". (p. 26)

" . . . a heuristic will be bald and bare; it is not an algorithm, nor a security blanket, nor a dogma". (p. 26)

(From *Conceptual Foundations of Scientific Thought*)
"The job of science . . . is to render subject to law what has been unpredictable in the past, thereby making it predictable. . . . Science, in this interpretation, smacks of that missionary urge to bring the heathen under the law, to civilize them". (p. 295)

(From *Models*)
" . . . *praxis* and *logos* are indissoluble, and in the genetic sequence of human development, the separation of *logos* from *praxis* is impossible". (p. 173)

" . . . many scientists are full of metaphysical hunches, but not many scientists, in my experience, can follow a metaphysical hunch across the street". (p. 71)

" . . . I am talking about metaphysics as that heuristic which serves the end of helping, guiding, suggesting how the scientist comes to *understand* what he is doing, and not simply how he comes to *do* what he is doing". (p. 73)

science and art: "modes of cognitive *praxis*" (p. xiii)

" . . . the crucial feature of human cognitive practice, namely the ability to make representations . . . this I traced to the primary production of artifacts – in the first place, tools and weapons, but more broadly, in good Aristotelian fashion, anything which human beings create by the transformation of nature and of themselves". (p. 13)

"The cognitive artifacts we create are models: representations to ourselves of what we do, of what we want, and of what we hope for". (p. 15)

"[Karl] Marx's striking aphorism, 'Language is practical consciousness', requires the elaboration that it is also social consciousness. . . . " (p. 18)

"*Anything* (in the strongest and most unqualified sense of 'anything') can be a representation of anything else. . . . It is *we* who constitute something as a representation of something else. It is essential to something's being a representation, therefore, that it be *taken* to be one". (p. xx)

> "The one-sided manifestos of classic philosophical materialism and idealism – 'being determines consciousness' or 'consciousness determines being' - leave out of account the crucial question: How? . . . Between an inert epiphenomenalism, and a hyperactive creationism, we are left with two unacceptable theories of mind; and in consequence, two distorted theories of science". (xxii)

> " . . . every model proposes a certain relation to the world, or to its object, and implicates the maker or user of the model in this relationship. We can therefore always read back or reconstruct the modeller from the model itself. . . . In this sense, all modes of representation can become themselves modes of self-knowledge as well". (xxiv)

> " . . . the ubiquitousness of handprints in paleolithic art: How shall we interpret this perhaps most ancient of all the visual artifacts of the human race? The simplest answer . . . that such prints are deliberate marks of presence: "I was here' . . . the handprint records a gesture, an action, an intention". (p. xxv)

> " . . . acknowledge the importance, in our quest for objective knowledge, of the human imperative to make one's presence known, to be recognized by the fruit of one's labor: theory is the graffiti of the intellect". (p. xxvi)

This selection is mine. Another reader will have another selection. Marx Wartofsky is my genial, lucid and relaxed philosophical companion. Happy Birthday, Marx.

HILDE HEIN

INSTITUTIONAL BLESSING: THE MUSEUM AS CANON-MAKER*

With characteristic acuity, Marx Wartofsky once dispatched George Dickie's Institutional Theory of Art by declaring it neither institutional nor a theory. Aestheticians, while commonly absorbed with theory, rarely have much to do with the actual institutions of the art world.[1] I propose here* to discuss one institution, the Museum, which, since the eighteenth century, has figured prominently in constituting the art world. I hold that the museum plays a major part in defining art and in evaluating it; that its institutional role is indeed culturally linked with the determination of what will count as art and, contingently, what is to be celebrated and preserved. As institutions, museums are mandated to collect and preserve objects identified as having a certain cultural value.[2] As social institutions, moreover, they depersonalize the judgment of value that they make, rendering it both public and normative. Museums purport not to be expressive of private tastes, but to be archivally retentive of objective value and to serve as legitimate celebrants of items of indubitable (if not universally acknowledged) merit.[3] Museums are thus implicated in the dissemination of cultural canons and, I contend, as well of their formation.

As John Guillory has noted, canonization is not possible in the absence of preservative institutions.[4] The concept of canonicity derived from ecclesiastical law refers to the selection of certain scriptural texts as not in conflict with Christian doctrine and hence as acceptable for inclusion in the orthodox Bible.[5] Ecclesiastical canonization also affirms conditions of beatitude for the elevation of persons to saintliness and confers that exclusive status upon them. Here the church or its official persona is the ratifying institution. Schools – primary, secondary and tertiary – are chiefly responsible for the perpetuation of the literary canon. They are supported in this endeavor by subordinate institutions – libraries, publishing houses, editors of anthologies, television producers, theaters, and a loose confederation of others. Canonicity in the musical and visual arts is similarly forged and maintained by preservative institutions – concert halls, recording companies, critics, program notes, catalogues, and museums. I will speak here principally of art museums, but also,

C.C. Gould and R.S. Cohen (eds.), Artifacts, Representations and Social Practice, 1–19.

secondarily, of museums of anthropology and ethnography (sometimes included within Natural History museums), craft museums, and museums of science and technology. The former are relevant because they identify objects, taken as non-art, that are produced by 'other' cultures, which may be aesthetically appealing, but are chiefly interesting because they illuminate those cultures.[6] The latter are pertinent because they display works of 'our own' culture that are disqualified as art because their primary function is intellectual or practical. These too may be aesthetically gratifying and have other features in common with art, but they do not become works of art in virtue of their possession of these features alone.

Art museums, as distinct from these other types of museums, purport to exhibit works chiefly for their aesthetic quality, divorced from whatever historical or cultural features such works might contingently have.[7] Critics of this view declare that the aesthetic exaltation of such works is the product of the cultural hegemony of a dominant social order, invisible to the degree that it is dominant. They hold that aesthetic value is neither timeless nor universal; nor does it transcend particular and specific interests. Indeed, they argue that aesthetic valuation is closely integrated and shifts with the entire economy of the evaluator's desires, interests and needs; these, in turn, being embedded in a social tapestry of similar and dissimilar desires, interests and needs. The naturalization and obfuscation of such social conformism works as an instrument of coercion to generate stability within a world of competing social claims. It also works to maintain social inequities and unequal power relations that are independently operative.[8] There is a noteworthy disposition on the part of the agencies of culture to conserve the canonicity of those objects to which it has been assigned, if necessary even by radical redefinition of the standards by which the status was originally awarded. Once admitted to canonic rank, it is more likely that a work will be valued for newly minted reasons than that it will be rejected along with the repudiation of obsolescent reasons. The canon thus lends itself to various modes of stretching and dislocation. It is not inflexible, but answers to the circumstances of the institutions responsible for its maintenance. The museum is an active site of the dynamics of canon formation, deformation, extension, revision or destruction in virtue of its explicit preserving and presentational function.

I. HOW THE MUSEUM CANONIZES

Unlike most canon producing institutions, whose explicit conferrals are more visible than their acts of exclusion, museums often display the excluded as prominently as the preferred. Their educational mission dictates the preservation of the despised 'other'. Anthropology museums, for example, display 'ourselves' to ourselves only indirectly and by implication as the source of the standard from which the exhibited 'other' is deviant. The 'other' is portrayed in tacit contrast, as strange or exotic and, allusively, as inferior to ourselves.[9] Lately, the constitution of the 'other' has been the object of a great deal of attention within and without the museum world. A number of major museums (especially of anthropology and ethnography) are in the process of reviewing their exhibit practice, and sometimes proposing entirely new exhibition philosophies. Museums that find themselves face to face in dialogue with 'others' who charge that they have been misrepresented and maligned, are challenged to examine their implicit assumptions. Where many canon forgers can carry out such deliberation in seclusion and behind closed doors (e.g. in the editorial office of the *Norton Anthology* publishers or within academic faculties), museums are denied such private discretion and may have to deal with public confrontations on the exhibit floor.

In art museums, where display of the 'other' is less common, the movement to enlarge the canon has been influenced as much by economic conditions as by cultural and political demands. Restructured tax laws in the 1980's in the United States and inflation in the art market meant that museums could no longer buy textbook masterpieces and were forced to change their acquisition patterns. This pragmatic need coincided felicitously with growth in the intellectual appreciation of art for its cultural and historical interpretation, and led to an expanded collection purview. College educated and well-travelled investors, less intimidated by social proscription than their predecessors, were readier to accept diversity and to appreciate it aesthetically. One opponent of this trend, Phillipe de Montebello, former Director of the Metropolitan Museum of Art, however, worried that "The danger is that the museum curator and ultimately the public will begin to venerate at the altar of lesser masters and more peripheral styles."[10] The monolithic view of the transcendent canon was clearly showing signs of weakness.

But how did museums come to be venerated as preservers of the canon? This was not their original function. The great museums of today hark back to a history of curiosity hunting, looting, military pillage, and grave robbery. Major general museums, of which the Louvre and the Metropolitan Museum are primary examples, are public descendants of private treasuries, and these museums played an active part in transforming the consciousness of obedient royal subjects to that of sovereign citizens of modern nation-states.[11] Repositories of material wealth, they alchemically transform it to concoct spiritual meaning.

Museums are indeed physical structures that serve as storehouses for valuable objects, but in addition they figure semiotically in an often ambiguous language of presentation. They can be deceitful, celebratory, respectful – but above all, they relate a story. At one time the story was meant to inspire awe for someone's wealth and power. Post-revolutionary public museums implanted national pride in the visitor or empathetic regard for individual genius. More recently, museum goers are stimulated to take pride in their own traditions and ancestry.[12] Some scholars have compared museums with other sites in which objects are displayed for public consumption, such as department stores and cafeterias.[13] Others fix upon more reverential analogies, treating museums as secular temples that endow the objects they contain with an aura and elevate the act of aesthetic contemplation to a sacred ritual.[14] Driven by the aestheticism of modernism, the modern art museum purports to tell a story that is autonomous and detached from all cultural ties. The museum is charged to collect and display all and only that which is aesthetically pure, referentially unburdened, without practical application, and transcendent of the particularizing rites and rituals of any given culture. White windowless walls and unadorned rectangular spaces are designed to impede local interpretation and thus to authenticate a universalized experience.[15] The art alone is meant to speak, and it speaks only for itself, and to a disembodied perceiver (an Eye).[16] Lifted out of their usual context and reference, visitors to the museum are supposed to enter a Platonic realm of pure, timeless being. The museum as canon forger is removed from the 'warm foothills of humanity'.[17] Just how intelligible, then, can the demand be that it reverse the 'inaugurating exclusions' that constitute canonicity? Are they not definitive of its own identity as well?

There is a certain contradiction in the demand for justice, democracy or equal representation as applied to the art museum. As a footnote to Plato, museums defy such egalitarianism. It seems that if museums are to ratify the test of time and to validate the canonic, they must repudiate the reality of time and the plurality of cultural perceptions in order to do so. But perhaps an alternative legitimation can be derived from an altogether different perspective – namely the immediate evidentiary impact of their sensory richness. Museums might be and often are appreciated for the beauty and authenticity of the objects they contain rather than for any abstract truths they may impart. A venue for direct, richly evocative, personal encounters, museums provide impressions that are repeatable in memory, reproducible in imagination and revocable through mnemonic devices (e.g. postcards, catalogues). Those who love them are drawn to museums phenomenologically, because they find there an experience that is aesthetically gratifying.[18] Typically, however, the genuineness prized by the art museum and its patrons is that of the provenance of the object displayed rather than that of the enraptured museum goer. The value of the object calculated in terms of its history of possession displaces that of the experience it is capable of generating. Its traceability is its truth. As Nelson Goodman has pointed out however, this authenticity does not guarantee a particular primary aesthetic effect, but only the legitimacy of the object's genealogy.[19]

Some innovative museums no longer give foremost esteem to the history or originality of their contents, but seek instead to induce an experience that is genuine. Their exhibits are carefully crafted stimuli – perhaps simulacra or environments – productive of a virtual reality. (The Holocaust Museum now on the Mall in Washington D.C. is an example.) Bordering upon the popular entertainment of the Theme Park, they have not achieved the respectability of the classical art museum nor the authority to canonize. As yet, they are parasitic upon that authority, but they are rapidly overtaking it, and conceivably their mode of preserving authenticity will prevail. Whichever the locus of authenticity, object or experience, museum exhibits fuse the intensity of personal experience with a reinforcing social narrative and the stamp of history that overwhelms private judgment. Indeed, it is sometimes difficult to avoid replacing one's own directly known reality with that persuasively authorized by the museum.[20]

Typically, museums have not authorized the reality of the represented, but that of the representors, who profess to define reality from a single, universal perspective. As trainers of aesthetic sensibility, art museums educated artists to hone their craft by imitation and to produce art that preserves canonic history either by perpetuating it or by dissent from it. In a pre-technological age, artists had little access to the art of earlier times. Unable to travel and without the photographic reproductions that have brought all art into a timeless present, artists depended on the authority of their individual teachers. The exposure to past masters that museums made available to them was liberating, but now museums controlled the repertoire of artworks that could affect and influence their judgment of what was to be emulated. In this fashion, museums presided over the production of art history as well as the history of taste. According to Arthur Danto's view of the artworld, art museums continue to exercise this control over art history, for artists still produce art with reference – if not always deference – to the art tradition. Art museums, above all other institutions, serve to keep it in place.[21]

Arguably, museums are as committed to completeness as to excellence. The former is a demand of scholarship; the latter of taste, and it dominates popular expectation. The several roles can be in conflict. From one perspective museums must collect and display only 'the best'. But this connoisseur function can interfere with the museum's educational and conservative function. The impetus to canon making clashes with the drive for completeness or even typicality, for example when a museum chooses to display the collected oeuvre of a single artist or genre.[22] Canonicity breeds another limitation. It can curtail a museum's willingness to take risks in acquiring new or difficult or controversial art which may turn out not to be 'the best'. Speaking of contemporary art, Stephen Weil comments that often museums would simply like to say "here, this interests us – we're putting it up because we thought you might be interested too."[23] But this is to step down from the magisterial timelessness of the 'museum piece' to a world of passing fancies. That world is now left to galleries, which are freer of Platonism and can experiment with the timely, the atypical, the untried and the unconventional.

The conservative display posture of the art museum is more like that of the taxonomic museum of natural history, although the former emphasizes the unique individual while the latter foregrounds its typicality as a specimen. Natural history museums are, nonetheless, more able than art museums to display the pathological. Pathology is a part of natural history, the converse of normality, and of interest for its own regularities. But pathology, equated with the aberrant, is only part of the romance of art history, the underside of genius. Its contribution to canonicity is causally contingent – *par hasard* it may yield a masterpiece. Scientific canonicity, in contrast to artistic, depends on lawfulness, whose exhibition requires generic reference and the subsumption of the individual to the type. The particular specimens that the natural history museum displays are thus to be viewed as instances of a kind and not for their own sake, while the works shown in the art museum, in order to be canonic, must transcend typicality as well as tokenism.[24] At the same time, and paradoxically, canonic works of art must crystallize about themselves a history of predecessors and successors that compromises their uniqueness.[25] Roughly speaking, natural history museums confer canonicity by assimilation to a standard, while art museums do so by rewarding imaginative differentiation.[26] This leaves unanswered the question, what is the normative standard against which that differentiation is made. The tacit understanding in the case of art is that works that withstand the test of time do so because they both transcend and reflect, and indeed prolong, their history. That characterization however, founders against the claim of critics who accuse the art museum of ideological partisanship and who demand immediate canonic access for an alternative set of objects – a contradiction.

II. A CHALLENGE TO CANONICITY

Some would exorcise the notion of canonicity altogether, especially since it excludes with the same stroke those items disvalued as bad art and those which, because of their exteriority to the canonic tradition, their otherness, do not register as art at all. Can we dispense with the canon along with the cultural disjunction between 'us' and 'them'? In a sense, the demise of the canon would be a tragedy for those aspiring to its blessedness. The irony is not lost on some who are cast in the thick of the institutional struggles. Henry Louis Gates, writing on the approaching realization of his long nurtured dream to create a black canon (by way of the assemblage and publication of a *Norton Anthology of African-American Literature*), negotiates a difficult path between critics from both the right and the left. Arguing (against Jacques Derrida) that formerly subjugated, voiceless and invisible people must explore and reclaim their own subjectivity, he says:

> Consider the irony: precisely when we (and other Third World peoples) obtain the complex wherewithal to define our black subjectivity in the republic of Western letters, our theoretical colleagues declare that there ain't no such thing as a subject, so why should we be bothered with that? In this way, those of us in feminist criticism or African-American criticism who are engaged in the necessary work of canon

> deformation and reformation confront the skepticism even of those who are allies on other fronts, over this matter of the death of the subject and our own discursive subjectivity.[27]

This would not be the first time that a well-fortified garrison, once penetrated, turned out to be empty. The victory leaves a hollow taste and provokes a crisis of self-confidence, for the question looms whether one's current success is only the obverse of the decline of the institution. Notwithstanding the triumph, one cannot help but ponder the significance of success.

Shall the canon be scuttled or extended to make room for those previously excluded by it? From time to time the controversy achieves crisis proportions. One such occasion took place in 1984, when the New York Museum of Modern Art mounted an exhibition, 'Primitivism in Twentieth Century Art'. Modernist essentialists who championed formalism and celebrated it in so-called primitive art were pitted against anti-canonists, who took that judgment as an act of cultural imperialism that would appropriate the non-canonic (primitive) into art while refusing canonic status to its original source.[28] Put more ironically, some skeptics wondered why a painting of a kimono or dressing gown (by Jim Dine, for example) qualified as canonic, while the crafted artifact did not. Such eruptions, viewed from a long perspective, may be cultural rearrangements required by global politics and socio-economic reshufflings. They mark cognitive dissonances, and having surmounted them, there is no return. Edward Said, in his monumental work, *Orientalism*, depicts just such a moment of punctuated cultural evolution.

Orientalism, he explains, is a style of thought that, for well over a century, did not simply describe, but ontologically constituted its subject – the Mysterious East. The Orient is not somewhere on the map, east of Suez, but is a mind-space, the product of orientalist literature and scholarship. Said's deconstruction of it, along with that of the scholars he commends (e.g. Clifford Geertz, Anwar Abdel Malek, Yves Lacoste, and Noam Chomsky), renders it impossible ever to go back to 'the Orient' with that same naive romanticism that aggrandized its counterpart 'the Occident' even as it estranged the one from the other.[29] Said's scholarship would have fallen on deaf ears twenty years earlier. It was made possible in the 1970's by cultural, economic and technological events that brought remote parts of the globe in proximity with one another and thereby undermined some of the mythologies that had sustained separate societies. As a result, some of the invidious distinctions of 'other' by which 'self' has been demarcated became untenable. The reconfiguration of the one draws in its wake inevitable changes in the other.

Museums confer the status of art, and like other purveyors of civil identity,[30] import with that designation a notion of selfhood as the opposite pole of otherness. The relativistic history that they proclaim serves both poles in the process of self-identification. Each defines itself in relation to the other; but their otherness is not symmetrical, for only one is absolutely other. The marked other comes to know itself as other and receives its reflected history as a sentence. Cultural liberation is achievable for it only through a revolutionary act that replaces the dependent history with one that is indigenous and self-made. To contrive such a history, however, is to engage in a form of self-annihilation, at least the negation of self-as-other. It is to repudiate the canon-created history's authority and thereby to destroy the relationship that heretofore established one's identity. If the defined-as-other are successful in their endeavor to undermine that designation, they necessarily uproot as well the pole relative to which their identification was made. Thus, the identity of that pole is also affected as its dependent breaks loose from its mooring. Museums have been other-marking agents, but they are now being appealed to to join in the liberatory movement to erase the stigma of otherness. They are enjoined to become agents of social change.

This charge to change has been actively resisted especially by people who deny the museum's social role to begin with, and would regard such participation as degrading. Holding out for the museum's transcendent function, these guardians of insularity object to any descent into a temporized political course, whatever the merit of its aims. They believe that museums are not and should not be politically engaged, and they reject the allegation that their own position is a politically committed one that favors preservation of traditional patterns of domination and an exclusive canon. Others, they say benignly, have only to set up their own museums if they are not content with the existing ones.

At a deeper level, the rebellion of the represented other calls into question the very nature of the museum. Are not museums inherently objectifying? Do they not necessarily present something for the scrutiny of someone, and do they not therefore depend upon an epistemology that enforces a hierarchical distinction between the subject/knower and the object known?[31] Could museums, if the representees became representors, be so radically restructured as to obviate otherness? The question does not pertain to superficial stylistic reforms that might be implemented by more sensitive exhibition design. It is ontological: What is a museum if not a display of otherness?

III. THE MOVEMENT FOR REFORM

The extent to which museums are complicit in the construction of 'otherness' poses a problem not alone for repentant insiders who would like to dis-establish the establishment. It is equally a problem for a segment of the excluded 'others' who see merit in the museum system and hope to appropriate it to their own ends. What happens when those who were previously objectified by them use museums to articulate and display their own traditions and culture to themselves and others? Can they avoid contamination by the instruments they appropriate? Will the institutional spaces and cultural forms that previously negated them now serve to reverse the indifference and sometimes contempt engendered even among their own young people who know themselves only as the other relative to the dominant culture and whose acquaintance with their own history is dilute? Audre Lorde has cautioned against the use of the 'master's tools' to dismantle his house.[32] Henry Louis Gates, on the other hand, argues that these are the *only* tools that are likely to work.[33]

The museum is a tool designed for a Cartesian task. It is a speculum that permits close examination of objects extracted from one environment and introduced into another where they are more readily controlled and surveyed. Such intervention is not possible without making intellectual compromises and these are bound to reflect the practices and values of one's culture. European culture, with its long history of analysis and self-examination and its glorification of objectivity initiated the museum as celebratory of itself.[34] But, as noted above, the illumination of self is at the cost of the obscuration of the other.

From the perspective of those who are seeking to reform or deform the canon, especially those most injured by it, simply gaining control of the instruments of its perpetuation is an important step. While one can doubt the merits of conceding to external standards that one does not respect, there is no doubt of the satisfaction that comes from surpassing those by which one has been diminished. If it accomplish nothing else, coming up to standard serves to build confidence and helps formerly objectified people overcome some of the debilitating misapprehensions and deliberate distortions that have been fostered by the dominant culture.

IV. CAN THE OTHER BE SUBJECT?

Certain features seem indispensable to museums. The durability of exhibits that makes them accessible to visitors seems incontestable. How can a museum fulfil its function if its exhibits do not withstand observation by the public? But what if the object depicted is, in its normal environment, a moment in an evanescent process? Far from truthful presentation, such museum displays factitiously distort and falsify a cultural conviction that things are and ought to be transient, ought not to be artificially fixed. Conventional museum exhibits stabilize their subject matter. They are designed to freeze time and space and to make observable, to whoever has the price of admission, objects (or events) that may be short-lived or fleeting in nature and, in some instances, only selectively available. In some cultures certain objects are approachable only seasonally or for special occasions, or exclusively by certain individuals. To display these in customary (western) museum fashion is thus to violate them. This is not to say that no form of exhibition is possible, but it may entail radical reconsideration of the capabilities of the tools that have been appropriated. As with the deconstruction of Orientalism, the appropriation (some would say subversion) of museum practice will vaporize some apparently solid entities, not to mention the typological distinctions between kinds of museums. Words shift their meanings; property is dematerialized; icons become individuals; ordinary things become works of art, and the museum building itself can become a ceremonial house that enshrines a spirit of a different nature.[35]

Those among the represented now negotiating to assume control of their representation will have to decide which compromises are to be made. They must strike the balance of institutionalized remembering against manipulated forgetting, knowing that both are inscribed in the medium on which their hopes are pinned. They must understand that the museum is not a transparent instrument, but has a charged history, and their transaction with it must be undertaken warily and with ambivalence.[36]

Museums cannot help but be affected as their staff and audiences are infiltrated by members of the object-group they formerly displayed.[37] Political pressures and affirmative action programs have already brought about a shift in their functional identity and in the perception of the core enterprise to which they are ostensibly committed. They will continue to collect, preserve, study, interpret and exhibit objects, assigning variable priority to each of these services and performing them differently depending upon their specific situation.[38]

Education has always ranked high on the list of museum services, but without a uniform conviction as to who is to be educated and to what end. Art museums have been especially vague in this regard, more or less operating on the assumption that the exposure to Beauty is necessarily enlightening. Science and history museums, maintaining collections for research, have served scholars. But the same political wind that caused museums to open to new audiences has swept away their concentration on scholarly research. Willingly or not, museums are giving up on the 'trickle down' theory of education and accepting their obligation to the public as primary. Of course, this is partially a function of resources. Both in the U.S. and abroad, funds for primary research are diminishing in all institutions, and museums have been cutting back their scientific/research staff in favor of more programming, more public relations and more administration. The very idea of pure research, academic as well as in museums, is in poor repute, perceived as self-indulgent and patrician (if not patriarchal), and applications for research funding must now justify their proposals in terms of some envisaged application.[39] For museums, that might take the form of scientific or visual literacy or multicultural diversity. Clearly the museums are striving to be responsive to the tenor of the times, but they are left in a state of confusion as they seek to redefine themselves from within and without.

As those who were excluded by the canon move to inscribe themselves within, they change the nature of the community. Their demand for self representation is not to be confused with political representation, i.e. a proportionate electoral voice.[40] It is not satisfied with having a vote in a system that someone else has constructed. The tardily appointed delegate, obliged to work within a set of rules that she or he had no part in designing and to situate her/himself within their framework, will invariably be disadvantaged even when she or he meets the standards. But to be truly representative, even politically, is to be creatively involved in forging the system and its canons.

The institutions of culture are more complex than those of politics and the rules of adherence more fluid. Gates refers with some embarrassment to Pierre Bourdieu's locution – "the redistribution of cultural capital."[41] The phrase is crassly quantitative, but not entirely out of order in speaking of canon deformation; for the reference is to an inventory of values that may be traded off, one against another, albeit the unit of their measure is obscure. Bourdieu's expression is embarrassing because it does not convey the epistemological magnitude of the good to be dispensed. By whom? To whom? By what means of exchange? And to what end? Is it, after all, only for the sake of maximizing profit?

The risk that storming the canon will explode it seems real. The very threat exposes the absurdity of the notion that the canon can be both democratic and transcendent. At the very least, the canon is trivialized; for superiority universalized is meaningless.

For museums in particular, even a momentary distinction between 'looker' and 'lookee' seems essential to the viability of the institution. A zoo in which no one is caged and no one free to walk about, is more like a jungle or a beach. As Arnold Berleant proposes, museums might be sites for the promotion of aesthetic (or other) pleasures.[42] But this would render them, on the one hand, indistinguishable from the world in general, or, would on the other, require protective and regulative devices that would make them once again selective and interpretive.

V. CONCLUSION

Must museums cease dispensing beatitude? Deeply compromised by their part in the construction of otherness, museums are caught in the antagonisms of the communities that are its consequence. Since communities determine what is of value, museums which purport to be repositories of valued objects, must exemplify the communities they serve. But if anything has been learned from the present climate of multiculturalism, it is that different communities assign value differently, sometimes to the same objects, sometimes disagreeing even as to what is an object or how value is to be expressed. We have also learned that civil society is composed of "a constantly changing mosaic of multiple communities and organizations."[43] It is therefore not to be assumed that a single value designation can be authoritative and will forever withstand the test of time. Most of us are (at some time) members of a number of (sometimes incompatible) communities, and so hold even our own values inconstantly. Philosophers have mystified this rather normal social phenomenon in the endeavor to fix value as contextually independent, if not as eternal ideals.

Museums as heirs to the same philosophical inclination have proudly served as footnotes to Plato, propagating cultural hierarchy as a fact of nature. They are now under pressure to change. Some critics would like them to be more experimental, or more negative and self-critical. Some fear that a sacrifice of authority would undermine their representational capacity. Some are concerned lest their preoccupation with presentational style and context become self-defeating and generate a museum-world cut off from any other community. Whatever their direction – and there is no reason to believe there should be but one – museums are likely to remain in the business of exhibiting objects that somehow embody somebody's values for someone. There is thus sufficient reason to attend both to objects and to those who value them.

Simone Weil speaks of a type of attention, converging upon love, that is needed to listen to both affliction and truth and that is drawn to the beauty in things. So to attend is to commune with an object (or person) at the level of the Impersonal – seeing it with acuity, understanding, and affection.[44] In my estimation, museums have the potential and should strive to foster that sort of attention. By way of their multi-dimensionality and textural depth, museums are uniquely qualified to reach out to vastly different audiences and to approach them simultaneously through variously responsive channels. Where the canon excludes, museums can include, and can do so without reductionism or simplifying uniformity. Where the canon is narrowly hierarchical, museums can be broad and generous and can encourage that spirit in the public. They can be culturally enriching not by coercively exposing us to the so-called best that has withstood the test of time, but by generating appreciation of the good that is everywhere and at all times so that we can find it for ourselves. They might enculturate a form of understanding that helps us to know with compassion, to feel with analytic clarity, and to enjoy the normality of such communion. In doing this, museums would fill their educational mission and need neither abandon judgment nor trivialize the canon. They would justify their conventional institutional functions and reveal with sympathetic insight what the genuine attraction is that draws people to take the canon seriously. That might put an end to the canon wars.[45]

College of the Holy Cross
Worcester, Mass.

NOTES

* This article has been accepted for publication by the *Journal of Arts Management, Law and Society* under the title "Philosophical Reflections on the Museum as Canon Maker" and is now in press.

[1] I use the two-word expression "art world" as distinct from Arthur Danto's single word "artworld" by design. The former, and more conventional, locution refers to institutions related to the arts materially and socially, which Dickie then designated as theoretically definitive of art. Danto's second order artworld is cultural and historical. It enters into the metaphysical act of art creation by shaping the classifying consciousness of producers and appreciators of those entities that come to be designated art. Both worlds reflect ideology and are infused with politics.

[2] The American Association of Museums officially accredited as museums only institutions which possess and preserve collections and which employ a staff for their maintenance and educational or aesthetic display. The specificity of this characterization has been somewhat relaxed in the past decade, since it caused some embarrassment with respect to science centers, which do not maintain collections in a strict sense.

[3] This is so despite the fact that many museums are composed almost entirely of the collections assembled by a particular donor or curator. The assumption is that this person's taste is not idiosyncratic but reflective of an absolute standard of quality.

[4] 'Canon', in *Critical Terms for Literary Study*, ed. Frank Lentricchia and Thomas McLaughlin, Chicago, U. of Chicago Press, 1990. Guillory says that the institution responsible for the biblical canon is the church, whose procedure of inclusion and exclusion is rigorously based on dogmatic grounds. The acts of judgment that compose the literary canon, he says, follow a different and less rigorous social agenda which is closely tied to the phenomenon of literacy and access to the means of literary production. He maintains that the school is the major social institution that regulates the practices of reading and writing and therefore the means of canon formation devolve to schools and, more broadly, to the educational system.

[5] From its inception, evidently, canonicity represents exclusion, the forging of orthodoxy by means of the suppression of alternative positions as well as dissent. viz. Christine Froula 'When Eve Reads Milton: Undoing the Canonical Economy' *Critical Inquiry 10*, no. 2, December 1983; see ref. to Elaine Pagels' discussion of second-century gnostic writings. See also Guillory *op. cit.*

[6] Such objects may come to be regarded as 'works of art' in their own right and may in fact be exhibited in art museums, but in terms of their production, they were created in contexts that did not include the concept 'work of art' under which they are subse-

quently classified. That notion is of comparatively recent and predominantly European vintage.

[7] The generic distinctions between museum types listed here are schematic. Of late, such logical and other differences have been subject to challenge and have been eroding under pressure to diversify and contextualize their exhibits. It is not uncommon now to find what would formerly have been considered an anthropological or historical exhibit in an art museum or, conversely, an exhibit of art in a natural history museum. I have discussed this shift in exhibition philosophy elsewhere.

[8] See Barbara Herrnstein Smith, 'Contingencies of Value' and other contributions to a special issue on Canons of *Critical Inquiry 10*, no. 1, September 1983.

[9] Rarely are we invited to contemplate western ear piercing or excessive dieting as forms of body mutilation. Nor are we shown gentlemen's clubs and business associations as parallels to the men's long house of tribal societies.

[10] Cited in 'Expanding the Canon of Art Collecting', Michael Conforti, in *Museum News*, Sept./Oct. (1989) *68*, no. 5, p. 36.

[11] Carol Duncan, 'Museums and Citizenship' in *Exhibiting Cultures: The Poetics and Politics of Museum Display*, ed. Ivan Karp and Steven D. Lavine, Washington D.C. Smithsonian Institution Press, 1991. Duncan describes the deliberate transformation of the princely gallery with its display of military exploit and enviable wealth to become a repository of spiritual wealth wherein even lowborn citizens share in national glory.

[12] A source of possible conflict here is that the objects that these ancestors produced may not meet the current conventional standards of canonicity; hence the dilemma of the museums that seek at once to instil ethnic pride and to perpetuate 'excellence'.

[13] Neil Harris, 'Museums, Merchandising and Popular Taste: The Struggle for Influence' in *Material Culture and the Study of American Life*, New York, Norton, 1977.

[14] Walter Benjamin, 'The Mechanical Object in An Age of Reproduction' in *Illuminations*; Dillon S. Ripley, *The Sacred Grove: Essays on Museums*, 1969, New York, Simon & Schuster.

[15] No notice is given to the fact that these sterilizing procedures are themselves conventional and speak to a cultural tradition that associates truth with stripped down essentialism.

[16] Brian O'Dougherty, *Inside the White Cube*, 1976 San Francisco, The Lapis Press. Indirectly, the message is a deeper metaphysical lament; for it bespeaks the existential isolation of the human condition. The lonely, non-contextual, and thus vacuously universal content of the artwork is the solace of mankind, but not yet its transfiguration.

[17] Clive Bell, in *Art*, spoke with contempt of those who carry life associations with them to the appreciation of art and expressed with moral fervor his preference for the cold, ecstatic heights of aesthetic exaltation.

[18] Arnold Berleant maintains that a museum is an environment whose distinctive purpose is to promote the experience of aesthetic engagement. He deplores the more usual aesthetic of objects that prompts most museums to organize exhibits in a manner that often impedes aesthetic response. See his 'The Museum as a Participatory Environment' in *The Aesthetics of Environment*, 1992, Philadelphia, Temple University Press.

[19] *Languages of Art*, Indianapolis, Hackett Publishing Co., 1976. Goodman does claim that such legitimacy has a secondary aesthetic effect pertaining to a body of work by an artist or a genre, but his point is that no particular tingles or frissons follow from the genuineness of the object as such.

[20] I do not mean to suggest that museums alone have this capacity. Film and perhaps television appear to have comparable impact on the public concept of history. Their unrepeatability is a detriment, but is lessened by the availability of video. Still, the multisensory presentation and multidimensionality of the museum exhibit and its implied provenance give it an ontological edge.

[21] *The Philosophical Disenfranchisement of Art*, N.Y. Columbia University Press, 1986.

[22] A simple solution would be to endorse a variety of museums with distinct missions; but on the contrary, museums are becoming depressingly alike, cast in the same formulaic mold, regardless of their size, location or their resources.

[23] Weil, p. 38.

[24] Jack Glickman argues that creation, unlike making, is of types. The type in question may or may not determine multiple instantiations. In that respect, it is not apposite to the classificatory genera of natural history, which are meaningless if empty. I mean to deny the logical appropriateness of the latter sense of typicality to art. See 'Creativity in the Arts' (1976) in *Culture and Art*, ed. Lars Aagaard-Mogensen, Eclipse Books.

[25] Anita Silvers suggests that a canonic work cannot stand alone. Its canonicity is therefore not a first order predicate or even a dispositional one. It resides in the art historical and historical scholarship that supervenes upon its production. 'Has Her(oine's) Time Now Come?' JAAC *48*, no. 4, Fall 1990. An implication is that, its canonicity arrived at, the work ceases to be an individual and becomes a paterfamilias, responsible in a sense for all its progeny.

[26] Diachronicity and synchronicity are common to both standards. Both reveal temporal ancestry and conspecificity over spatial barriers. Both exhibit paradigms – however in a sense that is equivocal. The art museum emphasizes the evaluative dimension of the normative – that which sets the standard by virtue of its perfection – while the natural history museum focuses upon the statistical – that which is standard by virtue of its numerosity,

[27] Henry Louis Gates, Jr. *Loose Canons*, 1992, New York, Oxford University Press, p. 38.

[28] This exhibition, curated by the modernist William Rubin, angered many people because it apparently denied the relevance of history or culture to aesthetic expression, or even of the artist's individual intention. Assimilating the work of African, Native American and Pacific Island producers to that of self-conscious western artists, the exhibition professed an essentialism that reduced the former to "a footnote to the development of art in the West." (Karp and Lavine, 1991, p. 376). Strictly speaking, neither side advocated abandonment of the canon: The formalists declared that primitive artists, as Picasso demonstrated, understood it intuitively and so merited inclusion. Their opponents scoffed at that claim, but argued for categorically expanding the canon in order to include 'tribal art'. This was to redefine 'art-making', but not to defend anti-art. See William Rubin, ed. *Primitivism in 20th Century Art: Affinity of the Tribal and the Modern*, 2 vol. NY, Museum of Modern Art (1984), Thomas McEvilley (1984), 'Doctor, Lawyer, Indian Chief: Primitivism in 20th Century Art at the MOMA' *Artforum*, 1984.

[29] Edward W. Said, *Orientalism*, 1978, New York, Pantheon Books.

[30] I use the term 'civil identity' as distinct from 'political' in the manner made famous by Antonio Gramsci in explaining the power of the intellectuals to dominate by 'cultural hegemony'. See *Selections From the Prison Notebooks*, ed. and translated by Quinton Hoare and Geoffrey Nowell-Smith, New York, International Publishers, 1971.

[31] Pieties of scholarship notwithstanding, the attraction of museums that display the culture and artifacts of others is essentially that of a human zoo.

[32] Audre Lorde, 'The Master's Tools Will Never Dismantle the Master's House' in *Sister Outside: Essays and Speeches*, Trumansburg, N.Y. Crossing Press, 1984.

[33] *Loose Canons*, p. 38.

[34] Elsewhere I have likened the museum to a laboratory, which also selects items from reality and isolates and simplifies them for close analysis. A culture that holds science in high regard does not perceive itself as degrading the object it scrutinizes, but only as satisfying the need for knowledge (which it considers as wholly laudable.) In the particular case that I examined, the Exploratorium, the objective is in fact democratic, to extend the facilities of the laboratory to a broader population, enabling everyone to probe with their own senses those features of the world that are ordinarily inaccessible to all but a few. See *The Exploratorium: The Museum as Laboratory* (1990) Washington D.C. Smithsonian Institution Press.

[35] James Clifford, 'Four Northwest Coast Museums', in Karp and Lavine. Another example of adaptive configuration of museum convention and spiritual expression occurred with respect to the 1984 exhibition *Te Maori*, in which the Maori elders were imported to ceremonially consecrate the building in which the artifacts were housed with each relocation of this travelling exhibition.

[36] James Clifford, 'Four Northwest Coast Museums: Travel Reflections' in *Exhibiting Cultures: The Poetics and Politics of Museum Display*, ed. Ivan Karp and Steven D. Lavine, Washington D.C. Smithsonian Institution Press, 1991.

[37] Arguably, there are similar identity shifts in the canonic structure of various artforms as they become appropriated by non-traditional users whose influence eventually becomes determinative. Gates (1991) refers to an identifiable style of African-American writing that has so-far not achieved that status. On the other hand, it seems that jazz has made canonic inroads upon original American music, and perhaps the rhetorical style of women's writing (if there is *one* such style) can be discerned to have an effect upon contemporary American literature (or at least upon its critics.)

[38] According to the American Association of Museums definition, collections are an essential ingredient of museums. However, this requirement has been challenged by science centers, which, although they do display objects, do not assess their own value according to the value of the objects they display. Science centers are a special case. They *use* objects in order to display ideas or to make them accessible. The exhibits themselves are successful insofar as they have a certain transparency that shows forth the concepts they embody. Science centers are thus very much unlike art museums in one respect,but very much like them in another: they are not in the business of collecting specimens or studying typical instances. They are, in a sense, celebrations of the ideal. In both cases, the true value of the content of the museum is not to be found in the physical stuff that can be touched and moved. For an eloquent defense of the preservative ideal of the art museum, see Sherman Lee, *On Understanding Art Museums*, Englewood Cliffs, N.J., Prentice Hall, 1975 (Introduction).

[39] With the animus against 'elitism', there is a tendency to reject all preoccupation with theory that is 'disinterested' or purely contemplative. This conflation of the speculative with the useless or egoistic is as prejudicial against some art as against some science, and when funds are in short supply, it can be a dangerous political weapon.

[40] There can be no doubt that the demand for representation in the museum as in the canon is political. Kwame Anthony Appiah, writing on Race, argues that literature is one of the primary instruments by which nationalism is shaped, and nationalism itself depends upon selective segregation of 'us' and 'them'. (See 'Race' in *Critical Terms for Literary Study*, ed. F. Lentricchia and T. McLaughlin, Chicago, Chicago University Press, 1990.) Control of the canon, in the form of access to literacy and dissemination (through formal schooling) as well as to creation and publication, is a component of self-determination. The dispute among African-Americans over the desirability of a separate canon of black literature (the Black Aesthetic) as distinct from inclusion in the 'great' canon reflects political differences within the African-American community.

[41] *Loose Canons*, p. 177.

[42] *Op. cit.*

[43] *Museums and Communities: The Politics of Public Culture*, ed. I. Karp, C. M. Kreamer, S. D. Lavine, Washington D.C. Smithsonian Institution Press, 1992, p. 3.

[44] 'Human Personality' (1943) in *Simone Weil: An Anthology*, ed. Sian Miles, New York, Weidenfeld and Nicholson. Iris Murdoch has made reference to this sense of attention in her essay 'The Sovereignty of Good' and Sara Ruddick refers to it in 'Maternal Thinking'. Both are striving to identify an experience that is intensely cognitive and sharply affective. The dichotomous philosophy that is canonic (sic) in our culture, deprives us of a language to express such a melange.

[45] I am grateful for critical assistance from Ivan Karp, Arnold Berleant, Aleta Ringlero and Predrag Cicovacki.

GREGG M. HOROWITZ

"SUDDENLY ONE HAS THE RIGHT EYES": ILLUSION AND ICONOCLASM IN THE EARLY GOMBRICH

I. INTRODUCTION

For a while it seemed that one of the central projects of modernist artmaking had fallen off the map. From Manet until the late 1950s, the goal of displaying the process of artistic fabrication, of showing the making of art out of its raw materials, had been taken up as the self-conscious goal of artmaking itself; aesthetic appearances were distorted and fractured by the attempt to force what resisted aesthetic figuration to appear as unfigured. A paradoxical, perhaps impossible project, teetering unstably between the jubilant overcoming of aesthetic appearance and its sublimely pleasurable incompleteness, but a project renewed over and again in the face of, perhaps even by means of, its paradoxical nature. However, with the interconnected losses of faith in the very idea of unmediated 'material' raw at the moment of artmaking and the heroic activity of the artist confronting it, postmodern art damped the eruptive fire at the heart of modernist pictoriality and replaced it with a cool detachment from appearance itself. Among its defining features, postmodern art rejected, also over and again, the modernist dialectic of matter and appearance. It is no wonder, then, that painting, the peerless art of appearance, went into eclipse in the 1960s and '70s.

For better or worse, though, and for reasons not yet completely understood, painting has returned from the dead. Abstraction, opticality, beauty, and other modernist themes are again being raised in and about painting, although to my mind without either the rigor of high modernism or the merciless ironizing of postmodernism. If this return of painting reveals anything it is that the characteristically modernist dialectic of pictorial appearance and its resistances has not, despite appearances, been overcome. So now may be a propitious moment to approach this complex theme anew.

In this essay I will discuss some early writings by E.H. Gombrich. My thesis is that Gombrich is a modern iconoclast engaged in an effort to break down a premodern regime of pictorial appearance. His icono-

C.C. Gould and R.S. Cohen (eds.), Artifacts, Representations and Social Practice, 21–38.

clasm is however, simultaneously destructive and productive, aiming at exhibiting pictorial appearance along with its resistances and so at calling forth novel modes of spectatorial subjectivity, new ways of seeing. My strategy here is indirect: I will be treating Gombrich as a synecdoche for projects that are clearly not his alone and I will use the painting of Paul Cezanne, also synecdochically, to allegorize Gombrich's theorizing and Gombrich to allegorize Cezanne's painting. Although more traditional arguments about influences and anxieties could be made I will refrain; my goal is to try to approach the modernist iconoclastic project of remaking human perception from two different, if not discrete, vantages. Since this project is still alive, if either approach provides a fresh perspective my goal in this essay will have been achieved.[1]

II. CEZANNE'S PAINTING

Describing Cezanne's achievement of a kind of neoclassical internal completeness and harmony despite his self-imposed restriction to impressionist, relational coloration has been, not surprisingly, a chore. But because Cezanne's style, beyond being another move in the history of painting is also, I think, a mutation in the history of appearance itself, it is not surprising that no one has captured the experience of watching a Cezanne more perspicuously than a poet. In a letter, Rainer Maria Rilke writes of a Cezanne still life:

> Although one of his idiosyncracies is to use pure chrome yellow and burning lacquer red in his lemons and apples, he knows how to contain their loudness within the picture: cast into a listening blue, as if into an ear, it receives a silent response from within, so that no one outside needs to think himself addressed or accosted. His still lifes are so wonderfully occupied with themselves.[2]

Rilke here expresses a key paradox of Cezanne's pictures. While they are indeed paintings, painted objects made to be looked at and so made to address a spectator, they do not appear to address the spectator. Although they are essentially relational objects, objects that have their being in being seen, they nonetheless seem closed in on themselves. Cezanne's paintings, in short, seem to exclude the spectator to whom they are addressed by exhibiting or performing for the spectator her own exclusion. Now, Rilke sees this as a feature of Cezanne's use of color, where every color calls forth a response but the response is provided by the painting itself. Let us note how odd it is, but also how pointed, that Cezanne's exclusion of the spectator operates through color. Color, after all, is the paradigmatic secondary quality, that which, in some influential ontologies, is an index of the subjective and so cannot even exist without a spectator to experience it. In a world without spectators there could be no colorful drawings, which is to say no paintings, yet it is just through the quality that needs spectatorial complicity to come into existence that Cezanne, according to Rilke, tries to make the spectator secondary. This is perhaps another version of Cezanne's notorious anxiety explained by Merleau-Ponty, citing Emile Bernard, as "aiming for reality while denying himself the means to attain it".[3]

What this paradox makes manifest is perhaps already obvious: Cezanne cannot really eliminate the spectator. He is after all, a painter, and every painter (at least every painter in the illusionist tradition) is also already herself a spectator of the spectacle she brings into being.[4] Rather, Cezanne exhibits for the spectator that her position before, in front of, the painting is secondary because her role is already inscribed inside it. Cezanne displays the relation between picturing and spectation in the painting itself; the painting does not await the arrival of the experiencing spectator but instead preempts her experience and then shows the preemption. Hence, Rilke's description of the still life (and remember how central that apparently sedate genre is to Cezanne) as self-contained, as in repose; however, since it also performs its repose, it is self-contained differently from the way a mere object, a real orange, is. Everything the painting needs, which includes someone to see it, it shows itself to already have.

Cezanne's painting is, I think, a genuinely revolutionary response to the doubleness of pictures, their being both material objects in the world and visual representations which appear to be of the world they inhabit. It is revolutionary because it does more than simply pose the problem by virtue of appearing, it also takes the problem as its subject matter. Cezanne's painting is about pictorial doubleness, which is to say it is self-reflexively about its own status as picturing activity. Not only is a Cezanne like all other paintings in being a doubled artifact, it is also a picture of the doubling of artifacts and so a picture of its own coming to be a picture. Cezanne never hides the paint, the material out of which he makes his pictures, but rather foregrounds it assertively. In the *Tulips in a Vase* of 1890–92, not only is the wall behind the tulips painted in patches which make it appear more paint than wall (a visual fact a simply mimetic interpretation might recuperate with reference to patchily painted walls), but the back edge of the table clearly outlined in black is overrun with paint (and here there is no possible mimetic analog). This vibration of outline and *un*contained paint is iterated throughout the painting, in the leaves of the tulips, in the oranges arrayed to the side of the vase, in the front edge of the table the frontality of which, were it not overrun with paint, might even be seen as a mark of respect for a more traditional view of the picture plane as a window. To look at the painting is to look into a world of paint. Crucially, though, it is a *world* of paint one looks into and not just paint one looks at. One sees outlines failing, but not without effort, to contain the colors they cannot bind. In other words, one sees not a world made of paint but a world struggling to make itself out of paint. One sees, that is, the process of painting itself.

In Cezanne's painting the doubleness of pictures appears, comes to the surface. His paintings are not mere painted things but neither are they optical artifacts one grasps by not noticing the paint. To see a Cezanne is to see paint worked and working to be something else, which means seeing paint and the limned out-lines of that something else, and so to see the movement from the one to the other put on display. Cezanne freezes the process of pictures without stopping it. This thematization of process within the picture is, I suggest, a defining feature of modern pictoriality. Not only does a Cezanne appear as something it is not, it also appears to so appear; it performs its own process of splitting in plain sight. The pictorial appearance which was presumptively opaque, occlusive of the paint which supports it, in the illusionist tradition is rendered transparent in Cezanne by being shown, being made visible as an appearance, a showing which requires the exhibition of the paint, the appearance, and the incomplete emergence of the latter from the former. This surfacing of appearance as an object for vision both is and requires a novel understanding of the arena of pictoriality.

That this is a new moment in the history of appearance might seem a melodramatic claim. Every picture has always been doubled, so while Cezanne may make a new kind of painting it may seem a stretch to say he makes a new kind of appearance. But I think it is not a stretch at all. Not only does Cezanne make pictures that look new, he also makes the look of pictures look new by visibly inscribing the traditional complementary role of the spectator in his painting. This creates a new look because the spectator thus presumed by the display of the painting is a new kind of spectator. Once again, this is Rilke's paradox: Cezanne's painting is to be looked at, but this act of looking is something new. ". . . For a long time nothing, and suddenly one has the right eyes".[5] Cezanne's painting is still to be engaged with the eyes, hence there is still an apparently traditional painting/spectator encounter. But the painting does not need the spectator to make the appearance emerge, hence this is not a traditional encounter at all. Cezanne calls into being a new kind of visual attention, a new kind of spectator with new eyes. To look at the pictoriality of Cezanne, to look without overlooking, is to be a new kind of seeing subject. A new mode of pictoriality with a concomitantly new mode of visual subjectivity – this justifies the claim that Cezanne installs a new regime of appearance.

There is an important sense in which this new regime of appearance reconfigures in its wake the history out of which it emerged, thus making its novelty hard to discern. Put bluntly: once we develop the eyes with which to see Cezanne, no picture will ever look the same again. The mode of subjectivity appropriate to Cezanne is alert to the doubleness of pictures and so is not able to engage uncritically (which is not the same as being unable to engage at all) in the kinds of visual attention presumed by earlier modes of appearance. Masaccio will not look the same to post-Cezannian spectators as he did to fifteenth century eyes. The Cezannian dialectic of appearance and subjectivity generates modes of attention incompatible with those installed by prior, illusionistic forms of pictoriality. In this sense Cezanne is an iconoclastic painter. However, Cezanne continues to make pictures to display, for vision, the relation between appearances and what resists appearing, and so his iconoclasm is crossed with iconophilia. The problem of exhibiting appearance is for Cezanne a problem precisely for painting, for that activity which manufactures modes of visual attention. Only by trying to look at our own looking can we grasp what painting in the illusionistic tradition has been for. Put simply, the proper iconoclastic critique of illusionistic picturing is to look again, which in Cezanne's case means – more painting.

III. GOMBRICH'S ICONOCLASM

Throughout his career Gombrich has addressed the perennial art-historical problem of distinguishing conceptual from perceptual, or illusionistic, images. In his earlier work he developed a rich and ramifying account of this distinction. Most to the point, this early account is iconoclastic in much the way Cezanne's painting is; to be precise, Gombrich develops an iconoclasm which calls for more, not less, looking at pictures.[6]

The distinction between conceptual and illusionistic images has long resisted precise clarification, so to avoid begging any questions let us start by aiming for intuitive plausibility. Roughly, a conceptual image is an abstract one, the understanding of which depends on a beholder using her cognitive stock to decode or read it, while an illusionistic image is a naturalistic one the recognition of which depends on the beholder simply seeing it. Standard examples of conceptual images include children's pictures in which, say, all four sides of a house are drawn because that is how the child thinks of a house, stick figures,[7] and Egyptian art. Gombrich explains Egyptian styles (excepting those of the Eighteenth Dynasty reforms) as follows:

> The point is that Egyptian art is not based on what the artist could see at a given moment, but rather on what he knew belonged to a person or scene. It was out of these forms which he had learned, and which he knew, that he built his representations, much as the primitive artist built his figures out of the forms he could master. It is not only his knowledge of forms and shapes that the artist embodies in his picture, but also his knowledge of their significance. We sometimes call a man a 'big boss'. The Egyptian drew the boss bigger than his servants or even his wife.[8]

In other words, conceptual art is formulaic, deploying codes according to which relative importance is signified by size, location by a token of place, and so on. Conceptual images do not appear to be of their motifs, but they were never meant to. The standard contrast to the Egyptian conceptual image is the Greek naturalistic one which so influenced Roman art and through it the Italian Renaissance. In a passage which also incidentially clarifies Plato's famous preference for Egyptian art, Gombrich says of an early Greek kouroi:

> It . . . shows that the artist who made this statue was not content to follow any formula, however good, and that he began experimenting for himself . . . It was no longer a question of learning a ready-made formula for representing the human body. Every Greek sculptor wanted to know how *he* was to represent a particular body. The Egyptians had based their art on knowledge. The Greeks began to use their eyes.[9]

Naturalistic art appeals not to readerly knowledge but to recognitional capacity, not to the beholder's knowledge but to her eyes. Whereas conceptual images are formulaic and so depend for their comprehension on shared formulae, illusionistic images are antiformulaic and so apparently depend for their effect (we can no longer call it simply comprehension) on – well, on what? We might want to say on the beholder's capacity to see, but it is in complicating that "mere" capacity that Gombrich's own theorizing begins.

A standard way of explicating the recognition of illusionistic images is in terms of their greater resemblance to what they picture. On this view, whatever it is that allows us to recognize the everyday furniture of the world also allows us to recognize its appearance in an illusionistic image. The more contemporary manner of putting the point employed by Gombrich is that illusionistic images are or are made of iconic signs. An "iconic sign . . . is any sign which is similar in some respects to what it denotes"[10] and, further, denotes what it does by virtue of this resemblance. Both the more traditional and the semiotic ways of defining the uniqueness of the illusionistic image thus avail themselves of the putative fact of resemblance between the image and its object. Whereas the illusionistic image looks like what it pictures, hence provoking the illuded perception that the thing is there in a manner similar to the way the thing itself would provoke the veridical perception that it is there, the conceptual image does not look like its object and so its routes of reference are more language-like.

Gombrich rejects this way of making the distinction. Demarcating conceptual from illusionistic images by reference to the iconicity of the signs used implies that the difference is objective, i.e. that it is a difference in the objective features of the signs themselves. However, even if we choose our examples uncontroversially, this implication is plainly false. Consider again the child's drawing with all four sides of the house crayoned in and compare it with the house in the upper right of Pissarro's *Boulevard Montmarte*. The latter is surely an illusionistic image to whatever degree the former is not, but whatever iconicity it possesses cannot derive from the smudges of paint that signify bricks and windows. The greater iconicity of the Pissarro cannot be a function of the objective iconicity of its constituent signs; indeed, on that criterion, the child's drawing would be the more iconic – at least its depicted walls have straight edges! – and thus the more illusionistic. Unless we wish to rearrange the membership of the two classes of image the difference between which we wanted to explain, the objective iconicity of the visual signs will not help us to grasp the distinction.

Using the example of Guardi's realistic paintings of Venice, Gombrich generalizes this point:

> Is the primitive manikin, complete with ten fingers and toes, less iconic than the patch of colour which may denote a gondolier on one of Guardi's paintings? Here again many difficulties may be avoided by concentrating less on a morphological classification of signs that on an analysis of their interpretation. Guardi relies on the beholder's capacity to read "iconicity" *into* his sign. The contextual, emotional, or formal means by which this type of interpretation is evoked or facilitated – in other words, the relation between objective "iconicity" and psychological projection – would have to form one of the main fields of study of a descriptive semiotic of the image.[11]

It is important to stress that Gombrich is not here rejecting the distinction between the illusionistic and the conceptual image, the iconic and the non-iconic sign. He is instead trying to save it by rejecting the attempt to analyze it in terms of the objective features of the signs themselves. Uninterpreted, the visual signs which constitute an illusionistic image are no more nor less iconic than those which make up a conceptual image. Put this way the point is obvious, perhaps even banal. Painters must manipulate the materials available to them in whatever way serves their purposes, but whether one chooses to describe those materials morphologically or materialistically, there is nothing in an ovoid patch of white or in a deposit of white pigment that as such signifies 'window'. What is remarkable is not that this is true, but that anyone ever believed otherwise.

However, in the passage in which Gombrich makes his general point there is a hint as to why not just anyone but indeed most theorists would have believed otherwise. Rejecting an explanation of iconicity in terms of objective properties of signs does not entail there being no difference between the iconic and non-iconic at the level of the *appearance* of objective difference. No morphological or material difference does not mean no difference at all, which is why Gombrich points us in the direction of a pragmatic analysis by saying Guardi incites beholders to read iconicity into his signs. All image makers incite beholders to read their signs but the illusionistic image-maker tries to get beholders to see the signs as objectively resembling what they depict. In short, that illusionistic pictures are not themselves iconic does not mean that they do not appear iconic to their users, and it is this appearance that is left to be explained by reference to the way the pictures are seen. If he can explain how such an appearance is successfully induced Gombrich will also have explained non-parochially why it has appeared to earlier theorists that the illusionistic image objectively resembles its object, and so why there would have been an effort to develop theories of objective iconicity.

Gombrich makes this displacement of the analytic burden from objective properties of the sign to psychosemiotics his explicit project by saying that such an investigation "will show that what has been called the history of 'seeing' is really the history of a learning process through which a socially coherent public was trained by the artists to respond in a given manner to certain abbreviated signs".[12] This is an overly condensed, even coy way of passing the burden to an investigation of response. However, all the elements of it are carefully chosen and I will unpack its meaning by concentrating on four of them: the idea of a public, the social coherence of that public, the notion of 'seeing', and the scare quotes that surround it. Taken together, these elements point to the reasons a false theory of iconicity would gain widespread acceptance and what happens to pictorial appearance when that theory is overturned.

First, the concept of a public. It is a fact too frequently neglected that although images are made in almost every human culture, in only a few do the users of the images relate to them as a public to presented artifacts. The range of uses of images is as wide as the range of human activities although the concept of image itself, connoting a mere to-be-looked-at-ness, tends to obscure the diversity. While it is perhaps trivially true that every image is made to be seen, since it could not serve any purpose otherwise (although, if Nietzsche's account of the veiled god of the Bacchic rites is true, even this would be too absolute[13]), not every image is made merely to be looked at. Illusionistic images are always of this latter sort, however, for otherwise they could not do their specific work of passing for what they are not. Such images appeal to the eyes and to no other sense organs, to the eyes at the expense of other sense organs, and as such require a kind of detachment from practical use. The attempt to see Henry VIII in Holbein's portrait, for instance, requires that the observer keep a distance, but not too great a distance, from the painting, meaning that she ought not to try to smell it or examine its back or try to confirm through any other sensory modality that the king is really present. Indeed, in the extreme case of *trompe l'oeil* illusionism, seeing the illusion requires standing quite still at an observation point appointed by the picture itself. Not every illusionistic picture is as restrictive as the *trompe l'oeil*, but every one to some degree demobilizes (or presumes the demobilization of) the picture user in the name of the appeal to the eyes, thus transforming her into a looker, or onlooker. This demobilization opens up a space of visual action at a distance, a space which, while crossed by the painter with tool in hand, is only crossed by the spectator with her eyes and so allows for a multiplicity of simultaneous users. It is this demobilized multiplicity that is connoted by the idea of a public.[14]

The public, of course, has been known to be rowdy. At times it refuses to accept the discipline of illusionistic pictures. Periodic eruptions of iconoclasm have led to the vandalizing and destruction of pictures, but it should be noted that physical destruction is an *excessive* reminder of the materiality of pictures. The iconoclast's mere approaching of the illusionistic picture is itself the destructive act since at a certain proximity the illusion collapses; the actual consignment to the flames is an orgy of violence intended to save others from having to repeat the destruction. The real danger posed by the iconoclast is her demonstration that seeing the illusion is optional, that it is always in the power of the beholder to refuse the pictorial invitation. The iconoclastic assertion of the spectatorial role, the forcing to the surface of the materiality of the picture, is precisely what the public of detached beholders is constructed to avoid. For the image to appear iconic it must not seem as if its so appearing is a function of the public attendant on it. In other words, the public of beholders must see itself as attendant on, not productive of, the illusion. This requires not just the exclusion of the iconoclast but her repression. The threat of iconoclasm must not be fought off, it must be forgotten, if the appearance of illusionistic art is to emerge. This is the social coherence of the public to which Gombrich refers.[15]

The socially coherent public contructed by and for illusionistic image-making must be so constructed for the sake of appearances. For legible representations like books the public is unlike that for paintings, in the important regard that books are not to be looked at but rather are to be seen, at least ordinarily, in order to be read; there is no level at which legible representations appear to represent – they just do represent. Put differently: books are to be seen in order to do something else but illusionistic images are just to be seen. Perhaps this too is obvious since there is little one can do with an illusion except see it, but we must ask – obvious to whom? It is only obvious once the socially coherent public has been constituted. once the game of reading iconicity into a marked surface is fully engaged. If one is not committed to such a reading into, if, that is, one is an incipient iconoclast, there are as many things one can do with an illusionistic image as one's imagination permits. The public committed to a reading-into suppresses alternative possibilities in order to see, to really and merely see. For the sake of appearances the public must be trained to merely see, to be a seeing public. Now, not all illusionism works in the same way; impressionism, for instance, requires a different kind of public from Baroque illusionism. The history of illusionism, then, is the history of the various learning processes, the history of the construction and reconstruction of the public. This is what Gombrich means by the history of seeing.

Finally, I come to the scare quotes. This is the coyest element in Gombrich's proposal; with them he gestures at taking back the claim that the history of seeing is really a history of seeing, suggesting instead that it is a history of something-like-seeing. Why the apparent withdrawal? On the one hand, if Gombrich is right that iconic signs do not in themselves resemble their referents but rather only appear to because a public has been properly taught, then their iconicity is not itself visible; the way a mark is treated by a contingently constituted public cannot itself be seen in the mark. (This is why Gombrich uses the concept of projection in explaining iconicity.) If the iconicity is not visible naturally it cannot be seen and so the history of 'seeing' cannot strictly speaking be a history of seeing. It is instead a history of seeming-to-see, which is to say a history of appearances.

On the other hand, while the scare quotes seem to take back what is advanced between them, they also oddly insist on the necessity of advancing it. Gombrich could simply replace 'seeing' with 'seeming to see', but in signifying his refusal to do so he signals that something important would be lost. What would be lost, I think, is the idea that seeming to see is really a kind of seeing. The argument Gombrich is advancing would seem to provide sufficient motivation to substitute 'interpreting' or 'using the signs of painting' for 'seeing', but to do so would be to submit to the idea that projection and vision are exclusive processes, that projection is the pulling down of a veil over what is really there and is now obscured. Instead, Gombrich suggests that there is a kind of seeing which centrally depends on a projection which is delayed or deferred in its return.

I will not discuss all the conditions under which seeing what has been projected becomes a kind of seeing as such. One of the conditions, though, is surely that the projection is hidden from the spectator, since if she believes herself to be projecting then she will also believe that what is in front of her, despite appearances, is a screen. (One of the goals of psychoanalytic treatment, after all, is to get the analysand to see the screen in order to see that she is projecting.) If Gombrich's argument is that projective seeing is a kind of seeing as such, he will have to argue that at least this condition obtains. He does so in "Meditation on a Hobby Horse", and his argument there is already in part predictable: what hides the fact of projection from the projector is her membership in a socially coherent public.[16] The projections which create iconicity are licensed, which is to say licensed to be forgotten or disavowed, by the collective behavior of the public engaged in projective seeing.

Two points can be made now. First, Gombrich's arguments make explicable why a false theory of objective iconicity would once have been plausible. If one of the conditions under which a culture of illusionism emerges and which differentiates it from the various cultures of the conceptual image is the forgetting of projection, then iconicity will appear objective therein. A culture of illusionism depends on a kind of unspoken consensus not to represent to itself the grounds of its consensus. From Vasari through Ruskin the pursuit of the *non*-conceptual image has been taken as the pursuit of the *anti*-conceptual image, precisely because illusion has been identified as the creation of *objectively* iconic signs. For a theorist in this tradition what would appear to need explaining would be the objective resemblance of certain kinds of representations to their objects. Instead of explaining their own projection - indeed, in order not to explain their own projection - such theorists would instead seek to justify it by thematizing the objectivity of the icon.

A theory of objective iconicity would appear not only plausible, however, it would also appear obvious. Given that the successful construction of the coherent public is symptomatized by the forgetting of projection, what is seen after projection, the pictorial appearances, would appear, so to speak, on their own. They would be brute perceptual data, facts of the perceived world; the movement from factum to fact, from the made to the given, would have already disappeared in the pictorial appearance. The second point that can be made, then, is this: Gombrich's capacity to explain iconicity without being committed to its objectivity, to distinguish the illusionistic from the conceptual image through psychosemiotic analysis rather than through (let us call it) ontology, is a sign of his external relation to the illusionistic tradition.

It is, in short, the non-parochialism of his defense of the distinction between the two types of images.

I would like to put this point one other way in order to highlight the nature of Gombrich's externality to the tradition of illusionism. It would be a misrepresentation of his point about the socially coherent public to assert that its members do not see the real marks, the material substrate, of illusionistic pictures. Indeed, Gombrich shifts the burden of explaining iconicity to psychosemiotics to escape the dichotomy of either seeing the appearance or seeing the "real" thing; reading iconicity into visual signs requires a visual grasp of the material of the sign, of how to use the sign with one's eyes. The appearance of a conflict between the visual appearance and the materiality of the mark can only arise if one starts with the premise that when a spectator looks at the mark she really sees something else, and this premise can only arise within an illusionistic culture committed to the objective iconicity of its visual signs. The conflict itself is a symptom of the forgetfulness of the illuded public which expresses that public's self-misrecognition. Gombrich's critique of the traditional explanations of illusionism thus cannot amount simply to pointing out that when we see an illusionistic image we really see a marked surface since his counterpresumption is that we already know that however hard we may try to forget. A better way to think of the critique would be by analogy to the observation (and I do not remember by whom) that not seeing the object in the illusionistic image is a kind of pictorial dyslexia, a deficit in the capacity to keep what are known to be meaningful signs in a meaningful array; in this light, we might see Gombrich as trying to induce pictorial dyslexia in the socially coherent public. He tries to render iconic signs unreadable as icons by foregrounding that they are signs. The spectator so incited will not experience the breakdown of the illusion into its substrate, but will experience the presence of the screen onto which she has been projecting. In seeing the signs as signs, the spectator will witness her own activity of projection, her own spectatorial activity.

What Gombrich is trying to get us to remember is our own making of iconic signs, our own making iconic of some signs, and so our making some signs to resemble their referents. This position is genuinely orthogonal to the forgetfulness of the illusionistic tradition; if the reminder works it undermines the disavowal of our own activity which would otherwise allow us to take a sign to be objectively iconic because it appears to be. The premise of so many earlier attempts to explain pictorial appearance as rooted in objective features of the icon expresses how earlier theorists have taken themselves to be observers attendant on the fact of iconicity. Such a premise, of course, just is the mark of membership in the socially coherent public; to take as given what has been produced, to experience iconicity as objective, is to duly forget the projection which underwrites the illusion. Those projects which attempt to explain iconicity from the point of view of its spectators thus can be seen as attempts to justify the forgetting. As Gombrich displays the projection along with the exhibited sign, the search for the grounds of iconicity in natural mimesis appears as a constitutive misrecognition of a human activity. By arguing that the illusion in illusionism needs to be traced to the psychological structure of the mimetic regard, the mimetic eye, Gombrich adopts the genuinely iconolastic posture. Rather than overthrowing a regime of pictorial appearance by destroying the icons (an act which actually embodies the misrecognition by attacking the object itself instead of the psychological structure which animates it), Gombrich pushes the anti-objectivist position where it needs to go, to the activity which is the root of the experience of iconicity. This reawakening return to look at the picture again is a much more radical posture, one much more external to the illusionistic tradition than any bonfire of the vanities.

IV. NEW EYES

The strength of Gombrich's psychosemiotic analysis of pictorial illusion is that it makes the conceptual/illusionistic distinction a functional one. We do different things with signs that are morphologically or materially identical and it is what we do that makes the difference. The burden of explaining the difference thus must be borne by an analysis of the different functions, the different ways of looking and making. If we functionalize the distinction between signs that need to be read and those that need merely to be recognized it comes to this: conceptual images are made to be read, to be interpreted by referring back to a code, but illusionistic images are made only to be recognized, made in the context of a visual code, the operation of which is concealed in the actual operating of it by the spectator. This concealment is not an accidental property of illusionistic pictorial appearance but rather its defining feature. Since to be an iconic sign is to be be seen as objectively embodying some characteristic of the represented motif, to be iconic is to be seen as uncoded. It is not so much that the icon resists the recognition of its own being operated, it is, in a sense, the product of that resistance.

The structural feature of the function of iconicity that its satisfaction requires the concealment of the function is what makes the psychoanalytic concept of projection attractive to Gombrich. While it may be an obvious truth about all signs that we make them to represent, we only make a sign iconic by forgetting that we make it, and to what discipline other than psychoanalysis should we turn to grasp the productivity of forgetting? This forgetting is constitutive only of illusionistic pictures since only they require the appearance of objective mimesis; only they resist being seen in the light cast by the history of their own making. Thus, psychosemiotics is iconoclastic only for illusionistic images, only, that is, for those pictures which appear in the wake of the disavowal of their histories. To point out the underlying semiotic functions of other sign systems just is to do semiology, but to point out the semiotic functioning of the sign of forgetfulness is to call the spectator back from her "objective" self-effacement.

The goal of the iconoclastic concern with signs that resist their own history is described in *Art and Illusion*, the lectures in which the psychosemiotics is actually done, as follows:

> The main aim I have set myself in these chapters is to restore our sense of wonder at man's capacity to conjure up by forms, lines, shades, or colors those mysterious phantoms of visual reality we call 'pictures'.[17]

That our sense of wonder has been lost reinforces the notion that Gombrich's iconoclasm is intended to *productively* break the spell on the spectator by bringing to her attention both the literal manufacturing of illusionistic pictures (the ways they are made to be for the eyes) and her own constitutively hidden complicity in their making. Now, iconoclasm usually connotes a hatred for pictorial illusion, understandably so given its rather dour champions, but if by wonder Gombrich means that profound cognitive and perceptual engagement which Bacon called the seed of knowledge, his aim is not to be an art historical Savonarola but rather to expand the possibility of spectatorial pleasure in the appearance of self-activity – pleasure and iconoclasm need not be forever disunited. Gombrich's goal in restoring history to the sign which resists it is to reanimate wonder at our making of pictorial appearance, to remind us that we are present in the signs which appear not to need us. In short, Gombrich aims at making manifest, visible, both pictorial illusion and the spectatorial function whereby it is made but without destroying the apparent space of pictures. In setting this new project for looking, Gombrich's iconoclasm, like Cezanne's, is an inflationary rather than a deflationary one.

The addressee of Gombrich's iconoclasm is thus a new spectator of pictorial appearance capable of watching, from the external perspective opened up by psychosemiotic analysis, the traditional acts of spectation. This addressee is, in other words, the same one as the new seeing subject of Cezanne's painting. To watch a Cezanne is to watch paint become a world, or, put differently, to see the emergence of the space of the picture from the picture making. Cezanne shows us the doubleness of the picture and thus calls us to witness our own spectation made objective. This is a delicate operation requiring a double resistance, a resistance both to the deflationary drive to collapse painting back into mere paint, and to giving in to the traditional spectator's desire for the appearance of readymade iconicity. Cezanne cannot accomplish this on his own since – to reiterate, but I hope not pointlessly – he is a painter, a maker of appearances, and so needs to call forth the right eyes. But the eyes he calls forth are self-reflectively aware of their function. It is this kind of seeing subject whom Gombrich's iconoclasm also wants to generate.

Cezanne and Gombrich share a common foe in the forgetfulness of the illusionistic tradition. Both address themselves not to the false transparency of illusionistic appearance but rather to its production as if transparent, to its production by means of an induced forgetting which produces the appearance of objective iconicity. In short, both pursue a deconcealment and exhibition of the spectatorial function, a project which is a far cry from demanding the destruction of that function. If the construction of the public for illusionistic pictures was and is a productive enterprise, so too is the construction of the self-reflective spectator of Cezannian and Gombrichean iconoclasm. To be able to see our perceptual worlds as made and keep making them – perhaps the seemingly paradoxical nature of this splitting of the subject was the threat held over those who might feel the pull of iconoclasm. To be an iconoclast and yet not succumb to that threat is the constructive project of modernist pictoriality shared by Gombrich and Cezanne.

Vanderbilt University

NOTES

[1] Cezanne, of course, plays a central role in most received histories of modernism; Gombrich, in part by his own design, in none that I know of. This seems to me unfortunate. If in the course of this essay Gombrich is rehabilitated for modernism that would be a substantial fringe benefit.

[2] Rainer Maria Rilke, *Letters on Cezanne*, translated by Joel Agee (New York: Fromm International Publishing), 1985: p. 87.

[3] Maurice Merleau-Ponty, 'Cezanne's Doubt', in *Sense and Non-Sense*, translated by Hubert L. Dreyfus and Patricia Allen Dreyfus (Evanston, Illinois: Northwestern University Press), 1964: p.12.

[4] See Richard Wollheim, *Painting as an Art* (Princeton: Princeton University Press), 1987: pp. 13–100, especially pp. 42–45.

[5] Rilke, *op. cit.*, p. 43.

[6] By Gombrich's earlier work I mean everything up through the 1956 lectures published as *Art and Illusion* (Princeton: Princeton University Press), 1960. Whether there is a turn away from the radicalism of the early posture, indeed whether the early posture was ever as radical as it seemed, is a vexed question. See the harsh exchange between Murray Krieger, 'The Ambiguities of Representation and Illusion: An E. H. Gombrich Retrospective', *Critical Inquiry* **11** (1984): pp. 181–94, and 'Optics and Aesthetic Perception: A Rebuttal', *Critical Inquiry* **11** (1984), pp. 502–08, and Gombrich, 'Representation and Misrepresentation', *Critical Inquiry* **11** (1984): pp. 195–201. I believe there is a turn but that it is not sharp and that the signs of it are already visible in *Art and Illusion*. In any case, since I am restricting myself here to the earlier work alone, I can leave a resolution of this matter for another time.

[7] John M. Kennedy has argued that stick figures are more nearly recognized than decoded. If true, they would then need to be reclassified according to this scheme. See 'Depiction Considered as a Representational System', in John Fisher, ed., *Perceiving Artworks* (Philadelphia: Temple University Press), 1980: pp. 131–65.

[8] E. H. Gombrich, *The Story of Art* (London: Phaidon Press), 1951: pp. 36–38.

[9] *Ibid.*, p. 52.

[10] E. H. Gombrich, 'Signs, Language and Behavior', in *Reflections on the History of Art* (Berkeley: University of California Press), 1987: p. 244

[11] *Ibid.*, p. 248.

[12] *Ibid.*, p. 248.

[13] Friedrich Nietzsche, *The Birth of Tragedy*, translated by Walter Kaufmann (New York: Random House), 1967: pp. 61–67.

[14] The demobilization referred to here is not, to be sure, inconsistent with all other practical uses. Some political uses, for instance, coincide with the distancing aspects of illusion; a great enough distance between observers and image can open up the space for the mass spectacle characteristic of fascist pageantry. Similarly, some religious purposes can be served by the way demands for stillness can coincide with a contemplative attitude and humility. The investigation of these 'secondary subjections' would form an important pendant to a full-scale study of the history of illusionism.

[15] The construction of the coherent public does not take place in a cultural vacuum. Cultures of illusionism provide both literal and figurative supports for it. Greek illusionism thus can only be understood in the context of stage design, of theatrical display aimed at an already existing commitment to an effect of the real in the representation. For the

analysis of Renaissance illusionism one would need to place artistic practices into the contexts of the illustration of hagiographic narrative, contemporary debates about the Incarnation, and so on. The coherence on which the public's existence as a public depends is a complex social product.

[16] "The blob in the painting by Manet which stands for a horse is no more an imitation of its external form than is our hobby horse. But he has so cleverly contrived it that it evokes the image in us – provided, of course, we collaborate", Gombrich writes in 'Meditations on a Hobby Horse or the Roots of Artistic Form', in *Meditations on a Hobby Horse and Other Essays on the Theory of Art*, 4th edition (Chicago: University of Chicago Press), 1985: p. 10.

[17] *Art and Illusion, op. cit.*, p. 8.

MICHAEL KELLY

DANTO, DUTTON, AND OUR PREUNDERSTANDING OF TRIBAL ART AND ARTIFACTS

How can we understand the art and artifacts of tribal societies, given that the ethnocentric label 'tribal' implies that those societies are radically different from our own culturally and conceptually? To begin with, how do we know whether tribal people make an art/artifact distinction? If they do, is it the same as ours? If not, how can we correctly understand what seem to us to be their art and artifacts?

These questions suggest a dilemma. If we disregard the historical differences between our practices of creating cultural objects and those of tribal societies, a universal art/artifact distinction is possible; but at the same time the claim to universality is difficult to defend either because it is not clear how it could ever be substantiated or because it is thought to be a weapon of cultural imperialism. Yet if we highlight historical differences too much, a universal distinction becomes impossible and relativism sets in, locking us in our own world, unable to understand tribal objects. Is there any way out of this dilemma, which has led, according to James Clifford, to "a pervasive postcolonial crisis of ethnographic authority"?[1]

Arthur C. Danto and Denis Dutton have recently reflected on these issues.[2] While Danto is right that "it is through a system of philosophical thought that we must construct the questionnaires that are to take us into the other culture,"[3] he is too quick to elevate this "system," and particularly our art/artifact distinction, to universal status. Dutton is judicious in balancing Danto's conceptualism with the perceptual factors needed to discern the differences between tribal objects, but he misguides us by suggesting that our perception is independent of our concepts of art and artifact. After examining Danto's and Dutton's strategies, I argue that we can rely on our art/artifact distinction if we do not insist that it be universal. It is rather to be seen as one of three major components of our *pre*understanding of tribal objects, the other two being our 'commodification of tribal objects' and 'aesthetic interest.'

By 'preunderstanding' I mean our historical understanding of our own art and artifacts, which forms the cognitive and affective starting point of our experience of tribal objects and which enables them to become

C.C. Gould and R.S. Cohen (eds.), Artifacts, Representations and Social Practice, 39–52.

objects of possible understanding.[4] Although this preunderstanding is constitutive of our understanding of tribal objects as art and artifacts, it alone does not transform them into art and artifacts any more than the label 'tribal' transforms a society into a tribal one. That is, how we come to understand tribal objects is a distinct, though not separate, matter from what makes them what they are. The 'what' is determined by historical factors in addition to our preunderstanding, beginning with the cultural conditions under which the objects are created and including how they are conceived and perceived by the tribal people themselves.

Now, given our preunderstanding of objects, and in light of the above dilemma, how can we acquire a 'correct' understanding of them? A critique of our preunderstanding allows us to do so, and thereby to dissolve this dilemma, though only if we remember that our preunderstanding is a historical condition of, but not equivalent to, our understanding of tribal objects. The criteria of correct understanding are not established by us or the tribal people alone, but in the process of our dialogue with them; such criteria are therefore neither prior to nor independent of this dialogue. My strategy here is thus a middle ground between Danto's formal ideal of universal concepts of art and artifact, and Dutton's ethnographic ideal of cross-cultural perception of tribal objects.

I. DANTO'S POT PEOPLE AND BASKET FOLK

Danto discusses the problem of how to understand tribal objects by devising a thought experiment. He conceives of two cultures, the Pot People and the Basket Folk, who occupy "distant regions of the same vaguely bounded area, but separated by some geographical feature that has enabled them to evolve in different ways."[5] Each culture makes pots and baskets, but with an important difference. As their names suggest, the Pot People's pots are "thick with significance," while their baskets are merely items of domestic life; and the Basket Folk's baskets are "objects of great meaning and possessed of special powers," while their pots are useful objects with no further importance.[6] Since the two cultures have never had any contact with one another, each is quite confident of its distinction between their own pots and baskets.

Matters become complicated when examples of the two types of pots and baskets are transported to two museums in a third, Western culture.

The works of art, the Pot People's pots and the Basket Folk's baskets, are exhibited in an art museum; while the artifacts, the Pot People's baskets and the Basket Folk's pots, are displayed in a natural history museum. So far so good. But then two dioramas are set up in the natural history museum to represent the everyday lives of the Pot People and Basket Folk. Each diorama includes examples of pots and baskets in their separate roles as art and artifacts. When some young school children visit the natural history museum, one girl, who is an aspiring philosopher, is puzzled about the Pot People's and Basket Folk's baskets because they look exactly the same, yet some are labeled artifacts and some art. She asks her teacher, who responds that there is a difference though one that she herself cannot explain. Unwilling to accept that there is a distinction she cannot see, the student persists with her inquiry about how to discern art and artifacts when there are no perceptual differences between them.

After rejecting several ways of making the art/artifact distinction other than on perceptual grounds,[7] Danto introduces his own proposal. He argues that the distinction is a conceptual or philosophical matter to be treated accordingly: the meaning, thought, and content of the Pot People's pots and of the Basket Folk's baskets distinguishes them as art from their look alikes – the Pot People's baskets and the Basket Folk's pots – which have only utility values. The particular meaning will of course vary from one culture to another (we emphasize reason, for example, while tribal peoples are thought to emphasize power), and there will be similar variances with utility. Yet the distinction itself is universal: "My sense is that the philosophical structure of African artworks is the same as the philosophical structure of artworks in any culture,"[8] and similarly for the structure of the respective artifacts. Thus, according to Danto, what makes a work of art a work of art is the same, in concept, for tribal cultures as it is for Western cultures, and what makes an artifact an artifact is likewise the same for both (and, in fact, for all) cultures. Hence the need for the young girl in the thought experiment to be an aspiring philosopher: she is finally able to discriminate between the perceptually indiscernible baskets by relying on a universal, conceptual art/artifact distinction.

Part of the attraction of Danto's argument is that his universal art/artifact distinction seem to solve the young student's problem; that is, it is persuasive on pragmatic grounds, albeit with conceptual tools, by showing that the resemblances between the different baskets are aesthetic illusions.[9] Yet although some such distinction is useful in solving the problem at hand, it is not for that reason universal. What argument does Danto then provide to justify his claim that the art/artifact distinction is universal? He invokes Hegel at this point, which I think only weakens his case.[10] Whereas Hegel has a metaphysico-logical system to support his contention that the Western concept of art is universal (since it reflects the highest stage of the world-historical development of art toward its Idea), Danto rejects such a system.[11] He also discusses here the similarities and differences between the Greek philosophical understanding of art and artifacts and our own; plus he recommends that we can come to understand the art and artifacts of other cultures through the meaning and utility they embody just as Wittgenstein suggests that we can understand other people's minds through the medium of their bodily actions. While these points clarify Danto's ideas, they do not support them.

Since Danto provides no other convincing argument for his claim about the universality of the art/artifact distinction, and since it is not clear to me what other he could give, the claim remains unpersuasive. This means that, on Danto's own terms, he has not yet solved the young student's problem; for he would not be satisfied appealing to an art/artifact distinction that is ours alone. In addition, as Susan Vogel argues in the same catalog in which Danto's essay appeared, a universal distinction would not solve the problem anyway; for there are many tribal objects that are meaningful though not beautiful and many that are useful and beautiful at once.[12] The point of this objection, which Dutton echoes, is that the problem cannot be solved without the help of the aesthetic, perceptual dimension of tribal art and artifacts which Danto systematically excluded.[13]

II. DUTTON'S JUNGLE PEOPLE AND TOURIST PEOPLE

Dutton raises several of his own objections to Danto's thought experiment.[14] His major objection begins with a question: By whom can the differences between the Pot People's and Basket Folk's baskets (or pots) not be perceived? To answer, Dutton introduces a new thought experiment, drawing on his experiences with art and artifacts from the Sepik River region of New Guinea.

He asks us to imagine two cultures, the Jungle People and the Tourist

People, who in general develop separately but have had some contact, which we know because their languages and mythologies overlap in significant respects. The two peoples produce very similar carvings, but the Jungle carvers make them as part of their ritual practices and cultural traditions. They are highly revered for making intricate, two-sided carvings with hard wood which contain spirits of the dead visible in the dark. The Tourist People, as you would guess, are merchants from the same region who make carvings exclusively in order to sell them to tourists. They carve with soft wood and only on one side, and they keep the designs as simple as is required to make the tourists believe they are getting authentic Jungle carvings.

Dutton continues his thought experiment by introducing what he takes to be a Danto-like supposition, namely, that nobody can ever tell the difference between the Jungle People's and Tourist People's carvings. While logically possible, this is barely conceivable once we step out of the philosopher's study, Dutton argues. First of all, when we make mistakes about the different carvings, we misperceive individual examples, not a whole genre, as Danto seems to assume. So "the irrelevance of perception," which is his conclusion, "cannot be said to follow" from our isolated mistakes.[15] Secondly, even if at times *we* cannot distinguish the different carvings, certainly the Jungle and Tourist People are able to differentiate them. Moreover, Dutton adds, we can learn to avoid such mistakes by following their lead: " . . . trained perception, the ability of tribal peoples themselves to see systematic differences between art and artifact – and the ability of the informed Westerners also to learn to perceive differences – is the key."[16] In effect, he thus suggests to the young student in Danto's thought experiment that she emulate the Pot People's and Basket Folk's perceptual acumen in order to distinguish their baskets in the natural history museum: " . . . it is a matter of gaining cultural knowledge in order to see aesthetic qualities which have intentionally been placed in the objects to be seen."[17]

Although I agree with the spirit of Dutton's objections, since he critically examines the relevance of Danto's thought experiment to our understanding of tribal objects, he also fails to solve the young student's problem. The 'by whom' question at the heart of his major objection, which seems disarmingly to the point, speaks to a different issue. Danto never claims that the Pot People and Basket People cannot recognize the differences between their respective pots and baskets. Rather, his inter-cultural problem is how *we* in the museum can distinguish them.

Dutton cannot solve it simply by appealing to how *they* (the People) make their art/artifact distinction on perceptual grounds and suggesting that we follow their example. This appeal merely shifts the problem *from* one of how to understand tribal objects from the perspective of our art/artifact concepts *to* one of how to understand them from the perspective of the Tribal People's perception. To resolve this new problem, he must explain both how they perceive the differences between their objects, and, more importantly, how we can understand their perception, especially given the difficult case of indiscernibles Danto imagines. These explanations, not the acumen of those who make the relevant distinctions, provide the key here.

The difficulty with Dutton's strategy is that we cannot adopt the perceptual perspective of the tribal people without first critiquing our own perspective in conceptual as well as perceptual terms. Such a critique is precisely what Danto opens up by stressing the role of our concepts in the understanding of tribal objects, but then he forecloses it, I think, by claiming that our art/artifact distinction is universal. Dutton, on the other hand, takes one step forward by reintroducing perception into the picture, but then he retreats, in effect, by underestimating Danto's perceptual problems. He does address these problems by discussing forgery, namely, the possibility that the Tourist People could eventually make carvings that are perceptually indistinguishable from the Jungle People's. He tries to solve them by claiming, again, that we could be duped only in individual cases, and that we could minimize the risk of forgery even in those cases by acquiring more knowledge of the Jungle People's cultural meanings and artistic practices.[18] But this strategy cannot work because we cannot learn to discriminate between Jungle and Tourist carvings without relying on how we understand (perceive and conceive) our own art and artifacts, that is, without relying on what I am calling preunderstanding.

III. OUR PREUNDERSTANDING OF TRIBAL OBJECTS

A nineteenth-century prejudice in art history, anthropology, and philosophy was that you could not have art without culture, that tribal people did not have culture, and that therefore they did not have art.[19] By contrast, a modern, twentieth-century prejudice is that, as our own understanding of art was expanded, we began to discover that tribal people have both culture and art.[20] Most recently, postmodernists have admonished us to bracket our concepts of tribal art and artifacts in deference to those of the tribal people whose objects we are trying to understand.[21]

These shifts in our view of other cultures seem momentous, until we look more closely. While the elevation of tribal objects to the status of art seems to be part of a world-historical process of decolonization, it actually reproduces one of the main discursive weapons of colonialism. Earlier, it was the Western definition of culture that excluded tribal people's objects from the pantheon of art, and now it is the Western understanding of art that includes them. In both cases, the West has what Sally Price calls "definitional prerogative," i.e., an advantage in making important distinctions for all the world's cultures.[22] As a result of this advantage, there is an asymmetry in our discussion of tribal objects: we set the standard to judge whether they are to be excluded or included, while insisting that the standard is not ours alone but universal (even if the tribal people are not aware of it).

In the shift to postmodernism, this asymmetry appears to be overcome because the tribal people are now allowed to speak in their own voice, which is clearly a positive step. Unfortunately, the asymmetry is replaced here by unintelligibility rather than symmetry if we are expected to bracket our preunderstanding of tribal art and artifacts. For we cannot possibly have a dialogue with the tribal people if we silence our own voice, if we eliminate one of the conditions that makes tribal objects into objects of possible understanding. On the other hand, we can have a dialogue with tribal people, *if*, while acknowledging our preunderstanding, we refrain from claiming that any of its components is universal.

There are, I think, at least three major components of our preunderstanding of tribal objects: (A) our commodification of tribal objects; (B) our aesthetic interest; and (C) our art/artifact distinction. In explaining these components, I would like to separate two of Clifford's insightful observations about the 1984 primitivism show at the Museum of Modern Art in New York:[23] (1) that the MoMA curators' concepts of art and artifact shaped their understanding and exhibition of the 'primitive' objects; and (2) that some of these curators claimed to have discovered a universal art/artifact distinction in the course of their experiences with tribal objects.[24] While the first point concerns what I think is a historically conditioned dialogical process of preunderstanding, experience, and understanding, the second is a philosophically unjustified elevation of our Western concepts.

(*A*) Most of us in the West experience tribal objects in the context of commercial trading and cultural exchange. Such trade and exchange are material evidence of the fact that we experience them in the way we conceive of our own art and artifacts, namely, as commodities to be bought, sold, and displayed in museums. Although this fact may seem merely negative because it implies an imposition of our cultural practice onto their objects, it also has a positive or enabling side: by establishing a mediating context between tribal cultures and our own, the commodification of non-Western objects makes them into objects of experience and possible understanding.

Once we experience the tribal objects, there is of course a problem about how we can attain a correct understanding of them. This problem arises, however, only after they are in this new context. So it is the dynamics of this context which set our possible understanding in motion and which determine, in part, the criteria of correct understanding. We soon find, for example, that we cannot conceive of them just as commodities for the simple reason that we do not conceive of our own art and artifacts as mere commodities. That is, although tribal objects have to become commodities to be understood by us as art and artifacts, they are not for that reason only commodities. Their objects, like ours (we assume), have complex identities as modes of cultural expression, spirituality, critique, etc. While these identities are certainly altered once the objects become commodities, they are not entirely lost any more so than are the identities of our art and artifacts.[25] There is thus something else about the tribal objects, especially the ones we think of as art, which prevents them from being reduced to commodities. What is that something? The answer to this question requires that we understand more about the historical context in which these objects are created. But before doing that we must examine the other components of our preunderstanding because they shape how we understand that context.

(*B*) The Kantian notion (and practice) of aesthetic interest (as distinct from moral, practical, scientific, and other interests) has historically been an integral part of our conception of art, helping us to distinguish works of art from mere objects, commodities, and artifacts. But aesthetic interest has also helped to blur the art/artifact distinction. For it is when we believe that we can willy-nilly take an aesthetic interest in cultural objects regardless of their historical context that we find ourselves confused about the art/artifact distinction; in short, we often find ourselves taking such an interest in artifacts as well as art. This happens especially when our aesthetic interest is combined, as it generally is, with commodity exchange. For example, it is by divorcing tribal objects from their historical context in order to deal with them as commodities that we first being to think of them in terms of aesthetic interest. But when these same tribal objects are perceived in aesthetic terms, they resemble our artifacts and so we become confused (along with the young girl in Danto's natural history museum) about what distinguishes art from artifact after all. Aesthetic resemblances thus blur the very distinction aesthetic interest was intended to clarify.

At the same time, however, our aesthetic interest is an enabling condition of our experience and possible understanding of tribal objects. Without it, we would not likely interact with them at all, or else we would tend to treat them as mere commodities, artifacts, or, as Danto has noted, art only in the analogical sense that they are thought to reflect earlier stages of our own art history.[26] So to be regarded by us as what we consider to be art today, tribal objects must be viewed, to begin with at least, in terms of aesthetic interest.[27]

(*C*) Finally, our art/artifact distinction also enables our understanding of tribal objects, for we simply cannot understand them without presupposing how we conceive of our art and artifacts. This does not mean, however, that we understand their objects only in terms of this distinction. The reason for this qualification is not only because of the other components of our preunderstanding and the historical context in which the tribal objects are created, but also because of the open-endedness of our distinction. That is, we have so much trouble distinguishing tribal objects, as well as understanding how the tribal people distinguish them, because we have so much trouble making our own objects conform to our distinction. Many people agree that most, if not all, works of art are also useful and thus are artifacts as well, though few, if any, artifacts have what else it takes to be works of art.[28] So art has something more than use value, even though it may have that, too. There is little agreement, however, about what the 'more' is, nor does there need be any. It is enough that people agree that there is something more, granting them the freedom to explain what it is on their own terms.

Now, the freedom we already extend to people in our own culture is just as easily and commonly extended to those in tribal cultures regarding their own objects. But it remains to be seen what they make of this freedom, namely, how they make the art/artifact distinction; and it is even an open issue whether they make the distinction at all. These possibilities – that the tribal cultures make the art/artifact distinction differently or that they do not make it – are just extensions of the fact that the distinction is so unstable in our own culture. From this point, in order to understand the tribal peoples's actual ideas about art and artifacts, we need to follow Dutton's advice and learn more about their culture. But we must acknowledge, in a way that Dutton does not do explicitly, how such learning is itself affected and even made possible by our preunderstanding. What we are most likely to find is that the tribal objects will confound as much as confirm our art/artifact distinction, which suggests that we are on the right path to attain a correct understanding of them.

In the long run, I think what we do is *construct* a common art/artifact distinction between ourselves and tribal peoples in the historical, dialogical process of coming to understand each other's cultural objects. Our common distinction, to the extent that it exists at all, is thus a result of our dialogue with them. Satisfied with this distinction insofar as it facilitates mutual understanding, we have no need to regard it as universal. Rather, it is bound by the context of our dialogue, which is determined in part by our preunderstanding and in part by the historical factors constitutive of the cultural creation of tribal objects.

Now, if the abstraction of tribal objects from the cultural context in which they were originally embedded is part of our commodity practice, aesthetic interest, and art/artifact distinction, it might seem that acknowledgement of this point should inspire us to return to the original or 'pure' context in order to understand the objects better. This is not possible, however, because that context disappeared long ago when we made contact with the tribal cultures; it is now merely an ethnographic ideal. Moreover, as Price and Howard Morphy have pointed out, when we have emphasized a 'pure' context in dealing with tribal objects, we have traditionally done so to discriminate against them, i.e., to regard them as merely useful objects circumscribed by a narrow cultural context that makes them unfit to be treated as art.[29] Rather than pursue a utopian or belittling notion of context, we need to critique our preunderstanding of tribal objects in order to comprehend how it shapes the present context in which we experience and understand them. It is this inter-cultural, dialogical context that is most relevant now, for it is here that a correct understanding of tribal objects (and their correct understanding of our art/artifacts) is determined.

IV. DIALOGUE AND CRITIQUE

How does my brief account of our preunderstanding of tribal objects dissolve the dilemma introduced at the start? It does, I think, because our preunderstanding – namely, our commodification of art and artifacts, aesthetic interest, and art/artifact distinction – makes it possible for us, first, to experience tribal objects and, then, to understand them. This preunderstanding is a major though not the sole determinant of what makes the objects we experience what they are; this 'what' is also determined by historical factors in the tribal culture. The part that our preunderstanding determines is more the *whether* and *how* we understand than the *what*.

We might imagine, along with Dutton, that without our preunderstanding we would have less, if any, problem perceiving the distinction between tribal objects; but without it either we would not experience them at all or, if we did, they would be perceptually unintelligible to us. By relying on our preunderstanding, on the other hand, we admittedly have a problem of correct understanding since our preunderstanding may limit us. To eliminate this presumed limitation we might then imagine, along with Danto, that the concepts constitutive of our preunderstanding are universal. But it is virtually impossible to substantiate such a claim. My alternative to Danto and Dutton has been (1) to acknowledge the historicity of our understanding of tribal cultures in the form of our preunderstanding of them; and (2) to show that we are enabled by, but not confined to, our preunderstanding as long as we critique it when we are in dialogue with the tribal peoples. While Danto over-values our concepts and Dutton over-rates the tribal people's perception, the focus of this dialogue should be the relationship between historical concepts and perception on both sides.

Since I have objected to the alleged universality of our art/artifact distinction, how can I answer the charge of relativism? It may be objected, for example, that I have told only one side of the story about tribal objects; for my account says nothing conclusive about whether the tribal cultures regard their objects as art or artifacts, whether they view them differently at all, or whether they view them differently only when they are taken out of context by us, either by force or by trade. Am I indeed a relativist trapped in our preunderstanding of tribal peoples and unable to explain what constitutes a correct understanding of tribal objects?

In response, I must emphasize that to understand tribal objects from the perspective of our preunderstanding does not exclude the possibility of having a correct understanding of them. It is precisely by engaging in dialogue that we escape our own perspective and it is by engaging in critique that we prevent our side of the dialogue from becoming imperial or imperious. Someone from the other side of the dialogue will have to tell us whether we have understood tribal objects correctly or not; it is they, not we, who have to make the relevant conditions of their cultural context part of the dialogue. This is already happening in disciplines – e.g., ethnography, anthropology, art, literature – where the tribal people are speaking for themselves. We will have to engage in this dialogue in more depth before knowing whether our understanding, based on our preunderstanding, is a misunderstanding or not. At the same time, we have to continue to critique our preunderstanding, recognizing its enabling as well as limiting roles in the understanding of tribal objects.

Columbia University

NOTES

[1] James Clifford, *The Predicament of Culture: Twentieth-Century Ethnology, Literature, and Art* (Cambridge: Harvard, 1988), p. 8.

[2] Danto, 'Artifact and Art,' in *ART/artifact: African Art in Anthropological Collections* (New York: Center for African Art, 1988), pp. 18–32; and Dutton, 'Tribal Art and Artifact,' *The Journal of Aesthetics and Art Criticism 51*, 1 (Winter 1993): 13–21.

[3] Danto, p. 31.

[4] The term 'preunderstanding' stems from Hans-Georg Gadamer's notion of "prejudice," which in turn stems from Martin Heidegger's concept of "foreunderstanding," both of which were conceived in contrast to Kant's "conditions of possibility of knowledge and experience." Cf. Gadamer's *Truth and Method* (New York: Seabury, 1975).

[5] Danto, p. 23. For a similar contrast from which a different conclusion is drawn, cf. Franz Boas, *Primitive Art* (New York: Dover, 1955), pp. 17–20.

[6] Danto, p. 23.

[7] The first involves scientific testing of the Pot People's and Basket Folk's pots and baskets to find traces of a different chemical or mineral unique to each culture's art; but such a procedure cannot work because, whatever art is, it is certainly not constituted merely by chemicals or minerals. Nor can we rely on aesthetic criteria alone since we can view any object aesthetically, as Marcel Duchamp taught us with his urinal. Lexical or linguistic evidence will not help either; for whether the respective cultures have the words for the art/artifact distinction cannot determine whether they have art and artifacts, since nobody doubts that the Greeks had examples of both without having this distinction (cf. Danto, pp. 26–8).

[8] Danto, p. 31. And cf. p. 18, where he says that the boundary between art and nonart (e.g., artifact) is inflexible (i.e., absolute – cf. p. 20).

[9] Danto, p. 32.

[10] Danto, pp. 29–30.

[11] Danto, pp. 24, 32.

[12] Vogel was also the curator of the ART/Artifact show.

[13] Danto does not exclude this dimension entirely; rather, he uses it to set up the problem of the indiscernible baskets in the natural history museum, and then argues that perception cannot solve a problem it has created.

[14] Two of Dutton's other objections (pp. 15–6) concern (1) Danto's analogy between, on the one hand, Warhol's Brillo Box and the industrial Brillo Box that inspired it and, on the other, the Pot People's and Basket Folk's pots; and (2) Danto's misleading connection between Duchamp's art/nonart distinction vis-à-vis readymades, and Pablo Picasso's and Roger Fry's art/nonart distinction in relation to African sculpture, i.e., Duchamp was challenging the concept of art that Picasso and Fry wanted only to expand.

[15] Dutton, p. 20.

[16] Dutton, p. 20.

[17] Dutton, p. 20.

[18] After acquiring some of this knowledge, Dutton defends the idea of universal definitions of art. Cf. Dutton's review of Sally Price's *Primitive Art in Civilized Places* (Chicago: University Press, 1989), and Marianna Torgovnick's *Gone Primitive: Savage Intellects, Modern Lives* (Chicago: University Press, 1990) – both in *Philosophy and Literature 15*, 2 (October 1991): 379–90.

[19] In both his philosophy of history and philosophy of art, Hegel is the best spokesperson for such a view.

[20] Danto discusses Picasso's 1907 experience with African sculpture in the Trocadéro Museum, which reflects this expansion and realization, as well as the views of Roger Fry and others at the beginning of the twentieth century.

[21] Torgovnick is a good example of this view.

[22] Price, p. 68; cf. also Torgovnick, pp. 81–4.

[23] *'Primitivism' in Modern Art: Affinity of the Tribal and the Modern*, 2 Vols., William Rubin, ed. (New York: MOMA, 1984). Cf. the critical reviews of this show by Price, Torgovnick, and Clifford, as well as Danto's comments (*op. cit*).

[24] Clifford says, more precisely, that the MOMA exhibition, which was centered on the notion of 'affinity' between the primitive and the modern, was primarily a way for the West to constitute non-Western arts in its modernist image (p. 193). He then adds that the show was also used to discover universal human truths, defined by the West of course, which are embodied in art worldwide.

[25] Cf., e.g., Kwame Anthony Appiah's *In My Father's House: Africa in the Philosophy of Culture* (New York: Oxford, 1992), where he argues, among other things, that the African culture*s* have *not* been overrun by Western influence.
[26] Danto, *Encounters and Reflections: Art in the Historical Present* (New York: Farrar Strauss Giroux, 1991), p. 164.
[27] As Clifford expresses the general point here, ignorance of cultural context is a precondition of artistic appreciation, for it makes space for aesthetic judgment; p. 200.
[28] Danto makes this point (pp. 28–9). But cf. Randall R. Dipert, *Artifacts, Art Works, and Agency* (Philadelphia: Temple, 1992) for a more extended discussion of this issue.
[29] See Price, p. 99; and Morphy, *Ancestral Connections: Art and an Aboriginal System of Knowledge* (Chicago: University Press, 1991), p. 24.

PETER KIVY

IN DEFENSE OF MUSICAL REPRESENTATION: MUSIC, REPRESENTATION AND THE HYBRID ARTS

I

The importance of Marx Wartofsky's contribution to our understanding of visual representation hardly needs advertisement on the pages of the present volume, being well known to all his admirers. It is in the spirit of that work, although not, obviously, as a direct contribution to it, that I offer, for this happy occasion, some brief remarks on the problem of musical representation.

It seems to me apparent enough, and hardly controversial, that Beethoven musically represented the risen Christ, in the *Et resurrexit* of the *Missa Solemnis*, with rapidly ascending scale passages. And it also seems to me apparent enough that the rapidly ascending passage, a so-called *coup d'archet*, with which Mozart opens his Symphony No. 31 in D (K. 297), the Paris Symphony, does not represent anything at all.

Mozart's Paris Symphony is a work of what the nineteenth century came to call 'absolute music,' and what I sometimes like to call 'music alone,' which is to say, music without text, program, programmatic title or any other such extra-musical paraphernalia.[1] It is my view that absolute music, music alone, has no representational or semantic content whatever. Many historical figures, and some contemporary ones, disagree. I have defended my own view of music alone, which might be described as 'enhanced formalism,'[2] elsewhere, and will not undertake to rehearse that defense here.[3] For present purposes, I simply assume that my view of absolute music, enhanced formalism, is true.

It is also my view that music with text or program or programmatic title sometimes contains musical passages palpably 'representational,' in more or less the same sense of that word in which we would want to say that a certain painting by Van Gogh represents sunflowers or one by Botticelli a woman standing in a seashell. The rising scale passages in the *Et resurrexit* of Beethoven's *Missa Solemnis* are such passages. Their number is legion.

But just as there are many, both past and present, who demur from

C.C. Gould and R.S. Cohen (eds.), Artifacts, Representations and Social Practice, 53–67.

the view that absolute music is absolutely non-representational, there are those, both past and present who, appearances to the contrary notwithstanding, demur from the view that some passages of music, those occurring in compositions with text, program or programmatic title, are truly representational in the full-blooded sense of the word. Again, I have defended the possibility – indeed the actuality – of musical representation elsewhere, both against the old as well as the new arguments to the contrary.[4] But musical representation apparently is in need of eternal vigilance. For new enemies, with fresh arguments, are ever on the march. And one such reinforcement to the anti-representationalist camp, with a new and intriguing argument, is the subject of the present paper.

My good friend Stephen Davies has, recently, come forth with an array of new arguments against the notion that passages such as the one from Beethoven cited at the beginning of my paper are truly representational.[5] I have dealt with all but one of them on another occasion.[6] But the one I have not yet dealt with has given me pause. In brief, and without any adornments, it is that if music cannot be representational without a text, program, or programmatic title, as, indeed, I claim, then it cannot become representational merely by the addition of one of them. Even unadorned, the argument has a certain intuitive appeal, although, clearly, it is not the case that in general, something lacking a property in one set of circumstances cannot gain the property in another set, without there being any change in *it*. I may not be rich in New York, but may be in Outer Mongolia. A mere change if venue here does the trick (assuming what is to become my 'wealth' is portable).

What drives Davies' argument, then, is not merely the general principle, which appears to be false, that something lacking a property in one context cannot gain it merely by a change of that context. Rather, it is his conception of what kind of an artwork a musical work is when it possesses text, program or programmatic title, or, in the case of opera, a *mise-en-scène* as well. Such works are what Davies calls, following Jerrold Levinson, "hybrid art forms." And to appreciate Davies' argument fully we must first, therefore, turn to Levinson's account of just what a hybrid art form is.

II

According to Levinson:

> An art form is a hybrid one in virtue of its development and origin, in virtue of its emergence out of a field of previously existing artistic activities and concerns, two or more of which it in some sense combines . . . . The components of a putative hybrid must be locatable somewhere in the preceding culture and must be plausibly seen as having come together in the result. In short, hybrid art forms are art forms arising from the *actual* combination or interpenetration of earlier art forms. Art forms that have *not* so arisen, though they may be intellectually analyzable into various possible or actual structural or mediumistic components, are not hybrids in the primary sense.[7]

Among such artistic hybrids, Levinson wants to distinguish three different kinds, "which can be labeled *juxtaposition* (or addition), *synthesis* (or fusion), and *transformation* (or alteration)."[8]

Juxtapositional hybrids, on Levinson's view, are art forms in which "the objects or products of two (or more) arts are simply joined together and presented as one larger, more complex unit."[9] In such an amalgam, each element maintains its individual character, as, for example, in song accompanied by an instrument or instruments; and thus the "elements [are] imaginable in isolation from the others to which they are joined and which, so isolated, would count as bona fide (if peculiar) instances of the arts entering into the hybrid."[10]

In synthetic hybrids, on the other hand, "two (or more) arts are brought together in such a way that the individual components to some extent lose their original identities and are present in the hybrid in a form significantly different from that assumed in the pure state."[11] An example of such a synthetic hybrid – a very controversial example, one might add – is, Levinson suggests, Wagnerian music drama.

Finally, Levinson describes the transformational hybrid as "closer to the synthetic model than to the juxtapositional one, but which differs from the former in that the arts combined do not contribute to the result in roughly the same degree."[12] An example of a transformational hybrid might be, Levinson suggests, kinetic sculpture, which he sees as "sculpture modified in the direction of dance. It is not an equal fusion of the two, but rather an incorporation of some of the special or distinctive *characteristics* of dance into what remains recognizably *sculpture*, though in an extended sense."[13]

As I said, Davies bases his denial of the possibility of representational music on Levinson's account of hybrid art forms. It would be well, therefore, to at least suggest, parenthetically, that as a foundation, Levinson's account is seriously flawed in one rather crucial respect. For if Levinson's historical criterion for hybrid art forms is adopted, it then turns out that all forms of vocal music with text, which is to say, all music sung to words, with the exception of opera and music drama (and, of course, 'melodrama,' which is not sung), will turn out, on Levinson's view, *not* to be hybrid art forms in any of its three variations. But as vocal music with a text is usually thought of, and is so thought of by Levinson, as a paradigmatic example of the hybrid, such a conclusion is highly unpalatable, to say the least, and, many would say, a *reductio* of Levinson's historical criterion, although not necessarily of the three-fold hybrid concept itself.

The criterion, it will be recalled, for being a hybrid art form, is not merely that two or more art forms be recognizable in the compound, but that they get there in a certain way: "The components of a putative hybrid must be locatable somewhere in the preceding culture and must be plausibly seen as having come together in the result." But such is *not* the case with instrumentally accompanied song, in the Western musical tradition, or of unaccompanied vocal music. It is certainly not the case that, historically, accompanied song is the putting together of a previously existing vocal music with a previously existing, independent instrumental idiom. On the contrary, it is quite the other way around. Instrumental music, in the West, as an autonomous musical art form, is a late development, and comes out of vocal music; in the Renaissance, this is apparent in the fact that a good deal of what instrumentalists played was arrangements of vocal music for instruments.

Furthermore, neither polyphonic *a capella* vocal music nor unaccompanied monodic song can be thought of as historically being put together out of one art form, music, and another art form, 'literature' or 'words.' For there *never* was a time, in the memory or recorded history of Western civilization, where music was not sung music with a text, with or without musical accompaniment. Of course opera and music drama, which certainly are on everyone's list of hybrid art forms, do qualify under Levinson's historical criterion. For they are, historically, the self-conscious result of just the kind of putting together Levinson has in mind. But they are not the three-fold putting together of the 'pure' art forms of text, music and dramatic representation, but, rather, the putting together of vocal music and *mise-en-scène*. Yet even though opera and music drama do survive as hybrid arts under Levinson's historical criterion, texted vocal music of other kinds does not. And this latter result seems too high a price to pay.

Of course Levinson's general classification of hybrid art forms, and the concept of the hybrid art form itself need not fall with the historical criterion. But the criterion itself is in need of either repair or replacement, if we are to fully understand just what an artistic hybrid really is. And to that extent Davies' attack on representational music has been launched with one seriously defective weapon. Nevertheless it may well have merits of its own. And with that in mind, and Levinson's notion of hybrid art forms in hand, I turn now to Davies' critique.

III

It is Davies' thesis that "an illustrative title or an accompanying text could not make a musical work representational if it could not be representational without them."[14] And he asks: "Why, then, does the debate about musical representation concentrate on program music?"[15] His answer, and his argument, based on Levinson's analysis of artistic hybrids, are as follows.

Most people, Davies points out, would take opera, for example, to be representational. But opera, he avers, is, to use Levinson's terminology, a "synthetic hybrid." And, he continues:

> In a synthetic hybrid the combination of constituents results in a new entity, rather than merely a mixture. It would involve a fallacy of decomposition to argue from the representational character of opera to the representational character of its constituents. It does not follow from the cake's being sweet that the salt which goes into it is sweet.[16]

Thus, even though opera is representational, it is not a bona fide example of *musical* representation because it is *emergent* from its parts, and if one of its parts, *music*, cannot be representational on its own, it certainly would be fallacious to infer that it is representational in the emergent art form which *is* representational, namely the synthetic art form of opera. Opera, not music, is the representational art here. That is how I understand Davies' argument. I will return to it in a moment.

But what of program music? Davies characterizes it a transformational hybrid in which "music plays the dominant role." And, he continues:

> Titles and accompanying texts may possess a narrative function, but I take it that they are not representational (as paintings are). If program music were to be representational, the representational component would have to be supplied by the music. If 'pure' music is not representational, then program music could not be representational, given that program music is not a synthetic hybrid.[17]

So program music, as I understand the argument, because it lacks any representational component whatever, and is not a synthetic hybrid with emergent representational features, fails to be representational itself at all. Furthermore, the musical part of program music, not being representational in its pure form, and not even being, in program music, part of a representational art form, doesn't even have the consolation of being 'representational' falsely so-called as the non-conclusion of a fallacious decomposition argument from whole to part. Nor can it gain representationality from the text, because the text is not representational. And how, good Simmias, can the non-representational gain representation from that which is not representation itself? Nay, good sir, it flees from its opposite.

So, to sum up Davies' argument, as I understand it: With regard to what might be called, stretching the point a bit, the musical arts of 'dramatic representation,' in which I include not only opera and music drama but cantata, oratorio, and most 'art song,' music in those art forms cannot be representational, even though the art forms themselves are. For, first, it does not follow from the whole of a synthetic hybrid being representational that any of its parts is; and, second, if music cannot be representational in its pure, absolute form, then it cannot be made so by the addition of text, *mise-en scène*, or anything else. Further, with regard to program music, in which I would also include tone poems and simply titled music like 'Scotch' Symphony or 'Pastoral' Sonata, here even the work as a whole is not properly speaking representational, although it may be narrative. For the text is not representational, properly so-called, nor is music, in its pure, absolute form. Nor can putting them together in program music make a synthetic hybrid, so there can be no emergent representation in the case. Program music is not representational, nor is either of its parts; and because it is not representational it can present no temptation to commit the decompositional fallacy on music's behalf. Such, as I understand it, is Davies' whole argument against the possibility of representational music in its texted forms.

IV

I believe that Davies is mistaken across the board. That is to say, I believe that both opera and program music present instances of representational music properly so-called, and that Levinson's account of hybrid art forms provides no reason to think otherwise. Let me turn to opera, and opera-like art forms first.

Now, as a matter of fact, Classical and Romantic opera present few real cases of musical representation. (Not surprisingly, Baroque opera is a richer source.) So in my previous work on the subject, *Sound and Semblance*, I adduced my examples mostly from Baroque cantata and oratorio. But since Davies has used opera as his paradigm of the synthetic hybrid, and has based his argument in this regard on it, I see no reason not to follow suit in my reply. So let me begin by adducing an example from the standard operatic repertory of what I take to be a bona fide musical representation. On it I shall base my subsequent argument.

In the second act, second scene of Beethoven's *Fidelio*, Leonore and Rocco, after a bit of huffing and puffing, succeed in moving a heavy boulder ("*einer grössen Stein*") from one place to another, for reasons it is not necessary here to relate. The task is accomplished, to be precise, in the *Andante con moto* section of No. 12: *Melodram und Duett*. The opera *Fidelio* represents, in the place just stipulated, two people rolling a heavy boulder. But, of course, this representation is made up of diverse representational elements. The act of rolling the boulder is represented by two singers pretending to push and shove what appears to be a heavy object. This large object, the boulder itself, is represented by a lump of painted papier-mâché. And, *pace* Davies, the *sound* of the boulder is *represented* by a 'grumbling' figure in the celli, double basses and contra-bassoon, over which figure in the score is the stage direction: "*Hier lassen sie den Stein über die Trümmer rollen.*"

Now I would like to call attention to two compelling reasons why I think it makes perfect sense, is indeed correct to describe the grumbling figure as musically representing the sound of the boulder being moved in the 'world' of *Fidelio*. First of all, this is the way I think listeners experience the figure, once they notice it. Of course, it is a very small detail in a very big art work. And I doubt not that it passes unnoticed among most listeners until some pedant like me points it out. But once it is pointed out, it is then experienced as a representation of the sound the boulder must make in the fictional world of *Fidelio*. To appropriate a concept from Richard Wollheim's account of visual representation, the listener 'hears in' the grumbling figure the sound of the rolling stone.[18]

Second, ever since such phenomena as the grumbling figure in *Fidelio* have been discussed, which, in my reading, anyway, goes back to the sixteenth century, the 'representation' words are the ones critics, composers and listeners have consistently used to describe them. 'Picture,' 'tone painting,' 'imitation' (in the Platonic and Aristotelian sense of 'mimesis') are, prior to the twentieth century, the more common words in use, 'representation' being, I think, the word of choice in more contemporary 'logical' and 'analytical' circles. In *Sound and Semblance* I made a distinction between what I called there musical 'pictures' and musical 'representations'; and I am certainly not denying that the words 'picture,' 'tone painting,' 'imitation' and 'representation' all possess individual nuances and connotations. Be that as it may, for the purposes of the present discussion I think we may say that the word 'representation,' in a broad enough sense to take in the historical periods in which 'picture,' 'imitation,' 'tone painting,' and the like, were current, captures the core concept that all of those words implied, the concept the current argument is about. So 'representation' (in this broad sense) is what, I am arguing, people seem to think they experience in cases like the grumbling figure in *Fidelio*, and 'representation' is how they refer to it; and furthermore, the experience and description have been more or less in place since at least the late Renaissance.

Now it is perfectly possible that people have been both mistaken about what they are experiencing, and logically inaccurate in their referring to things like the grumbling figure as 'representations,' 'tone paintings,' 'pictures,' and the like. So let us now see if Davies provides us any conclusive argument to think so.

It seems to me we are offered two basic reasons for rejecting the common notion that things like the grumbling figure in *Fidelio* are bona fide representations. First, the notion is reached by the fallacy of 'decomposition,' and, second, it is just evidently the case that, as Davies puts it, "an illustrative title or an accompanying text could not make a musical work representational if it could not be representational without them."

Let us take the fallacy of decomposition first. Here's how Davies illustrates it. "It does not follow from the cake's being sweet that the salt which goes into it is sweet." But nor does it follow from the cake's being sweet that the sugar which goes into it is sweet. And, of course, the sugar *is* sweet. Which just goes to show, as everyone knows, that an invalid argument can have a true conclusion.

But in any case, we do not reach the conclusion that the grumbling figure in *Fidelio* is representational by decomposition, any more then we reach the conclusion that the sugar in the cake is sweetening that way. We conclude that the cake is being sweetened by the sugar and not the salt from what we know of cakes and sugar and salt. And we know that thc grumbling figure in *Fidelio* is representational by what we know of such musical figures and what we know of just how this one functions in its place. So we can dismiss the fallacy of decomposition out of hand. It may be a snare for the feet of others, but not for ours.

If anything is going to dislodge me, then, from my belief that the grumbling figure is representational, properly so-called, and that that is the way I experience it, it is going to have to be the contradictory belief that, to refine Davies' axiom a bit, an illustrative title or text cannot make a melody token of a given melody type representational if other melody tokens of the same melody type can not be representational without them. But is this axiom true?

I am calling it an 'axiom' because, so far as I can see, no argument is offered for it. But as an axiom I must confess that it does not tug at my heartstrings. It is far from self-evident to me. Indeed, it is far *more* evident to me that the grumbling figure in *Fidelio* has become representational by the addition of text, where it would not have been otherwise, merely as a figure in a symphony or sonata, than it is evident to me that the axiom which denies this possibility is true. As far as I am concerned, I have just refuted the axiom.

But why should it be even initially plausible to think it impossible that a scrap of music not representational in a piece of absolute music can *become* representational with the addition of text or title? I can think right now of only two possibilities, the first trivial, the second at least of some interest, although, in the event, unconvincing.

Suppose a real nit picker should argue as follows. Closely considered, and precisely put, the grumbling figure did not become representational when it was not before; it is, indeed, no more representational in the second act of *Fidelio* than it would have been in a sonata or symphony. What *is* representational is the whole – that is the whole segment of time – of which it is a part: the segment that includes the scene, the action on stage, etc.

Well this way of 'closely' and 'precisely' putting the case just seems to me to be wrong: unfaithful to our experience. What is doing the work of representing the sound of the rolling boulder just *is* the grumbling figure in the bass instruments, and there's an end on't. The text, and *mise-en-scène* (perhaps) are *necessary conditions* for the representationality of the grumbling figure. *It* wouldn't be representational without *them*; but *it* is what is representational of the sound, not *they*. However, the above way of putting the matter does, I think, point up a more plausible, although in the event inconclusive reason for thinking the passage of music from pure to representational impossible. That reason requires some consideration.

It may be thought there is some alchemical sleight of hand going on here. How, it might be queried, can some 'object,' without there being any change in *it*, be non-representational in one place and representational in another? It sounds like black magic. (Or quantum mechanics.) Of course many 'Cambridge changes' may occur. But certainly the change from non-representation to representation cannot merely be that. So it looks as if we are being asked to accept mysteries. Only someone prepared to believe the doctrine of the real presence, it might be argued, could believe that the *same* musical figure is non-representational in one place, representational in another, while, like the wafer and wine, keeping the self-same perceptual qualities in both.

Well there is a kernel of truth in this objection, and it is, simply, that we still have an imperfect understanding of artistic representation *tout court*. Certainly I do not pretend to have presented a full account of the musical kind here or elsewhere. But that being readily granted, it must nevertheless be insisted that there is no divine mystery in the passage of an 'object,' musical or otherwise, from non-representational to representational. It is just too common a phenomenon, and too commonly thus described, not merely as regards music, but in the most ordinary circumstances, to be thought arcane or occult. It just happens.

A thumbtack lies on the general's desk and is just a thumbtack. He puts it on his map and it represents the Third Army. A squiggle is just a squiggle; but in a diagram it represents a coil, on a road sign a dangerous curve. An ordinary chair becomes the throne of a king in an informal run-through of *Lear*. A broom is a horse when it is between the legs of a child, a rapidly ascending scale passage a *coup d'archet* in Mozart's Paris Symphony but the risen Christ in the *Missa Solemnis*. These are all too common, too ordinary for black magic and too deep, no doubt, for facile explanations. But if they are not *all* 'representation,' properly so-called, then I don't know what they might be.

I find no plausibility whatever, then, in the 'axiom,' as I have called it, that a melodic token of a given type cannot be made representational by the addition of a text or title. It seems to me to happen all the time. And unless the 'axiom' can be shown to follow from some higher principle that I do accept, I am inclined to simply dismiss it as patently false, because in conflict with some of the most familiar and obvious experiences of my musical life, and of musical life in the Western classical tradition at least as far back as the sixteenth century. With that being said, I think I can now make quicker work of the question of whether program music is properly representational, and press on to my conclusion.

V

The *Harvard Dictionary of Music* (1st ed.) defines 'program music' as "Music inspired by, and suggestive of a program, i.e., an extramusical idea indicated in the title of the piece and sometimes substantiated in explanatory remarks or in a preface."[19] And the first two musical compositions that are adduced as examples are, not surprisingly, Beethoven's Pastoral Symphony and Berlioz's *Symphonie fantastique*, which, I would think, are the first examples that would occur to anyone having even a passing acquaintance with the standard concert repertory.

But these are examples, it is worth noting, of two *different* kinds of composition, both of which fit the broad definition quoted. For Berlioz's 'program' is a narrative, while Beethoven's is merely a series of titles that set pictorial scenes. But in *both* cases, I would insist, the music generally performs the same function: to 'represent' or 'illustrate.' As I have argued elsewhere,[20] it is not in the power of music to tell stories. What it can do is illustrate scenes from stories much in the way that pictures do in those deluxe, illustrated versions of novels which used to be popular, and of which Rockwell Kent's beautifully illustrated *Moby Dick* is a notable example.[21] And it can do it, to some extent, anyway, 'temporally,' which 'still' pictures cannot but which motion pictures can, silent film with titles thus providing another visual analogue to narrative program music.

But in any case, the 'problem' of representation in program music, even in its narrative form, is no different from and no more troublesome than the same 'problem' in opera, opera-like vocal music, or any other texted musical form.

Davies, it will be recalled, makes the following claims with regard to program music. First, programs and titles and texts, although they may be narrative, are not representational. Second, therefore, if program music were representational, the representationality would have to come from the music. But, third, since pure, absolute music is not representational, program music cannot be either, because program music is not a synthetic hybrid, and, thus, cannot 'emerge' as representational from non-representational parts, the underlying assumption of course being here, as elsewhere, that the mere addition of text cannot make the music representational, where it was not before.

With the first and second claims I have no quarrel at all. They are both, on my view, correct. From the third, however, I demur, as the argument that has gone before should have made readily apparent.

Since, as the previous argument makes clear, I have no trouble with the notion that the addition of a text, whether narrative, descriptive or otherwise can make a melodic token of a given type representational, where another token of the type, without the text, was not, and, thus, since I thoroughly reject the 'axiom' to the contrary, there is little need for me to say much more in defense of the view that program music, both in its narrative and non-narrative varieties, can, at times, in places be representational. It is so where the text makes apparent *what* the music is representing, and the music makes it apparent *that* the music is representing this by facilitating our experience of hearing the object of representation in the musical fabric. More examples of this would be otiose, as would further argument. What was said of the grumbling figure in *Fidelio*, in the previous section, applies, *pari passu*, to unnumbered examples in the program music repertory. And the nuts and bolts of how music facilitates our hearing representations 'in' it I have treated in detail in *Sound and Semblance*, both as regards the representation of sounds, as well as the representation of other phenomena.[22] For music is by no means limited to the representation of sounds, as my very first example, Beethoven's representation of the risen Christ in the *Missa Solemnis* makes clear. For other such examples I can only refer the reader to my previous work on musical representation and to his or her own musical experience.

VI

To conclude, let me point out and comment briefly on what I take to be a puzzling resurgence, in recent years, of the very old suspicions regarding the genuine possibility of representation in music. Three times in the past fifteen years philosophers in the modern analytic tradition have found logical or metaphysical reasons for affirming that so-called musical representation is not a bona fide instance of the phenomenon. In 1976 Roger Scruton argued that musical 'representation' was defective because it did not maintain a viable distinction between subject and medium of representation, and because, on his view, any piece of so-called representational music can be fully understood without our being aware of the supposed representationality at all.[23] In 1987 Jenefer Robinson argued that musical 'representation' was defective because it did not facilitate 'hearing in,' the musical version of Richard Wolheim's 'seeing in' in visual representation, which Robinson took to be a necessary condition for genuine representationality.[24] And now Stephen Davies (1992) has argued that musical 'representation' is defective because, essentially, it is said to exist (for the most part) only with the help of a text, and the mere addition of a text cannot make absolute music, which, by hypothesis, is not representational, into a representational thing.

I have tried, I hope with some success, to answer the arguments of Scruton and Robinson elsewhere,[25] and the argument of Davies in this place. And although it would be disingenuous of me not to express satisfaction with my responses, I still find myself puzzled by this recent resurgence of doubt, by philosophers of deep insight, about the real possibility of musical representation. The old, historical debate was very much a normative one, in the heat of battle, over whether representation in music was a desirable aesthetic goal. Furthermore, within that debate it is sometimes very difficult to know whether what is being claimed is that composers ought not to pursue musical representation because it is impossible, or that it *is* possible, but not worth pursuing. But in any case, it is clear that the debate was driven by a fierce disagreement over what the 'music of the future' should be.

I take it that no contemporary philosopher would be prone to confuse the normative with the 'possibility' question, or to be driven by the former to take a stand, one way or the other, on the latter. What then drives the debate today?

I can only conjecture that the exponential increase in, and understanding of representation in the visual arts, of which Marx Wartofsky's work is a prime example, has made it increasingly clear what representation in its most paradigmatic form *is*, and at the same time made it increasingly clear that the musical cases are *peripheral* ones. A hard-headed sense of parsimony will do the rest. It is very tempting, I suppose, to want to get rid of the peripheral cases by drumming them out of the corps altogether.

But I feel obliged to play the role of gadfly, in the present debate, and press upon those of a parsimonious bent in representational matters that peripheral cases are *cases* nonetheless. A philosopher of the distant past said "What is is." A philosopher of the more recent past said (with almost equal pith) "Everything is what it is." I fondly hope, and seriously doubt that if the skeptic contemplates these two excellent sayings, while listening intently to a cantata by Bach, it will settle for him or her forever in the affirmative the question of whether music *truly* 'represents.'

Rutgers University
New Brunswick

NOTES

[1] Mozart's Symphony No. 31 is called the 'Paris' Symphony because it was written to be performed in Paris, not, as in the case of Mendelssohn's 'Scotch' Symphony, because it was meant to evoke any geographic location.

[2] Philip Alperson so describes it because, unlike traditional formalism, it countenances expressive properties as part of the musical fabric. I gladly adopt Alperson's nomenclature.

[3] On this see, Peter Kivy, *Music Alone: Philosophical Reflections on the Purely Musical Experience* (Ithaca: Cornell University Press, 1990). See also, Peter Kivy, 'The Fine Art of Repetition,' 'A New Music Criticism?,' and 'Is Music an Art?,' in Peter Kivy, *The Fine Art of Repetition: Essays in the Philosophy of Music* (London and New York: Cambridge University Press, 1993).

[4] On this see, Peter Kivy, *Sound and Semblance: Reflections on Musical Representation* (2nd ed.; Ithaca: Cornell University Press, 1991).

[5] Stephen Davies, 'Representation in Music,' *The Journal of Aesthetic Education, 26* (1992).

[6] Peter Kivy, 'Listening: Responses to Alperson, Howard and Davies,' *Ibid.*

[7] Jerrold Levinson, 'Hybrid Art Forms,' reprinted in Levinson, *Music, Art and Metaphysics: Essays in Philosophical Aesthetics* (Ithaca: Cornell University Press, 1990), pp. 27–28.

[8] *Ibid.*, p. 30.

[9] *Ibid.*
[10] *Ibid.*, p. 31.
[11] *Ibid.*
[12] *Ibid.*, p. 32.
[13] *Ibid.*, p. 33.
[14] Davies, "Representation in Music."
[15] *Ibid.*
[16] *Ibid.*
[17] *Ibid.*
[18] On 'hearing in,' see Kivy, the second edition of *Sound and Semblance*, 'Afterword: Pictures, Representations and Hearing-in,' pp. 217–226.
[19] Willi Apel, *Harvard Dictionary of Music* (Cambridge, Mass.: Harvard University Press, 1951), p. 604.
[20] See *Sound and Semblance*, Ch. IX.
[21] Garden City, New York: Garden City Publishing Co., 1930. Many of Rockwell Kent's illustrations are, actually, more decorative than directly relevant to the narrative. Better examples – better not from the aesthetic point of view but from the 'logical' – are the illustrated versions of the 'classics,' *The Three Musketeers, Treasure Island*, etc. that were issued for children in my boyhood.
[22] *Sound and Semblance*, Ch. III.
[23] Roger Scruton, 'Representation in Music', *Philosophy, 51* (1976).
[24] Jenefer Robinson, 'Music as a Representational Art,' in *What is Music?: An Introduction to the Philosophy of Music*, ed. Philip Alperson (New York: Haven, 1987).
[25] See *Sound and Semblance*, second edition, pp. 146–159, and 217–226.

DOUGLAS P. LACKEY

TWO VIGNETTES IN THE HISTORY OF THE MENSURATION OF VALUE

I wish in this article to consider certain medieval examples which bear on the problem of the mensuration of value; that is, the attempt to give some sort of quantitative character to things judged good or bad. I am referring to processes of rationalization which start from a judgment that an item is good, then proceed to consider whether it is better or worse than other things, then finish with some inference about how much good the item contains. This thrust towards mathematicization is often described as characteristically 'modern'; it is allegedly spawned by 17th century physical science, then applied or misapplied by 18th century social science, from whence, for good or ill, it comes down to us. But the drive to mathematize is found in earlier phases of Western culture in unfamiliar and surprising contexts. My examples are the development of the theory of indulgences in certain medieval texts and the representation of Prudence in late medieval art. The examples, which indicate how social conditions can influence theology and visual representation, should interest someone who is a Marxist and an aesthetician, if not a Marxist aesthetician.

In my view what I have called mathematicization derives from or accompanies a more general process of rationalization, a process in which several phases can be distinguished, nicely illustrated, as it happens, by successive steps of medieval intellectual history. The first stage involves the mere additive collection of ideas, of the sort seen in the late Roman compilers, for example, Stobaeus; in the second stage, ideas are collected together in broad categories and presented in encyclopedias, as in Martianus Capella, or Isidore of Seville. In the third stage, ideas are compared, checked for logical compatibility, and organized into sub-sets of consistent ideas, a process of purgation popular in the twelfth century and exemplified in many texts of Abelard. In the fourth stage, the purged sets of consistent ideas are deductively organized, with more general and important ideas at the beginning and more particular ideas following; thus Aquinas, in the *Prima Secundae*, considers the general character of sin and virtue, and in the *Secundum Secundae*, considers

C.C. Gould and R.S. Cohen (eds.), Artifacts, Representations and Social Practice, 69–86.

specific sins and specific virtues. After all this, mathematicization can and might proceed.

I. THE CALCULUS OF INDULGENCES

My first example of the rationalizing process involves the Catholic doctrine of indulgences, defined in Canon Law as

> a remission before God of the temporal punishment of sin which is already forgiven, which a properly disposed member of the Christian faithful obtains under certain definite conditions with the help of the church, which, as the minister of redemption, dispenses and applies authoritatively the treasury of the satisfactions of Christ and the saints.[1]

To understand the evolution of this canon we must step back and consider the broader theory of the good and the bad in the tradition of Catholic orthodoxy, the ideas of grace and virtue, reward and sin, and punishment and penance.

I begin with sin because, as we know, it is more interesting. Sin in both the Old and New Testaments is identified as a rebellion against God; in the Old Testament this includes insults to God and transgressions of His laws; in the New Testament, it is recharacterized as an offense against love, a violation of the first and dominant commandment to love God. Now, to break any of God's commandments, knowing that they are God's commands, qualifies one as a rebel, from which it follows that all sinners are alike, qua sinners and rebels, and all suffer equally from separation from God. The equal perdition of sinners is a standard view of the very early church, in which sins are cleansed by baptism and there is no second cleansing for those who relapse. Each and every mortal sinner suffers the equal pain of permanent excommunication.

As early as the Shepherd of Hermas (c. 120) the notion of a second baptism or second cleansing circulates through Christian communities. With second cleansings, Christians now can be grouped into three classes: sinners, cleansed sinners, and those who have not sinned. The notion of the absolute character of mortal sin, however, is retained in the post-Nicene system of public and canonical penance, by which every sinner worked his way back into the congregation of communicants through the same successive public stages of Weeper, Hearer, Kneeler, and Co-stander. But when the church became firmly established in the fourth century, and Christian bishops took over many of the functions of Roman judges, the system of canonical penance was modified. Just as different crimes merited different prison terms under Roman law, different grades of sin merited different assigned periods in the different grades of penitude. In the canons of St. Basil, those guilty of homicide, sorcery, or incest were excommunicated for twenty years, four as weepers, five as hearers, seven as kneelers, and four as co-standers. How there could be different grades of sinners given the standard definition of sin as rebellion is not explained in the early patristic discussions, and the inconsistency between the absolute character of sin with the relative rehabilitation of sinners led Jovinian in the fourth century to revive the view that all sins are equal, a view sharply rebuked by Augustine (Ep. 167) in the fifth.

Around the sixth century the new system of private penance began to supplant the old system of canonical penance. In the new system, sinners were not graded but sins were, each kind of sin receiving a characteristic penance. The old problem that all sins are equally rebellions against God is reflected in the new idea of priestly absolution, which removed the guilt for each sin, i.e. which removed the barrier that the sinner had placed between himself and God. Absolution is not harder to obtain for some sins than others; all that is required is sincere contrition. There remained, of course, differences of penance, which was from early on curiously described as a kind of debt which remained to be paid to God by the newly reconciled sinner. So we have in working form even in the penitential period the high scholastic distinction between the *culpa* of sin and *poena* of sin, and the implication, made explicit by Aquinas (ST I-II 72.5), that the *culpae* of all mortal sins are equal though the *poenae* are not.

If God is owed a debt even by an absolved sinner, there must be some way these debts be paid if the sinner dies before his earthly penance is finished. If not, then absolved sinners obtain heaven without penance, a violation of God's retributive justice. So in the penitential period there is a slow evolution of the doctrine of purgatory, absent in scripture and the early patristic period. By the end of the millennium it is widely believed that the souls of most repenting sinners spend time after death suffering intense but finite quantities of pain. Once established, the idea of purgatory could be pressed into service for quite a different task, providing punishment for non-mortal, or venial sins. The very idea of a non-mortal sin develops slowly and with considerable labor, until the 13th century consensus develops that a venial sin is not an act which is against the law (*contra legem*) but an act outside the law (*praeter legem*), a doctrine that was glossed by Cajetan as implying that mortal sins are acts which have an evil goal but venial sins are acts with good goals but not good means.

The establishment of purgatory, from which sinners are released after finite time, provides the possibility of earthly succor of departed souls, and the notion of suffrages for the dead is as old as the doctrine of purgatory itself. But as soon as the idea of suffrages took hold, the question arose as to the relative efficacy of suffrages, and the problem of how much temporal suffering is relieved by a given suffrage. A nice illustration of the mathematicizing urge is found in a seventh century discussion reported by Professor Lea in his *History of Auricular Confession*

> A question arose whether it is better to have a daily mass sung for a year or seven masses over one week by 52 priests. The answer is that the former is more meritorious, but the latter brings speedier relief.[12]

By the 13th century, then, we have a doctrine that (i) mortal sinners are equally guilty, (ii) mortal sins are unequally grave; (iii) there are non-mortal sins, unequally grave, (iv) the worst venial sin is less grave than the least mortal sin, (v) no number of venial sins equals in gravity a single unabsolved mortal sin, (vi) the punishment for an unabsolved mortal sin is infinite. (vii) the penance for an unabsolved venial sin may equal the penance of an absolved mortal sin (viii) the penance for an absolved mortal sin or an unabsolved venial sin is proportionate to the gravity of the offence.

The penances assigned in the penitential manuals are often quite severe, and it is natural that means were gradually developed for side-stepping the full weight of penance. Two devices were increasingly pressed into service: commuted penance, and vicarious penance. With commuted penance, which becomes common in the 10th century, some quantity of the penance was excused in return for good works or donations to the church: the debt for sin is not paid but excused. With vicarious penance, the full debt is paid, but is paid by someone other than the sinner.

The development of vicarious penance is still, to my knowledge, an uncharted historical problem, and the professional historians may determine that its origins lie with such customs as the hiring of professional mourners to weep at funerals, still, I believe, a Mediterranean practice. But the usual suggestion is that its origins lie in the customs of feudalism. Feudalism imposed heavy duties on vassals, and frequent arrangements had to be made for substitution of one service for another, or one payment in kind for another payment in kind. Eventually the practice of substitution percolated into the legal system. In trials by combat, a sufficiently wealthy litigant could hire a substitute champion to defend his honor. Even more interestingly, a convicted subject could get another man to stand in for him and take his punishment. Naturally, if one can hire a substitute to serve the punishment for earthly crimes, one could hire a substitute to serve the penance for sins. Manuals were devised to work out the processes of substitution: for example, one manual explains how a sufficiently rich man, assigned a seven years fast, could complete his penance in six days, as follows: "the penitent will pay 12 men to fast for 3 days on bread, water, and vegetables. Then he will get 7 times 120 men to fast for three days. The total is 2556 days, equals seven years."[3]

Now indulgences begin as commuted penances in the 11th century and end as vicarious penances in the 13th. The informal practice of commuting penance for some particular service is generalized into a publicly available recourse in which some act or payment of a certain kind reduces the amount of penance by a certain fraction, or reduces the suffering of the absolved sinner by a certain number of days in purgatory. In 1095, Pope Urban announced that all who served in the first crusade would receive a complete reduction of all owed penance – the first of the plenary indulgences.

With the introduction and spread of plenary indulgences, lavishly distributed by Innocent III to obtain recruits for the war against the Albigensians, a certain uneasy feeling develops among the theorists that God's justice is subverted if indulgences are viewed simply as a commuted penances. After all, a penance is not a debt owed to God because God needs something, a penance is a debt owed to God because the sinner deserves punishment, on grounds of retributive and vindictive justice. To commute a penance, then, is to go against justice, and on this and other grounds they were condemned by Abelard. In the 13th century, Albertus Magnus commented that indulgences are pious frauds to encourage the faithful to good works, but they cannot in fact reduce spiritual debts. Bonaventure [In *Sent.* IV. d.70] complained that the sale of indulgence creates "too great a market in God's mercy." The situation was rescued by Alexander of Hales, who construed indulgences not as commuted penances but as vicarious penances. Apparently it was Alexander (or perhaps Hugo of St. Cher) who introduced the idea of a spiritual treasury which has dominated the theory of indulgences ever since, though the idea of a store of merit on which a sinner can draw was hinted at in the previous century by Anselm in his *Cur Deus Homo*.

The idea of the spiritual treasury is described in modern catechisms as a storehouse containing the superabundant merits of Christ and the saints. But the real coins of the spiritual treasury are coins of meritorious suffering, the same sort of suffering that constitutes the penance of sinners. Thus the full debt of each sin is paid in suffering, and God's justice is not subverted by indulgences. But the suffering of the sinner is vicarious, since he does not suffer but someone else does, and by this other suffering the sinner's debt is paid.

In the third quarter of the 13th century the issues involved in the measurement of sins, penances, and indulgences are reviewed by Aquinas, who confronts old difficulties and introduces new ones.

As regards the equality of sins, Aquinas is in worse trouble than most of his predecessors. Aquinas had deftly dispatched the problem of evil with his theory that evil is a privation, from which it follows that sin is a privation, not a substance or a quality. But by Aristotelian logic all privations are equal in what they lack, and there can be no degrees of privation. It follows, apparently, that all sins are equal: they are all equally unlawful, they are equally a transgression from the rule of reason, "like a deviation from a straight line," and they are all equally opposed to virtue. We are back with Jovinian, in the fourth century. (ST I-II 73)

Aquinas's response is worth quoting at length, since it shows unusual probing and intellectual discomfort:

> The matter must be considered carefully – the Stoics took this stand because they considered only the privative element in sin, that is, the departure from reason. Assuming simply that privation does not admit of degrees of difference they concluded that all sins are equal.
>
> But two kinds of privation are apparent. One kind of privation is pure and simple and contains nothing of its opposite. Thus death is the privation of life and darkness

> is the privation of light. This kind of privation does not admit of degrees of difference, for nothing remains of what has been taken away. Hence a man is not less dead on the first day than on the third or fourth or at the end of the year. The same holds for darkness. If a single shade shuts out all the light the room does not get darker if more shades are added.
>
> The second kind of privation is not simple but retains something of its opposite. It consists more of the lessening of a quality than in its complete absence, for example, sickness which involves a bodily indisposition that leaves the vital balance of functions intact so that the animal continues to live. The same holds for such things as disgrace. These privations admit of a difference of degree with reference to the remaining element. And it makes a great difference in sickness and in dishonor just how much of the basic organization or acceptance remains.
>
> Sin or vice outright ought be considered in the same way, for in them there is a certain deviation from reason But not a total absence of rationality otherwise . . . neither the substantial reality of the sinful act nor the actual bent of the sinner would remain . . .
>
> Therefore it matters much to the gravity of the sin whether one departs more or less from the rightness of reason. (ST I–II 73.2)

Now this is not argument but assertion. The correct analogy is between mortal sin and death and venial sin and sickness, rather than between degrees of sickness and degrees of mortal sin. The fact is that privations are not qualities and therefore not intensive qualities, which admit of greater or lesser degrees. If one must be five foot six to become a policeman, then the five foot five man fails and the five foot four man fails, and they are both equally non-policemen, and one is not less of a non-policeman than the other.

Fortunately Aquinas has another method of differentiating sins, which is to identify different sins as privations of different virtues. Then the gravity of the sins will vary as the goodness of the virtues. There is a little stumbling block stemming from Aristotle's view that every virtue is a maximum, and maxima are incapable of variations or degrees, but this problem is dispatched by the observation (ST I–II 66) that there are no degrees within one virtue but there may be comparative differences between virtues. The differences are that the theological virtues are more important than the cardinal virtues, that intellectual virtues are more important than moral virtues, and that moral virtues that involve the will are more important than moral virtues that involve appetite. The full sequence in order of value is Charity, Faith, Hope, Prudence, Justice, Fortitude, and Temperance. Defining the seven deadly sins as privations of these seven virtues is left as an exercise for the reader; there are hundreds of medieval diagrams devoted to this problem, all of them more or less futile.

Different sins lead to different penances, and difference of penances leads us to Thomas on indulgences. Indulgences must have merit, Aquinas observes, because the Church uses them and the Church does nothing in vain. (ST Supp q. 25) Nevertheless the question persists how one man can pay satisfaction for another; after all, Aquinas has committed himself to the metaphor of sin as sickness, and I cannot be cured of sickness if someone else takes my medicine. As Aquinas quaintly notes, "one man's fast cannot mortify another man's flesh." (ST Supp. 13.11) Nevertheless, the substitution is possible, because what we are dealing with here is not sin but the debt due on a forgiven sin, and "one man can pay the debt of another." Furthermore, one who bears the suffering for another engages in a fundamental act of charity, and to deny the possibility of assuming another's suffering is to block a route to charity (an argument that leans on Anselm's view that Christ assumed the penance due from mankind as a supreme act of charity). This argument only works, obviously, if penance is construed as a kind of debt; this is the language the theologians had used for penance for centuries, but the warrant for the use of the term 'debt' is not provided by Aquinas. A linguistic practice that calls penances debts does not demonstrate that penances *are* debts or that they can be paid by third parties.

But Aquinas does have an ace in the hole with which to rescue the legitimacy of indulgences. The ace is the doctrine, long developing, of the union of all Christians in a single body, the body of Christ:

> The reason why [indulgences] so avail is the oneness of the mystical body in which many have performed works of satisfaction exceeding the requirements of their debts . . . so great is the quantity of such merits that it exceeds the entire debt of punishment due from those who are living at this moment, and this is especially true of the merits of Christ . . . the merits of the saints are the common property of the Church (ST Supp Q 25)

With the doctrine of the mystical body, the old notion of one person paying the debt of another is replaced by the notion of all Christians paying the debt of all Christians, and indulgences no longer seem like as evasions of God's justice.

The linked notions of the Spiritual Treasury and the Mystical Body rendered the concept of indulgences sufficiently intelligible and doctrinally credible that the doctrine remained unchallenged until the assault of Luther in 1517. In the treatment of Aquinas the theory has reached a remarkable degree of abstraction and organization: the deposits of some are balanced off the withdrawals of others, like the late lamented internal checking and banking system of the U.S. Congress. The metaphors of the theory: Treasury, debt, payment, and so forth, are distinctly commercial, even before there was commerce, and in the later Middle Ages the system of indulgences did come to form a small but substantial part of the economic system, a way of raising capital before the devising of the modern repertoire of instruments of credit.

But though the system of indulgences bears some resemblance to a modern system of checks or reserve notes, there were limitations that were distinctly medieval. The system was not fully tractable to the laws of arithmetic: Aquinas argues, for example, that the purchase of three indulgences each of which remits one third one's total penance is not equivalent to a plenary indulgence or a total remission of penance due. (ST Supp Q 25.2 ad Obj 3). Furthermore there does not seem to have been a secondary market for the resale of indulgences; each indulgence was consumed by the purchaser, and each indulgence is transmitted directly from the issuing authority to the penitent: Chaucer's pardoner has his indulgences "hot from Rome." The suspicion of paper money before the 18th century was too strong to overcome these obstacles, even if the paper were backed by the superabundant merits of Christ and the saints.

The limitations of the system of indulgences were not special but typical of late medieval systems. Despite these limitations, the system of indulgences is a tribute to the generalizing power of the European mind. Luther's attack on the abuse of indulgences, his resurrection of the doctrine of the equality of sins, and his rejection of any notion of vicarious punishment, represents a backward step in the rationalization process. But it is difficult to imagine what the next forward step might be. On the Catholic side, the field of sin, penance, and indulgences fell prey to the degenerating research programme of 17th century casuistry. In that system, stealing 1–5 ears of corn is no sin; stealing 6–10 ears is a venial sin, and stealing more than 10 ears is a mortal sin.[4] Perhaps the doctrine of the Spiritual Treasury provided all the precision the subject could bear.

St. Ambrose and St. Augustine from *Visconti Hours*. Giovannino dei Grassi (c. 1390). Reprinted by permission of George Braziller, Inc. © 1993.

II. AN IMAGE OF PRUDENCE IN THE VISCONTI HOURS

In the late fourteenth century Ginagaleazzo Visconti commissioned a book of hours, now in the National Library in Florence, illuminated by Giovannino dei Grassi. In the lower border of his illumination for the Te Deum [L-F 17], the main panel of which shows Saints Ambrose and Augustine, Giovannino has depicted the virtues of Justice and Prudence. Justice wields her traditional broadsword, a true Visconti. But the illustration of Prudence is innovative and surprising. Instead of the usual mirror and snake, Prudence holds a calipers, an instrument of measurement.[5] The suggested affinity of prudence and measurement is an innovation in the iconography of Prudence; no earlier representation of Prudence with an attribute signifying measurement can, I believe, be found in medieval art.[6]

Giovannino's innovation did not die with him. His successor, Belbello, who completed the Visconti hours in the early 15th century, gave Prudence the same calipers in his complex painting of the Fall of the Rebel Angels [L-F 11v]: Belbello's Prudence also carries the traditional mirror, traceable back at least to Giotto's Prudence in the Scrovegni Chapel (1305)[7]. In his great sculptures for the tomb of the Henry II, Duke of Brittany, in the Cathedral at Nantes (1507), Michael Colombe presents Prudence carrying a mirror in one hand and an instrument identified by Emile Male as a compass in the other:[8] dei Grassi and Colombe at least agree that Prudence requires instruments of measurement. The calipers or compass appear in a representation of Prudence on the tomb of Gauvain de Dreux at Saint Nicholas en Loye en Normandie, and Prudence carries a compass in a tomb sculpture in the cathedral at Amiens dating from 1543.[9] Giovannino's calipers initiate a iconographic tradition that lasted over 150 years and died out only with the general decline of allegorical/moralistic painting in the last half of the 16th century.

What are we to make of the appearance of the calipers in a representation of Prudence in the late 14th century? Is this individual whim or some symptom of a deeper cultural fact? Unravelling this problem takes us not only to the history of representations of Prudence but the history of prudence itself.

Prudence, as Giovannino knew, is one of the four cardinal virtues, but it is not generally known how long it took prudence to assume this rank. In Xenophon and Plato (*Rep.* 436 ff.), justice, temperance, and courage are recognized as primary virtues, but Plato's fourth virtue is reason [*nous*], which knows ends, not prudence, which is the skillful selection of means to ends. Likewise in Aristotle the moral virtues of justice, temperance, and courage are well established, but there is no fourth primary virtue. Instead Aristotle presents (*NE* vi.) an array of five co-equal intellectual virtues, of which only one, *phronesis*, is identifiable as prudence. The Stoics argued for the unity of virtue and the identity of virtue with reason, so they too have not yet enthroned prudence. In the Apocryphal Book of the Wisdom of Solomon (c. 50 B.C.), written in Greek, the author speaks of prudence [*phronesis*], fortitude, justice, and temperance in a single phrase (*Wis*. 8:7), but all of these virtues are said to derive from the primary virtue of wisdom [*sophia*]. Prudence comes into her own as a primary species of virtue, underived from any higher category and co-equal to the other moral virtues, only with Cicero, who achieves this result by translating Aristotle's '*phronesis*' as '*prudentia*' and then listing '*prudentia*', not the more intellectual '*sapientia*' along with *justicia*, *fortitudio*, and *temperantia* as the four basic human virtues[10], and the influential discussion of Macrobius follows suit.[11] The Latin Fathers, for whom claims concerning the ultimate ends of human life were matters of faith, not reason, happily join Cicero in ranking *prudentia* over *sapientia*, and neatly supplement the Ciceronian four with the Pauline three – faith, hope, and charity – producing the canonical list of seven virtues, familiar in Ambrose, Augustine, and Gregory the Great.[12]

Despite the authority of the Latin Fathers, prudence plays only a small role in the imaginative and spiritual life of late antiquity and the early middle ages. When the Shepherd of Hermas, in the second century, inaugurated the tradition of personifying the virtues as maidens, his list of virtues did not include prudence,[13] and when Tertullian, in the early third century, initiated the literary tradition of dramatizing moral conflict as a battle between personified virtues and vices,[14] he did not include prudence in his list of combatant virtues. In Prudentius's fourth century poem *Psychomachia*, the poet neglects his own namesake and does not mention prudence as one of the virtues contending for the human soul.

The pictorial representation of the virtues, in early medieval painting, derives mainly from Prudentius and accordingly neglects prudence. Likewise prudence, always a tame virtue, plays almost no role in the tension-ridden dramatizations of virtues and vices in Romanesque sculpture, even those not based on Prudentius. In these early periods,

Prudence puts in an appearance primarily in representations of the virtues of kingship, in which the centrally placed monarch appears surrounded by four virtues, as in the ninth century painting of King David in the Bible of Charles the Bald. In slightly later pictures, starting with Cambrai Gospels (c. 850) and the Bible of San Callisto (876), Prudence is shown holding a book, an iconographic choice that exhibits the Carolingian fetish for literacy but cuts off from prudence that majority of human beings who could not read.

Reconciling the presence of prudence in the Latin Fathers with the absence of Prudence in the art of the early middle ages is a deep scholarly problem; I can only point to elements which require deeper analysis. One element is the anti-intellectualism of early Christianity, with its short run earthly future, its stress on the sufficiencies of the lilies of the field, and its pride in being 'foolish,' i.e. imprudent, in the eyes of wise and prudent Greeks. A second element, for an age imaginatively focussed on spiritual conflict, is that there is no deadly sin which is the opposite of prudence; indeed, the opposite of prudence, *stultitia*, seems not be a sin at all but a kind of inborn failing for which a person is not morally responsible.[15] The content of spiritual life in these centuries derives mainly from Paul and the confessional side of Augustine, and in this tradition prudence has no role. The third element is obviously social; in strict feudalism, the Church selected ends, the nobles made policy, and all the rest needed to do was obey. In such a society there is no need for selection of means by each and every individual, and therefore no need for universal prudence.

With the Renaissance of the 12th century, the iconography of Prudence shifts. Prudence still holds her book in an 11th century Rhenish sacramentary; but in an early 12th century lectionary at Cologne Prudence is shown with a book and a snake.[16] Later in the 12th century, the book almost disappears[17] and Prudence is everywhere accompanied by a snake and a dove, after Matthew 10:16: "Ye shall be as wise as serpents, and as gentle as doves." The changed iconography of prudence democratizes the virtue: prudence is no longer an obligation confined to literate monarchs but an obligation of every Christian, enjoined by Christ. This is the new status of Prudence in the didactic manuals of the mid-12th century, like the *De Virtubus et de vitiis* of Alain de Lille (c.1160), or the *de fructibus carnis et spiritus*, falsely attributed to Hugo of St. Victor, truly attributed to the Pseudo-Hugo of St. Victor![18]

The democratization of Prudence inspires those great cycles of virtues and vices which one finds in relief sculptures at Notre Dame de Paris (1210), Amiens (1230), and Chartres south (1240). Set at eye level and near a main entrance, the cycles provide instruction for every Christian passing through. The pattern is set at Paris; twelve virtues are shown personified and with attributes; below each virtue a naturalistic scene depicts the contrary vice. The selection of the twelve virtues (Faith, Hope, Charity, Chastity, Prudence, Humility, Strength, Patience, Gentleness, Concord, Obedience, and Perseverance) remains something of a mystery[19], but Prudence is found in all the cycles, and not just on the strength of being one of the cardinal virtues, since not all the cardinal virtues are represented. What I have called the democratization of prudence is completed by Aquinas, who is so removed from the Carolingian attitude that he has to devote a separate article to showing that there even is such a thing as kingly prudence.[20]

The great classicizing weight of Aquinas's treatise on prudence (*ST* II–II Q 47–56) gives Prudence an unimpeachable role in 14th century representations of the virtues. There are several iconographic innovations. The 12th century dove becomes, in the 13th century, a disc with a picture of a dove, as at an Eilburtus altar from 1250,[21] and the disc becomes, in the 14th century, a mirror at which Prudence gazes, as in Giotto's Scrovegni fresco. The head of Prudence, naturalistically represented in the 13th century, acquires a double face and then a triple face in the 14th century, as the artists follow Dante (*Purg*. XXIX:132), who gave prudence three eyes, for seeing past, present, and future.[22] These grotesque heads, often with an aged face pointing backwards and a youthful face pointing forwards, show a developing interest in foresight as an essential element of prudence, and a declining interest in counsel, discretion, circumspection, and the other more distinctively Christian elements of prudence that dominate the element of foresight in the didactic manuals in the 12th and 13th centuries. By the 16th century, foresight is the only surviving component of medieval prudence. Thus Titian's three-headed Allegory of Prudence, now in the National Gallery in London, contains the inscription, "*Ex praeterito, praesens prudentur, agit ni futura actione deturpet*" [Instructed by the past, the present acts prudently so as not to be disrupted by the future.][23]

The depiction of Prudence gazing into a mirror creates a semantic problem for the late medieval artists, since mirrors are commonly associated with the vice of vanity. The calm and dignified young lady with a mirror depicted in Memling's *Vanity*, now in the museum at

Strasbourg, could easily pass for a late medieval Prudence, if only she would put on a few clothes.[24] The self-knowing and self-absorbed mirror gazing Prudence of the fifteenth century represents yet another assault on the snake-holding 13th century Prudence, which, according to Thomas, can take as its goal either the common good or individual good.[25] The tension between the old communitarian virtues of Thomas and the new egoistic conceptions resonates through Valla's *De Voluptate* (1433) and produces the 'new iconography' of the fifteenth century, on full display in a remarkable manuscript (c. 1470) of Jacques d'Armangnac, duc de Nemours, now in the Biliotheque Nationale.

Prudence in the Nemours cycle holds a mirror but she also carries a colander and a thin coffin rides on her head. At her feet is an open jewel case, and the floor is littered with discarded jewels. This is a Christian prudence, who knows the last things. But the mirror strikes a non-Christian and ego-centric note, and the mirror connotes self-interest as much as self-knowledge.

The representation of Temperance in the Nemours ms. is even more extraordinary: Temperance carriers a clock on her head, a bridle in her mouth, eyeglasses in her hand; she wears shoes with spurs and steps on a miniature windmill. Faced with this regalia, even Emile Male threw up his hands, explaining it all as artistic license:

> Je suis convaincu qu'une oeuvre si froidement extravagante n'a pu etre concue que par quelquc illustre pedant, quelque future laureat de palinods ou des chambres de rhetorique. Ce qui est certain, c'est qu'aucun livre de morale ecrit par un theologien, aucun traite populaire redige par un clerc ne nous present les Vertus sous cet aspects Pour pouvoir comprendre les attributs les estranges Vertues, j'ai parcouru, je crois, a peu pres toute la literature morale due xv siecle, sans rien trouver de satisfaisant; j'allais y renouncer, quand un heureux hasard me fit mettre las main sur les vers inedits qui rendaient raison de tous les detailes de l'oeuvre a expliquer. Cela prouve que se figures des Vertus ne doivent rien a la tradition et a l'enseignment theologiques, mains sont nees de la fantasie individuelle. Je croirais voluntiers (s'il faut risquer une hypothese) que c'est quelque bel espirit de Rouen qui a imagine cette mascarade.[26]

But if the imagery of the Nemours ms. is just personal fantasy, it is hard to explain the attraction of these images for so many subsequent artists. Temperance with a clock in her hand and a bridle in her mouth appears in bomb sculptures by Michael Colombe from 1507, and, some ninety years after the Nemours painting, the same array of attributes appears in the engraving of Temperance by Bruegel. The relating of temperance to the measurement of time apparently struck a responsive chord among northern European artists, and the old conception of temperance as the right balance between asceticism and gluttony seems to be giving way to a new conception of temperance which involves the technical governance of life with the instruments of science. In the Nemours ms. the natural rhythms on which the virtues supervene in the ethics of Aristotle and Aquinas give way to the mathematical rhythms of clocks, seventy years before Copernicus and a century and a half before Galilean mechanics. Male was wrong in thinking that the new iconography sprung from some freak of individual imagination. The artists were responding, not to didactic sources, but to their own environment, full of new gadgets with strange resonances that put pressure on the traditional conceptions of virtues and the good life.

The same tomb for which Michael Colombe sculpted a Temperance with a clock has a Prudence with a compass in her hand. Whatever impulse attracted Colombe to clocks also attracted him to compasses, and it seems that all the calipers and compasses and clocks from dei Grassi on down are part of the same sensibility. I have already mentioned the egoism connoted by handing Prudence a mirror, and the development of foresight into the dominant component of prudence. If we add to this future directed egoism the element of rational calculation, expressed by calipers and compasses, we are on our way to a recognizably modern conception of prudence. The decently attired and bustling peasants of Bruegel's engraving of Prudence (1559) – storing grain, repairing dikes, counting money, drying meats – are not stewards for God's possessions but proto-types of the possessive individualists of Hobbes and Locke. The inscription on Bruegel's engraving reduces prodence to calculative foresight: "*Si prudens esse cupis, in futurum prospectum ostende, quae possunt contingere, animo tuo cuncta propone.*" [If you would possess prudence, consider all future contingencies, and weigh your soul's decision.]

By 1651, in Hobbes' *Leviathan* (viii.) prudence could be described as the selection of means by which desires are satisfied. By the end of Hobbes' century, even professedly Christian authors had forgotten the medieval conception of prudence, and the calculation of profit and loss and wise investment became all that prudence was thought to be. The description of prudence in Thomas Traherne's *Christian Ethics* (1675) is so distant from Aquinas that the connection is almost lost:

> For the designs of Prudence are to secure one self in the exercise of every virtue . . . as neither to hurt a mans self in his Life, Estate, Honor, Health, or Contentment

> . . . Prudence has an eye to every Circumstance and Emergence of our lives. Its design is to make a mans self as great and glorious as is possible: to reconcile our Devotion, Obedience, and Religion, to our interest and Prosperity in the World, to shun all extreams, to surmount all difficulties, to overrule all Disadvantages, to discern all opportunities, and lay hold of all occasions of doing Good to ourself.[27]

Such a prudent man needs a calipers and compass and much else besides.

Can we credibly attribute to dei Grassi, in a deeply felt religious work of the 14th century, a preview of the secular future of the 17th century? Perhaps we can: the man was, on the surviving evidence, a great artist and a great visionary. Furthermore, the image which concerns us lies in the border of a book of hours, and it is on the borders of these great books of the later middle ages that signs of the new birth are most often found. Giovannino may have been mainly concerned with his picture of Ambrose and Augustine, and put the calipers in the hand of Prudence without much thought. But he worked in the most advanced market economy in Europe, and the measure of profit and loss and the accumulation of capital was the talk of Milan as well as Florence. The belief that prudence requires calculation and measurement was even in the 14th century not just an assumption of cathedral building engineers but a common realization of the merchant class. And it was their virtues that would, over the next three hundred years, creep from the borders to take center stage.[28]

Graduate Center and Baruch College,
CUNY

NOTES

[1] Code of Canon Law, Latin-English Edition (Washington, DC: Canon Law Society of America, 1983), at 992.

[2] H. Lea, *A History of Auricular Confession and Indulgences* (Philadelphia, 1986), Vol III, p. 325, n.

[3] William A. Bausch, *A New Look at the Sacraments* (Mystic, CT: Twenty-third Publications) p. 175.

[4] Paul L. Williams: *Everything you Always Wanted to Know About the Catholic Church But were Afraid to Ask* (New York: Doubleday, 1989) p. 293.]

[5] The identification of the object in the hand of Prudence as a calipers is given by E. Kirsch and M. Meiss in *The Visconti Hours* (New York: Braziller, 1972). The size of the object would indicate a calipers, although a compass cannot be ruled out. The identification of this virtue as Prudence is confirmed by Belbello's Prudence in the fall

of the rebel angels, where the identification of Prudence is certain and the imagery is the same as the earlier Virtue.

[6] Prudence and other virtues are occasionally depicted with instruments of measurement in works earlier than the Visconti hours. But in all these works prudence is shown with the other virtues engaged in some common work of construction: for example, in an illustrated ms. of the Somme le Roy dating from 1289 the virtues all wield implements as they work to construct the temple of the Seven Pillars of Wisdom. The Visconti work is the first in which Prudence appears *alone* with some sort of implement.

[7] *Giotto: The Arena Chapel Frescoes* ed. James Stubblebine (New York: W. W. Norton, 1969), fig. 68.

[8] Emile Male, *L'art religieux de la fin du moyen age* (Paris: Armand Colin, 1949 [first published in 1908]), p. 324. Male includes a bizarre and unsubstantiated footnote that the compass is "*un attribut francais*," and the mirror of Prudence, also depicted by Colombe, is "*a la fois un attribut italien at un attribut francais*."

[9] *Ibid.*, p. 325, n.

[10] *De. off.* I. 43; *de. fin.* II. 22; *de. nat. deo.* III. 15, but especially *de. inv.* II. 53.

[11] Macrobius, *in. somm. scip.* I. viii. 7.

[12] Ambrose, *de. off. min.* I. 24; *in. lux.* V. 62, *de. par.* III' Augustine, *de. Lib. arb.* I. 13; Gregory *in Job* III. 9.

[13] Shepherd, III. viii, 2–5. The Shepherd's virtues are Faith, Continence, Simplicity, Knowledge [*episteme*], Innocence, Reverence, and Love.

[14] *de. spec.* xxix.

[15] The reader might feel that the appropriate opposite of prudence is sloth, but in these centuries sloth or *acedia* is a kind of giving in to despair, and the prescribed spiritual remedy for despair was not prudence but hope.

[16] Adolf Katzenellenbogen, *Allegories of the Virtues and Vices in Medieval Art* (London: the Warburg Institute, 1939), p. 33.

[17] Images do linger on. Prudence holds a book in sculptures at Chartres south, (c. 1240), and in Giotto's Arena fresco, which has both the old book and the new mirror.

[18] For the didactic context see Jennifer O'Reilly, *Studies in the Iconography of the Virtues and Vices in the Middle Ages* (New York: Garland, 1988), chs. 3–4.

[19] On the selection of the twelve virtues see Emile Male, *The Gothic Image* (New York: Harper and Row, 1958), pp. 108–130.

[20] *ST* II–II, Q 47, art. 12.

[21] Katzenellenbogen, *op. cit.*, p. 45.

[22] Dante is probably here following Aquinas, who is following Cicero (*de. inv.* II.53) in listing memory, perception, and foresight as three components of prudence.

[23] For analysis of the Titian allegory, see Erwin Panofsky, *Problems in Titian* (New York: New York University Press, 1969), 103–108.

[24] For Memling see Raimond van Marle, *Iconographie de l'art profane au moyen age et a la renaissance* (New York: Hacker Art Books, 1971), fig, 95.

[25] *ST* II–II, Q 7, art. 11.

[26] Male, *op. cit.* (Note 8) p. 316–217.

[27] Thomas Traherne, *Christian Ethicks* int. Carol Marks; ed. George Guffey. (Ithaca, NY: Cornell University Press, 1968).

[28] My thanks to Professor Pamela Sheingorn for help in the preparation of this essay.

IRONY, LTD., AND THE FUTURE OF ART

Unless my title is ironic – and it isn't – it must be tendentious, since it presupposes that irony is Limited (that is, *has* limits) and that art does indeed have a future.[1] The only questions then remaining would be what the limits are and what art will look like in that future. But neither of those first assumptions is self-evident or for that matter evident at all. We have heard talk (and shall hear more here) about the end or the death of art – *something* of a future, although not one that promises much comfort; and by far the greater emphasis in modern views of irony (the importance attached to irony *is* largely modern) has been on the absence in irony of all and any limits. I propose to discuss these two views (and their alternatives) in light of what I claim to be an underlying connection between them; that is, in the relationship between art and irony. Even if art and irony were not related in any deep sense, the questions of what the limits of each were individually would warrant discussion; *with* a connection, that question becomes more pressing still.

1

In its usual contemporary view (both advocated and disputed), the phenomenon of irony appears limit*less*. That is, with every ironic statement opening the way to another one, and that in turn to still another – denying any stopping-point except for the ironic consciousness with *its* suggestion only that there is no bottom to reach. "Nice day", I say to you, with the rain gushing down on us while we wait for a bus already ten minutes late – and neither the rain nor the irony is going to stop there (our only comfort is in sharing the two). The irony expressed does not, on this view, assert anything in particular; it is only a reaction against the miserable rain – as it might react against anything else that so dampened the speaker's expectations. Irony characteristically turns a 'given' against itself; it hardly matters what the particular 'given' is.

Irony thus initiates (more precisely, takes place in) an infinite

C.C. Gould and R.S. Cohen (eds.), Artifacts, Representations and Social Practice, 87–103.

movement – perhaps regress, perhaps progress – leaving behind not only the one context in which its individual statement is set, but *all* contexts: it is irony that is asserted, not just this or that irony. Kierkegaard's estimate of irony's "infinite absolute negativity"[2] may seem too heavy (or light) for merely human practice. But master-ironist himself, his diagnosis commands attention, and not only because with it he finds himself in momentary agreement with Hegel who would thereafter appear more often as his arch-antagonist.

To be sure, irony does not openly declare this 'infinite' ambition; indeed, it hardly speaks about itself at all. But that is because reticence is a characteristic part of its mechanism and so also of its representation: it has no tell-tale signs, except perhaps for understatement itself. In recognition of this, the critic and aesthetician Jean-Paul proposed the introduction of an 'irony mark' – which, like the question-mark or the exclamation point, would identify ironic sentences whenever they occurred. But the idea never caught on, in part no doubt because the force of irony is diminished by all self-reference; it is a feature of irony that its appearances should *not* name themselves. In this sense, irony is the more present for being the more absent – an irony in its own constitution.

Admittedly, other literary figures or tropes – metaphor, synecdoche – also avoid advertisements for themselves; there, too, blatancy is self-defeating or, more basically, anti-figurative. (Although the figurative impulse might turn forthrightness too into a figure of speech – perhaps called 'candor', or even 'truth'.) All figurative language, moreover, involves deviation from a norm – but none other implies or exemplifies as irony does the exact reversal of what had been, only a moment before, 'literally' asserted or implied. Thus, the oxymoron – e.g., 'loud silence' – openly displays a contradiction not in order to advance the cause of contradiction but to intensify a *particular* contrast. Also catachresis – assigning a figurative term to something which has no other, literal designation (as in the 'leg' of a chair) – and the aporia, which in so many words tells its audience that the subject referred to is beyond words, ineffable – although they hint at a realm transcending discourse, do not designate or argue for a general consciousness beyond the particular phrase. By contrast, the reversals of irony enter and reflect not only the eye but the disposition of the reader or viewer. We speak of persons as well as lines or phrases as ironic – but we could hardly imagine a person who was metaphoric or metonymic. Even in referring to a tragic life, we consider that life as the object, not as the medium or mind of tragedy.

When Oedipus, conscientious and feelingful, acts to avoid the fate prophesied for him, he succeeds only in bringing that fate more directly to him. Here, too, moreover, where irony is the subject and not only the means of dramatic representation, it is the viewer who first contrasts Oedipus's efforts to avoid moral violation with the undoing to which those efforts lead him. To be sure, Oedipus himself becomes aware of the same irony or at least of the reversal at its source. (Does the 'object' of irony also *experience* what occurs *as* irony? It seems more likely that irony is limited to second-or third-person views – the first-person never being able to detach himself sufficiently from what is happening to him.) It is in any event the reader or viewer, not Oedipus, who first and then last follows the subversion of his efforts and who – at once inside and outside the text – sees that undoing with the double vision of irony.

This double vision and its presupposition of a transcendent consciousness would in the 19th Century – expanded and dramatized – become what has since been known as Romantic Irony, where the ironic consciousness, in both author and reader, remains always a step ahead of the disparity between appearance and reality that marks the ironic crux. It is worth noting that this consciousness prescribes a manner of resolution similar to that found in comedy – as we observe in them both someone's undoing through the very means by which the same person had before that asserted himself: the pickpocket who, practicing his trade, has his own pocket picked; the dreamer who, giving life to the characters he dreams, then discovers that he himself has been dreamt.[3] The tragic irony of *Oedipus* is thus not tragic *as* irony but tragic beyond irony which by itself is neither tragic nor not. (Otherwise, every occasion that reveals a disparity between appearance and reality would be tragic – which, happily, is not the case.)

The viewer or reader of transcendent irony thus finds refuge in the doubled form of consciousness that constitutes it. From the nineteenth century on, moreover, this conception of the ironic consciousness has moved in two directions which are not, however, so different as to obscure a common transcendent impulse. On one of these versions – the Romantic irony mentioned – the role of ironist is like that of a God who observes detachedly all human (i.e., finite) concerns and efforts – the outcomes of which are anticipated by the omniscient observer beforehand and which invariably turn out differently from their original and merely human design. The latter efforts, then, are only apparent, an immortal longing on the part of an agent who is fatally mortal (a requirement for all objects of irony). This ironic view, except as it is *actually* God's, is a fiction – most vividly displayed in art itself as the artist breathes life into characters who then seem, but only *seem*, to move and act on their own. Having been set in motion *ex nihilo* there is no reason why that motion should ever stop – an intimation of immortality that apparently holds even for those characters who, fictionally, meet their deaths. (Madame Bovary dead, for example, remains more alive than the characters who survive her in the novel; she remains more alive for that matter than Flaubert himself, who on his own deathbed cursed her for this injustice.)

In this way, the ironic artist and viewer stand outside the event–ridden public world, encompassing the fragments and tensions of human experience that they depict and observe. Are the limits and parts of these fragments also the limits of the artist or viewer? Not at all – and a proof of this is in the mind's disposition to oversee, to reconcile the disparity of tragic irony, stirring the hope that the future will mimic and sustain that transcendence, going beyond that and perhaps all other reversals. As God can recognize contradiction without suffering it, so, too – at least in aspiration – does the human creator or observer in Romantic Irony.

Is there, then, in this imperial conception of irony, any *non*-ironic literature, any texts that do not overcome a set of conflicting fragments to present the reader with, if not a contradictory, at least a divergent view? For some commentators – not always Romantics themselves – the answer to this question is "No". On their view, art is always a denial – because it is a revision – of reality, and this is, in their thinking, a sufficient basis for finding it ironic. Even so conservative a critic as Northrop Frye arrives at this conclusion, with none of the hedges one might expect in his sweeping claim: "The literary structure is ironic", he writes, "because 'what it says' is always different in kind or degree from 'what it means'".[4] But unless one saddles Frye with a Derridean metaphysics that distinguishes saying from meaning in *all* discourse, this particular claim seems too broad an application of what is in fact a restricted view of irony. Certainly *some* literary forms seem to oppose the reversals of irony, aiming at sincerity and immediacy, not at displacement or indirection. Thus, for example, the lyric (arguably the most romantic of all forms) – in which the reader is meant not only to avoid double vision but to merge his or her own 'I' with the 'I' of the text itself.

Transcendent Irony, in contrast, invokes a binary consciousness at once in and beyond the text. And Romantic Irony is perhaps the most obvious representation of this evocation of transcendence, epitomized in 19th century poets and novelists but appearing also among its philosophers, political leaders, and barons of commerce – all making claims for a unified self that comprehends (both surrounds and makes intelligible) the apparent fragmentation of history, much in the way that God had done in the past, only with still greater emphasis now because obsolescence had undone that God, his supposed omnipotence notwithstanding.

But if Romantic Irony is the most explicit version of Transcendent Irony, still another version, originally intended to displace the first, comes to the same transcendent end. This conception – what I call 'Skeptical Irony' – sets out to be tough-minded where Romantic Irony had been tender, this-worldly rather than other-worldly, pluralistic rather than monistic – all of these disclaimers directed *against*, not in behalf of a transcendent design. Notwithstanding such deflationary intentions, however, the effect of Skeptical Irony comes to exactly the same transcendent conclusion – in one sense, even more sharply, since now irony appears not merely as a means, but as an end in itself, a principle according to which limits not only do not but *could* not exist.

In Skeptical Irony – I refer here to the varieties proposed in post-structural and post-modern accounts – the lure of infinity, which in Romantic Irony leads to the resolving consciousness of the Transcendent Ironist, is itself ironized: here any one moment of irony appears only to be undercut by another – which then suffers the same fate, and so on in a great chain of 'Divided Being' that is irony all the way down (except, of course, that there is no 'down'). Skeptical Irony criticizes Romantic Irony as a vestige of non-ironic nostalgia, a longing for privilege and security which is fictional even beyond irony's original fiction. So it would dispute the aspiration of Thomas Mann (perhaps Romantic Irony's most deliberate 20th Century adherent) to "an all-embracing crystal-clear and serene glance, which is the very glance of art itself, that is to say: a glance of the utmost freedom and calm and of an objectivity untroubled by any moralism".[5] By contrast, how much more daring – and *truly* ironic – the free-fall of Skeptical Irony, the '*mise in abyme*'. " . . . It is necessary . . . that writing literally mean nothing",

Derrida can say – presumably meaning *this* literally enough to give irony a constant and efficacious push forward, beyond itself.[6]

Instead of distancing himself and appearing then as an omniscient overseer, the Skeptical Ironist thus thinks to sublimate the transcendent impulse that moves him by *becoming* irony; there is nothing, including himself – more precisely, his 'self' – that he will not see double in his then infinite journey: the double itself becomes double, and that in turn, and so on. Instead of first positing transcendence and looking down or back from it, the ironist in this second version starts with the particular and looks up or forward: Irony is destiny and the future, not history and the past. And although there is nothing intrinsically transcendent *or* ironic about wanting to eat your cake and have it too, the cake eaten here, meant to be irony plain, nonetheless has transcendence as its main ingredient; here, too, the impulse for escape or transcendence motivates the ironist, and here, too, the double vision of irony is the thread on which transcendence hangs.

On these two versions of transcendent irony, then, the answer to the question of whether irony has limits is – not simply perhaps, but plainly – "No". Transcendence is a motive force for them both precisely because it challenges the role of limits. In Romantic Irony the founding principle of a non-ironic ironist is openly acknowledged; in Skeptical Irony, that Unmoved Mover is replaced by a process with even more explicit designs on transcendence, aiming to escape the here-and-now which is where irony necessarily begins.[7] On neither of these do limits have a grip. To attempt to set limits either within or outside these versions of irony is (in their own terms), moreover, to invite a variety of charges, the mildest of which is the accusation of naiveté. For readers who would look non-ironically backwards, that monocular vision will – to the *Skeptical* Ironist – demonstrate the blinding effect of nostalgia; for readers who look non-ironically forward, the prospect, as judged by the Romantic Ironist, wanders between randomness and totalitarianism. In both versions, *apparent* limits such as moral conscience, aesthetic form, the consistency of scientific inference, have no purchase. Any appeal to such reactionary constraints represents contamination – psychological or even physiological blockage; it discloses not authentic limits but a failure of nerve.

2

In my title, I associate irony with the topic of the future of the arts, and I want now to consider this connection – mainly because of irony's standing as the dominant trope or figure both of modernist art and of modernism in general. Certainly in the most original – and originative – modernist 'philosophers of suspicion', the subversion of appearance by reality is a dominant motif (so, for example, in Marx and Freud, with – respectively – the deflation of superstructure by foundation and the displacement of manifest by latent content.)

In these terms alone, irony, insofar as it impinges on all cultural expression, would by implication have a central role also in determining the future of art. But its presence is even more strongly grounded than this – reflecting, beyond irony's general claims, a specific implication in the character of art itself. The latter implication appears most dramatically in the Hegelian view of the future of art, where the movement of consciousness represented by irony is taken also to herald the end or the *death* of art. This is, admittedly, not much of a future, but it is, in Hegel's terms, all the future that art *can* have; and it is important for our purposes here to understand the connection of irony to art from which that prediction (more exactly, that pronouncement) emerges.

The latter connection holds, I shall be suggesting, even if we dispute Hegel's specific claim – or indeed, even if we question the conception of irony on which it draws. But let us start here with Hegel in his own terms, viewing the question of the future of art through the lens of a conception of irony as limitless – 'infinite', as Hegel himself also inscribed it. Hegel was, I believe, the first writer to think or at least to speak about the death or the end of art, and it is worth considering how he came to that conception. To be sure, death was itself a favored Romantic theme, with the phenomenon itself often exerting a powerful attraction (eventually including, of course, the death of God). Not only did Goethe's 'Young Werther' commit suicide, for example, but a number of the novel's readers did so as well (one, if not a conclusive, argument for the efficacy of art). But for Hegel, this attraction, like every other, had a historical ground, and it was from that basis, not personal disposition, that he then made the startling announcement of the end or death of art.[8]

What exactly was Hegel committing himself to in this declaration? His claim, it seems to me, is usefully divided into two parts. The first of these – a superficially modest although still, a century and a half later, unsettling proposal – finds in art the taint of mortality. Here Hegel asserts that art not only *has* a history but that it is itself *in* history – that like individual art works with their beginnings and ends, art itself, as a vehicle of cultural or individual expression, is contingent. In practical terms, he here views art as having come into existence in response to a particular set of human needs or purposes; as those needs or purposes alter historically (in part, as their creators and viewers react to the presence of art itself), thus the role that art initially had would also change – possibly to the extent of losing its place entirely. In this part of his thesis, Hegel announces not the death of art but its contingency, judging only that it *can* end or die as the result of a process of aging to which it is no more immune than other aspects or instances of mortal being.

The second part of Hegel's thesis speaks, beyond possibility, of actuality – asserting that not only does art, as contingent, occur within history, but that this possibility is now actual (that is, when he wrote – and *a fortiori* at the time of *this* writing, 150 years later): the time of art which had once come has now also gone. This second claim is, of course, stronger than the first one, and it is important to understand its connection to the character of irony discussed earlier. Hegel's obituary for art, it will be recognized, is impelled by his philosophy of history and by his conception of reason which is closely intertwined with it. Certainly he speaks in this context from no *animus* against art (art, he writes, captures "the profoundest intuitions and ideas of [nations'] hearts").[9] Indeed, art's obsolescence is but one in a line of other cultural 'overcomings' in the course of world history and thus part of the evolution within that framework of '*Geist*' (which is most concisely, although circularly, defined as what it is that world history is the development of).

The claim that Hegel makes about the nature of art as it undergoes this combination of evolution and devolution is, moreover, quite precise. As art once served certain important social and human functions – for the Egyptians and Greeks, for medieval and renaissance and then modern Europe; so, no less assuredly – he claims – those functions have become (as they must inevitably) obsolete. The consciousness that found itself at home with the human figure in Greek sculpture or with the landscape in a Dutch painting was itself in passage – impelled by those very moments. This passage, moreover, has been linear and progressive – consciousness becoming at each stage fuller and more explicit than it had been at the stage before. That process led, in the 'Romantic' art of Hegel's own age, to an increasing detachment of the Idea that animates artistic expression from its form. The latter process of dissociation – what is finally the dialectical movement of reason – is not peculiar to the history of art; Hegel views it as the motive force of history as such (at least for the history of consciousness which is, for Hegel, all of history that matters). The principle of motion here is straightforward: wherever mind makes an entry, mind also, later, by the same compulsion, forces an exit. Thus religion – with its own claims on *Geist* – must also pass (although after art); thus, too, the stages *within* art supercede each other, impelled at one level by cultural inventiveness or artistic originality, but more basically, by the historical logic of artistic progress, the growing *self*-consciousness that eventually exceeds the capacity of art itself. At that point – that is, when the Idea overpowers the vehicle – it then enters a new form; the old is passé, irrelevant – much as a snake shucks its outworn skin. This does not mean, to be sure, that even after its time has passed, art cannot serve as an 'entertainment' (here Hegel stole a march on Graham Greene's literary genre). But it does mean that art in this new condition no longer answers to the *basic* needs of its audience – what people require and are in a condition to receive, apart from or prior to their pleasure in being entertained.

Arthur Danto, at once philosopher, artist, and art critic, has been imaginative enough to put Hegel to an empirical test.[10] For Hegel's thesis should, it seems, have concrete implications by which it may be confirmed or disconfirmed. Danto's argument, reasonably enough, draws on the present condition of the arts (mainly, the visual arts). In them, his exemplary item of evidence is the advent, broadly construed, of 'conceptual art' – that array of objects or events which depend, unlike much traditional painting or other of the visual arts, directly on verbal formulations joined, either immediately or as presupposed, to a more traditional painterly past of the artistic structure. Such works are in effect instances of verbal no less than visual representation; they are unintelligible apart from the acknowledgement of that connection.

For some examples of this step taken by art, the mutation of visual form into idea is obvious. Marcel Duchamp's 'Fountain' was, before he titled it, a urinal. It was the title and the irony thus conveyed that effected the transformation and that brought the newly-pronounced fountain into a museum. To be sure, 'Fountain' remained a physical object and not only a nameplate or title; but once moved into the museum, it would be acknowledged as a physical object mainly in deference to a past used to commodities and to the need of museums to fill their floor and wall space with something bulkier than prose. Clearly, it was the ironic *concept* (the porcelain made word) that was installed in the museum; the urinal apart from that concept might have been, indeed *was*, one of many such fixtures, none of them produced by Duchamp himself. "Ideas alone", Sol Lewitt has said, epitomizing this conception of the artistic project, his own and others', "Can be works of art". There is, of course, no quarreling about usage – but Lewitt's reference here to "works of art" implies quite a different meaning from the one it has more traditionally had.

This is, I take it, what Danto's suggestion about the conceptual transformation of art amounts to, and its connection to my earlier discussion of irony may now come clear. Insofar as irony is without limits and insofar as art employs irony in *that* sense, then art itself is bound to become something other than what it had been: it turns into verbal or conceptual discourse which is intended literally and referentially and which thus is also distinct even from 'verbal art' (i.e., literature) as well. For Danto, art at this point has become philosophy – and although for Hegel, there is an additional mediating step (and perhaps more than one) before *that* happens, the eventual outcome, and the reasons for it, are the same. Art flourishes, has its place, only when a certain diffidence or naiveté persists in the creator and his audience, an acceptance of sensible pleasure in the consciousness that creates or appreciates it. This capacity is not a feature of consciousness that can be willed into existence – since the act of willing it would already transform it. Where consciousness posits only and always a further consciousness against itself – as it does in Transcendent Irony – the outcome will not be art in any traditional sense, but a self-conscious assertiveness that can never rest easy; this new appearance seeks rather to disrupt or to reinvent form, in effect to move *beyond* form, and with this also to move beyond the merely aesthetic sensibility. There is here a will for consciousness that speaks the idiom of reversal or opposition, with the latter – that is, irony – then becoming an end in itself.

The effect of this change is an approach, in art's own terms, to the end of art. That possibility is to some extent already familiar to us: in the visual arts, with the conceptual art already mentioned but also with the related self-consciousness of minimalism and hyper-realism; in music, with the efforts to incorporate random variation, nearly-random instruments, or at a last extreme (in John Cage's well-known example), the medium of silence itself which is as close to pure irony as irony can come; in writing, with the search for a discourse that has as its subject only writing itself – and fails in this search only because it cannot entirely invent the words it uses. (Samuel Beckett typifies this search: he wrote in French rather than his native English, he said, because he hoped this would enable him to write "without style".)

The issue here is not whether what comes of these efforts has been significant; the individual works have indeed been of great interest and not only because they summarize and by contrast reflect back on the past history of the individual arts and of art as such. The issue persists, however, as to what they promise – in addition to their look at the past – by way of a future, *beyond* the breach of limits thus effected. What can the ironic consciousness propose – as art would ordinarily hope for and do – *on its own behalf*? And here its own means seem in fact to incapacitate it, since irony – an oppositional phenomenon – can hardly say much of anything by itself; that is, when there is no ironic object to push against. What is likely to occur under these circumstances, in fact, once we go beyond the assertion of self-consciousness itself, is what we now find increasingly in what is already, after all, 'post-conceptual' art (Stanley Fish proposed to settle once and for all the issue of such nomenclature by invoking the 'post-contemporary'). Here art works gather which look like traditional art in the way that shadows look like their originals – imitative, governed by eclecticism, unconsciously disclosing their conscious intentions (that is, art failing to conceal art), driven less by genuine originality than by a will only to be *different* from other past work. In Danto's terms, such structures, abstracted beyond art, verge on, if they do not become, philosophy, political science, social criticism, and so on. In all this, moreover, there is the additional threat of a commercialism that flattens such first-level motives still further. A book of poems by a known poet does reasonably well with a publishing run of 2500 copies; a Michael Jackson recording will easily sell 25,000,000 copies. Odds of 10,000 to 1 are bound to have consequences beyond the ratio itself.

Is this then *inevitably* the future of art – an end that, part-bang, part-whimper, mirrors the divided vision of irony itself? Certainly that is a possible and plausible future, intriguing as a narrative and supported by evidence of current appearances. It has the additional attraction, moreover, of showing history itself in motion – an aspect often missing even in historical accounts.

And yet there are also reasons for suspecting that this may not be the whole story, that *as* story, it is not quite right even as far as it goes – or more precisely, that an alternate future is no less possible or plausible. One reason for this suspicion has nothing to do with irony. This is the fact that even if a live and active future *were* in store for art, its form would be virtually impossible to predict. This constraint is familiar from art's own past, even from its most productive moments. What could Raphael or his contemporaries have 'predicted' of the baroque flourishes of Bernini? For Bernini *himself* to predict them, after all, required their actualization; it is the artist himself, its creator, who imagines a future for art. Prior to or otherwise apart from such imagining, it may *always* seem that art has no future at all – that is, until this is disproven by art itself. As it repeatedly has been.

On the other hand, ignorance about art's future and the silence that may cloak it are obviously no arguments *for* its viability either. One course to follow in finding *that* evidence, it seems to me, would be to challenge Hegel's historicist thesis at its origins – opposing to his conception of an evolving consciousness or human nature an essentialist view of that consciousness or nature as unchanging. On this alternate view, the arts would have originated and evolved in response to human needs which do not themselves alter; the arts thus might *have* histories – there could be new genres or forms – but they would not, collectively, be *within* history. So, for example, when Aristotle writes in explanation of the 'birth' of poetry, that "all men take pleasure in imitative representations" (*Poetics*, 1448b), he is evidently thinking of an essential human trait which then expresses itself in poetry. As long as the trait remains – that is, as long as humans are human – so also the idiom of expression will be constant (at least in a general way; Aristotle was well aware that individual genres have histories).

The disagreement between Aristotle and Hegel on this point leads far afield from the topics we are discussing; in any event, a second possible future for art is closer at hand – in a loophole left in the discussion of irony from which I set out. On the view of irony as limitless or 'infinite', reflecting in this the movement of consciousness beyond – more precisely, against – the idea and even the form of art, it seems to me that the Hegelian thesis of the 'end' of art is itself unavoidable.

Like the news of his death that Mark Twain, at *one* time, found greatly exaggerated, we might quarrel with Hegel about the actual date of art's death, but not about its inevitability. But it is also possible – this is a more subversive opening – to conceive of irony itself in different, even contradictory terms to those presupposed in the Hegelian thesis: to imagine irony *with* limits, and then to predict the future of art which would then be linked with *that*.

I will not attempt here to develop or to argue for this alternate view of irony – a foundational view, as it would be, irony *with* tears – although I believe that this can be done, and done in a way that makes fuller sense of the same evidence that led to the contrary view of Transcendent Irony. More directly pertinent for the relation between irony and art, however, are the consequences that this alternate view of irony would have for the question of the future of the arts. What we might do for the moment, as a thought-experiment (which would, I believe, more thoroughly argued, disclose itself as fact), is to imagine irony as contained, bound (in the sense of limited) by external and thus independent criteria; in other words, as asserting or implying truth-claims rather than as bringing truth or fact always into question. Viewed from this perspective, moreover, irony would be *within* art, not the other way round – and with this, a very different picture of the future of art also would emerge.

The strategy here must be to look sidewards or obliquely – and *to be able* to do this as irony itself permits, that is, to look at irony ironically. "Nice day", I say in the midst of a downpour (repeating my earlier example and discomfort). But now I hold up the hope and possibility of such a day – its *reality* for us – and not only the defeat or denial that the rain seen by itself through one (wet) eye would represent. There can be little question, it seems to me, that the history of art, with the weight of self-consciousness that it has come to impose on artists, has, since the Renaissance, made an unavoidable difference to artists who, looking constantly backward over their shoulders, act within it. But what, we might ask, if we anticipate that history, looking rather to what art was *before* its history imposed the self-conscious ironies that now buffet it? Or what if we look at it even in the present as it might be now *apart* from those ironies?

Cues that bear on the result of such experiments come, it seems to me, from several directions. The first of these is the fact that the very concept of art is an anachronism when applied to the greater part of what we now speak of, rather too quickly, as the history of art. The Greek tragedians did not think of themselves as 'artists' in the modern sense of creators – any more than did medieval painters or the builders and artisans responsible for the Gothic cathedrals. This does not mean that it is a mistake to view such works now *as* art, but it means that in doing so, we have, as a matter of historical fact, 'ironized' our view of those works through the lens of that concept as *we* impose it. Nor is this process of aesthetic ironizing directed only to a remote past (one which because it is past could hardly dispute anything now said about it). For example: well within recent memory, the very idea of regarding film as an art form would have seemed outlandish, at best a not very funny joke. There were, for example, no published starting times for 'movies'; this would have been to give the medium unjustified weight. Viewers entered the 'theatre' (*sic*) whenever they arrived and stayed until they had seen as much as they wanted to. The classical conception of beginnings and ends – applicable in drama, for instance – was irrelevant. Only by a gradual and self-conscious process of change over a period of decades was film 'inducted' into the category of art.

From this and other examples, we infer that the *intention* to produce art (an intention that in putting a 'frame' around practical immediacy is intrinsically ironic) is no necessary feature of art itself; the same effect can be realized without the intention – and still more effectively because more ironically (an effect that is *almost* uncaused. . . .). A summary view of the conditions that could make art possible in the absence of artistic intention comes under the not very technical heading of 'surprise' – as the phrase might then be: 'surprised by art'. The phenomenon this term refers to should not itself be surprising; it is familiar from the experience of art itself – for one thing, as certain works of art, no matter how many times one may have seen or heard them, continue to catch and hold our attention – in part at least, by surprising us *again and again* in the causes and effects that constitute the artistic structure.

In respect to the '*first-time*' experience – that is, seeing individual works or genres for the first time – the phenomenon of surprise, of taking pleasure in novelty – in 'what happens' – is a commonplace. But at least something of the same characteristic seems to have a role where art is concerned even in 'second-' or 'third-time' experiences: the fact that a reader or viewer knows what is going to happen next does not preclude this. So far as this is the case, moreover, no explanation of the attraction exerted in such instances would be possible that omits the factor of surprise – which seems, in fact, a likely formula for the opposite or antonym of irony, at least for an irony without limits: where there is no regularity, there could also be nothing unexpected. If, on the other hand, irony is conceived of with*in* limits, its oppositional character provides assurance that it will also be reactive and thus predictable, thus also laying the ground for the 'surprise' of art. The very occurrence of surprise in the experience of art, then, argues for a version of irony with, rather than without limits. Even if we resisted this implication, moreover, the role of surprise in the encounter with art would remain to be explained – and more than that, to be taken account of in predicting art's future or more precisely, in addressing the question of whether it *had* a future.

I here venture only one, fairly obvious suggestion about what might in the future appear as an unanticipated art form. It is no accident, I believe, that viewers of television (as various polls suggest) watch television *commercials* more attentively than they do the programs between them. More time, effort, and talent – and as evidence of this, money – are invested in 'making' the commercials that in making the programs themselves. And this shows in the results: not necessarily in the commercial efficacy of advertising, but in its power to hold the viewers' attention – that is, in its aesthetic appeal. I can imagine – indeed there already exist – 'anthologies' of such commercials which may eventually have the status of the first anthologies of poetry (the anthology is itself, we recall, an artistic genre with a history of its own).

This is, I recognize, a conventional and hardly startling comment. But on the view I have proposed, predictions that try to go much deeper would require the authority of art itself; we must ourselves *be* the artists, I have claimed, if we are to predict what – or that – art will be. What this means in effect is that the most telling prediction we can now make is that what will in the future appear as art (if anything appears at all) will surprise us – first in its immediate presence, but also later, after the 'first-time' experience, when we find ourselves treating it *as* art, incorporating it into the history that had before been fashioned out of older, more familiar stuff.

Admittedly, this is not proof that art *does* have a future – although in its own terms, it is as much proof as the issue allows. It argues only that just as for art in its traditional form and history there was no way to predict the character or even the fact of its future, we must – now as before – concede the possibility of its continued existence. Not because the future as such is uncertain, but because, beyond that, uncertainty – even improbability – is intrinsic to art. The only basis for denying this view, and thus for predicting the end of art, would be if somehow the capacity for innovation in the sense cited were destroyed or otherwise lost, in artists and/or in their potential audience. This would indeed be the outcome of an irony-without-limits of the sort asserted in Romantic and Skeptical Irony, where any such sense, even if it existed, would have no function, where it would – like the human appendix – become merely vestigial. But since there are reasons to doubt that account of irony – indeed to doubt that it is coherent even in its own terms – we are compelled then to think of art as *possible*: now as before, and in the future as now. At issue here is not the difficulty of predicting the future as such (although there is something of that as well), but the difficulty of predicting a future in an expressive form that is intrinsically unpredictable. Prediction *retrospectively* is, of course, only too easy; we do better here to agree to 'Wait and see' – a recommendation that need not be always as prosaic or trite as it seems. Especially not when the alternative is the lure, but also the mystification of the infinite – a promise that neither art nor irony is capable of keeping.

State University of New York at Albany

NOTES

[1] An early version of this paper was presented at a conference on 'The Future of the Arts' at Buffalo State College in November, 1991.

[2] Kierkegaard, *The Concept of Irony*, trans. Howard V. Hong and Edna H. Hong (Princeton: Princeton University Press, 1989), p. 26.

[3] The latter example comes form Borges's story, 'The Circular Ruins'.

[4] Northrop Frye, *Anatomy of Criticism* (Princeton: Princeton University Press, 1957), p. 81.

[5] Thomas Mann, 'The Art of the Novel', in Haskell M. Block and Herman Salinger, eds., *The Creative Vision* (New York, 1960), p. 88.

[6] Jacques Derrida, *Positions*, trans. A. Bass (Chicago: University of Chicago, 1981), p. 14.

[7] It is not easy to find a definition of Romantic Irony that has won general acceptance; one view of it, moreover, would in effect equate it with what I call here 'Skeptical Irony'. (See, e.g., Lillian Furst, *Fictions of Romantic Irony* (Cambridge: Harvard University Press, 1984), 227–8.) For the purposes of the argument here, however, the distinction

between the two types of irony is more important than the names given them. In any event, grounds are present in the acknowledged 'founders' of Romantic Irony (e.g., Friedrich Schlegel) both for claiming a connection between Romantic and (e.g.) Greek irony – and for drawing a subsequent and sharper distinction between them both and 'Skeptical Irony'.

[8] See, e.g., T.M. Knox, ed., *Hegel's Aesthetics* (Oxford: Clarendon Press, 1975), Vol. 1, pp. 7–12. The extent to which Hegel is committed to the 'Death of Art' thesis (or if so, in what sense) has been disputed (see, e.g., Curtis L. Carter, 'A Re-examination of the "Death of Art" Interpretation of Hegel's Aesthetics', in Warren E. Steinkraus and Kenneth L. Schmitz, eds., *Art and Logic in Hegel's Philosophy* (Atlantic Highlands, N.J.: Humanities Press, 1980), and William Desmond, *Art and the Absolute* (Albany: SUNY Press, 1986), Chs. 1 and 2. The role of the 'Death of Art' thesis as raised in the present essay is primarily conceptual, not historical – although I would also argue on textual grounds that such 'revisionist' readings as those mentioned that claim to get Hegel off the hook of the thesis (why should it be considered a hook at all?) are mistaken.

[9] 'Introduction' to *The Philosophy of Fine Art*; see Berel Lang, ed., *The Death of Art* (New York: Haven Publications, 1984), p. 262.

[10] Arthur Danto, 'The End of Art', in Berel Lang, ed., *The Death of Art*, pp. 5–35.

A GENEALOGY OF 'AURA': WALTER BENJAMIN'S IDEA OF BEAUTY

No doctrine of Benjamin's has been more influential in contemporary aesthetic discourse than that of the decay of the aura of the traditional work of art, which he developed in his essay on "Das Kunstwerk im Zeitalter seiner technischen Reproduzierbarkeit."[1] Yet justifiable concern has recently been expressed that the conventional reading of the essay in the 1970s has "tended to obscure its more incongruous and ambivalent features."[2] Whereas such attention has been directed at elements of the essay tied to the political circumstances of its origin as well as to Benjamin's lifelong meta-project of a theory of experience, I shall maintain that its philosophically most intriguing ambiguity follows from his introduction and deployment of the term 'aura.' Furthermore, I shall maintain that this term's only decisive incongruities are the legacy of Benjamin's explorations of a semantically related category some fifteen years earlier, the concept of beauty.

No other category of such prominence in Benjamin's thought has received so little attention. This often subterranean leitmotif surfaces at crucial points in his writings and – like his concept of myth[3] – bears upon his employment of a wide range of traditional philosophical categories (including truth, nature, critique, semblance,[4] and the sublime). Once we have accepted the conceptual propinquity of the grammars of the concepts of aura and beauty, or more precisely, aura and *schöner Schein*, two points will follow: first, that Benjamin's essay on Goethe's *Elective Affinities*[5] – Benjamin's most difficult if not esoteric effort – can be read as developing an elaborate metaphysics of beauty in its relation to truth and semblance, and finally in its critique through the sublime in, and hence beyond, the literary work of art. Second, we will see that Benjamin's introduction of the term 'aura' serves to present a specifically modern but semantically filiated correlate to his early idea of beauty,[6] which in his essay on Goethe's *Elective Affinities* had already exploded the canonical confines of literary beauty.

This latter point relies on more than merely the suggestiveness of the analogy between what are clearly two categories of appearance, aura and *schöner Schein*. In his various definitions of aura,[7] Benjamin

C.C. Gould and R.S. Cohen (eds.), Artifacts, Representations and Social Practice, 105–119.

makes both coded and explicit references to its aesthetic lineage. In doing so, he is resituating not just beauty but the sublime as well vis-à-vis their classical, Romantic, and German Idealist counterparts.[8]

It is safe to say, if surprising, that Benjamin is indebted to that influential tradition in aesthetic theory that conceives of the history of aesthetics as identical with the history of the idea of beauty. This tradition lost its authority in the eighteenth century, with the shift in focus to the subjectivity of beholding or *aisthesis*. Benjamin's disagreement with this shift, exemplified in his attempt to work with a pre-Kantian notion of beauty in his early works, is an underlying presupposition of all of his philosophizing about art. An investigation of beauty's elaborate philosophical genealogy as well as Benjamin's own differentiated exploration of this issue is much too intricate to handle within the scope of this paper. Yet our comparison of the respective philosophical grammars of these concepts can be aided by refuting those who would claim that Benjamin's concept of 'aura' has no such elaborate philosophical genealogy;[9] if anything, there is a strong methodological reason for suspecting a profound affinity between these two terms, that is, the disdain Benjamin shared with Kant for inventions in philosophical terminology.

In his book on the German Baroque drama or mourning play (*Ursprung des deutschen Trauerspiels*), Benjamin offers a definition of philosophy as "a struggle for the representation of a limited number of words which have always remained the same" [I, 217]. Benjamin's attention, as countless passages in his letters and writings testify, is to the semantic complexity which results from the accrual of diverse meanings to canonical philosophical terms in the course of time rather than to new terminology. His ambition was to recover what he considered the primordial meaning of a term,[10] an intention which follows from his theory of language. Benjamin's view, that it is preferable to recontextualize an existing concept rather than to invent a new term[11] derives from his view of the history of philosophical terminology and is a quintessentially Kantian sentiment, especially if we recall Kant's discussion of 'Ideas in General' in the *Critique of Pure Reason*. There Kant asserts:

> To coin new words is to advance a claim to legislation in language that seldom succeeds; and before we have recourse to this desperate expedient it is advisable to look about in a dead and learned language, to see whether the concept and its appropriate expression are not already provided there. Even if the old-time usage of

> a term should have become somewhat uncertain [. . .] it is always better to hold fast to the meaning which distinctively belongs to it.[12]

Given this precept, I would like to suggest that those very problematic features of the notion of beauty – or specifically the relationships between beauty and truth, beauty and semblance, which he explores in the concluding section of his *Elective Affinities* essay – contribute to Benjamin's abandonment of the notion of *schöner Schein* for the new notion of 'aura'.

Benjamin first signals aura's debt to beauty with his epigram to the 'Work of Art' essay, a quote from Valéry's 'The Conquest of Ubiquity.' In this passage Valéry asserts that the modern development of techniques "make it a certainty that profound changes are impending in the ancient craft of the beautiful." Technical innovations are bringing about a transformation of artistic 'invention' as well as "an enchanting change in our very notion of art" [*Illum.*, 217]. Benjamin elaborates upon this pronouncement later in his essay, where he asserts that for three centuries following the Renaissance, until the advent of "the first truly revolutionary means of production, photography," the **secular** cult of beauty (which preserves the artwork's ritual function) prevailed. With photography, however, the notion of authenticity was thrown into doubt; no longer the measure of artistic value, the function of art lost its ritual grounding, which it had retained "even in the most **profane**[13] forms of service to beauty" [*Illum.*, 224].

"PRYING THE OBJECT FROM ITS SHELL"

Benjamin's Redetermination of the Relationship between Truth and Beauty

In one of Benjamin's definitional explorations of the notion of aura, he hints at the underlying epistemological dilemma this category shares with beauty. In the third section of his 'Work of Art' essay, Benjamin sets forth the thesis that the mode and organization of human sense perception is socio-historically contingent. The ultimate social causes for the decay of aura are twofold; the desire to overcome "the uniqueness of every reality by accepting its reproduction" and "to bring things 'closer' spatially." The "mark" of this desire for nearness and for overcoming singularity, according to Benjamin, is "prying the object from its shell" (*die Entschälung des Gegenstandes aus seiner Hülle*) [*Illum.*, 223].

With this phrase, Benjamin reinvokes the metaphysics of *Hülle* (husk, shell) and *Verhüllten* (the concealed), two key terms of his grammar of beauty in the *Elective Affinities* essay. Benjamin's explorations of these terms are furthered by his consideration of related figures in Goethe's works, especially that of the *Schleier* (veil). Goethe's *Dichtung*, Benjamin writes, "remains turned towards the interior in a veiled light." Benjamin also cites a line Goethe sent to Zelter soon after completing the *Elective Affinities*: "I am convinced that the both translucent and opaque veil will not hinder you from gaining view all the way up to the actually intended figure" [I, 197]. Indeed, as Benjamin notes, Goethe's portrayal of the beauty of his three principal female figures – Ottilie, Mignon, and Helena – turns on his use of the veil motif. [I. 197].

For Goethe this term betokened more than the imagery of the surface – "it is the husk, which moved him again and again where he struggled with insight into beauty" [I, 197]. Recall his invocation of the grammar of the veil throughout *Wilhelm Meister*, for example, as in Wilhelm's first encounter with the beautiful Amazon: Her figure was concealed by a man's white greatcoat, which she takes off in order to hand it down to the wounded and unclad young Wilhelm. "Astonished" and "bewildered" by her beauty, he perceives her "head encircled by rays; and a glancing light seemed by degrees to spread itself over all her form." This saintly or "angelic" presence disappears, her beauty – her affective aura – dissipates with the mystery of the visionary experience.[14] Later in Book IV, Goethe's probing of the veil extends to his *recollection* of this visionary encounter, whose central detail is, once again, the vanishing of her figure in the moment that the cloak falls:

> Incessantly he kept recalling that incident, which had made an ineffaceable impression on his heart. He saw the lovely Amazon again come riding out of the thickets; as she neared him, dismounted, went to and fro, and attended to his injuries. He saw the veiling raiment fall from her shoulders; he saw her countenance, her figure vanish in their radiance.[15]

What is the nature of the veil as the beauty's surface and what is being veiled? Although the veil necessarily conceals, it appears and thus reveals. Does the veil serve to mediate the truth or as its treacherous illusion? Benjamin finds it necessary to distinguish between several kinds of beauty, in part due to this subcutaneous indeterminacy, in part due to the intricate relations between myth, truth, and beauty. Part of the solution to this enigma rests on Benjamin's redetermination of the concept of truth and the relation between truth and myth. Thus Benjamin distinguishes between 'essential' or 'philosophical' beauty, both of which are necessarily veiled, and 'ephemeral' or 'symbolic' beauty, which are grounded upon myth. Like truth but unlike myth, philosophical beauty "does not manifest itself by being exposed." This phrase both raises the issue of the truth's *Sichtbarkeit*, discussed in the penultimate section of this paper, and points to the issue of Benjamin's disagreement with Plato, for whom the truth is beautiful. Moreover, it underscores the primacy of the *relationship* between the veil and the veiled. Their very inextricability reveals what Benjamin considers the fundamental enigma of *schöner Schein*: The secret of a work's beauty cannot be discovered by lifting its veil, by destroying its husk, that is, by penetrating a surface or appearance. All such approaches alter – and hence Benjamin would say 'destroy' – the work in the process. The interpreter's strategy must be instead to comprehend and then articulate the grounds *why* such an approach cannot succeed, or, as Adorno mandates in another context, "To solve the enigma is equivalent to setting forth the grounds of its insolubility." Thus Benjamin argues convincingly that what becomes visible in such beauty is not its *idea*, as Platonists would have it, but rather its *secret* [I, 196].

Hence art criticism (*Kunstkritik*), whose task would seem to be to lift or penetrate the veil, has *not* "to remove the shell, [but] far more [to] raise itself to the true intuition of beauty by a precise cognition of it *as* shell."[16] Critique must make the intuition of beauty possible while defending its integrity. Benjamin elaborates upon this notion in his announcement of the planned journal *Angelus Novus* [II, 242], in which he terms *critique* the "*Hüter der Schwelle*" – "guardian of the threshold", namely a threshold which, like the veil and that which is concealed in Benjamin's metaphorical vocabulary, brings together myth which is '*scheinhaft*' (i.e., it appears) and truth which does not.

Between Recollection and Mémoire involontaire

The most intuitively perplexing association Benjamin makes with aura is in his *Baudelaire* book, where he develops the notion with reference to Proust's deployment of *mémoire involontaire*. There is more to this conceptual association, however, than the furthering of his own definition *vis-à-vis* a concept more familiar and accessible at the time, due to the widespread ideas of Proust and Bergson.[17] This reference is strikingly reminiscent of an earlier, canonical use of memory with relation to beauty. And here we can begin to comprehend his *Elective Affinities* essay, in which Benjamin demarcates his position *vis-à-vis* his understanding of the quintessence of the Platonic view of the apprehension of beauty, as represented by the following lines from the *Phaedrus*:[18]

> But he who is newly initiated, who beheld many of those realities, when he sees a godlike face or form which is a good image of beauty, shudders at first, and something of the old awe comes over him, then, as he gazes, he reveres the beautiful one as a god [. . .] his memory is borne back to the true nature of beauty, and he sees it standing with modesty upon a pedestal of chastity.[19]

Several features of this passage invite the interpreter to associate that very image of beauty, standing modestly "upon a pedestal of chastity," with the figure of Ottilie in Goethe's *Elective Affinities*. Such imbrications of divinity and beauty, for example, are familiar features of Goethe's rendering of Ottilie in the novels: We are reminded of Ottilie's role as the Virgin Mother in the Christmas frieze ("who can describe the presence of this recreated queen of Heaven?" II, 6). And whereas the filiation of beauty with chastity seems to bear upon the figure of Ottilie, Benjamin convincingly contests such readings by demonstrating the ambiguity of her innocence.[20]

Indeed, Benjamin's interpretation coincides with other critics insofar as he agrees that Goethe endeavors to evoke beauty in the figure of Ottilie. For Benjamin, "Being convinced of Ottilie's beauty" constitutes "the fundamental condition for our participation in the novel" [I, 178f.]. And hence Benjamin follows Plato's *Symposium* in specifying the figure of the loved one, for Benjamin the "corporeal, living beauty" [I, 194], as the paradigm of beauty. Yet the question of *how* Goethe evokes the beauty of Ottilie remains. If it is uncontroversial to construe Plato as asserting, on the basis of his theory of recollection, the beautiful object to be beautiful because it evokes our recollection of the Form of Beauty from that time before we existed in this sensual world, then we can safely take Benjamin to be presuming no more than this. The decisive distinction he makes *vis-à-vis* the fictional character of Ottilie does rest on this very comprehension of visible beauty as the recollection of its idea. But Benjamin explicitly states that *the figure of Ottillie does* **not** *awaken such recollection* and thus rules out that the reader finds Ottilie beautiful in what he regards as the Platonic sense. For Benjamin, then, Ottilie's beauty becomes the foil of an elaborate critique of the idea's *Sichtbarkeit*.

The problem Benjamin poses – why Ottilie fails to awaken recollection of the idea of beauty – has several answers. One answer is that in her being [*Dasein*], beauty remains "first and foremost" [*das Erste und Wesentliche*]." As a consequence, *the truth does not*, as it does in the Platonic metaphysical hierarchy. Ottilie, Benjamin argues, does *not* partake of the truth. Hers is not the semblance of beauty, but an essential beauty, and as such, mysterious, a secret: her "innermost nature [remains] closed." As such Goethe's figure goes up against the conventions of the classical portrayal of beauty. This is the point of Benjamin's citation of a criticism of Ottilie made by Julian Schmidt, that she is "not a genuine offspring of the poet's *Geist* but illegitimately spawned."[21] Benjamin is applying distinctions most memorably examined in Lessing's *Laökoon* when Benjamin charges Goethe with crossing "the borderline separating the epic from painting" and rejecting "the famous Homeric prototype for the epic representation of beauty" [I, 179]. In terms of literary portrayal, Benjamin is referring to the tradition whereby other characters in the novel voice a sequence of brilliant epithets about the beautiful character, in this case Ottilie. This strategy of representation is controverted in the Homeric epic tradition – at least according to Lessing – in that Homer avoids making such direct statements that Helen is beautiful, how she is beautiful, and why she is beautiful.

The mode of literary representation of beauty chosen by Goethe, however, is *indeed* to make such direct statements about the beauty of Ottilie. The author states that she is beautiful in perhaps a dozen places;[22] but it is of paramount importance, however, that the reader never lose sight of her beauty. Her beauty is quintessentially *visual*: Ottilie is "a feast for the eyes" (*ein wahrer Augentrost*).[23] Even Ottilie's coffin is left open, such that her physical beauty – its corporeal nature – never disappears. Even in death, significant not least for its traditional association with the sublime, the mythical domain resists relinquishing Ottilie's "ambiguous innocence and semblance-like beauty" (*zweideutiger Unschuld und scheinhafter Schönheit*) [I, 179].

Benjamin as a Theoretician of Visuality

Aura and beautiful semblance are both crucial components of a Benjaminian theory of visuality (*Sichtbarkeit*). Aura is, after all, a *visual* metaphor. Analogous to his treatment of the notion of aura, Benjamin's critique of Ottilie is a critique of her pure visuality. Her visual beauty remains silent in the acoustic realm and is ultimately destroyed by her (sublime) silence.

Benjamin's critique of *schöner Schein* thus essentially becomes a de-valorization of optical appearance, of *Augenschein*.[24] Whereas as early as Plato the appearance of beauty was inextricably bound up with the wearing of the face,[25] Benjamin's theory of *Scheinlosigkeit* represents an attempt to ground art at a 'higher' level than through an optically dominated concept. The category he pairs with *Sichtbarkeit*, the *Ausdruckslose* or 'expressionlessness', evokes, to be sure, the Jewish prohibition against the image, but most significantly does *not* imply a dismissal of the auditory;[26] silence, here both sublime and the rhetorical analogue of the *Ausdruckslose*, should be heard.

Beauty is also bound to the illusion of the work's perfection and totality, both of which have been traditionally offered as constitutive for beauty. Hence the destruction of the *Schein* also becomes the extinguishing of the illusion of totality, whereby the artwork becomes a ruin, an allegory. In destroying this false totality, however, the artwork is perfected by the very act of being smashed into a fragment. This helps illuminate why Benjamin's notion of the *Ausdruckslose* must be violent: truth requires the disruption of *schöner Schein*. It is curious, consequently, that although Benjamin's metaphysical architectonic remains intact in his *Trauerspiel* book, the category of the *Ausdruckslose* as such has been abandoned. Its place, however, has been taken by the notion of "the light of divine learning":[27]

> In the field of allegorical intuition the image is a fragment, a rune. Its beauty as a symbol evaporates when the light of divine learning falls upon it. The false appearance of totality is extinguished. For the *eidos* disappears, the simile ceases to exist, and the cosmos it contained shrivels up. [OGTD, 176][28]

In Benjamin's late works, moreover, he distills the substance of these categories into a phrase which is both more precise and more immediately accessible philosophically: the *Auslöschung des Scheins* [cf. I, 1138 & 1144]. In the latter case, an annotation to Baudelaire's poem *"L'amour du mensonge*," Benjamin draws a direct analogy to the example of the *Ausdruckslose* presented in the earlier essay, when he writes: "*Auslöschung des Scheins durch seine Verherrlichung in der Lüge.*" And in the more familiar, aphoristic style of "Central Park," it is antimythical allegory which effects the same act of extinguishing:

> Majesty of the allegorical Intention: destruction of the organic and living – the extinguishing of semblance [*Auslöschung des Scheins*].[29]

Moreover, here as well, the dissolution of semblance is connected with death: Ottilie's "auslöschender Schein" dissipates with her passing.

Benjamin's Rhetoric of Silence and the Sublime

One part of what is novel in Benjamin's view is, in general terms, that beauty's claim on the truth is indirect; it requires a third element to partake of or participate in truth: critique. Furthermore, allegorical critique is immanent: it takes place within the artwork itself, i.e., *vis-à-vis* the critique from inside, through the work's category of the *das Ausdruckslose*, which Benjamin defines as "the sublime critique of the true." Its analogue in Hölderlin's poetics is interruption, caesura; in tragedy, it is the tragic hero's silence. Thus Benjamin concludes, in a central section of his *Origin of the German Lamentation Play*, that "the paradox of the birth of the genius in moral speechlessness, moral infantility, constitutes the sublime element in tragedy."[30]

Accordingly, the *Ausdruckslose* is a category of language. The *locus classicus* for the sublimity of the rhetoric of silence is Longinus's *Treatise on the Sublime*; but there are many other, less canonical junctures in the history of poetics where Benjamin developed his perspective on this notion. The stillness of Stifter's fictional worlds, for example, provided Benjamin with a fertile expanse to develop surprisingly elaborate reflections on the imbrications of the visual and the acoustic:

> Stifter can only create upon the basis of the visual. This does not imply that he merely renders the visible in other terms. [. . .] The problem of his style is, then, how he conceives of the metaphysical visual sphere in all things. Connected with this fundamental peculiarity is first of all that any feeling for such revelation which must be *examined* is lacking: stillness. Stillness is namely first absence and above all any acoustic sensation. Language [. . .] according to Stifter [. . .] is a display of feelings and thoughts in a deaf room.[31]

Benjamin's abiding preoccupation with Mallarmé's poetics of *blanc*, *absence*, *silence* and *vide* finds its gestative grounds in his early reading of *Un coup de dés* near the end of the war.[32] In Hugo von Hofmannsthal, who published Benjamin's *Elective Affinities* essay after it had been turned down by the fledgling *Deutsche Vierteljahresschrift für Literaturwissenschaft und Geistesgeschichte*, Benjamin found a natural patron. Hofmannsthal's Chandos letter, his powerful statement of the crisis of language, is presented from the rhetorical perspective of a "propriety of silence."[33]

The richest laboratory for Benjamin's study of silence, however, remains the writings of Goethe. The inexpressibility of beauty through saying as opposed to showing is thematized in salient statements from *Der Sammler und die Seinigen* to the *Farbenlehre*.[34]

Why does Benjamin choose another term than *schöner Schein*, especially given the disdain he shared with Kant for inventions in philosophical terminology? One answer can be found in the moral dimension of beauty, which is subsumed under the political dimension of the newer concept. Ottilie is not a moral figure, for, as Benjamin argues, her suicide is not a choice but fate. Ottilie's "plantlike existence" does not lead to moral categories; by spurning a conscious life she is rejecting the possibility of *Sittlichkeit*. Her actions only appear to be moral, her fatedness and death are of mythical character. But just as in his *Elective Affinities* essay Benjamin has shown the moral and aesthetic domains to be inextricable – contra German Idealists such as Hegel and Schelling – by demonstrating a moral dimension of the sublime category of silence, in his 'Work of Art' essay he explores the deceitful terrain where the political and aesthetic converge.

What provisional conclusions can be drawn from the aspects of Benjamin's aesthetic adumbrated above? The modern ultimately demands a category more differentiated than beauty, at least the concept of beauty it inherited from its immediate precursors, the German Idealists. The Idealists fail even to discern the disparity between the artwork's ritual and exhibition values; yet this does not imply an insensitivity to the fact of these phenomena, as Benjamin demonstrates with two telling passages from Hegel which he quotes in a significant footnote to the 'Work of Art' essay. Beauty has nonetheless become a footnote in Benjamin's late works. The other salient mention of beauty in connection with his theory of aura also occurs in a lengthy note, in this case to the essay 'On Some Motifs in Baudelaire.' The artwork's cult value appears in the beautiful. *This* semblance is "the aporetic in the beautiful." Moreover, just as Benjamin offers us a natural and a historical definition of aura, he distinguishes here between beauty's "*historical* being"[35] and its "relation to *nature*." Once again Benjamin quotes himself: "The beautiful in its relation to *nature* can be defined as that which remains 'essentially' the same only beneath its concealment."

A second conclusion has been confirmed by a recently published fragment related to Benjamin's 'Work of Art' essay. In the late 1930s Benjamin wrote that:

> The significance of beautiful semblance for traditional aesthetics is grounded in the rapidly closing era of perception. The corresponding doctrine received its final formulation in German Idealism. Yet it already bore epigonic features. His well-known maxim that beauty is semblance – phenomenal appearance of an idea or sensuous semblance of the true – was not only vulgarized in classical antiquity but relinquished its experiential grounds. These reside in the aura.[36]

Thus Benjamin's introduction of the term 'aura' attempts to restore both the semantic depth and experiential ground of *schöner Schein* in classical aesthetics, which was lost in the austerity of the well-known German Idealist dogma of beauty as of the "sensual semblance of the idea." (Benjamin similarly sees himself differing with Plato, who, he says, "develops truth – the realm of ideas – as the essentiality [*Wesensgehalt*] of beauty.") Aura is the equivalent of beauty's veil in the more complex classical formula. This conclusion is strengthened by the quotation Benjamin appends to this passage, a self-citation from the *Elective Affinities* essay: "The beautiful is neither the husk nor the encased object but rather the object in its encasement."[37]

Second, we should note Benjamin's move from the literary work of art in his early works to other media. The specificity of the tradition of the evocation of beauty in literature and poetry – which Benjamin transforms through an aesthetic of the ideal of the human body – could not be mapped onto the domain of the reproducible arts.

The incongruities of Benjamin's treatment of aura reflect neither the term's confluence of semantic legacies nor internal inconsistencies in an *ex nihilo* definition, but rather a less than seamless transfer of the grammar of beauty's relations to truth and the sublime onto the specifically modern category of aura. Yet although Benjamin stands almost alone among modern theorists of art in taking beauty seriously as an aesthetic category, he concludes in his late essay, 'On Some Motifs in

Baudelaire,' that because reproducibility leads to decay of the artwork's 'aura', "the beautiful has no place" in technical reproduction.

Einstein Forum, Potsdam

NOTES

1 Walter Benjamin, 'The Work of Art in the Age of Mechanical Reproduction,' in W.B., *Illuminations* (London: Collins/Fontana, 1973), pp. 219–253.

2 See especially Miriam Hansen's excellent reading of the essay, 'Benjamin, Cinema and Experience: "The Blue Flower in the Land of Technology",' *New German Critique*, Nr. 40 (Winter 1987), pp. 179–224 (Special Issue on Weimar Film Theory); this citation, p. 180.

3 See Winfried Menninghaus, *Schwellenkunde, Walter Benjamins Passage des Mythos*. Frankfurt am Main: Suhrkamp, 1986.

4 I shall usually refer to *Schein* as 'semblance', even though I find Knox's argument for 'pure appearance' as opposed to 'semblance' persuasive [cf. *Hegel's Introduction to Aesthetics*, trans. T. M. Knox. Oxford: Clarendon, 1979, p. 4 n. 1]. Nonetheless, I make allowances for divergences in Benjamin's employment of *Schein* from Hegel's more precise usage by heeding various nuances, such as mere appearance (Findlay), illusion, luminousness, and glimmer.

5 In: Walter Benjamin, *Gesammelte Schriften*, vol. I, ed. R. Tiedemann and H. Schweppenhäuser (Frankfurt a.M.: Suhrkamp, 1974), pp. 123–201. References to texts in this seven-volume edition are hereafter referred to by volume and page number only.

6 In general terms, both concepts represent new paradigms and both are negatively determined by that which they do not cover.

7 These definitions can be found primarily in the essays on 'The Work of Art . . .' and 'Charles Baudelaire'; see W. B. *Illuminations* . . . Hereafter cited as: '*Illum.*'

8 Benjamin's dialogue with these legacies is a major thread of my forthcoming monograph, entitled *Walter Benjamin's Idea of Beauty*. Benjamin's reappropriation of the discourse of beauty places him virtually alone (significant exceptions being Georg Lukács, Ernst Cassirer, and Martin Heidegger) among early twentieth century writers on aesthetics.

9 Rolf Tiedemann, for example, wants to give Benjamin and Adorno full credit for first giving this ago-old term a philosophical connotation, in his dictionary article on 'Aura' in the *Historisches Wörterbuch der Philosophie* (ed. Joachim Ritter. Vol. 1: A-C. Darmstadt: Wissenschaftliche Buchgesellschaft, 1971, col. 652–53). Birgit Recki catalogues several philosophical senses of 'aura', without making the customary facile claims of influence in her monograph, *Aura und Autonomie. Zur Subjektivität der Kunst bei Walter Benjamin und Theodor W. Adorno* (Würzburg; Königshausen & Neumann, 1988). These are not relevant to the present investigation, however, and in the end, they have little bearing on Benjamin's employment of this term.

10 This is Benjamin's principal point in a letter he wrote at the beginning of 1924 to Hugo von Hofmannsthal, who had just declared his willingness to publish Benjamin's essay 'Goethes Wahlverwandtschaften' in his journal, the *Neue Deutsche Beiträge*.

Benjamin draws a parallel to a short (unpublished) essay, 'Fate and Character,' which he had written five years earlier, where he addresses; "[die] in der Philosophie erfahrenen segensreichen Wirksamkeit einer Ordnung, kraft welcher ihre Einsichten jeweils ganz bestimmten Worten zustreben, deren im Begriff verkrustete Oberfläche unter ihrer magnetischen Berührung sich löst und die Formen des in ihr verschlossenen sprachlichen Lebens verrät. [. . .] *So versuchte ich vor Jahren, die alten Worte Schicksal und Charakter aus der terminologischen Fron zu befreien und ihres ursprünglichen Lebens im deutschen Sprachgeiste aktual habhaft zu werden.*" Walter Benjamin, *Briefe*, ed. T. W. Adorno and G. Scholem (Frankfurt a.M.: Suhrkamp, 1966), p. 329.

[11] This point is made by Winfried Menninghaus, in his inquiry into 'Walter Benjamin's Theory of Myth'; cf. Gary Smith, ed. *On Walter Benjamin, Critical Essays and Recollections* (Cambridge, MA: MIT Press, 1988), p. 293, and explored in depth in Smith, *Walter Benjamin's Idea of Beauty.*

[12] Cf. Kant, *Critique of Pure Reason*, trans. Norman Kemp Smith (New York: St. Martin's Press, 1965), p. 309.

[13] I endeavor to set forth the details of Benjamin's notion of the profane in *Walter Benjamin's Idea of Beauty.*

[14] Translation emended from: [Johann Wolfgang von Goethe,] *Wilhelm Meister's Apprenticeship and Travels*. Trans. by Thomas Carlyle. Vol. I of III. London: Chapman and Hall, 1874 [1824], pp. 199–201. (Book IV, Chapter VI)

[15] Goethe, *Wilhelm Meister's Apprenticeship*, p.207 (Book IV, Chapter IX).

[16] "Die Kunstkritik hat nicht die Hülle zu heben, vielmehr durch deren genaueste Erkenntnis als Hülle erst zur wahren Anschauung des Schönen sich zu erheben." [I, 195]

[17] By pursuing the following line of thought, I do not wish to ignore the interpretive dilemmas caused by another concept Benjamin puts forth in this context, that of *Eingedenken* (memoration). Above all, the ambiguity of this concept's voluntary and involuntary determinants should be explored. Nonetheless, it is this concept which allows Benjamin to extend Proust's Bergson critique, consistent with his own conceptual scaffolding and analogous to Cassirer's argument against Bergson. For an excellent analysis of how this category is bound up with Benjamin's notion of temporality, see Stéphane Mosès, 'Zu Benjamins Begriff des Eingedenkens', in: *Bücklicht Männlein und Engel der Geschichte. Walter Benjamin. Theoretiker der Moderne*, ed. Werkbund-Archiv Berlin. Gießen: Anabas, 1990, pp. 100–101.

[18] The lines Benjamin cites from the *Phaedrus* reveal that his source was not Plato directly, but Julius Walter's monumental study of *Die Geschichte der Ästhetik im Altertum* (Leipzig: Reisland, 1893. Reprint: Hildesheim Georg Olms, 1967), p. 286f. He erroneously follows Walter in presenting two separate passages from Plato's dialogue [251 A, 254 B] as a single one. Benjamin's reliance on this particular source is significant not only because it alludes to his confidence in Walter's reading of a 'history of aesthetics' as a certain conceptual development of the 'idea of beauty' from Hesiod to Longinus but for its learned explorations of beauty's *Scheinhaftigkeit* and related notions.

[19] Plato, *I. Euthyphro Apology Crito Phaedo Phaedrus*, trans. H. N. Fowler. Loeb Classical Library (Cambridge, MA: Harvard, 1982), pp. 487, 497. "Wer nun erst frisch von der Weihe kommt und einer von denen ist, die dort im Jenseits viel erschauten, der, wenn er ein göttliches Antlitz, welches die Schönheit wohl nachbildet oder eine

Körpergestalt erblickt, wird zunächst, der damals erlebten Bedrängnis gedenkend, von Bestürzung befallen, dann aber recht zu ihr hintretend, erkennt er ihr Wessen und verehrt sie wie einen Gott, [. . .] denn die Erinnerung zur Idee der Schönheit erhoben schaut diese wiederum neben der Besonnenheit auf heiligem Boden stehend." [I, 178]

[20] Ottilie's passivity is not morally commendable. Citing Werner's sonnett, Benjamin observes "die Keuschheit dieses Kindes hutet kein Bewußtsein" [I, 173].

[21] "nicht ein echtes Kind von des Dichters Geist, sondern sündhafter Weise erzeugt."

[22] Several such utterances are concentrated in Goethe's portrayal of Ottilie's arrival, in which Goethe invokes several terms of the grammar of beauty Benjamin will later explore (Part one, Chapter six). Other such direct statements occur at crucial junctures of the novel and are duly noted by Benjamin. In addition, note the crucial passage in the next chapter, in which Ottilie relinquishes her miniature portrait of her father to Eduard, removes it from its place next to her breast, presses it to her forehead, and places it into his hands. In this moment of yielding and of uniting. Eduard presses her hands to his eyes: "Perhaps the loveliest pair of hands that had ever been clasped together" (English, p. 74; German, p. 51); English passages are quoted from Johann Wolfgang von Goethe, *Elective Affinities*, trans. with an introduction by R. J. Hollingdale. Harmondsworth, etc.: Penguin Books, 1986; German passages are from: *Goethes Werke*. lm Auftrag der Goethe-Gesellschaft hrsg., Vol. 4 (Epische Dichtungen 1). Wiesbaden: Insel Verlag, 1949–1952.

[23] English, p. 64; German, p. 43.

[24] Indeed, Goethe repeatedly invokes the *eyes* in his rendering of Ottilie. Not only is she an *Augentrost*, as previously mentioned, but Eduard finds her amusing before she even speaks (62; 42); she is so captivating that "the trees and bushes should have [. . .] been given eyes so as to admire and to delight in her (259; 195); and her silences throughout are such that she speaks through actions – gestures – rather than words.

[25] This feature of the Platonic aesthetic is exemplified not least by the passages from the *Phaedrus* cited above.

[26] The imbrications of Benjamin's reflections on this pair of categories – the auditory and the visual – with those of the younger Gerhard Scholem in the years prior to the publication of Benjamin's *Wahlverwandtschaften* essay shall prove more complex and crucial than heretofore suspected. I intend to explore this theme briefly in two forthcoming essays: 'The Magic Jews. Benjamin, Scholem, and the Rhetoric of German-Jewish Esotericism Between the World Wars' and 'Gershom Scholem's Poetics of Judaism.'

[27] 'Das Licht der Gottesgelahrtheit' [I, 352].

[28] 'Das Bild im Feld der allegorischen Intuition ist Bruchstück. Ru[i]ne. Seine symbolische Schönheit verflüchtigt sich, da das Licht der Gottesgelahrtheit drauf trifft. Der falsche Schein der Totalität geht aus. Denn das Eidos verlischt, das Gleichnis geht ein, der Kosmos darinnen vertrocknet." [I, 352]

[29] [I 670]. Cf. Benjamin, 'Central Park,' trans. Lloyd Spencer, *New German Critique*, no. 34 (Winter 1985), p. 41.

[30] Translated by John Osborne as: *The Origin of German Tragic Drama* (London: New Left Books, 1977), p. 110.

[31] "Stifter kann nur auf der Grundlage des Visuellen schaffen. Das bedeutet nicht, daß er nur Sichtbares wiedergibt [. . .] Das Problem seines Stils ist nun, wie er an allem die metaphysisch visuelle Sphäre erfaßt. Zunächst hängt mit dieser Grundeigentümlichkeit

zusammen, daß ihm jeglicher Sinn für Offenbarung fehlt, die *vernommen* werden muß, d.h. in der metaphysisch Grundzug seiner Schriften: die Ruhe. Ruhe ist nämlich die Abwesenheit zunächst und vor allem jeglicher akustischen Sensation. Die Sprache [. . .] bei Stifter [. . .] ist ein Zurschaustellen von Gefühlen und Gedanken in einem tauben Raum." Walter Benjamin, *Briefe I*, ed. Theodor W. Adorno and Gershom Scholem. Frankfurt a.M.: Suhrkamp, 1966, p. 197.

[32] Scholem recalls seeing this tome on Benjamin's desk "in a special quarto edition whose graphic form was clearly in keeping with the title" during his last visit to Berne. Gershom Scholem, *Walter Benjamin, The Story of a Friendship*, trans. Harry Zohn. Philadelphia: The Jewish Publication Society of America, 1981, p. 86. Especially significant in this context is the *Préface pour un Coup de Dés: cf. S. M., OEuvres complètes*, texte établi et annoté par Henri Mondoret G. Jerau-Aubry (Pléiade), Paris 1974. See also Benjamin's mention of Mallarmé in: *Charles Baudelaire. A Lyric Poet in the Era of High Capitalism*, trans. Harry Zohn. London: NLB, 1970, p.106.

[33] Hugo von Hofmannsthal, *Gesammelte Werke in Einzelausgaben*, ed. Herbert Steiner, Frankfurt a.M.: S. Fischer, 1959, p. 215.

[34] I hope to elaborate these ideas in a study of the rhetoric of silence in Goethe's *Elective Affinities*.

[35] "geschichtlichen Dasein"

[36] "Die Bedeutung des schönen Scheins für die überlieferte Ästhetik ist in dem seinem Ende sich zuneigenden Zeitalter des Wahrnehmung tief begründet. Die dementsprechende Lehre hat ihre letzte Fassung im deutschen Idealismus erfahren. Aber sie trägt schon epigonale Züge. Seine berühmte Formel, daß Schönheit Schein sei – sinnliche Erscheinung einer Idee oder sinnliche Erscheinung des Wahren – hat die antike nicht nur vergröbert sondern ihren Erfahrungsgrund preisgegeben. Dieser liegt in der Aura." [VII, 667]

[37] "Weder die Hülle noch der verhülte Gegenstand ist das Schöne, sondern dies ist der Gegenstand in seiner Hülle" [I, 195]. Hence the beautiful can be grasped less through the idea of disclosure than that of the impossibility of disclosure (*Unenthüllbarkeit*)." [I, 195]

ROSHDI RASHED

ANALYSIS AND SYNTHESIS ACCORDING TO IBN AL-HAYTHAM*

Among problems at the border between philosophy and mathematics, *analysis and synthesis* have occupied a central place for two millennia. Indeed, few are the problems of philosophy of mathematics which have existed for so long and which have stimulated so many writings. Present in filigree in Aristotle's writings,[1] they are there in person in the writings of his commentators[2] and in the writings of philosophers and logicians up to the beginning of the last century. One easily imagines the diversity of meanings and the multiplicity of formulations of this problem of analysis and synthesis, which designated a domain vast enough to include at the same time an *ars demonstrandi* and an *ars inveniendi*.

In history of mathematics, the situation is slightly different: analysis and synthesis present themselves more in the mathematicians' applications of them than in their theoretical formulations. Thus, in Archimedes[3] the 'analysis-synthesis' pair indicates, at first sight, two ordered stages of geometrical demonstration: but nothing assures us that he was the first to introduce such usage, and nothing reveals to us the intended meaning of the mathematician. It is a fact that this usage then became systematic, at least in the writings of Archimedes' successors, as in some of Apollonius' books,[4] and that it was never found to be lacking in geometry until the end of the eighteenth century and the beginning of the next century. But, on the other hand, there are few theoretical texts by mathematicians dedicated to this question – two short texts of unequal importance: a page by Pappus, and a few lines by pseudo-Euclid, which both raise a certain number of historical and philological problems. These texts, especially that of Pappus, did not fail to provoke a certain number of questions, debates, even controversies. Do analysis and synthesis designate, in ancient geometry, a method of demonstration of propositions and of solution of given geometrical problems? According to some commentators, this is and can only be the case, and so the problem underlying analysis and synthesis is purely logical: that of the conditions of reversibility of analysis in order to attain the synthesis and, thus, the demonstration. But the rules of this method are neither to be

C.C. Gould and R.S. Cohen (eds.), Artifacts, Representations and Social Practice, 121–140.

found in Pappus' text, nor, for that matter, in any known text of a mathematician from Antiquity. Adding further to this difficulty is the fact that, at that time, no one ascribed a prescriptive value to analysis and synthesis. So, if these terms point to a method, we must first unearth the meaning behind this word and the function provided by this new meaning. It had also been asked if analysis and synthesis would not designate a discipline; this thesis was defended by some historians who wanted to exclude the idea of a method:[5] according to them, it is the case of a 'corpus of mathematical techniques'. But, no one is able to tell what is included in this corpus and, if we discuss the hypothesis of a set of recipes intended to resolve geometrical problems, we see many questions cropping up: what would this discipline be? Would it be a mathematical discipline, other than geometry and superior to it? What would be, finally, its relations to geometry, but also with arithmetic, which also uses the terminology of analysis and synthesis, as witnessed by Diophantus' *Arithmetica*? As it presents itself, Pappus's text is no help in clarifying these questions but rather adds further problems. Indeed, Pappus speaks of a 'domain of analysis', 'αναλυόμενοζ, occupied mainly by three men, Euclid, Apollonius and Aristaeus 'the elder', where one proceeds by analysis and synthesis.[6] We know, however, that the order of exposition of the *Elements* is synthetic, similarly for the *Conics*; the seven books of the *Conics* which have come down to us are indeed written in a clearly synthetic style, even the most analytical of them, such as the fifth. We must therefore turn to other books by these authors such as the *Data* and *On the Cutting off of a Ratio* – reconstruct the eventual concept of this discipline. But, what assurance do we have against arbitrariness?

To this question, as with many others, it seems to us futile to seek an answer by once more triturating Pappus' text. We are suggesting that we start with the writings of the heirs of Hellenistic mathematics which tackled the problem of analysis and synthesis. Throughout the history of classical mathematics, two periods were essential for research on this theme. The first is not well-known, and is as of yet not studied: in the tenth and eleventh centuries, the most distinguished geometers took up this question, some of them even devoting voluminous treatises to it. The second took place six centuries later. I wish to investigate here the most important of the contributions on this theme which were developed in the tenth and eleventh centuries, that of the mathematician and physicist Ibn al-Haytham. He is the author of two treatises which deal directly with this subject: one is entitled *Analysis and Synthesis* and the other, mentioned in the latter's introduction, is *The Known* [*al-ma'lūmāt*; litt.: *the known things*]; to which one must also add other opuscules which have the same theme. However, before commencing our investigation of these treatises, we shall stop and look at some features of Ibn al-Haytham's mathematical practice, which will enable us to understand his contribution.

CONSTRUCTION AND EXISTENCE

One of the most striking features of Ibn al-Haytham's geometrical writings is the reinforcement of the distinction between the construction of a mathematical object and the demonstration of its existence. Hellenistic mathematicians having previously caught a glimpse of it, this distinction was to be stressed more and more by their Islamic successors in the tenth century: al-Qūhī, Ibn Sahl, Al-Sijzī, among others – because of the changes in mathematical activity itself: the growing number of solid problems, the beginning of algebraic translation of some of them, the reactivation of research on the area of curved surfaces and conical curves, which are essential to the study of the optical properties of mirrors and lenses, etc. With Ibn al-Haytham, we observe a growing concern about following the construction with a proof of existence, especially in the case of problems which cannot be constructed with the ruler and the compass.

In order to gain a better understanding of the scope of this distinction, we have to remind ourselves that for Greek and Arab mathematicians, existence was, as it were, defined by construction as long as the given object could be constructed with the ruler and compass. We cannot better Zeuthen's word, when he writes: "the essential value of geometric construction resides in that it must serve to demonstrate that that to which the determination of the construction leads, really exists".[7] It may therefore seem, for this class of objects – which can be constructed by ruler and compass – that abstract existence and effective existence, resulting from these instruments, overlap inasmuch as the construction provides in the end an existence theorem. Yet, even in this case, it remains to be proved that the construction determines exactly what it was meant to determine and, in order to conceive the latter demonstration, the mathematician is forced to invoke the fundamental properties concerning the relations of position between the straight lines and the circles, thus coming under a geometry of situation.

When, on the other hand, we are dealing with objects which cannot be constructed by ruler and compass but only with the help of conical sections or again of transcendent curves, one must transform, explicitly this time, the construction in a logical proof of existence. Here again, the mathematician is forced to invoke the fundamental properties of curves, which this time come under a proto-topology. In order to speak of the existence of inexpressible objects, he needs at any rate a more general language than that of ruler and compass; this task is evidently not simply linguistic but also mathematical and methodological, since these fundamental properties must be either properly postulated or deduced from other already postulated properties. But, before indicating how the analysis-synthesis pair is linked to this problem, we shall begin by quickly showing this distinction between construction and existence in the mathematical works of Ibn al-Haytham.

An important part of the mathematical research of Ibn al-Haytham is in fact allotted to solid problems: two treatises on the regular heptagon, a short memoir on the lemma to the fourth proposition of the second book of Archimedes' *On the Sphere and the Cylinder*[8] and a solid numerical problem which he asked himself. Like all his predecessors, Ibn al-Haytham was constructing solution-points with the aid of intersections of conical sections. This technique – and that of the intersection of conic and circle – is still frequently used in a treatise where the mathematician was trying to reconstruct the lost eighth book of Apollonius', *Conics*. But, in all his writings, Ibn al-Haytham never fails to demonstrate that the point thus determined really exists, or to specify the necessary and sufficient conditions for its existence. So, for example, to solve the solid numerical problem which he brings back to $X = (a - X)^3$, $0 < X < a$, he deals first with the lemma:

Find four numbers a_1, a_2, a_3, a_4 such that $a_i > 0$ for $1 \leq i \leq 4$; $a_{i+1} > a_i$ for $1 \leq i \leq 3$, $a_1/a_2 = a_2/a_3 = a_3/a_4$ and $a_4 - a_3/a_1 = k$ the given ratio. The solution, if it exists, will depend on a parameter. In order to solve this problem, the author writes $k = b/c$ in the reference $(Nx, Ny) = (NC, NU)$, he takes the points $A(c, b)$, $B(c, 0)$, $D(2c, 0)$, $E(2c, b)$, and considers

$$\mathbf{H} = \{(x, y);\ y(x - c) = bc,\ x > c\}\ (E \in \mathbf{H}),$$

$$\mathbf{P} = \{(x, y);\ y = x^2/c,\ x > 0\}.$$

The author shows that **H** and **P** necessarily intersect in G (x_0, y_0). Indeed, if $P \in \mathbf{P}$, we have $d(P, AB) \to +\infty$ when $x_p \to +\infty$, and if $M \in \mathbf{H}$, we have $d(M, AB) \to 0$ when $x_M \to c$, since AB is asymptotic to **H**.

The existence of G is proved by Ibn al-Haytham in the following terms, where **P** is designated by NO and **H** by EGH: "the section NO cuts the section EGH, since the more we extend the section NO in the direction of O, the more it moves away from the axis NU and it moves away from the straight line AB; and the more we extend the section EGH in the direction of E, the more it moves closer to the straight line BA, as was shown in proposition 14 of Book 2 of the *Conics*. The section NO cuts therefore the section EGH; it cuts it at the point G".

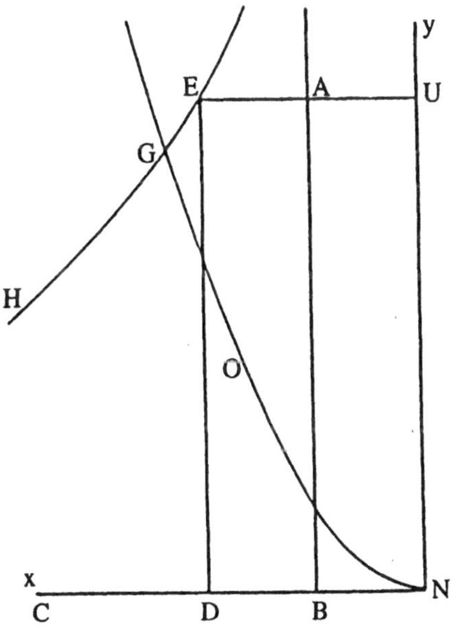

Fig. 1.

We see here that Ibn al-Haytham utilizes the continuity of the curves and their behavior at infinity. Thus, independently of the calculation of x_0 and y_0, after having proved the existence of $G(x_0, y_0)$, the author gives the solution:

$a_1 = c$, $a_2 = x_0$, $a_3 = y_0 = x_0^2/c$, $a_4 = y_0 + b = x_0^2/c + kc$, solution depending on the parameter c. The number x_0 is the root of the third degree equation[9] : $x^3 - cx^3 = kc^3$.

There is nothing exceptional with this example. For half a century, solid problems had multiplied considerably and they were no longer confined to those inherited from Antiquity: the trisection of any angle, the two means proportional, the duplication of the cube, the regular heptagon . . . As for this type of existence proof, it is present in other works by Ibn al-Haytham. Thus, in the treatise which he devoted to the reconstruction of the lost eighth book of Apollonius' *Conics*,[10] Ibn al-Haytham does not apply this demonstration technique, with all the notions which are tied up with it, only to solid problems; he also applies it to plane problems which are capable of construction. This is an important remark: it shows that the concern to demonstrate existence is there even if the problem is capable of construction, and that resorting to the intersection of curves becomes a technique which is relatively independent from its original background, i.e. the solid problems. Let us recall here, for example, problem 20, which reads: "if we have a section of a given cone [parabola or hyperbola] and if we consider two points on its axis, to draw two straight lines from these two points, which meet on the circumference of the section, such that their sum is equal to a given straight line": it is reducible to the study of the intersection of a parabola – or an hyperbola – and an ellipse with the same axis. Here again, Ibn al-Haytham does not forget to demonstrate the existence of intersection points with the help of notions such as interior, exterior.

The situation is similar for another class of problems in this book – from 1 to 17 – where it is a question of finding a tangent to a conic which has a property characterized by a given length or a given ratio of two lengths. Each of these problems is reduced to the determination of the point of contact of the tangent sought after or of a point from which this point of contact will be deduced. Ibn al-Haytham does not undertake, however, a geometrical construction with ruler and compass but shows, with the aid of the intersection of conic and straight line or, accordingly, of conic and circle or two conics, and appealing to the notions of interior, exterior, concavity, asymptotic behavior, that this point exists, with or without supplementary conditions.

We could find in the works of Ibn al-Haytham many other examples which all confirm that the study of the intersection of curves really constitutes, in the geometry of the time and especially in Ibn al-Haytham's works, a privileged and advanced domain of research. Even if one can see emerging from behind the rapid expansion of these studies the interests and prompting of algebraists, it remains true that geometers, Ibn al-Haytham first among them, had definitively recognized in these studies the elements of a geometry of coordinates and are naturally, we might say, appealing to the notions of interior, exterior, concavity and asymptotic behavior in order to deal with the existence of points.

In the preceding examples, as in many other problems studied by Ibn al-Haytham, the properties which are essential to the argumentation are not explicitly mentioned. These are the properties of curves which are, in final analysis, necessary to the establishment of intersection points and which, to some extent, all amount to the notion of convexity. It would indeed have been sufficient, in order to firmly establish existence proofs, to postulate explicitly such properties as: *if a straight line has points on either side of a circle or, more generally, a closed curve, it cuts the latter* or: *if a continuous curve has points on either side of a convex curve, it cuts it*. But these properties, like so many others which should either be axioms or should derive directly from them, are based on notions such as continuity. Their implementation in the solution of problems, as witnessed by the works of Ibn al-Haytham, his contemporaries and his successors, requires that we appeal to coordinates.

Concerned with demonstrations of existence, Ibn al-Haytham could not be indifferent to this shortcoming: he had thus to secure these properties and the notions, such as continuity, on which they are based. Is the continuity of curves guaranteed, in Ibn al-Haytham's works, by a disguised postulate, or is it derived from a more primitive property? In short, what secures the continuity of curves and surfaces? To this question, Ibn al-Haytham's response is, in part, traditional: it is motion which secures continuity.

For him, as for his Greek and Arab predecessors, a curve is obtained kinematically by a continuous motion or two, of which it suffices to conceive their coordination. It is thanks to the uniformity of motion that Ibn al-Haytham succeeded in conceiving this coordination: that is to say that the continuity is secured by the coordination of uniform motions. If, moreover, the curve is convex, it is impossible to go from its interior to its exterior without cutting it. Similarly, an already conceived curve will sweep a surface with a uniform motion, this surface being able to envelop a solid. Therefore, there are continuous curves or surfaces only where there is uniform motion.

There is nothing particularly new in all of this. For the Greek mathematicians, cone, cylinder, sphere, cissoid, conchoid, helix, spiral, quadratrix,[11] etc. are all obtained by motion. Even when the curve was constructed by points – for example, the parabola in Diocles – it earned its name only after verifying the *symptoma* of the curve obtained by motion (here, by motion and plane section). In the *Elements*, the only exceptions are the straight line, the circle and the polyhedra. It is a fact that in Euclid's treatise, motion is excluded from the books on plane geometry and is only introduced – moreover, quite discretely – in Book XI; except if we consider the use of the compass as a motion – as was envisaged by Thābit ibn Qurra (who died in 901).

So, Ibn al-Haytham opts for a kinematic conception of curves and writes, following the example of predecessors such as Ibn Sahl,[12] a treatise on the mechanical construction of curves. Nevertheless, this conception had two difficulties to face: motion was absent from the geometrical lexicon, which was defined essentially by the *Elements*; for the other part, the distinction between construction and existence raised another obstacle: how does one resort to a continuity which is guaranteed mechanically when one demonstrates existence, without eventually having to consider the possibility of construction? In order to side-step these difficulties, the most effective method is to introduce motion into the fundamental notions of geometry; this is what Ibn al-Haytham assigned himself to.

In his commentary on the *Elements*, written before the treatise on analysis and synthesis and before that on *the Known*, Ibn al-Haytham took up again the principal Euclidean definitions of the straight line, the circle, the angle, etc. to introduce motion. The straight line is thus defined as "the line such that if we fix arbitrarily two of its points and if we make it turn, its position does not change". Motion intervenes also in the discussion of figures as in the case of the fifth postulate. Ibn al-Haytham was to be criticized a century later by his successor al-Khayyām for using this notion in such a way. In his treatise on *The Known*, which is directly linked to analysis and synthesis, Ibn al-Haytham brings motion into geometry, as we shall see, as much to guarantee continuity as to justify geometrical transformations. For the time being, notice that the requirement to which he is submitted in order to demonstrate the existence of solutions to solid problems, led Ibn al-Haytham to examine the properties pertaining to a geometry of situation or topological properties. The language necessary to evoke the existence of such objects, which is slightly more general than that of the ruler and compass, had to be able to express, in some way, these properties.

EXISTENCE OF TRANSCENDENT MAGNITUDES

It is not only the case of algebraic magnitudes but also in that of transcendent magnitudes that Ibn al-Haytham considered the problem of existence. An important part of his research is indeed dedicated to lunes and the squaring of the circle.[13] We should see, by examining very briefly the latter case, how he conceived the distinction between existence and construction.

Ibn al-Haytham begins by recalling two results about lunes, known to Hippocrates of Chios. Let AC be the diameter of a given circle, and let there be two semicircles, respectively, AB and BC. We have

$$L_1 + L_2 = \Delta(A\ B\ C). \tag{1}$$

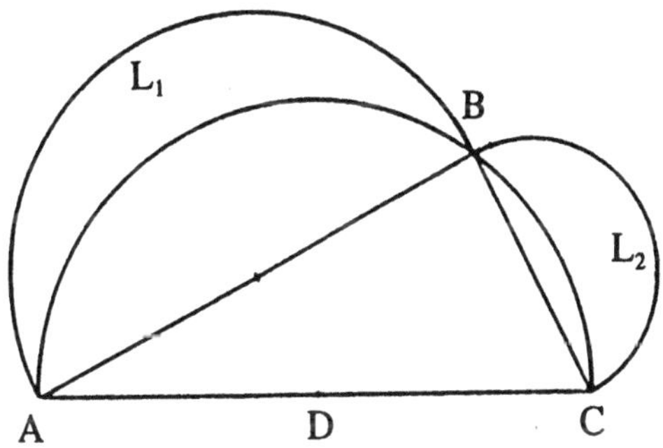

Fig. 2.

Consider now the particular case where B is the middle of the arc AC, and D the middle of the segment AC. We have

$$L = \Delta(A\ B\ D). \tag{2}$$

Let now DK be the perpendicular bisector of AB, which cuts the arcs AB of the two circles in their respective middle H and E; and let C(HE) be the circle tangent to the two circles. We have

$$C(HE) < L$$

and

$$C(HE)/L = k\ [= \pi(\sqrt{2} - 1)^2/2].$$

So, let DU be such that DU/DA = k, then $\Delta(BDU)/\Delta(BDA) = k$, therefore

$$\Delta(BDU) = C(HE).$$

But we know how to build a square SPQO equivalent to $\Delta(BDU)$, therefore

$$C(HE) = \text{square SPQO}.$$

We then construct a square of side QX, such that QP/QX = EH/AC; we then have

$$C(AC) = \text{square TQXY}.$$

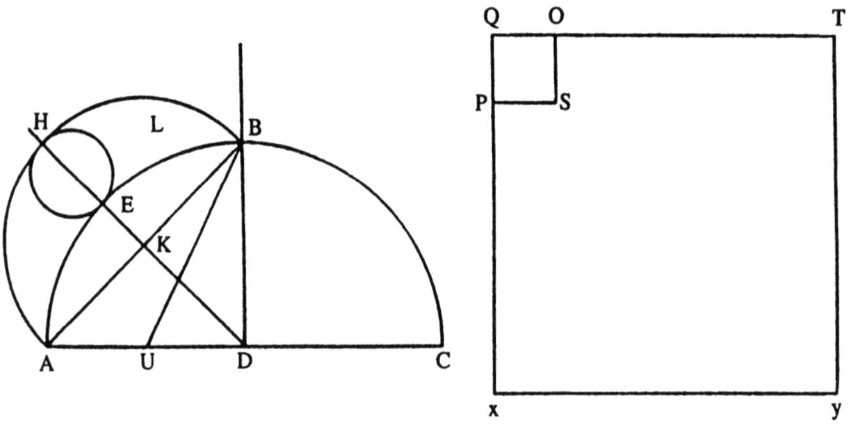

Fig. 3.

It is clear that the whole reasoning relies on the existence of k, the ratio of two plane surfaces. The existence of the segments DU, QP and QX is deduced from that of k, even though their construction is possible only when k is known.

Then, Ibn al-Haytham devotes a long passage to the existence of this ratio independently of the possibility of its construction. Here is what he wrote:

> This circle is therefore itself a part of this lune. But every magnitude has a ratio to every other magnitude of which it is a part, even if we don't known this ratio and even if we can't construct it, because the ratio between two magnitudes is neither due to men's knowledge of them nor to their power to determine or know them. The ratio between magnitudes is a notion peculiar to magnitudes of the same kind. So, if two arbitrary magnitudes of the same kind are both limited, finite, invariable and they both keep their magnitude and do not change at all, neither increasing nor decreasing, nor do they change in kind, then one has, relatively to the other, one and only one ratio, which does not vary and which does not modify its form at all.

Ibn al-Haytham's way to show the existence of this ratio is to bring the problem back to 'known' elements – that is, in his own words, to what is "fixed in its state and does not change, because for mathematicians the known is that which does not change" – such as AC, AB, the arc AB, EH, the lune AEBH, the circle HE, etc. Demonstrating the existence of this ratio – which is for us transcendent – led Ibn al-Haytham to the study of the invariant elements of the figure, magnitudes, forms, positions. We understand, under these conditions, the reasons leading Ibn al-Haytham to opt for a mathematical realism which is here lacking nuances. As for the construction of this ratio, Ibn al-Haytham promises us to treat it in another memoir, which never reached us, if it was even ever written. But the 'known' are precisely that to which analysis must get to, before synthesis gets under way.

A NEW GEOMETRICAL DISCIPLINE: *THE KNOWN*

We just saw that in the tenth century, research on a vast range of problems, notably problems for which construction is not possible, led mathematicians to extricate that which had been showing through the surface in late Hellenistic mathematics, and to ask the question of existence in a more abstract manner, independently, therefore, of construction. All through the century, this new demand gained in strength, playing finally the role of a norm in Ibn al-Haytham's works. This is, according to us, the main reason for this unprecedented interest in the problem of analysis and synthesis.

In order, however, to conform as much as possible to this norm, it seemed necessary to increase the knowledge of general properties of figures, that is of those properties which subsist when the figure is transformed and even when we express it differently, as in the case of a change of the system of coordinates: these situations became more and more frequent in mathematical practice. For example, it was common in the study of problems in which construction is not possible, to bring the different parts of a figure, which are described in different systems of coordinates, under the same system of coordinates; or even, in the study of curved surfaces and volumes, to proceed by affine transformations; or finally, in the study of the exact representation of the sphere on a plane, to define and examine the cylindrical and conical projections.[14] These general properties are, however, of diverse origin: they may result from the position of the elements of the figure, as they may come from the form.

It is precisely to the study of these properties that Ibn al-Haytham devoted a voluminous treatise entitled *The Known*. This book, written after the treatise on *Analysis and Synthesis*,[15] is nevertheless already mentioned in the introduction of the latter as a propaedeutic for analysis and synthesis. Therefore, we must stop and have a look at this book on *The Known*.

According to Ibn al-Haytham, a notion is said to be 'known' when it admits no change, even if it is thought of, or not, by someone. The 'known' designate these invariant properties, which are independent from our knowledge of them. Yet, we can read in his treatise on *Analysis and Synthesis* that the goal of the analyst is precisely to reach these known notions, and that it is only when he has reached them that his task is over and synthesis can get under way.

The book devoted by Ibn al-Haytham to these notions opens on a long and profound introduction, where, in an Aristotelian-Euclidean language, more traditional notions blend with less traditional ones. It is in this way that Ibn al-Haytham begins by recalling the Aristotelian subdivision of quantity. He expounds first the subdivisions of discrete quantities, although he does not study any problems that come under them in other parts of the book, since only geometrical problems are tackled. Continuous quantity contains all magnitudes. Like his predecessors. Ibn al-Haytham considers as a set of magnitudes a set of ordered elements, with addition, such that $A + B > A$ and that $A > B$ entails the existence of an element C such that $A = B + C$. This set is, moreover, Archimedean. For Ibn al-Haytham, as for any Archimedean, this set could be formed by segments of straight lines, areas, numbers, rectilinear angles, arcs of curves, skew areas, times, weights. These are in fact most of the magnitudes taken into account in his book.

But, another, less traditional, perspective is superimposed on this traditional beginning, a set-theoretical bias. When he deals with an element of a figure. Ibn al-Haytham considers it not only as a magnitude but also as a variety of the family of which it is a member. So, the knowledge of this element will be about its magnitude, its position, its form and about its relation to others; in short, the properties of space. The step thus accomplished is quite important and Ibn al-Haytham dedicates the whole introduction to the elucidation of these notions; let us take as an example the henceforth central notion of position.

Position is defined by Ibn al-Haytham with the help of three notions: motion, order and relation. So, the position of a point – considered as the extremity of a line – is known when its distance (or its distances) to another point (or other points) remains invariant. Many cases must be taken into account: the point P is fixed and so are the other points; the point P has a circular motion around a fixed point but without any change in the distance between these two; the point P and other points are all driven by the same motion, leaving the distances between P and each one of these points unchanged.

Similarly, the position of the line is defined in relation to fixed points; in this case, the line never has any motion, with the exception of increasing and decreasing, and the distance between its points and two points or more does not vary. This line will be said to be *of an absolutely known position*. The position of the line could also be located in relation to only one fixed point; in this case, the known notions will be the invariant distances between every point of the line and this fixed point, no matter whether the line itself is fixed or moving. One also locates the position of the line in relation to another line, no matter whether the latter is fixed or moving. Again, one locates the position of the line in relation to a moving point, or to a set of moving points, and in this case the known notions will be the invariant distances between each point of the line and each of the moving points; the line has to be driven by the same motion, in the same direction, as that of the points under consideration. Finally, one locates the position of the line in relation to a fixed line, and in this case the known notion is that of the angle formed by the intersection of these two lines, or of their continuation, whether the line, the position of which one is trying to know, is fixed or moving, with the condition, however, that the angle thus formed remains invariant. If the line or its continuations does not cut the line in relation to which its position will be known, it will be, at any rate, if these two lines are cut by a straight line which will form with each of them a known angle.

Ibn al-Haytham pursues his enumeration further, locating the position of the line in respect to a moving line, to a fixed surface and, finally, to a moving surface. He takes up again an analogous task in order to define the position of a surface and that of a solid, and to examine the other notions: known form, known magnitude, known ratio.

We see initially, by examining this long introduction to his book, that Ibn al-Haytham has integrated motion as a primitive notion of geometry, necessary to the definition of the position and form of every geometrical magnitude, and in order to guarantee continuity. The same examination then shows that this heir of Archimedes and Apollonius explicitly distinguishes the properties of position from the metric ones. Even if a property of position can be presented with the help of distance and angle measurements, that is under a metric form, Ibn al-Haytham insists nevertheless on describing what is distinctive to a position. At this point, it is essential to locate the position – of, say, a point – without the use of any system of coordinates, but only with respect to points or lines, fixed or moving; it is, so to speak, a descriptive geometry in the proper sense of the word. The goal set by Ibn al-Haytham in *The Known* is clear: to identify invariant relations which enable one to describe the position, form, magnitude and ratio. Each group of relations will constitute a chapter of a geometry to come, or of this science christened 'The Known'.

This introduction, where powerful intuitions and penetrating insights abound, may mislead the reader; it is better, therefore, to go back to the two parts of Ibn al-Haytham's book. In the first part, this distinguished mathematician and physicist claimed complete originality; he begins with a proposition which says that the set of points situated at a given distance from a point A, is a circle of center A whose radius is the given distance. His demonstration is effected with the help of the preceding definitions. In the following three propositions, he shows that, if a point with a given property P is associated to every point of this set, this second point pertains to a figure of known position and magnitude which is deduced from the set of points, that is from the circle, by a transformation which is characterized by Ibn al-Haytham. So, he characterizes a similarity (center, ratio, angle) in propositions 2 and 4, and, in proposition 3, a homothetic transformation (center, ratio); in proposition 11, he shows that two equal circles correspond in a translation defined by the two centers, and he uses this property in proposition 12; in proposition 13 and 14, he starts from three non-aligned fixed points A, B and C, and tries to determine for $D \in [AB]$, the set of points $E \in [CD]$ defined in 13 by

$$DC/DE = DA/DB$$

and in 14 by

$$DC.DE = DA.DB.$$

In both cases, E is the image of D by a transformation t and the set **E** of the points E is $t[AB]$. In each case he determines the support of **E**: a straight line in 13 and a circle in 14.

In propositions 15, 16 and 17, he uses the power of a point with respect to a circle. In another group, formed by propositions 6, 8, 9, 10, 23, he studies in each case a set of points M defined from two points A and B, with a property P(M). In each case he shows that the support of the set of points M is a known straight line or circle. P(M) is given respectively by:

$$AMB = \alpha,\ \alpha \text{ a given angle;}$$
$$MA = MB;$$
$$MA/MB = k,\ k \neq 1, \text{ given ratio;}$$
$$\text{area }(AMB) = S,\ S \text{ given area;}$$
$$MA^2 + MB^2 = l^2, \text{ given sum.}$$

In the second part of his book, Ibn al-Haytham announces that he will discuss propositions such as those which are found in Euclid's *Data*, but which were not dealt with by the latter. These are for the most part construction problems. So, in a group of eleven problems – from 1 to 5 and from 14 to 20 – he considers a straight line passing through a given point and verifying a property P relative to the given elements. He shows that, if such a straight line exists, it will pass through a second point, defined by the intersection of two lines: a straight line and a circle or, depending on the case, two circles. One of these two lines is given with the problem, while the other is obtained from the property P, by applying one of the propositions from the first part of the book. With this type of problem, Ibn al-Haytham tries to show that, if such a straight line exists, its position is then known. In another group of five propositions

– 6, 7, 8, 21 and 22 – the question raised is that of studying two straight lines, each passing through a given point and verifying a property P. These two straight lines are determined by their point of intersection, as defined above: the point of intersection of two lines. In general, most of the propositions of this second part are reduced to the determination of a point by the intersection of two lines, or to the determination of two points, the first of which is obtained by the intersection of two lines and the second being deduced from it.

Without lingering any more over these, let us notice that the two parts of the book are in fact complementary: in the first part, Ibn al-Haytham deals with sets of points and punctual geometrical transformations; he looks, in the second part, for the simplest geometrical means of determining positions of points and their relations, from the known elements.

Even this brief account brings to light the assignment of Ibn al-Haytham's book: he wishes to obtain geometrical results which can effectively shorten the analysis and which can, moreover, contribute to the development of a geometrical discipline which deals with these 'known'. But, we have shown that the objects of this discipline are the invariant properties of figures obtained by motion and plane sections; one is interested not only in metric properties, but above all in properties of position and form which remain stable after punctual transformations of the figures. Even if, in this book, Ibn al-Haytham borrows his examples from plane geometry, in his introduction he considers these notions in themselves, that is, as notions which are common to different domains of geometry. The geometrical discipline which deals with the 'known' appears, therefore, to be assigned to the linkage of particular chapters of geometry; this renders it more general and more substantial than any other chapter.

According to Ibn al-Haytham's own admission, the art of analysis and synthesis finds its laws in this geometrical discipline. The analyst has done his job only once he has reached the invariant properties of the geometrical figure under consideration – the known notions – in order then to show – under which conditions, already fulfilled or to be added – how these properties are indeed the properties of this and only this figure. It follows therefore that, if the art of analysis and synthesis designates a method, this method takes its meaning and extension only from this discipline. It must indeed refer to it; its generality is echoing that of the objects of the geometrical discipline, and it does not express a universality which would characterize it as a method.

One still has to examine the stages which mark out the application of this method to different sorts of problems. It is to this task that Ibn al-Haytham assigned the other treatise – the first one. Let us notice first that, if the modern reader puts together all of Ibn al-Haytham's writings in which analysis and synthesis intervene, he will see in this geometrical discipline notions taking shape which will, much later, become part of different disciplines: descriptive geometry, geometry of coordinates and proto-topology. Thus Ibn al-Haytham's words, in his claim to originality and in detecting traces of this discipline in Euclid's *Data*, are clarified.

* * *

According to this first meaning which is, it seems to us, a fundamental one, analysis and synthesis are neither two steps of a method, nor even a method common to all of mathematics. Analysis and synthesis are determined by this geometric discipline conceived by Ibn al-Haytham in order to fulfill the new requirements of mathematical practice. If these concepts describe a procedure of discovery and demonstration, it is inasmuch as we applied them to a problem of this discipline or which contains one of its notions.

However, on this first determination of the meaning of analysis and synthesis another one is superimposed, of a different origin. This second meaning appears when Ibn al-Haytham takes up, while generalizing it, the traditional problem of the procedure needed in order to find geometrical concepts. Thābit ibn Qurra had already written a memoir on this subject,[16] taken up later by other mathematicians, such as his grandson Ibrāhīm Ibn Sinān[17] and al-Sijzī. True to his style, Ibn al-Haytham then asks the question of the procedure, or method, which is the most common to the different mathematical disciplines. This time, the domain of application of analysis and synthesis is identified to that of quantity: discrete as much as continuous. Then, the problem is that of knowing what this extension of analysis and synthesis is based on. Here again, the 'known' or this geometrical discipline founded by Ibn al-Haytham reappears, since only analogies with this discipline would allow us to provide a foundation for this extension. Ibn al-Haytham fashions, indeed for every discipline which deals with quantity, definitions similar to those conceived for geometry. So, in arithmetic, he would speak of "a number of known quantity" for magnitude, of "known properties" – a perfect number, for example – for position and form. The care taken in introducing these definitions contrasts, however, with their fate: once the analogy is secured, these notions do not serve anymore, since the book on the known contains only geometric problems. Although this analogical extension somewhat conceals the geometrical origin of analysis and synthesis, it certainly inflects its methodological and formal meaning. Ibn al-Haytham then faces the problems which are always at the forefront when dealing with analysis and synthesis, that is, those which they raise, as an art of discovery and as an art of demonstration: the different kinds of demonstration, the role of intuition as learned intuition, the reversibility of analysis, the role of supplementary properties necessary to render analysis reversible, the distinction between propositions and problems, the meaning of analysis and synthesis in each case, and the distinction between problems according to the number of solutions, hypotheses, etc. Ibn al-Haytham contributes substantially to the study of these problems, but this will be the topic of another paper. Let us notice, in order to conclude, the double determination of analysis and the synthesis in relation to mathematical practice itself.

CNRS – Paris

NOTES

* This text was first published under the title 'L'analyse et la synthèse selon Ibn al-Haytham' in R. Rashed (ed.), *Mathématiques et philosophie de l'Antiquité à l'Âge Classique*. Paris, Éditions du C. N. R. S., 1991, pp. 131–149. Translated by Mathieu Marion.

[1] Many of Aristotle's texts testify to the presence of 'analysis and of synthesis'. The most cited of texts is that of the *Nicomachean Ethics*, 1112b. Following are those of the *Prior Analytics*, 46b and 51b, and the *Physics*, II, 9, 200a. To these, one must also add the *Prior Analytics*, 43b.

For the commentators of these passages – cf. B. Einarson [1936, 36–37] the problem remains that of the exact relation between analysis in the sense of Aristotle's syllogistic, and mathematical 'analysis' as we find it in Pappus' famous text. N. Gulley [1958] and G. Granger [1976], for example, claim that it is the same schema. Thus, Granger writes [1976, 313] on Aristotle's conception of mathematical demonstration: "the fundamental schema surely appears to be that of *analysis*, exactly described as Pappus would do some six centuries later". N. Gulley had written: "We have then, in Aristotle, the recognition of a method of analysis in geometry, corresponding to Pappus' description (as an upward movement to prior assumptions from which our prior assumptions follows), and illustrated by examples where the relation between ἀρχή and συμπέρασμα is recognized to be irreversible" [1958, 8–9]; this bringing together of Aristotle and Pappus derives from a logical interpretation of Pappus' text. Another problem, not of lesser importance, raised by historians, is that of the existence of such a conception of mathematical 'analysis' in Plato, even if the word is lacking, and of its relation to philosophical analysis. Ch. Mugler, for example, has defended the thesis according to which both analysis and synthesis had already began to take root in the *Republic* [1958, chapter 5, 290 sqq.], a thesis later rejected by H. Cherniss [1951, 415–416].

2 On the commentators' standpoint, see N. Gulley [1958].

3 Archimedes, *On the Sphere and the Cylinder*, II prop. 4, 5, 6, for example.

4 Although he cites the term in the introduction to Book IV of the *Conics* [1974, 4, I, 16], it is mainly in books such as *On the Cutting off of a Ratio* that Apollonius resorted to analysis and synthesis.

5 Cf. M. Mahoney [1968].

6 Hultsch [1876, vol. II: 634–36]. The translation of this passage raises some serious problems. We do not discuss them here. Cf. Th. Heath [1921, vol. II: 399–401], and recently Hintikka & Remes [1974, 8–10].

7 H. G. Zeuthen [1902, 74]. Cf. also O. Becker's commentary [1954, 90 sqq.].

8 Ibn al-Haytham, *Oeuvres mathématiques*, in preparation.

9 Notice that al-Khayyām, like al-Ṭūsī, studies this very equation later, with the help of an hyperbola and a parabola, but that the choice of curves is different. Cf. Sharaf al-Dīn al-Ṭūsī [1986, vol. 1, CLXVI sqq. and 66 sqq.].

10 Ibn al-Haytham, *Oeuvres mathématiques, op. cit.*

11 We discuss these curves somewhere else [Ibn al-Haytham, *Oeuvres mathématiques, op. cit.*]. For the moment, notice only that Dinostratus' quadratrix is generated with the help of two motions: a vertical motion of constant speed, and a relative motion on a mobile horizontal. So, according to Pappus' description, one draws a quadrant of a circle BOA; the radius OM moves uniformly from OB to OA; and simultaneously the horizontal P, which cuts the radius in N, moves down uniformly in such a way that the extremity P moves from B to O.

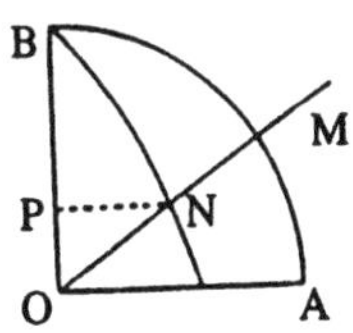

12 *Géométrie et Dioptrique au X e siècle: Ibn Sahl, al Qūhī, Ibn al-Haytham* [Les Belles Lettres, 1993]. Ibn Sahl constructs the three conics by a mechanical procedure.

13 R. Rashed: *Les Mathématiques infinitésimales du IXe au XIe siècle*: Ibn al-Haytham, vol. II, ch. 1. In print, éditions Peeters, Louvain – Paris.

14 *Géométrie et Dioptrique au X e siècle: Ibn Sahl, al Qūhī, Ibn al-Haytham, op. cit.*

15 Cf. R. Rashed: MIDEO, 20, 1991: 31–231 and R. Rashed: MIDEO, 21, 1993:87–275.

16 Ms. 4832 Aya Sofia – Süleymaniye, Istanbul, ff. 1^v–4^r.

[17] Ibrāhīm Ibn Sinān wrote a treatise entitled *On the Method of Analysis and Synthesis and other Procedures in Geometrical Problems.*

REFERENCES

Becker, O.: *Grundlagen der Mathematik in geschichtlicher Entwicklung*, Freiburg-Munich [1964].

Cherniss, H.: 'Plato as a Mathematician', in *Review of Metaphysics*, **4** [1951, 395–425].

Einarson, B.: 'On Certain Mathematical Terms in Aristotle's Logic', *American Journal of Philology*, **57** [1936, 33–54 and 151–172].

Granger, G. G.: *La théorie aristotélienne de la science*, Aubier, Paris [1976].

Gulley, N.: 'Greek Geometrical Analysis', in *Phronesis*, **3** [1958, 1–14].

Heath, Th.: *A History of Greek Mathematics*, Oxford [1921].

Heiberg, I. L.: *Apollonius Pergaeus edidit I. L. Heiberg*, Teubner, Stuttgart [1974].

Hintikka, J. & U. Remes: *The Method of Analysis*, Reidel, Dordrecht [1974].

Hultsch, F.: *Pappi Alexandrini Collectionis ... quae supersunt e libris manu scriptis edidit latina interpretatione et commentariis instruxit F. Hultsch*, Berlin [1876–1878].

Mahoney, M.: 'Another look at Geometrical Analysis', in *Archives for History of Exact Sciences*, **5** [1968, 318–348].

Mugler, C.: *Platon et la recherche mathématique de son époque*, Heitz, Strasburg-Zürich [1948].

Rashed, R.: "La philosophie des mathématiques d'Ibn al-Haytham. I: l'analyse et la synthèse", *MIDEO*, 20, 1991 [critical edition and French translation of Ibn al-Haytham's treatise].

Rashed, R.: "La philosophie des mathématiques d'Ibn al-Haytham. II: les connus", *MIDEO*, 21, 1993 [critical edition and French translation of Ibn al-Haytham's treatise].

Rashed, R.: *Géométrie et Dioptrique au Xe siècle: Ibn Sahl, al-Qūhī, Ibn al-Haytham*, Les Belles Lettres, Paris, to appear.

Sharaf al-Dīn al-Tūsī: *Oeuvres mathématiques, Algèbre et Géométrie au XIIe siècle*, edition, translation and commentaries by R. Rashed, I & II, Les Belles Lettres, Paris [1986].

Zeuthen, H. G.: *Histoire des mathématiques dans l'Antiquité et le Moyen-Age*, translation by J. Macart, Paris [1902].

JOHN STACHEL

CHANGES IN THE CONCEPTS OF SPACE AND TIME BROUGHT ABOUT BY RELATIVITY

In this paper, I shall confine myself to the physical concepts of space and time, that is, the concepts as employed in the formulation of various physical theories. These concepts are intimately involved with two other concepts: the concept of the motion of some object, i.e., the change over time of its position in space; and the concept of the change over time of some property of an object (or field) at the same point in space. So the concepts of time, space, and motion must be discussed together. As we shall see, the concept of the change of some property – quantitative or qualitative – at the same point leads to a concept of time that we may call 'local', since it only applies to events at that point; while the concept of motion leads to a concept of time that we may call 'global', since it requires the comparison of times at different points.

I shall start with the concepts of time and space associated with Galilei–Newtonian physics, and then proceed to discuss the changes in these concepts necessitated by the advent of the special theory of relativity; and finally the even more striking changes brought about by the advent of the general theory of relativity. My approach will not be primarily historical, but what might be called historical-critical. I shall not hesitate to violate the historical order of development in my account, introducing more recent concepts and/or viewpoints into it whenever this facilitates the understanding of some question.

Two points of view towards the nature of time and of space have dominated discussions of this topic in modern times (let us say since the time of Newton, Leibniz and Huygens). These have been called the relational and the absolute concepts. For space, these two concepts were well summarized by Einstein:

> Two concepts of space may be contrasted as follows: (a) space as positional quality of the world of material objects; (b) space as container of all material objects. In case (a), space without a material object is inconceivable. In case (b), a material object can only be conceived as existing in space; space then appears as a reality which in a certain sense is superior to the material world.[1]

A similar contrast can be made in the case of concepts of time:

C.C. Gould and R.S. Cohen (eds.), Artifacts, Representations and Social Practice, 141–162.

(a) time as an ordering quality of the world of material events; (b) time as a container of all material events. In case (a), time without a material event is inconceivable. In case (b), a material event can only be conceived as existing in time; time then appears as a reality that in a certain sense is superior to the material world.

Absolute Time

It seems logically possible to combine either point of view about space with either point of view about time; but in fact most natural philosophers either adopted the absolute viewpoint (like Newton) or the relational viewpoint (like Leibniz and Huygens) about both. We shall return briefly to the question of absolute and relational concepts of time later on, but for the moment shall focus on a presupposition about temporal order common to both views before the advent of the special theory of relativity. Whether inherent in the nature of the events themselves or drawn from their immersion in the temporal continuum, this order was presumed unique and unproblematical. For a continuous sequence of events occurring 'at the same place' – (provisionally taking for granted the meaning of this expression), this order does indeed appear to be unproblematical, as does the associated concept of local time.[2] But for events occurring at 'different places', a convention is needed to allow the introduction of a global time. The possibility of their unique temporal ordering rests on the possibility of introducing a universal or absolute relation of simultaneity between events at different places.[3] Assuming that such an absolute simultaneity relation exists, then it must be an equivalence relation. That is, it is reflexive (any event is simultaneous with itself), symmetric (if event a is simultaneous with event b, then b is simultaneous with a), and transitive (if a is simultaneous with b and b is simultaneous with c, then a is simultaneous with c). An equivalence relation on a set divides the set into equivalence classes: Each member of the set belongs to one and only one class, all members of which are equivalent under the relation. Hence, an absolute simultaneity relation among events divides them into equivalence classes of simultaneous events. We can use this relation of simultaneity to define simultaneous events as occurring at the same global time – an obvious-seeming convention, but a convention nonetheless. We thereby transfer the local temporal ordering of events at one place (which place is chosen does not matter because of the existence of the absolute simultaneity relation) to a global temporal ordering of events at different places. Hence, in Galilei–Newtonian physics, the concepts of local time and of global time, although logically distinct, effectively coincide in the concept of a universal or absolute time.

Relative Space

Now we must return to the concept 'at the same place'. Newton believed that, just as his dynamical theory of mechanics presupposes the existence of a universal or absolute time, it also presumes the existence of a universal or absolute space. If this were so, it would be clear what 'at the same place' means: at the same place with respect to absolute space. But Newton himself was aware of an 'operational' problem with the concept of absolute space: While it is rather easy to give an experimental prescription for deciding whether or not a body is in absolute rotation (Newton's bucket experiment) or (if we neglect gravitation) in absolute linear acceleration, there is no prescription that enables us to decide whether a body is in uniform linear motion at a constant velocity with respect to absolute space. That is, laws of mechanics do not allow us to single out, by means of the result of any mechanical experiment, such a state of absolute rest from the class of states in uniform motion. If one believes, as did adherents of the mechanical world view, that all of physics can be explained by the effects of mechanical interactions, this inability implies the need to abandon the autocratic concept of a single absolute space, and to replace it by a democracy of so-called inertial frames of reference. An inertial frame can be defined as one in a state of uniform motion, that is, one in which Newton's First Law is valid: an object acted upon by no external forces moves in a straight line with constant velocity with respect to an inertial frame of reference.[4] This equivalence of all inertial frames may be called the relativity principle of Newtonian mechanics. There is a three-fold infinity of inertial frames, each of which is in a state of uniform motion with respect to all the others; in each of them Newton's laws of mechanics are valid.

With the abandonment of absolute space, the concept 'at the same place' loses its absolute significance, but it may be replaced by the concept 'at the same place relative to some inertial frame of reference.'[5]

In summary, whether or not adherents of the mechanical world view were aware of it, Newtonian mechanics properly understood does require a universal or absolute concept of the time of an event, but leads to abandonment of the absolute concept of position in favor of a relative concept of position in space.

These concepts of absolute time and relative space may be given operational significance in terms of the measurement of spatial and temporal intervals with measuring rods and clocks, respectively. For example, the temporal interval (time) between two events, as measured by two clocks is independent of the inertial frame of reference in which the clocks are at rest (and therefore, they need not both be at rest in the same frame). On the other hand, the spatial interval (distance) between two non-simultaneous events, as measured by a rigid measuring rod for example, is not absolute, but depends on the inertial frame of reference in which the measuring rod is at rest. Since there is no preferred inertial frame of reference ('absolute space'), none of these relative distances can claim the title of 'the (absolute) distance'. For simultaneous events, of course, the distance between them is independent of the inertial frame in which the measuring rod is at rest: It makes no difference whether the rod is at rest or in motion relative to the two events. The reason that we can speak of 'the length' of an extended object, is that we mean the distance between two simultaneous events, one at each end of the object, and this is an absolute concept.

The triumph of the wave theory of optics in the mid-nineteenth century, and its later union with electrodynamics in Maxwell's theory of electromagnetism, led to the introduction of a hypothetical medium, the luminiferous ether, as the seat of light waves and later of the electric and magnetic fields. Extensive debates during that century about the nature of the interaction between the ether and ordinary matter ultimately led to a preponderant viewpoint: The ether fills all of space even where ordinary matter is present, and it remains immobile even when matter moves through it. Ejected from mechanics, Newton's absolute space seemed to have returned in the form of this immobile ether, which provided a unique frame of reference for electrodynamical and optical phenomena. In particular, the speed of light (or other electromagnetic waves) was supposed to have a fixed value with respect to the ether, but its value should differ when measured with respect to any other frame of reference, such as that defined by the earth, which moved relative to the ether. Yet all attempts to measure this change of speed, or any other predicted effect of motion through the ether on the behavior of light or any other electrodynamical phenomenon, failed. Various efforts to resolve this dilemma within framework of classical kinematics raised their own problems, until Einstein cut the Gordian knot in 1905. He pointed out that:

> The theory to be developed – like every other electrodynamics – is based upon the kinematics of rigid bodies, since the assertions of any such theory concern relations between rigid bodies (coordinate systems), clocks, and electromagnetic processes. Insufficient consideration of this circumstance is at the root of the difficulties with which the electrodynamics of moving bodies currently has to contend.[6]

Relative Time

A re-analysis of the foundations of Galilei–Newtonian kinematics led Einstein to pinpoint the concept of universal or absolute time as the source of the incompatibility of the relativity principle of Newtonian mechanics with Maxwell's electrodynamics. He found that, once this concept was abandoned, the ether could be eliminated, and the relativity principle – the full equivalence of all inertial frames – could be reinstated for *all* phenomena, electrodynamical and optical as well as mechanical. Elimination of the ether also eliminated the need to find an explanation for the failure of all attempts to detect motion through the ether. Of course, taken together with the relativity principle, no ether frame implies that the speed of light must be the same in all inertial frames of reference, a result clearly incompatible with the old kinematics. Einstein developed a new kinematics, in which temporal intervals – local or global – are no more universal or absolute than spatial intervals: Both (global) time and space are relative to an inertial frame of reference. Since there is no universal or absolute time, clocks can no longer be synchronized absolutely, and the concept of simultaneity must be analyzed very carefully. As noted earlier, an element of arbitrariness (convention) enters into any definition of global time and hence of (relative) simultaneity. But, in each inertial frame, the choice of one convention is clearly superior since it results in the greatest simplicity in the expression of the laws of nature. This convention, applied in each inertial frame of reference, results in that frame having its own relative simultaneity. The convention can be defined operationally, for example, with the help of a system of equivalent clocks at rest in that frame. These clocks may be synchronized using any signal that has a speed in that frame which one may justifiably take as independent of the direction of the signal.[7] The speed of light in vacuum meets this criterion, and furthermore has the same value in all inertial frames if this synchronization convention is used in each. So, it provides a convenient signal for use in synchronization. The global time interval between two events now depends on (is relative to) the inertial frame of reference, as did the spatial interval between two (non-simultaneous) events in Galilei–Newtonian kinematics. Since simultaneity is now relative, the spatial interval between *any* two events is also relative.

There is now, in many ways, much more symmetry between the properties of time intervals and of space intervals. For example, just as the proper length of an object is defined as the spatial interval between two simultaneous events, one at each end-point of the object, in an inertial frame of reference in which the object is at rest, the proper time between two events can be defined as the time interval between the events in a frame of reference in which both events occur at the same place. A curious feature of the new concept of time is now manifest. According to the old concept, for any two non-simultaneous events, there is always a frame of reference in which both occur at the same place, because it is always possible to travel between the two events with sufficient speed. According to the new concept, the speed of light is a limiting velocity, which it is impossible to even reach, let alone exceed, by any material process. Hence, there are pairs of events that do not occur at the same place in any inertial frame of reference, but for which there is an inertial frame of reference with respect to which they occur simultaneously. Such pairs are said to be space-like separated. Similarly, there are pairs of events that do not occur at the same time with respect to any inertial frame of reference, but for which there is an inertial frame of reference with respect to which they occur at the same place. Such pairs are said to be time-like separated. Finally, there are pairs of events, neither space-like nor time-like separated, but which can be connected by a light signal. Such pairs are said to be null or light-like separated.

The local time is also relative in the special theory. For a continuous series of events taking place at the same point, but one that is not permanently at rest in any one inertial frame – that is, one which is in accelerated motion – we can define the *proper time* of the series by picking a sequence of events E1 (the first), E2, . . . , E(n–1), En (the last) out of the series, calculating the proper time between the pairs of events E1–E2, . . . , E(n–1)–En in the sequence, adding these proper times, and taking the limit as we make the sequence denser and denser. The result is the local time interval of the series of events, usually called the proper time interval in relativity. This relativistic local or proper time is quite distinct from the global or inertial-frame time. Most notably, it is relative to the path of the object, i.e., path-dependent. We are quite used to the idea that the (proper) distance travelled between any two points in space depends on the (space-like) path taken between them (for example, the distance travelled between Naples and Boston depends on whether you go by way of Paris or Cape Town), and that the straighter the path between them the shorter the distance. A similar effect now holds for time: The local (proper) time that elapses between two events depends on the (time-like) path taken between them; but here the straighter the path, the *longer* the proper time elapsed. In spite of the analogies between the two in special relativity, there is still a big difference between space and time! This effect is the essence of the famed twin paradox.

Fixed Kinematic Structure

One could expatiate on many other curious features of the special-relativistic times (local and global) as compared to the absolute time. Dramatic as are the differences between these concepts of time, however, the global concepts share an important common feature as compared with the general-relativistic concept. Both the Galilei–Newtonian and special-relativistic concepts of time (and space) are based upon the possibility of establishing a kinematic framework, the structure of which is independent of any dynamical processes. Whether or not one regards this kinematical framework as existing prior to these dynamical processes depends on whether one adopts an absolute or a relational account of space and time. (In special relativity, with its intimate commingling of space and time, it is hard to see how one could combine a relational account of one with an absolute account of the other.) But in either case, the kinematical structure is fixed once and for all by a group of symmetry transformations of space and time, a group under which the dynamical laws governing any closed system remain invariant: in the Galilei–Newtonian case, this group is the inhomogeneous Galilei group; in the special-relativistic case, it is the inhomogeneous Lorentz group (also called the Poincaré group).

A number of elements take the same form in both groups: the spatial and temporal translations, which express (respectively) the homogeneity of the relative space of each inertial frame and the uniformity of the time – absolute or relative to each inertial frame; and the spatial rotations, which express the isotropy of relative space of each inertial frame. The two groups differ in the form of the so-called boosts, which relate the spatial and temporal coordinates of two inertial frames in motion relative to each other. The Galilei–Newtonian boosts involve the absolute time, which does not change from frame to frame; so they involve only a transformation of the spatial coordinates of an event with respect to the two frames. The special-relativistic boosts involve a transformation of the temporal as well as the spatial coordinates of an event. The well-known spatial (Lorentz) contractions and time dilatations can be deduced from the relativistic boost transformations.

Four-Dimensional Formulation

Perhaps it is worthwhile to begin with a word of caution about a prevalent pathology in theoretical physics, which I shall call 'the fetishism of mathematics'. We invent physical theories to enable us to better comprehend and cope with the world, or rather with limited portions of the world. We employ mathematics as a vital tool in such attempts. Mathematical structures help us to correctly encode numerous, often extremely complicated, relations among physical concepts, relations of both a quantitative and a qualitative nature. If these structures have been judiciously chosen, they do more than encode the relations that led to their introduction: formal manipulation of the structures leads to the discovery of new relations that can then be tested successfully. However, there are almost always redundant elements in the mathematical structure that have no obvious correspondents in the physical theory – and perhaps indeed no relevance at all to the physical content of the theory. What is more, all-too-often there comes a point at which the mathematical structure leads to predictions that fail the test of experiment; or experiments yield relevant results that the theory seems unable to encompass. Then we may say that the mathematical structure and/or the physical theory, has reached its limits of validity. But before such a limit is reached, there is often a tendency to forget that the mathematical structure was introduced originally as a tool to help us encode known and discover new relations among physical concepts;

that is, to forget about the constant dialectic between attempts to encompass new relations within the given structure and attempts to modify the structure itself in response to new relations that cannot be so encompassed. Instead, the mathematical structure is considered to be (or to represent) a more fundamental level of reality, the properties of which entail the concepts and relations of the physical theory – and indeed those of the phenomenal world. Drawing on the language of Marx, who speaks of "the fetishism of commodities," I designate as "the fetishism of mathematics" this tendency to endow the mathematical constructs of the human brain with an independent life and power of their own. Perhaps the most flagrant current examples of this fetishism are found in the realm of quantum mechanics (I need only mention the fetishism of Hilbert spaces), but the fetishism of four-dimensional formalisms in relativity does not fall too far behind. In my exposition, I shall try not to fall into fetishistic language; but if I do, please regard it as no more than a momentary lapse.

Poincaré and Minkowski showed how to represent the special-relativistic space-time transformations in a mathematically simple, elegant and fruitful form by the introduction of a four-dimensional formalism, in which space and time coordinates are represented together in a four-dimensional space-time. In this formalism, a 'point of space' is represented by a so-called world-line: a one-dimensional curve in space-time representing the history of this point over time. It follows that a three-dimensional space is represented by a congruence of such (non-intersecting) world-lines filling the entire space-time. (This is often called a fibration of space-time.) An instant of time is represented by a hypersurface intersecting each curve in the congruence once and only once. It follows that a (non-intersecting) family of such hypersurfaces filling all of space-time represents a (global) time variable. (Such a family of hypersurfaces is often called a foliation, or more informally a slicing, of space-time.) To summarize: In four-dimensional formalism, a (relative) space is represented by a particular fibration, and a global time (absolute or relative) by a particular foliation of space-time. The proper time along any world-line is the local time associated with a series of events taking place along that world-line.

Of course, the introduction of time as a fourth dimension to facilitate the description of motion in space long predates relativity. But in Galilei–Newtonian kinematics, due to the existence of the absolute time the four-dimensional space-time can be uniquely sliced (foliated) by a family of hyperplanes of simultaneity (i.e., of equal absolute time). In this sense, the four-dimensional unification of space and time remains formal. Yet it does allow a simple representation of the relativity of space. An inertial frame of reference is represented by a congruence of parallel, time-like straight lines, each line intersecting each hyperplane of simultaneity once and only once.

In the four-dimensional space-time of special relativity, each inertial frame is represented by a different slicing (foliation) of space-time into hyperplanes of equal time, together with the corresponding three-space defined by a congruence (fibration) of parallel time-like lines that are 'pseudo-orthogonal' (a concept I will explain soon) to the hyperplanes. Since space and time are both relative, the special-relativistic unification is much less formal.[8] The Lorentz boosts, which express the relation between two inertial frames (each represented by a foliation and fibration), are represented by so-called 'pseudo-rotations' taking one foliation and fibration into the other.

If we examine the kinematic structure of the four dimensional formulation of space-time more closely, we find that it consists of two distinct but interrelated structures, chronogeometrical and inertial. The chronogeometrical structure models mathematically the spatial geometry and the measure of global time, as these are exemplified respectively, for example, by the behavior of a system of rigid rods and clocks all at rest relative to each other. The inertial structure models mathematically the behavior of physical bodies not acted on by external forces, i.e., behavior characterized by Newton's First Law (the law of inertia). If we can neglect the spatial dimensions of such a body, we refer to it as a particle, and speak of the behavior of a freely-falling particle. Each of these mathematical structures is described by a geometrical-object field: the chronogeometry by the so-called metric tensor field, the inertial structure by the so-called affine connection field. These two structures are not independent of each other, but must obey certain compatibility conditions. Roughly speaking, these conditions ensure that sets of freely falling particles can be used to construct measuring rods and clocks; or equivalently, that a suitably chosen set of freely falling measuring rods and clocks is at rest in some inertial frame of reference, and so can be used to correctly measure distances and global times, respectively, relative to that frame.

As indicated above, in both Galilei–Newtonian and special-relativistic kinematics the kinematic structure is fixed and given *a priori*; this structure includes the chronogeometrical and the inertial structures. Stemming as it does from the common law of inertia, the inertial structure is common to both kinematics: This inertial structure, that of a (linear) affine space, allows (among other things) the definition of parallel straight lines, parallel hyperplanes, and the equality of parallel vectors in space-time. A fibration of space-time by parallel, time-like straight lines represents a particular inertial frame.

The Galilei–Newtonian chronogeometrical structure consists of a foliation by the hyperplanes of simultaneity (representing all events at the same absolute time), plus a degenerate (rank-three) four-dimensional metrical structure representing the Euclidean geometry that holds in every inertial frame.

The special-relativistic (or Minkowskian) chronogeometry consists of a non-degenerate four-dimensional pseudo-metric of signature two. A metric is a quadratic form used to compute the 'length-squared' of vectors, which is always a positive quantity. If the tensor does not assign a positive 'length-squared' to all vectors, i.e., if some of them have zero and negative 'length squared', the tensor is called a pseudo-metric. The Minkowski metric is such a pseudo-metric of signature two. This means that, when diagonalized, it has three plus terms and one minus term, or the opposite – at any rate three terms of one sign and one of the other, which represent space and time, respectively. If the 'length squared' of a vector computed with this pseudo-metric is positive, the vector is called space-like; if negative, the vector is called time-like; if zero, the vector is called null or light-like. Signature two for the four-dimensional Minkowski pseudo-metric implies that the null vectors at each point form a three-dimensional cone with vertex at that point, the so-called null cone. A time-like and a space-like vector are said to be 'pseudo-orthogonal' if their scalar product, computed with the pseudo-metric, vanishes. A fibration of parallel lines is time-like if it has time-like tangent vectors. Only a time-like fibration can represent an inertial frame. A hyperplane pseudo-orthogonal to the fibration (that is, with all space-like vectors in the hyperplane pseudo-orthogonal to the tangents to the fibration) represents the set of all events of equal global time relative to that inertial frame; and the slicing (foliation) consisting of all hyperplanes parallel to this one represents the passage of relative global time in that inertial frame. In both cases, the compatibility conditions require that the 'length-squared' of equal parallel vectors (space-like, time-like, or null) be the same.

Gravitation

In general relativity, both parts of the kinematic structure, the chronogeometrical and the inertial, lose their fixed, *a priori* character: both become dynamical structures. There are two basic reasons for this: General relativity is a theory of gravitation, and gravitation dynamizes the inertial structure. Furthermore, in general relativity the chronogeometrical structure is uniquely related to the inertial structure, so if the latter becomes dynamical, the former must. We shall now look more closely at each of these two assertions. First of all, by examining the Newtonian theory of gravitation from the four-dimensional point of view, we shall see that gravitation always dynamizes the inertial structure of space-time. As we have seen, the concept of inertial structure is based on the behavior of freely falling bodies, that is the trajectories of bodies (particles) not acted upon by any external force. In the absence of gravitation, force-free motions can be readily identified (in principle); but the presence of gravitation effectively nullifies the concept of such motions. How could we realize such a motion in principle (that is, ignoring purely practical difficulties)? First of all, non-gravitational forces (electrical, magnetic, etc.) can be neutralized or shielded. But gravitation is universal: it can be neither neutralized nor shielded. This would still not constitute a fatal difficulty if we could correct for the effect of gravitation on the motion of an otherwise force-free body. But the effect of gravitation is universal in a second sense: As Galileo is supposed to have demonstrated using the leaning tower of Pisa, it has the same effect on the motion of all bodies (this is often called the weak principle of equivalence). This would still not constitute a fatal difficulty if it were possible to single out the class of inertial frames of reference in a way that is independent of the concept of force-free motion. That this is possible is tacitly assumed when gravitation is described in the style of Newton as a force pulling objects off their inertial paths. But inertial frames cannot be defined independently of inertial motions, which are just force-free motions. So our attempt to distinguish between inertial and gravitational effects ends up in a vicious circle.

The only way out is to admit that we cannot distinguish in any absolute sense between the effects of inertia and gravitation on the motion of a body. There is a single inertio-gravitational field, and 'free falls' are the motions of bodies subject only to this field. (In principle, we can eliminate, or correct for, the effects on a body's motion of all non-gravitational fields.) Gravitation is no longer to be conceived of as a force pulling a body off a 'free' inertial path, but rather as a factor entering into the determination of the 'free' inertio-gravitational paths.

As a consequence of this new outlook on gravitation, the class of preferred frames of reference must be enlarged: It is now impossible in principle to distinguish between an inertial frame of reference with a (possibly time-dependent but) spatially homogeneous gravitational field in some direction, and a (possibly time-dependent but otherwise) uniformly accelerated frame of reference with no gravitational field. So, when gravitation is taken into account, all linearly accelerated frames of reference are equally good: linear acceleration is relative. (Rotating frames of reference can still be distinguished from non-rotating ones, so rotational acceleration is still absolute.) Within this enlarged class of preferred frames, the Galilei–Newtonian chronogeometric structure (absolute time, Euclidean geometry in each relative space) can still be established in a way that is independent of all dynamics, including that of the inertio-gravitational field.

Dynamized Inertial Structure

It is different with the Galilei–Newtonian inertial structure. To account for gravitation, as we have seen, it must be generalized to the inertio-gravitational structure, which is still modelled mathematically by an affine structure. But now, in virtue of the inclusion of gravitation, this structure is no longer fixed *a priori* (to wit, as the structure of a linear affine space), but must be given subject to certain dynamical field equations specifying just how the presence of matter affects the inertio-gravitational field.[9] What about the compatibility conditions between the still-fixed, kinematical chronogeometrical structure and the now-dynamized inertio-gravitational structure? In Newtonian gravitation theory, these conditions are still valid, but now they do not serve to fix uniquely the inertio-gravitational structure when the chronogeometrical structure is given. They allow just enough leeway in the choice of the former to impose the four-dimensional equivalent of Newton's law of gravitational attraction on the inertio-gravitational (affine) structure:

$$\mathbf{a} = -\text{grad}\ \varphi,$$

where φ is the Newtonian gravitational potential. So even at the Galilei–Newtonian level, it is the nature of gravitation, when interpreted as part of the inertio-gravitational (affine) structure, that dynamizes this structure. The breakup of this affine structure into inertial and gravitational parts is no longer absolute, but depends on (is relative to) the state of motion – in particular, the state of acceleration – of the frame of reference used. This implies that, in going from one allowed frame of reference to another that is accelerated with respect to the first, φ is no longer an invariant, but is subject to a more complicated law of transformation.

What is the exact nature of the change in the inertio-gravitational structure? Before gravitation is introduced, the linear affine space modelling the inertial structure has a property described mathematically as flatness or lack of curvature, which corresponds to the vanishing of a certain tensor formed from the affine field and called the Riemann or curvature tensor. In the presence of a Newtonian gravitational field this tensor no longer vanishes, but is subject to a certain set of field equations.

Physically, this curvature tensor has several interpretations. For example, in an affine-flat space-time, the spatial distance (at the same absolute time) between any pair of freely-falling test particles (modelled by time-like straight lines in the linear affine space) varies linearly with the absolute time. In other words, the two particles have no relative acceleration. In an affine-curved space-time, the path of a freely-falling particle is no longer a straight line, but a time-like geodesic, the straightest possible path in such a space. A neighboring pair of freely-falling particles (modelled by a pair of time-like geodesics) now generally have a relative acceleration (also called a tidal force) between them, the magnitude of which in various directions is proportional to corresponding components of the curvature tensor of the space-time. Measurement of such relative accelerations (or tidal forces) thus constitutes a direct measurement of components of the curvature tensor.

The curvature tensor not only manifests itself in the behavior of pairs of freely-falling particles. As mentioned above, it also enters in a very simple way into the four-dimensional formulation of Newton's field law of gravitation, which uses Poisson's equation to describe just how matter modifies the inertio-gravitational field.

General Relativity

Up to now we have been discussing the Newtonian theory of gravitation, based on the concept of absolute time. Now we must apply the lessons learned to the task of formulating a relativistic theory of gravitation, without an absolute time. The most important lesson of the four-dimensional formulation of Newtonian theory is that gravitation is best incorporated into an inertio-gravitational structure, i.e., described mathematically by a non-flat affine connection that is subject to certain dynamical (gravitational) field equations involving the curvature tensor of that connection. But an important feature distinguishes (non-gravitational) special-relativistic theories from Galilei–Newtonian theories: In special relativity, the compatibility relations between the chronogeometrical and inertial structures are completely restrictive: The chronogeometrical structure uniquely determines the inertial structure. When incorporating gravitation by converting the inertial structure into an inertio-gravitational structure and dynamizing it, we are confronted with a choice: either give up the unique relation between chronogeometrical and inertial structures, which is the road taken by the various special-relativistic theories of gravitation, or preserve the unique relation by dynamizing the chronogeometry too.

The latter is the path chosen in general relativity.[10] The metric tensor field provides both the chronogeometrical and inertio-gravitational structures and is subject to a set of field equations relating its curvature tensor to a tensor describing the sources of the gravitational interactions, the so-called stress-energy tensor. Formally, these field equations for the affine curvature tensor look almost exactly the same as the four-dimensional Newtonian equations. But in the case of general relativity, as emphasized above, the affine structure, including its curvature tensor, is completely determined by the metric tensor field. So, in general relativity, everything depends on the metric.

Up to now, I have not discussed the mathematical structure of the space with which the metric and affine field of general relativity are associated. In the case of Galilei–Newtonian and special-relativistic space-times, we saw that the structure of these spaces is determined by the respective kinematical symmetry groups of these theories: The inhomogeneous Galilei group leads to Galilei–Newtonian space-time, while the Poincaré group leads to Minkowski space-time. Both are linear affine spaces with a flat affine structure. In general relativity, no kinematic space-time structure is given *a priori*, so there is no preferred kinematic symmetry group to represent symmetries that restrict the class of allowed point transformations on the underlying four-dimensional mathematical space. (In general relativity, we cannot even call this mathematical space a space-time, since it has no physical properties until a metric tensor has been selected.)

So, it might seem that any four-dimensional topological space would do as a mathematical starting point: its symmetry group consists of all the homeomorphisms, that is all one-one, bicontinuous point transformations of the space. But to do physics, we do need tensor and other geometrical object fields on the space, on which we may carry out various differential operations that require the use of coordinates. To carry out these operations in a way that is independent of the particular choice of a coordinate system, we need to have a differentiable structure on the underlying topological space. So we are led to take four-dimensional differentiable manifolds as our underlying mathematical spaces, which have the group of diffeomorphisms (that is, differentiable homeomorphisms) as their symmetry group. We are thus led to require general covariance of any field equations on such differentiable manifolds; that is, invariance under this diffeomorphism group. This requirement constitutes an important aspect of what is called the general covariance of the entire theory (see below).

It must be emphasized just how revolutionary are the steps leading to the general theory of relativity: identification of the chronogeometrical and inertio-gravitational structures, dynamization of both, and the associated requirement of general covariance. In many ways, these steps involve a much greater break with traditional physics than those leading from Galilei–Newtonian physics to the special theory of relativity. I shall discuss four such radical differences between general-relativistic and pre-general relativistic concepts of space and time.

First of all, there is no longer such a thing as an 'empty' region of space-time. Where there is space-time there is chronogeometry and hence at the very least there is an inertio-gravitational field. Returning for a moment to the controversy between absolute and relational concepts of time and space, it seems to me difficult to sustain the absolute position with regard to general relativity. As Einstein put it:

> [The metric tensor components] describe not only the field, but at the same time also the topological and metrical structural properties of the manifold. . . . There is

no such thing as empty space, i.e., a space without a field. Space-time does not claim existence on its own, but only as a structural quality of the field.[11]

Secondly, the space-time structures no longer form a fixed stage, on which the drama of matter and fields is enacted; space and time have become actors in the drama. Curious possibilities thereby arise, such as the possibility of solutions to the field equations for the metric tensor representing space-times that contain both open and closed time-like world lines. A time-like world line represents the possible history of a particle. So in such a space-time, the history of some particles (with open world lines) includes a doubly-infinite (past and future) amount of proper time, while for others (with closed world lines) only a finite amount of proper time elapses before they 'looped back upon themselves', so to speak.

Thirdly, not only is the local structure (local, in the sense of a finite but limited region) of space-time dynamized; the global structure (global in the sense of the entire topology) is no longer given *a priori*. For each locally-given solution to the gravitational field equations, one must work out the global topology of the maximally extended manifold(s) (the criteria for such an extension are not uniquely given) that is (are) compatible with the local space-time structure of that solution. Solutions are possible that have finite but unbounded spatial fibrations and/or finite but unbounded temporal foliations. In the former case, someone always marching straight ahead into the universe could end up back where s/he started. In the latter case, the entire history of the universe would repeat itself after a finite time had elapsed.

Fourthly, if we choose to mathematically abstract the metric tensor field (which, as said, gives rise to both the chronogeometrical and inertial structures) from the underlying differentiable manifold with which it is associated, there is literally no physical content associated with such a 'bare' manifold. Many otherwise excellent textbooks on general relativity often suggest that the points of a bare manifold are physically individuated (that is, represent 'events') without a metric tensor field, which merely defines the chronogeometrical relations between such events given *a priori*. Such an assertion is incorrect for two reasons. Globally, as just noted, without the specification of a particular metric tensor field, we cannot specify the topology of the differentiable manifold that this metric converts into a space-time.

But, even locally, there is no means of physically individuating *a priori* the points of a region of a bare manifold. If other, non-gravitational fields or particles are present in the region, their properties may suffice to individuate the points of the region. Even then, without any chronogeometry, it is generally impossible to fully interpret the physical properties of such non-gravitational entities.

If we confine ourselves to otherwise empty regions, in which only the chronogeometrical (and inertio-gravitational) field is present, then the points of such a region cannot be physically individuated by anything but the properties of this field.[12] Among the properties of any physical event are its position in space-time: the 'here' and 'now' of the event. So one cannot give any meaning to the 'here' and 'now' of an event in such regions of space-time without use of the chronogeometry (metric tensor). No meaning can be attached to the concepts of physical space or time in general relativity that does not depend on the metric tensor field. This represents a second relativization of the concept of time, more drastic even than that required by the special theory of relativity.

To remind you of what has been said earlier: In Galilei–Newtonian kinematics, the absolute or universal time functions as a global time, the same for all rigid reference frames, regardless of their states of motion, even when gravitation is taken into account; and also as a local time, the same for all time-like paths. In special-relativistic kinematics, the unique concept of absolute time divides into two, both of which are relativized. Time in the global sense (the time used to compare events at different places) is relative to the inertial frame, i.e., to one member of a preferred class of reference frames. Time in the local sense (the proper time that elapses between events along a time-like path) is relative to the path.

In both pre-relativistic and special-relativistic kinematics, the use of different frames of reference leads to different descriptions of the same process undergone by some dynamical system: The expression for the initial state of the system at some time and for the changes in that state, as it evolves in accord with the dynamical laws governing the system, depend on (are relative to) the choice of inertial frame. But these descriptions are all concordant with one another. They all lead, for example, to the same prediction for the outcome of any experiment performed on the system.[13] Mathematically, this is because different descriptions amount to no more than different slicings (foliations and fibrations) into a three-space and a one-time of the same background space-time, in which the unique four-dimensional description of the dynamical process is given.

The Galilei–Newtonian absolute time and the special-relativistic global (inertial-frame) relative time have this in common: they are defined kinematically in a way that is independent of any dynamical process. (Of course, the *measurement* of such times depends on certain dynamical processes, but that is a different question.) In general relativity, no time – global (frame-dependent) or local (path-dependent) – can be defined that is independent of the dynamics of the inertio-gravitational field. In general relativity, that is, time is relative in another sense of the word: relative to dynamics, in particular relative to a particular solution to the gravitational field equations. The proper (local) time is of course relative to a path in the manifold, but also depends on the particular solution: whether a path in a manifold is even time-like cannot be specified *a priori*. Introduction of a frame-relative global time depends upon a slicing (foliation) of the space-time into three-dimensional space-like slices; again, whether such a slicing of the manifold is even space-like cannot be specified *a priori*. While a space-like slicing is always possible locally, a global space-like foliation may not exist in a particular space-time. In general, there is no preferred family of slicings in a space-time (like the hyperplanes of Minkowski space) because in general the metric tensor of a space-time has no symmetries. The symmetries of a particular metric tensor field may entail preferred slicings. For example, there is a class of preferred slicings of a static metric such that the spatial geometry of the slices does not change from slice to slice. Hence, we can say that there is a preferred global frame time relative to a static solution. But in general, there is no class of preferred slicings, and a different global frame time is associated with each possible slicing, no one of these times being preferred over the others. As in the pre-general-relativistic case, different choices of frame time lead to different descriptions of the same dynamical process, but all these descriptions are concordant with one another because they are all just different slicings of the same space-time.

Of course, in general relativity, different dynamical processes cannot take place in the same space-time.[14] The gravitational and the non-gravitational dynamical equations must be solved as a single, coupled system. A space-time that is a solution to the gravitational equations is associated with a dynamical system that is a solution to the non-gravitational equations. Here no separation between space-time kinematics and the dynamics of other fields is possible.

As a result of these changes, there is a basic difference between the type of questions that one can ask in pre-general-relativistic physical theories and in general-relativistic theories. A typical pre-general-relativistic question takes the form: *Here* is a point in space and *now* is a moment in time (or, if you prefer, *here* is a point in space-time): What is (are) the value(s) of some physical quantity (quantities) 'here' and 'now'? We answer such questions theoretically by solving some set of dynamical equations for the quantity (quantities) in question evaluated 'here' and 'now'.

In general relativity, as we have seen, 'here' and 'now' cannot be defined before we have a chronogeometry, which presupposes that we have solved the dynamical equations for the metric tensor field. So 'here' and 'now' cannot be part of the question. Thus, in general relativity, we must start with the specification of some solution to these field equations – plus any other dynamical fields that may be involved in the questions we want to ask. Then either:

(1) we don't ask any questions that depend on specification of a 'here' and 'now': e.g., questions that depend only on globally defined quantities, such as the total mass or charge of a system; or
(2) we construct 'here' and 'now' from the given solution to the field equations; then they are part of the answer, not part of the question.

The construction could involve just the pure gravitational field, i.e., be based upon just the metric tensor field; or, it could involve the other non-gravitational field variables if such are present. I shall not go into further detail here, but end by noting that this problem reappears in all attempts to construct a quantum theory that incorporates general relativity – a quantum gravity.

A typical quantum question takes the form: If one physical quantity has been found to have a certain value 'here' and 'now', what is the probability that another (possibly the same) physical quantity will be found to have another value 'there' and 'then'? Again, a quantum gravity question cannot take this form because 'here' and 'now', and 'there' and 'then', have to somehow be constructed as part of the answer, not part of the question.

Boston University
Dept. of Physics

NOTES

[1] Albert Einstein, 'Foreword' to Max Jammer, *Concepts of Space*, Cambridge: Harvard University Press, 1954, pp. xi–xvi. Quotation from p. xiv.

[2] We shall not here enquire too closely into the significance of the concept of the continuity of a sequence of events, involving as it does the possibility of intervals of time and space that grow smaller and smaller without any lower bound.

[3] Note that the word 'absolute' here is being used in a quite different sense than above, where absolute and relational concepts of time and space were contrasted. It would be better to use the term 'universal' here, but absolute is so generally used in this context that I shall follow this usage in spite of the possibility of confusion. As will be seen below, in this context the contrast is between absolute or universal (i.e., independent of frame of reference) and relative (i.e., dependent on frame of reference).

[4] For the moment, we shall not examine the phrase: "acted on by no external forces". When we come to consider gravitation, we shall have to return to this concept.

[5] Of course, one might generalize, and say: "with respect to any arbitrary frame of reference". But in both Galilei–Newtonian and special-relativistic theories, the inertial frames maintain their privileged role: The laws of physics take their simplest form when expressed relative to these frames – as long as gravitation is disregarded. As we shall see, when gravitation is taken into account, the inertial frames lose their privileged role even in Galilei–Newtonian theory.

[6] Albert Einstein, 'Zur Elektrodynamik bewegter Körper', *Annalen der Physik, 17*, 1905, pp. 891–921. Translation from *The Collected Papers of Albert Einstein*, vol. 2, *The Swiss Years: Writings, 1900–1909*, editors John Stachel et al. (Princeton: Princeton University Press, 1989), p. xxiii. German text: "Die zu entwickelnde Theorie stützt sich – wie jede andere Elektrodynamik – auf die Kinematik des starren Körpers, da die Aussagen einer jeden Theorie Beziehungen zwischen starren Körpern (Koordinatensystem), Uhren und elektromagnetischen Prozessen betrifft. Die nicht genügende Berücksichtigung dieses Umstandes ist die Wurzel der Schwierigkeiten, mit denen die Elektrodynamik bewegter Körper gegenwärtig zu kämpfen hat".

[7] Two clocks are said to be synchronized (relative to the inertial frame in which both are at rest) if the time for a signal to travel from the first to the second is the same as the time for a signal to travel from the second to the first. Clearly, an element of convention enters into the definition of any one-way signal time. On the other hand, only the preferred convention makes the resulting relative simultaneity an equivalence relation.

[8] The presence in the special theory of an invariant speed c, usually referred to as the speed of light, enables time and space coordinates to be expressed in commensurable units. The existence of such an invariant speed is a consequence of the existence of a group of space-time transformations that treats all inertial frames on an equal footing. From this point of view, Galilei–Newtonian kinematics represents a degenerate limiting case, in which this speed becomes infinite.

[9] Einstein originally believed that the entire inertio-gravitational field could be attributed to the effects of matter – an idea to which he gave the name "Mach's principle" – which would make this field little more than an epiphenomenon of matter. After he adopted a field ontology, he gave up this position, which is still the subject of controversy. Many workers in the field consider it to be incompatible with the general theory of relativity.

[10] Of course, we might both dynamize the chronogeometry and give up the unique relation. Various tensor-scalar theories of gravitation, for example, have attempted to do this. But so far the minimal assumption, that it suffices to dynamize the chronogeometry without changing its unique relation to the inertio-gravitational field, has survived all experimental challenges by such theories.

[11] Albert Einstein, 'Relativity and the Problem of Space', Appendix V to the 15th ed. (1952) of *Relativity: The Special and the General Theory*, New York: Crown Publishers, 1961, p. 155.

[12] The assumption that no metric-independent individuating fields exist forms part of the content of the principle of general covariance of the general theory of relativity. For details, see John Stachel, 'The Meaning of General Covariance: The Hole Story', to appear in a *Festschrift* for Adolf Grünbaum.

[13] Classically speaking. Quantum mechanically, they should all lead to the prediction of the same probability for a given event.

[14] Barring exceptional cases, in which different dynamical processes happen to have the same stress-energy tensor.

ANDREW BUCHWALTER

HEGEL AND THE DOCTRINE OF EXPRESSIVISM*

Contemporary criticisms of modern thought and culture often invoke the concept of expressivism. Although variously construed, expressivism generally is a view of reason, humans and their world opposed to Enlightenment dichotomies. According to the expressivist doctrine, man and world are not abstractly juxtaposed but integrally interrelated: individuals exist as parts of a broader whole, just as the world is the place for their self-realization and self-discovery. Similarly, norms of rationality are not abstractly contraposed to particular forms of life but deemed to have meaning and reality only in expressing existing social practices.

In recent years, many writers have identified the expressivist doctrine with the philosophy of G.W.F. Hegel. Hegel is said to believe that self-realization requires external embodiment, that norms are meaningful only in expressing the exigencies of a form of life, that individuals have identity only in community, that human experience expresses a larger order of being and can find fulfillment only in nurturing ties to that order. In his 1975 *Hegel*, Charles Taylor presented a view of Hegel that incorporated features of all these themes.[1] Since the publication of this important and influential work, others have advanced interpretations of Hegel that highlight individual components of the expressivist doctrine.[2]

In what follows I challenge the expressivist reading of Hegelian thought. While not disputing that Hegel sought to surmount Enlightenment dichotomies, I argue that his specific solution cannot be construed through the concept of expressivism. My aim is to clarify the nature and import of what in fact is Hegel's *non-expressivist* approach to modern dualisms.[3] I also seek to promote a clearer understanding of Hegel's treatment of the problems and prospects of modern reason. In this way I hope to reaffirm one of the aims of Marx Wartofsky's rich and variegated work: to better understand 19th century German thought and to explore its contemporary significance.

C.C. Gould and R.S. Cohen (eds.), Artifacts, Representations and Social Practice, 163–183.

I

1. I begin by considering what may be called the *subjective* version of Hegelian expressivism. Subjective expressivism is the thesis that something finds realization by manifesting itself in a given medium.[4] According to Taylor, this doctrine is particularly important in Hegel's account of self-realization. In Hegel's view, Taylor claims, individuals realize themselves by embodying their purposes in external reality. Such externalization is understood, however, as the means not just to concretize given purposes but to clarify purposes themselves. In this way, expressivism for Hegel is the backbone of an emphatic account of self-realization, one in which externalization constitutes the very nature and identity of the self.

This view seems uncontroversial, for Hegel does adopt a concept of self-realization defined in terms of self-objectification. But his position is misconstrued when self-objectification is understood, as it is in the expressivist reading, as externalization and manifestation in a given medium. This view is above all problematic because it misrepresents the specific dimension of Hegelian self-realization. Self-realization for Hegel is realization of the self's essential nature. The nature of self is freedom. Freedom is defined as complete self-dependence and self-sufficiency – that which "does not find its content outside itself, but makes itself its own object and its own content."[5] In this sense, Hegel defines the self in opposition to all externality, indeed as "inwardness" (*Innerlichkeit*).[6] Accordingly, he presents self-realization not as a process of externalization but as the systematic purgation of all externality, the elimination of which inhibits absolute self-sufficiency. Self-realization is the process by which the subject "withdraws out of the external into its inwardness with itself."[7]

Hegel is, to be sure, opposed to a romantic notion of self-realization,[8] one based on a "withdrawal into pure inwardness."[9] He acknowledges that external expression is necessary for any genuine self-actualization. Precisely because the self realizes itself only in attaining absolute self-sufficiency, it must be able to find and maintain itself in the various external conditions of its existence. External manifestation is a condition of achieved selfhood. Still, externalization is not the goal of self-realization but a necessary stage in a process that only achieves its end in overcoming all externality. The self attains full self-sufficiency only when it can represent all external conditions of existence wholly as a product of its own activity. While Hegel asserts, against romantics, that external manifestation is central to "absolute inward autonomy," he does so only because the self cannot attain full self-dependence unless it explicitly "refuses to allow something that has the character of outwardness to be preserved in, and be valid for, it" (*EL* §60A[10]). If Hegel champions self-realization as externality, it is only because he understands externalization in the literal sense – as the alienation, expulsion or *Ent-äußerung* of external reality itself. Only this sense of externalization comports with a notion of autonomy that requires the self to "throw off all that is natural in it."[11]

On occasion Hegel does appear to espouse a notion of self-realization defined chiefly in terms of external embodiment. This is particularly the case in his political and historical writings, where freedom is realized through the concrescence of a concept that initially possesses only a potential significance. The will, for instance, is free not as a general capacity but only when embodied in historically achieved conditions, those that present the willing subject with objects and choices that accord with dignity and autonomy. As Hegel writes: "the will is free not only in itself but for itself also."[12]

Even here, however, externalization does not capture the full nature of what for Hegel is self-realization. This is so not just because any genuine externalization of freedom – i.e., reworking of volitional objects so that they accord with autonomy – is itself an affirmation of "inwardness itself" (*PR* § 22). What merits emphasis here is that objectified freedom is not synonymous with freedom fully realized. Complete autonomy requires that something not only be free but be able to comprehend this freedom. To be free without such comprehension is to have one's nature specified exogenously, through external reflection, and thus to be subject to a determination alien to the notion of freedom. Fully realized freedom requires that the subject possess an understanding of the nature and meaning of both its object and activity – an understanding achieved only through self-knowledge and self-reflection. In other words, here, too, self-realization is an interiorizing rather than exteriorizing affair. While externalization describes a process by which what is true in principle also becomes true for itself, full autonomy is only achieved when this "for itself" also becomes "for itself" (*für sich selbst für sich ist*).[13] As Hegel also writes: "The most important point for the nature of spirit is not only the relation of what is *in itself* to what it is *actually*, but the relation of what *knows itself* to be to what it actually is."[14]

In this context it is not coincidental that Hegel locates the highest expression of freedom in the domain of absolute rather than objective spirit.[15] What characterizes the absolute spheres of art, religion, and philosophy is precisely their inward autonomy. Whereas objective spirit attends to spirit as realized in the externally conditioned realm of space and time, absolute or "self-apprehending" spirit is conceived precisely through its freedom from such externality. In the cultural domains, spirit is fully self-dependent and at one with itself: the object of reflection is nothing but the self-apprehending activity of the subject, just as this activity is nothing but the object's proper mode of being.[16] To be sure, here, too, Hegel is uninterested in any romantic juxtaposition of inner and outer. Precisely because art, religion, and philosophy proceed from socially held and historically evolved beliefs and values, they presuppose the forms of external realization appropriate to the domain of objective spirit. The "self-consciousness of absolute spirit" is possible only "through the mediation of finite spirit."[17] Still, freedom finds its complete or "infinite" realization only when received assumptions are cleansed of adventitious admixture and explicitly represented as cultural objects. Only in the self-apprehending domains of art, religion and philosophy do modes of human self-definition assume a reality appropriate to their own nature. As in his logical writings, Hegel maintains here, too, that self-realization is best construed as a process of interiorization (*Erinnerung*), and indeed in the literal sense of a re-membering that, like Plato's *anamnesis*, conduces to true being.[18]

None of this should suggest that Hegelian self-realization is not also a process of self-objectification. Hegel's thesis is simply that the self does not find objectivity by manifesting or expressing itself in an external medium.[19] In keeping with the priority on interiorization, self-production is preeminently a matter of self-knowledge. "[E]verything depends on spirit's self-awareness; if spirit knows that it is free, it is altogether different from what would be without this knowledge."[20] In Hegel's ontology, concreteness only emerges from the conjunction of reason and reality, substance and subjectivity, thought and being. Inasmuch as freedom subsists and attains reality only through self-knowledge, self-reflection is the condition for spirit's very objectivity. "[B]ecause spirit is essentially consciousness, this self-knowing is a fundamental determination of its *actuality*."[21] For Hegel, self-realization is the activity in which the subject "brings about its own actuality," in which it "realize[s] itself without the help of a material that exists outside it."[22] Hegel may occasionally describe concretization as a process of external expression: "The power of Spirit is only as great as its expression" (*Äußerung*).[23] At issue, however, is spirit's capacity not to shape an external medium but to give itself reality through its own self-reflective resources. Self-objectification is a matter of interiorization. "[T]hat which withdraws itself into the simplest depth is the mightiest and most all embracing" (*Übergreifendste*).[24]

2. We see, then, that the expressivist thesis misconstrues Hegel's notion of freedom and self-realization. But it is also problematic because it leaves him vulnerable to unwarranted criticism. It has been claimed, for instance, that because Hegel defines autonomous action as the embodiment of subjective purposes in external reality, he adopts a practical philosophy modeled on work or purposive activity. In this way he not only denies social interaction its rightful place in political life; he reduces normative questions to an instrumentalism, voluntarism, or even decisionism which bars normative reflection on the ends of public life. This criticism has been skillfully formulated by Seyla Benhabib, who not only ascribes to Hegel a work-oriented conception of political action, but construes that conception in expressivist terms.[25]

Much can be said for this interpretation, for Hegel does understand action more in terms of production than social interaction, more *poiesis* than *praxis*. At the same time, it is doubtful that his account of productive activity gives short-shrift to questions of normative validity or presents normativity itself in instrumental terms – as a strategy for implementing ends whose validity is presupposed as given. This view is problematic because it misrepresents Hegel's notion of self-realization. Although self-realization is for Hegel a matter of production, this should not be understood on the model of work or labor. Such interpretation would be valid only if Hegelian objectification could be conceived expressivistically – as an agent's exteriorizing purposes in an external medium. For Hegel, however, self-realization is most fundamentally a matter of interiorization – a process of self-formation and self-reflection. It is no coincidence that Hegel presents self-objectification as the activity not of, say, making oneself into an object but of that which "makes itself *as itself* into an object" (sich *als sich* zum Gegenstande machen).[26] Implicit in the act of will is a reflection on the good life for the acting subject and, correlatively, the good life itself. Though Hegel understands action as a form of labor, he does not restrict its rationality to the instrumental determination of means for realizing given ends. He focuses rather on what he calls spiritual labor, the work of an agent who "acts essentially,"[27] one whose activity includes reflection on the nature and validity of goals it pursues. In the tradition of the German Enlightenment, Hegel holds a notion of productive activity rooted not in instrumental action but in a final causality that ascribes purpose only to work that attends to the ends of human life.[28]

Many questions surround Hegel's effort to define productive action so as to include the dimension of normative self-reflection. Does Hegel's account of productive activity represent an effort to surmount ruling dichotomies between *poiesis* and *praxis*, labor and interaction?[29] Or is it just an elaborate effort to remove communal will-formation from the sphere of political action? Does Hegel seek to surmount modern productivism from within the framework of production itself?[30] Or is his aim merely to supply backhand justification for what in fact is a voluntarist and decisionist account of political action? Whatever the answers to these questions, they cannot be properly obtained unless one first considers the nature and viability of his decidedly non-expressivist effort to locate a dimension of normative self-reflection within the structure of productive action itself.

3. It may be argued that a Hegelian form of expressivism has little to do with instrumentalism and thus is unaffected by the difficulties that beset that notion. Indeed, Hegelian expressivism can be understood precisely as a critique of instrumentalism. Whereas instrumental reason objectifies external reality, expressivism approaches reality with an attitude of respect and mutuality. For expressivism, external reality is not an acreage to be fenced and mastered but an environment for spiritual discovery and fulfillment. Hegel's position has been interpreted along these lines by Charles Taylor, who construes expressivism in terms of the quest for holism characteristic of early 19th century romantic aesthetics.[31]

This version of the expressivist argument can likely be associated with some of Hegel's contemporaries, yet, again, it should not be ascribed to Hegel himself. For Hegel, the very assumption that the self could find complete self-fulfillment in external conditions was mistaken. Far more sober than his romantic contemporaries, he maintained that the tensions and dissonances of modern life could never be fully eliminated. The disparate elements of modern social life could never be so harmonized to accommodate an individual's quest for wholeness. Indeed, such harmony is not even desirable, for it would eliminate the particularism and subjective individualism that is the strength of modern society. To the extent it is possible at all, an expressive *promesse du bonheur* under modern conditions is achievable only in the artistic imagination of a beautiful soul – "the sentimental *bel esprit*."[32] Precisely as the domain where "the grief would not endure,"[33] however, the realm of aesthetic redemption is inadequate for a world structured by bifurcation. Indeed, efforts to restore holism through aesthetic self-expression can only result, as writers like Simmel, Nietzsche, Adorno, and Foucault later argued, in self-dissipation, a state of affairs that intensifies the dichotomies in question.

None of this implies that Hegel is uninterested in surmounting modern bifurcations. His is not the position of Friedrich Schiller, who sought "only" to humanize and subject to communicative control a world irrevocably bifurcated.[34] Against this "pragmatic" approach Hegel remained committed to a "speculative" solution.[35] He simply maintained that modern bifurcation cannot be consequentially surmounted with expressive-aesthetic means. In his view, other tools are required – the logical-methodological tools of a dialectical science. Only dialectics can reconcile reason and reality in a way appropriate to the bifurcated reality of modernity. Only a science able to discern the co-implication of identity and difference can both confront the reality of modern dissonance and forge unity from opposition itself. While romanticism properly notes the increasing fragmentation of modern life, the expressive-aesthetic solution bypasses existing oppositions and thus undermines the possibility of a genuinely holistic alternative. By contrast, a dialectical logic, capable as it is of "looking the negative in the face and tarrying with it,"[36] can proffer a comprehensive totality, one that subsists not in spite but because of modern dissonance. It may be that romantic expressivism migrates from "*logos* to *poesis*,"[37] but Hegel traces the steps from *poesis* to *logos*.

To be sure, reconciliation, for Hegel, is not just a conceptual production. It is also a product of reality itself. "Its being as a product must be comprehended as a producing."[38] Anything else would ratify the opposition between subject and object, reason and reality, besetting the aesthetic-expressivist solution. Besides, a genuine reconciliation of individual and community requires that individuals knowingly embrace the universal and actively pursue its realization. In an authentic political totality, an individual perceives the whole as "the end and product of his own activity" (*PR* §257), as "that which he brings about through his own activity" (*PM* §514). If Hegel nonetheless invokes the need for theory, it is to expose the misconceptions which prevent individuals from recognizing the entwinement of their interests with those of public life. Claiming that popular consciousness is dominated by an analytic understanding or intellect (*Verstand*) that ratifies and perpetuates dichotomies, Hegel ascribes to theory the task of demonstrating for popular consciousness, *inter alia*, that rights and responsibilities are mutually implicative, that modern bifurcations facilitate rather than impede genuine political totality, and that such totality can represent something more than the coincidental byproduct of individuals pursuing private interests. "The task of philosophy is to construct the absolute for consciousness."[39] Still, the absolute itself must be regarded, not as external construction, but as a development immanent of "interior" to reality itself – as a product of society's self-comprehension and self-construction. Essential though philosophy is in uprooting dogmatic assumptions, a genuine political totality can emerge only as individuals recognize the compatibility of their interests with those of political community and proceed to reorder existing conditions accordingly. Only such an endogenously generated solution – rather than the exogenously produced one offered by aesthetic-expressivism – can surmount the dichotomies in question.

II

Thus far I have challenged the subjective form of a putative Hegelian expressivism: the notion that self-realization consists in the exteriorization of subjective purposes. Yet it may be argued that such challenges do not touch the core of Hegelian expressivism, what can be called *objective* expressivism. In this variant, human thought and experience express some greater whole and find fulfillment only in emphasizing the primacy of the whole. In what follows I evaluate this interpretation as it is attributed to Hegel's account of (1) rationality, (2) the relationship of individual and community, and (3) the relationship of man and nature. I argue that here, too, the expressivist model misconstrues Hegel's actual position.

1. One 'objective' form of Hegelian expressivism has been discerned in his account of rationality, especially as it pertains to moral norms

and principles. A particularly notable version of this view has been advanced by Sabina Lovibond, whose *Realism and Imagination in Ethics* fashions a non-cognitivist account of morals that draws significantly on Wittgenstein and Hegel.[40] Calling on Taylor's work,[41] Lovibond argues that Hegel holds an embodied or 'concrete' concept of moral rationality, where reason is not, as it is with Kant, abstractly juxtaposed to existing forms of life, but is seen to express the exigencies of a particular life context. Asserting that moral conduct is possible and meaningful only in the context of shared life experiences, Hegel is said to maintain that moral rules and norms themselves are intelligible only as expressions of shared social practices.[42]

On its face, this thesis, too, seems uncontroversial, for Hegel did formulate an account of concrete rationality which accommodates the substantive considerations lost in the abstract formalism of Enlightenment and in particular Kantian universalism. In this respect Lovibond may be correct to identify neo-Hegelians like Bradley and Green as proponents of the expressivist concept of rationality. But she errs when attributing this position to Hegel. Though Hegel does champion the claims of concreteness, he is not opposed to the Kantian formalism. On the contrary, he fully accepts the requirements of a formal conception of rationality. Such emphasis on the priority of form is certainly a *sine qua non* for any account of philosophical idealism.[43] It is particularly important in a world which, no longer able to rely on the substantive certainties supplied by aesthetically and religiously defined cosmologies, must appeal to the 'logical' structure of a self-grounding reason to account for truth, meaning, and rationality. When criticizing the abstract formalism of Kantian rationality, Hegel directs his animus not at formalism *per se*, but at those abstract conceptions of formalism which assume that substantive considerations are alien to those of form. In what has been called Hegel's "metaphysic of form,"[44] concreteness – a technical term denoting the unity of thought and being – is achieved only by further developing a concept of form so that substantive considerations are not inimical to formalism but a feature of it. Kantian formalism is inadequate, Hegel writes, not for "lack of that presumptive reality given in feeling and intuition but rather in the fact that the concept has not yet given itself a reality of its *own*, a reality produced from its own resources."[45]

Hegel, to be sure, recognizes the extent that claims to rationality are predefined by substantive assumptions. He criticizes Kant precisely for failing to realize that his account of formal rationality presupposes substantive assumptions which vitiate any wholly formal account of rationality. On this basis, however, Hegel does not champion substantive over formal considerations; he does not embrace an account of rationality which defines cognitive considerations in terms of the requirements of life-practice. This view assumes that Hegel criticizes Kant for his excessive formalism, when in fact he claims that Kant is not formal enough. As he noted when discussing the abstractness of formal theories generally: "it is not content that they lack, for they have a specific content; they lack rather form, which is their essential nature."[46] To acknowledge that substantive considerations insidiously cling to Kant's theory of rationality is only to demonstrate the need to move from an abstract or finite to a substantive or infinite account of formalism, one in which the substantive considerations tacitly presupposed in any account of rationality are rendered an explicit product of reason itself. Hegel strives to demonstrate that reason's presuppositions (*Voraussetzungen*) are posits (*Setzungen*) of reason itself.[47]

This point is central to Hegel's practical philosophy. In his theory of Objective Spirit Hegel seeks to surmount the dualisms of Kant's practical philosophy, to formulate a concept of concrete ethical life that conjoins universalism and particularism, norms and history, rules and community. But Hegel does not do so expressivistically. His aim is not to develop a concept of practical rationality rooted in the contextual exigencies of a particular form of life, but to develop an account of concrete ethical life from the (transformed) principles of practical reason themselves. He challenges the abstract universalism of a practical philosophy based on the principle of the universal will, not by repudiating the principle itself, but by expanding it so that it fulfills its claims to universality – by extending its reach so as to encompass not just the restricted realm of unconditioned morality but the realm of conditioned experience itself, including the economic, social, familial, legal, political and historical domains. The task of the *Philosophy of Right* is to reinterpret the 'conditioned' domain of laws and institutions in terms of the principle of unconditioned freedom implicit in the idea of universal, autonomous will. Hegel's "system of right is the realm of freedom made actual, the world of mind brought forth out of itself like a second nature" (*PR* §4).

Naturally, Hegel is uninterested in contraposing a constructed concept of rationality to existing sociopolitical reality. The uniqueness of the doctrine of Objective Spirit is precisely its incorporation of the content of existing historical realities into the framework of practical philosophy. Philosophy is its own time apprehended in thought. Yet this turn to existing reason is conceived not as a denial but as an elaboration of the universalist and emphatically normative concerns that traditionally typified the canon of practical philosophy. In Hegel's view, the claims to autonomy that inform above all the Kantian concept of practical reason can be comprehensively and definitively fulfilled only if philosophy attends to a concept of reason that exists not just for philosophy but is discernible in existing conditions and accessible to public consciousness. Philosophy is "the exploration of the rational," and "is for that very reason the apprehension of the present and the actual, not the apprehension of a beyond, supposed to exist."[48] Moreover, even though the doctrine of Objective Spirit proceeds from a historically received concept of reason, its normative principles derive not from existing reality itself but from principles that emerge only in the systematic reconstruction of existing reason. Hegel does elaborate a 'concrete' account of practical philosophy, yet concreteness for him is achieved not by submitting to a reason embodied in existing practices, but by attending to actuality (*Wirklichkeit*) or true being (*wahrhafter Sein*), which only emerges when assumptions expressive of mere existence (*Dasein*) are purged of their adventitious attributes and made to express their essence and nature, their concept (*Begriff*). However critical of Kant's 'abstract' universalism, Hegel follows Kant in insisting that practical philosophy must adhere to a more than merely historical account of rationality.[49]

2. A related form of Hegel's objective expressivism has been detected by Charles E. Larmore, whose *Patterns of Moral Complexity* focuses on political expressivism. Political expressivism connotes an organic account of the relationship of individual and community, one which views the individual as a part or expression of a broader social-political whole. Larmore variously characterizes political expressivism. On some occasions expressivism is the view that humans only realize their nature and identity in community. On others, political expressivism is the thesis that our highest political ideals should fully coalesce with our highest personal ideals. In general this form of expressivism challenges disjunctive accounts of the relationship of man and citizen, *homme* and *citoyen*, central to liberal political theory.

Although not without plausibility, Larmore's thesis in fact does not capture the Hegelian position. For one thing, the assertion that Hegel sought to bind personal ideals to those of political community is contravened by his contention that the highest human ideals are most fully expressed only in the cultural domains of art, religion, and philosophy. But the expressivist view is deficient even if we restrict our focus to social and political relations. While Hegel did, undeniably, seek to surmount the opposition between individual and community typical of liberal accounts, he never argued that individuals find meaning and reality only in relating to a broader political whole. Such views may have been held by German romantics, but not by Hegel, whose account of the political individual accords far more with the aims of liberal political theory.[50] This form of political *non*-expressivism is evident in Hegel's commitment to many of the basic principles of a liberal constitutional state, e.g., basic rights, separation of powers, structures of judicial review, federalistic conception of power.[51] It is evident in the fact that his account of political community is preceded by an explication of the 'right' of the legal person and the moral subject. It is also evident in his explicit founding of the *Philosophy of Right* on the rational and autonomous individual, "the standpoint with which right and the science of right begins."[52] While an expressivist account regards the self as constituted only in political community, Hegel follows liberals in ascribing rights and duties solely to already constituted legal subjects.

Certainly Hegel does not espouse the asocial concept of human nature associated with liberal political theorists. In line with subsequent developments in social theory, he holds that selfhood and autonomous ego-identity are necessarily social, that an individual acquires a sense of him/herself as individual only in social relations. Still, questions of empirical identity formation are for Hegel distinct from the normative concerns of political philosophy.[53] If individual self-formation is inconceivable outside of community, the political ascription of rights and duties, presupposing as it does individual autonomy, is meaningful only by already assuming the idea of free and independent subjects. To be sure, because political philosophy does presuppose the concept of the autonomous subject, it must also presuppose the social or communal processes of ego-formation in which individuals are constituted (*PR* §71). Still, questions of socially mediated ego formation, situated by Hegel in his theory of Subjective Spirit (psychology and anthropology),[54] occupy no systematic place in his *theatrum mundi juridici*. To assume otherwise, to assume that ego-identity is itself achieved in political community, is to ignore the distinction between man and citizen, *homme* and *citoyen*, that lies at the normative heart of the *Philosophy of Right*.

None of this implies that Hegel espouses a liberal conception of state and community. While opposed to any expressivist derogation of subjective rights and liberties, he is equally opposed to the liberal view that considers community merely as a means to safeguard the ends and interests of private individuals. Against this contingent-instrumental view, Hegel maintains that the state is an end in itself, one that constitutes an individual's "substantive groundwork and end" (*PR* §§137, 268). Even here, however, Hegel cannot be said to claim in expressivist fashion that the *individual* has meaning and reality only as part of a broader whole. His point is rather that *the rights and liberties* claimed by autonomous individuals have no meaning and reality outside a framework of justly ordered institutions (*PR* §268). If Hegel accords a certain substantiality to the state vis-à-vis the person, it is as regards what John Rawls calls the "*political* conception of the person."[55] At issue is not a public or communal definition of the self, but recognition that individuals can meaningfully assert rights only as members of a political community. And it is in this context that Hegel's identification of public and private ends should be understood. At issue is not an undifferentiated homogenization of personal aims and social values, but a thesis about the conditions for individual liberties. Citizens embrace the aims of the state because they understand that their assertion of rights is dependent upon the structures of a constitutional state, that their "particular aims can be only through the universal."[56] What Hegel calls political sentiment is rooted in "the consciousness that my interest . . . is contained and preserved in another's (i.e., in the state's) interest and end" (*PR* §268). Here we disregard Hegel's additional and possibly stronger thesis: the political realm itself has legitimacy and reality only inasmuch as citizens knowingly and actively embrace public ends as their own.[57] What should be emphasized is only that in challenging the *political* opposition of individual and community, Hegel is not also adopting the expressivist intention to define the individual as such in communitarian terms.

It should be further noted that while Hegel espouses an organic concept of state, his is not the expressivist organicism attributed to him by Larmore. Expressive organicism is an account of the relationship of parts and whole where parts exist only through the whole, where "each part expresses the essence of the whole."[58] While this is certainly one aspect of a genuine theory of political organicism, it neglects what for Hegel is the essential dimension: that the whole itself expresses its parts (*PR* §276). For Hegel, this is particularly important with regard to individual subjectivity and the right of subjective reflection. In expressivist accounts independent subjective reflection is given little consideration, for individuals have identity only in their dependence on the whole. By contrast, a genuine whole exists only by accentuating the right of subjectivity. This is so not just because a genuine totality requires differentiation, but because the organic bond between individual and community is forged only when individuals *perceive* that their interests are represented in the whole. Without such consciousness, the bond remains contingent and external; it may exist in thought but not in reality itself. The desired internal and necessary connection is achievable only from the standpoint of self-consciousness, which allows the individual to relate to laws and institutions as to his "own essence, the essence in which he has a feeling of his selfhood, and in which he lives as in his own element not distinguished from himself" (*PR* §147). Only by accentuating the standpoint of subjectivity is it possible to secure that "more immediate identity" central to a living account of the relationship of individual and community. Though Hegel attacks the disjunctions in liberal accounts of the relationship of individual and community, his own 'organic' alternative is fashioned not from an expressivist perspective but from within the ambit of liberal political theory itself.[59]

3. The last version of objective expressivism to be considered here is one advanced most clearly by Charles Taylor himself. Objective expressivism would reunite human beings with nature, reestablish "communion with nature."[60] Inasmuch as humans exist only in expressing a larger natural order, they can flourish and fulfill themselves only by cultivating their ties with nature. As Taylor writes:

> We ought to recognize that we are part of a larger order of living beings, in the sense that our life springs from there and is sustained from there. Recognizing this involves acknowledging a certain allegiance to this larger order. The notion is that sharing a mutually sustaining life system with other creatures creates bonds: a kind of solidarity which is there in the process of life. To be in tune with life is to acknowledge this solidarity.[61]

Although expressivism of this sort can be associated with some of Hegel's contemporaries, it cannot be appropriately identified with Hegel, his critique of Enlightenment dichotomies notwithstanding. While Hegel certainly sought to reestablish the self as a whole, he did not hold that this could occur through the self's reintegration into nature; nor did he suggest that nature itself was "the source of self." Such views ascribe to Hegel a "spiritualized" or "poeticized" approach to the natural world at odds with his strictures on enchanting nature.[62] "Nature is not to be deified."[63] Such 'speculative' pantheism, favored though it was by the likes of Baader, Hölderlin, Goethe and Schelling, had for Hegel been rendered illicit by modern science and mechanistic accounts of nature. But it was also untenable on metaphysical grounds. In Hegel's idealist ontology, a thing has concrete reality only insofar as it expresses an organic unity of concept and existence, the unity of a thing with the concept of itself. In this definition nature has no true being, for its concept exists outside itself – in the domains of self-consciousness, or *Geist*, where alone conceptuality has its place and reality. Indeed, nature, far from representing a higher order of being, is in Hegelian ontology an entity devoid of all being whatsoever; it is a "*res nullias*" (*non-ens*) (*PN* §248). This was the basis for Hegel's attraction to the Protestant mystic Jacob Böhme, who, far from sanctifying nature, identified it with Lucifer. And it was the same consideration that led him to dispute the then current *hen kai pan* Neospinozism, which deemed nature a source of spiritual renewal and reintegration. As "the Idea in Externality," i.e., the idea alien or external to itself, nature for Hegel perforce repels such efforts (*PN* §247, *PM* §381Z).

For his part, Hegel maintains that self-reconciliation is a development occurring wholly indigenous to the domain of spirit itself. It denotes the process by which the subject becomes fully what in principle it already is: that which has itself as its own object. The self surmounts fragmentation not by locating itself in a putatively redemptive order of natural being, but by systematically traversing the spheres of being specifically dedicated to the unity of substance and subjectivity. Certainly, self-reconciliation also involves interaction with nature, the physical source of all life. Yet for Hegel the interaction of *Geist* with Nature is predicated not on integrating the self itself into nature but in progressively liberating it from all that is natural. While nature may be the spatiotemporal source of all spiritual being, spirit itself has its ontological *fons et origo* only in spirit, which is fully attained in the complete extirpation of all elements that do not accord with a concept of subjective self-reflexivity (*PM* §381). Spirit finds reconciliation only when the substantive conditions of its existence can be fully represented as a product of subjectivity itself. What in his political writings Hegel said of the reconciliation of freedom and necessity applies generally to that of substance and subjectivity: it "is produced not through nature but through freedom."[64]

Hegel's paean to the autonomous power of spirit, to be sure, entails little of the disdain for nature discernible in, say, Descartes, who saw man as nature's *maître et possesseur*. Like his romantic contemporaries, Hegel also sought to liberate nature from the evisceration it suffered at the hands of Enlightenment thought[65] and from what he himself perceived as its alienated condition. Indeed, anticipating Ernst Bloch and Walter Benjamin, he allowed that his philosophy aspires to the "reestablishment of a true nature" (*die Wiederherstellung der wirklichen Natur*), one sensitive to a nature free of instrumental manipulation.[66] Moreover, such reestablishment would flow from reconciling the realms of nature and spirit. On the other hand, Hegel's restorative approach does not require spirit to bow to truths presumed to lie hidden within nature's breast. This would assume that nature's alienation derives from external exploitation and from its instrumental domination by human beings. For Hegel, however, nature's alienation is ontological rather than historical, intrinsically structured rather than externally produced. Unable to generate a concept of itself, nature has its essence in *Geist* and thus is alien or external to itself. In this view, nature can be liberated not by ending its subjective manipulation but by emancipating itself from itself – by moving from nature to spirit. If nature has its concept in spirit, it can achieve organic union only in *Geist*, only in the domain where conceptuality is a feature of reality itself. In this sense, Hegel claims not only that spirit attains true being when liberating itself from nature; nature itself comes into being only in this development (*PM* §384). Nature is resurrected only insofar as humans "elevate themselves beyond a fallen nature."[67]

Many questions can be asked of Hegel's view of the relationship of spirit and nature. If Hegel surmounts an instrumental approach to nature, does he do so only through a hyper-subjectivism that denies natural nature all reality whatsoever? If he did seek to restore appreciation for nature's intrinsic truth, what is the substance of his wholly spiritual concept of true nature? Is Hegel's project a desperate effort to retain some appreciation for an enchanted nature in a world where existing nature is thoroughly and irrevocably disenchanted? Did he embrace a mystical concept of a resurrected nature because he appreciated the paradoxes inherent in human efforts to restore natural harmony? Is his wholly generated concept of nature just a philosophical elaboration of Christian notions of creation? Does it exemplify what Feuerbach and Marx saw as Hegel's misguided and mystified tendency to reconcile reason and reality wholly at the level of thought? Does Hegel provide a metaphysical reconciliation of nature only to mask his failure to address the possibility of a real historical one? And if he did assert that nature can be reconciled only in a process of human self-cultivation, what follows from the fact that this process concludes in the cultural domains of art, religion, and philosophy?

Here is not the place even to attempt to answer these and related questions. My aim has only been to indicate that Hegel's concept of the relationship of nature and subjectivity should not be construed on a model of objective expressivism. While Hegel seeks to restore a unity in the self, he claims that this is achievable not by nurturing ties with nature but by liberating the self from nature itself. Similarly, while he does formulate a spiritualized concept of nature, his aim is not to recover a spiritual order of being in nature but to render nature itself a product of spirit. Taylor rightly notes that for Hegel nature and spirit are reconciled not directly, but only in the arduous process of human self-reconciliation.[68] Yet whereas Taylor sees this *Bildungsprozess* as culminating in a new appreciation of man's dependence upon nature, Hegel's account flows from a belief that both man and nature exist only through a dependence upon spirit. While expressivism regards nature as a source of being, Hegel recognizes only *Geist* as the source.

III

This paper has challenged the expressivist reading of Hegelian philosophy. While rightly noting Hegel's effort to surmount Enlightenment dichotomies, this reading fails, in both its subjective and objective versions, to grasp the essentially non-expressivist nature of Hegel's proposals. Hegel conceived self-realization not as the embodiment of subjective purposes in an external medium, but as the internal self-unfolding and self-production of a self-dependent subject. He understood concrete rationality not as a paean to historical particularity, but as further development and vindication of universalist claims central to the Kantian conception of reason. He challenged political atomism not by invoking a communal conception of human nature, but by demonstrating that political community is the precondition for political rights and liberties. He restored man's unity with himself and with nature not by integrating the self into a spiritual order of natural being but by presenting the substance of both nature and spirit as a product of subjectivity itself.

Appreciation of the non-expressivist dimension of Hegelian thought is important simply for the proper understanding of Hegel's philosophy. It is also important as it clarifies Hegel's place in continuing debates about the problems and prospects of the project of Enlightenment. Proper appreciation of Hegel's position demonstrates that in his view the critique of Enlightenment thought and culture, however urgent and justified, is best pursued only by further developing the project itself. For Hegel, the hand that inflicts the wound is also the one that shall heal it. At a time when discussions of the gains and losses of modernity are framed in static polarities, there is much to recommend consideration of Hegel's non-expressivist supersession.

University of North Florida

NOTES

* I want to thank Willem deVries for comments on an earlier version of this essay.

[1] *Hegel* (Cambridge: Cambridge University Press, 1975). See also Taylor's *Hegel and Modern Society* (Cambridge: Cambridge University Press, 1979) and *Sources of the Self: The Making of the Modern Identity* (Cambridge, MA: Harvard University Press, 1989), especially 368–390.

[2] Seyla Benhabib, *Critique, Norm, and Utopia* (New York: Columbia, 1986); Martin Jay, *Marxism and Totality* (Berkeley: University of California Press, 1984); Charles E. Larmore, *Patterns of Moral Complexity* (Cambridge: Cambridge University Press, 1987); Sabina Lovibond, *Realism and Imagination in Ethics* (Minneapolis: University of Minnesota Press, 1983); Eugene Lunn, *Marxism & Modernism: An Historical Study of Lukács, Brecht, Benjamin and Adorno* (Berkeley: University of California Press, 1982), pp. 28–32; John H. Smith, *The Spirit and Its Letter: Traces of Rhetoric in Hegel's Philosophy of Bildung* (Ithaca, NY: Cornell University Press, 1988). In some writings Jürgen Habermas has seconded the expressivist reading, while on other occasions he has been more critical. For the favorable, *The Philosophical Discourse of Modernity* (Cambridge, MA: MIT 1987), pp. 75–82. For the critical: 'A Reply to my Critics,' in John B. Thompson and David Held (eds.), *Habermas: Critical Debates* (Cambridge, MA: MIT, 1982), p. 224 and 'Questions and Counterquestions,' Richard Bernstein (ed.), *Habermas and Modernity* (Cambridge, MA: MIT Press, 1985), p. 209.

[3] For one of the few to emphasize this dimension of Hegel's position, see Michael

Rosen, *Hegel's Dialectic and its Criticism* (Cambridge University Press, 1982), pp. 122–142.

[4] Taylor, *Sources of the Self*, p. 374. For Taylor's discussion of subjective expressivism, see *Hegel*, p. 16ff.

[5] *Lectures on the Philosophy of World History*, trans H.B. Nisbet (Cambridge: Cambridge University Press, 1975) [hereafter *Lectures*], p. 47.

[6] *The Encyclopedia Logic*, trans T.F. Geraets *et al.* (Indianapolis: Hackett, 1991) [hereafter *EL*] §60.

[7] *Werke in zwanzig Bänden*, ed. Eva Moldenhauer and Karl Markus Michel (Frankfurt: Suhrkamp, 1969) [hereafter *Werke*] *14*, p. 129.

[8] See Michael Theunissen, 'Produktive Innerlichkeit,' *Frankfurter Hefte* 29 (November/Dezember 1984): 103–110.

[9] *Phenomenology of Sprit*, trans A.V. Miller (Oxford: Clarendon, 1977), p. 490.

[10] Translation modified.

[11] *Lectures*, p. 51, translation modified.

[12] *Philosophy of Right*, trans T.M. Knox (London: Oxford University Press, 1967) [hereafter *PR*], §21.

[13] *Phenomenology*, p. 14.

[14] *Science of Logic*, trans A.V. Miller (New York: Humanities, 1976) [hereafter *SL*], p. 37.

[15] See Emil Angehrn, *Freiheit und System bei Hegel* (Berlin: de Gruyter, 1977), pp. 291–406.

[16] *Werke 16*, p. 151.

[17] *Werke 16*, p. 197f. See also *Philosophy of Mind*, trans William Wallace and A.V. Miller (Oxford: Clarendon Press, 1971) [hereafter *PM*] §385Z.

[18] See Edith Düsing, 'Zum Verhältnis von Intelligenz und Wille bei Fichte und Hegel,' in Franz Hespe and Burkhard Tuschling (eds.), *Psychologie und Anthropologie oder Philosophie des Geistes* (Stuttgart: Frommann-Holzboog, 1991), p. 124.

[19] See Taylor, *Sources of the Self*, p. 374.

[20] *Lectures*, p. 48, modified.

[21] *SL*, p. 37.

[22] *The Logic of Hegel*, trans William Wallace (London: Oxford University Press, 1904) §§235 & 163.

[23] *Phenomenology of Spirit*, p. 18.

[24] *SL*, p. 841.

[25] *Critique, Norm and Utopia*. In his *The Philosophical Discourse of Modernity*, Habermas sees a connection between expressivism and productivism in the Marxist materialistic appropriation of Hegel, but not necessarily in Hegel's thought, committed as it is to the "philosophy of reflection." For a critique of Hegel's concept of action along the lines suggested by Benhabib, see Manfred Baum, 'Gemeinwohl und allgemeiner Wille in Hegels Rechtsphilosophie,' in *Archiv für Geschichte der Philosophie 60* (1978): 193-198.

[26] *Jenaer Systementwürfe*, ed. Rolf-Peter Horstmann (Hamburg: Meiner 1987), p. 186 (emphasis added); see also *PR* §7A.

[27] *Werke* 12, p. 99.

[28] See Otto Pöggeler, 'Das Menschenwerk des Staates,' in Christoph Jamme and Helmut Schneider (eds.), *Mythologie der Vernunft* (Frankfurt: Suhrkamp, 1984), pp. 175–225.

In this essay Pöggeler considers Hegel's distinction between the "work of humanity" and "the miserable work of man."

[29] See Guy Planty-Bonjour, 'Hegel's Concept of Action as Unity of *Poiesis* and *Praxis*,' in Lawrence S. Stepelevich and David Lamb (eds.), *Hegel's Philosophy of Action* (Atlantic Highlands, NJ: Humanities Press, 1983), pp. 19–30; Emil Angehrn, *Freiheit und System bei Hegel*, pp. 174–177; Rüdiger Budner, *Theorie und Praxis – eine nachhegelsche Abstraktion*, p. 24; and Henning Ottmann, 'Arbeit und Praxis bei Hegel,' in *Hegel Jahrbuch*: 28–35. For Ottmann, "Hegel taught a philosophy that still sought to preserve as a unity that which later was for the most part sacrificed at the altar of the new God of work, namely, the unity of labor and praxis" (28).

[30] See my 'Hegel, Adorno, and the Concept of Transcendent Critique,' *Philosophy and Social Criticism 12/4* (1987): 297–328.

[31] *Hegel*, pp. 3–49 and *Sources of the Self*, pp. 368–390. See also Eugene Lunn, *Marxism & Modernism*, pp. 28–32.

[32] *Faith and Knowledge*, trans Walter Cerf and H.S. Harris (SUNY: Albany, 1977), p. 65.

[33] *System of Ethical Life*, trans H.S. Harris and T.M. Knox (Albany, NY: SUNY, 1979), p. 146.

[34] See Jürgen Habermas, *The Philosophical Discourse of Modernity*, pp. 45–50.

[35] The distinction between pragmatic and expressive approaches derives from M.H. Abrams. For an application of the pragmatic interpretation to Hegel, see Laurence Dickey, *Hegel: Religion, Economics, and the Politics of Spirit* (Cambridge: Cambridge University Press, 1987), p. 267f.

[36] *Phenomenology*, p. 19.

[37] Charles Taylor, *Hegel*, p. 18.

[38] *The Difference between Fichte's and Schelling's System of Philosophy*, trans H.S. Harris and Walter Cerf (Albany, NY: SUNY, 1977), p. 91.

[39] *The Difference*, p. 94.

[40] *Realism and Imagination in Ethics.*

[41] *Hegel*, pp. 168–170; 191f.

[42] Lovibond is, to be sure, not the only to emphasize what she sees as the situated dimension of Hegel's concept of rationality. But she is, as far as I know, the only to interpret this dimension as the evidence for the expressivist character of Hegel's thought.

[43] Emil Angehrn, *Freiheit und System bei Hegel*, p. 122n.

[44] This is the expression of Peter Rohs, *Form und Grund* (Bonn: Bouvier, 1969).

[45] *SL*, p. 591.

[46] *The Phenomenology of Mind*, trans J.B. Baillie (New York: Harper and Row, 1967), p. 330. See also *EL* §45.

[47] *EL* §238ff. For one of the few interpreters who accentuated this aspect of Hegel's critique of Kant, see Eugène Fleischmann, 'Hegels Umgestaltung der Kantischen Logik,' *Hegel-Studien* 2 (1965): 181–207.

[48] *PR*, p. 10.

[49] See my 'Hegel, Marx, and the Concept of Immanent Critique,' *Journal of the History of Philosophy 29/2* (April 1991); 253–279.

[50] See Karl-Heinz Ilting, 'The Structure of Hegel's 'Philosophy of Right,' in Z.A. Pelczynski (ed.) *Hegel's Political Philosophy* (Cambridge: Cambridge University Press, 1971), p. 92.

[51] See K.-H. Ilting, 'Liberale Demokratie und 'sittlicher' Staat,' *Archiv für Geschichte der Philosophie 69/1* (1986): 2–21.

[52] *PR* §57. It may be true that Hegel defines the individual will in the essentialist terms that violate Larmore's strictures on substantive political theorizing. Yet Hegel's particular definition of the essence of right – the autonomous, self-determining personality – is presented not as a denial of the liberal individual but a clarification of what is presupposed in the theory of individual rights – that the individual should be treated as ends and never as means (*PR* §36). See Steven B. Smith, 'What is "Right" in Hegel's *Philosophy of Right*,' *American Political Science Review 83*/1 (March 1989): 3–18.

[53] This at least is the view of the mature Hegel. For a recent discussion of an alternate position sketched in the early writings, see Axel Honneth, 'Moralische Entwicklung und sozialer Kampf,' in Honneth *et al.* (eds.), *Zwischenbetrachtungen. Im Prozess der Aufklärung. J. Habermas zum 60 Geburtstag* (Frankfurt: Suhrkamp, 1989), pp. 549–573.

[54] See Ludwig Siep, 'Intersubjektivität, Recht und Staat in Hegels *Grundlinien der Philosophie des Rechts*.' in Dieter Henrich and Rolf-Peter Horstmann (eds.), *Hegels Philosophie des Rechts. Die Theorie der Rechtsformen und ihre Logik* (Stuttgart: Klett-Cotta, 1982), pp. 255–276.

[55] 'Justice as Fairness: Political Not Metaphysical,' *Philosophy and Public Affairs 14*/3 (Summer 1985): 232n.

[56] *Philosophie des Rechts. Die Vorlesungen von 1819/20*, ed. Dieter Henrich (Frankfurt: Suhrkamp, 1983), p. 227.

[57] See my 'Hegel's Concept of Virtue,' *Political Theory 20*/4 (November 1992): 548–583.

[58] Larmore, p. 99.

[59] For Hegel's political organicism, see my 'Hegel's Concept of Virtue,' pp. 565–568. See also Michael Wolff, 'Hegels Staatstheoretischer Organizismus: Zum Begriff und zur Methode der Hegelschen *Staatswissenschaft, Hegel-Studien 19* (1984): 147–77; Ludwig Siep, 'Hegels Theorie der Gewaltenteilung,' *Praktische Philosophie im Deutschen Idealismus* (Frankfurt: Suhrkamp, 1992).

[60] *Hegel*, p. 27.

[61] *Sources of the Self*, p. 384.

[62] *Hegel*, pp. 22–29, 350–355. For a judicious study of Hegel's philosophy of nature – one emphasizing Hegel's greater affinity to Kant than to Baader, Goethe or Schelling, see Gerd Buchdahl, 'Hegel's Philosophy of Nature and the Structure of Science,' in Michael Inwood (ed.), *Hegel* (Oxford: Oxford University Press, 1985), pp. 110–136.

[63] *Philosophy of Nature*, trans A.V. Miller (Oxford: Oxford University Press, 1970) [hereafter *PN*] §248A.

[64] *Vorlesungen über Rechtsphilosophie* 1818-1831, edited by Karl-Heinz Ilting (Stuttgart: Frommann, 1973) 1:239.

[65] As he writes in *The Difference*, Hegel seeks to develop "a philosophy that will recompense nature for the mishandling that it suffered in Kant's and Fichte's system" (83).

[66] *Werke* 11, p. 384.

[67] *Ibid.*

[68] *Sources of the Self*, pp. 385–389.

PETER CAWS

TRANSLATING FEUERBACH

The domain in which writers and thinkers live and move and have their being is stocked with names and texts, and structured by multiple relations of intertextuality and production and power. But it is also structured by affective relations of one sort and another. One sort of affective relation that may be in play is the sort that leads to *Festschriften*, where warmth and esteem and affection in the plain ordinary-language sense link the contributors to the person honored, the person once referred to rather quaintly by a German friend of mine, in another context, as "the jubilee" – that is, the one jubilated or rejoiced in on the festive occasion in question. Marx Wartofsky has always seemed to me a jubilant sort of person, in a nice way, and it's a pleasure to acknowledge my own affective involvement in this common enterprise.

But there is another sort of affective relation, less obvious, that operates in our domain, one whose status as affective may not always be perceived, let alone acknowledged: I mean the relation between one scholar and another that induces the first to 'work on' the second, to take his or her life or intellectual development or textual production as an object of research or translation or commentary. There may of course be quite banal reasons for this, especially in the case of beginners – the assignment, the assistantship, the fellowship – but it's nearly always a bad idea to choose the object of attention on such grounds, as generations of frustrated graduate students can attest. And there may be overtly affective forms of it, as when an expositor or translator is swept away by the force of a creative personality and becomes what is known in less intellectual circles as a 'groupie,' with the result that criticism may be hampered by enthusiasm or translation spoiled by sentiment. Mature scholars are less impressionable and more autonomous, and hence have freer choice. The question is, what guides it? Purely intellectual considerations, perhaps: this work is important, it should be available in my language, it should be better understood. But my thesis for the moment is that there are often other forces at work, that levels of affect may be engaged of which the scholar who does the work may be only partially conscious, if at all.

C.C. Gould and R.S. Cohen (eds.), Artifacts, Representations and Social Practice, 185–193.

The question I'm *not* going to try to answer here is why Wartofsky chose to write on Feuerbach. Some elements of an implicit answer to it are obvious enough, for example the influence of "the late Dr. Karl Marx of Berlin, Paris and London" the stimulation of whose work is so appreciatively noticed in the acknowledgments at the beginning of Wartofsky's *Feuerbach*; others may emerge anyway, but conclusions about that will be left to the reader. (What's not so obvious even in that obvious case is the possible deeper reason why someone called Marx would work on Marx in the first place.) Instead I shall concentrate on another and earlier case of attention to Feuerbach, that of the novelist George Eliot, who as Marian Evans (the only time she used her real name in a published work) translated his *Das Wesen des Christenthums* as *The Essence of Christianity*. I shall be speaking about George Eliot's attachment to things German in general and to Feuerbach in particular, but also by way of preface about the process of translation itself.

There is a certain class of translators – those who work in glass booths at international assemblies, or who accompany important persons on missions abroad – who are given the name of interpreter. The root sense of interpretation is the exchange of price for price: give these professional interpreters utterances in Russian, for example, and they will give you, in principle at least, equivalent utterances in English. I will call this type 1 translation. That is of course just what other translators are supposed to do, with extended texts rather than ephemeral utterances. People who translate books, however, aren't usually called interpreters. The translators of genuinely textual material – as distinguished from the prose of manuals or 'textbooks' which lack the dimensionality of the text[1] – inevitably do engage in interpretation to some degree, though a form of interpretation that 'interpreters' are precisely *not* supposed to practice. It's still giving price for price, but no longer merely on the level of the utterance: not what the speaker said, but what the writer meant (or what the text means). I will call this type 2 translation.

Translation or interpretation? Translation means a purposeful carrying-across; it's one of a series of terms (relation, oblation, etc.) that derive from the supine form (*latum*) of the Latin verb *fero*, to bear or carry. (The root form gives in the equivalent positions transference, reference, offering, etc.) There's a suggestion that what is carried across is in some way unchanged, as opposed to interpretation where we know it's been exchanged. A really good translation gives you something like the experience of the original, without too much interposition on the translator's part.

This is a point I can illustrate anecdotally. Some years ago I had occasion to go to a meeting in Venezuela. Years earlier I'd had pretty functional Spanish, functional enough to lecture in it on the philosophy of science (though there may still be a few former students of mine in Costa Rica who remember the logical trout, the *trucha lógica* that was the best I could come up with in reaching for a description of Hempel's paradox, which if I'd had a better command of the language I would have called a *truco lógico*, a logical trick). A good way of getting a language back under such circumstances is re-immersion, and that meant having a book in Spanish to read on the plane, or before going to sleep, or at breakfast. I chose *Cien años de soledad* of Gabriel García Marquez. By the end of my meeting I was about halfway through, and back in New York I really didn't have the leisure to continue in reading in Spanish, which is a lot slower for me than English, so I picked up Gregory Rabassa's translation. The point of the anecdote is this: when I looked back later it was completely impossible to locate the place at which I'd switched from García Marquez's Spanish to Rabassa's English. This was for me striking evidence of the brilliance of the translation. (Lest the reader think that such transitions are invisible in principle I may say that I had quite the opposite experience with the Moncrieff-Kilmartin translation of Proust.)

The ascent from utterances to texts lends itself to extrapolation. Suppose what is to be carried over is not a particular text but a body of work, a point of view, a philosophical life – that of a French philosopher such as Sartre, for example, or a German philosopher such as Feuerbach, who are to be made accessible to English-speaking readers. Or suppose the carrying-over here is not linguistic or geographical but temporal, the rendering of a nineteenth-century philosopher (perhaps Feuerbach again) in terms of the late twentieth century. How close to the ideal of a non-distorting translation (call it a type 3 translation) might such an enterprise come? This is a question I shall not pursue further on this occasion, but I introduce type 3 to complete a spectrum, as it were, to suggest a continuity from the most limited and literal exchange of word for word to the most comprehensive and speculative interpretative response to an entire corpus. At one point in the translation of *The Essence of Christianity* George Eliot sent to her friend Sara Sophia Hennell a draft of part of the Appendix with the remark: "I have written it very rapidly and have translated it quite literally so you have the *raw* Feuerbach – not any of my cooking."[2] It is the degree of this 'cooking' that marks the place of any particular piece of translation on the spectrum from type 1 to type 3; at the more elaborately worked end one scholar's reading of another may simmer for decades (though it must be said that this does not necessarily mean an improvement in the product – as in the more mundane case the risk of overcooking is always present). Cooking in fact may not be a bad metaphor for at least some kinds of interpretation, taking something and turning it into something else, in a dark enclosed space, and under the influence of fire and the aegis of those gods who, as Heraclitus pointed out to his visitors, are in the kitchen also.

The finished translation will inevitably be 'cooked' to some degree and consequently not exact or definitive. George Eliot is sharply aware of this, but she also sees that the idiosyncrasy of translation is a positively interesting feature of the intellectual scene. She twice tells the story[3] of the seventy translators who gave its name to the Septuagint, the Greek version of the Old Testament prepared under the direction of Ptolemy in the third century B.C.E., who according to legend were shut up separately to work on their translations and yet produced identical texts, but comments that "it would have been a dreary issue for mankind, if the division into nations had ended in such an identity of mental products, even though the standard had been English."[4]

She also sees that type 2 translation is humble work – though doing it certainly need not commit one to humility:

> It is perfectly true that, though geniuses have often undertaken translation, translation does not often demand genius. The power required in the translation varies with the power exhibited in the original work: very modest qualifications will suffice to enable a person to translate a book of ordinary travels, or a slight novel, while a work of reasoning or science can be adequately rendered only by means of what is at present exceptional faculty and exceptional knowledge.[5]

The Feuerbach translation fits the description. One thing that is true of George Eliot is that she did not set her sights low, and if her previous contribution to the art (David Friedrich Strauss's *Life of Jesus*) had been on the whole less challenging, what she went on to do afterwards, and before her career as a novelist finally took off, was certainly more so, since in Weimar she undertook the translation of Spinoza's *Ethics* and *Tractatus Theologico-Politicus*.[6]

And what was George Eliot doing in Weimar? This brings me to Germany and will eventually lead back to Feuerbach. German was one of her four modern languages, and German intellectual life was a leading interest; in her writing for the *Westminster* it was a recurring theme. Thomas Pinney's selection of her journalism,[7] which represents about half of what she wrote in that form, includes seven essays that deal with German subjects. In one of them she summarizes the national character, in its intellectual aspect at least:

> The German mind possesses in a high degree two tendencies which are often represented as opposed to each other: namely largeness of theoretic conception, and thoroughness in the investigation of facts. So undeniable is it that the typical German has these tendencies, that their excess is the very vice he is reproached with by those who know him and don't like him. Your German, it is said, can not write about the drama without going back to the Egyptian mysteries; he sees that everything is related to everything else, and is determined to exhaust you and the subject; his doctrine is all-embracing, and so is his detail. Quite true. No man is less disposed than our German to accept a too slight induction, to let pass an inaccuracy of statement, or to report a conclusion from imperfect observation or experiment; on the other hand, no man is more likely to be contemptuous towards desultory labours which are not *wissenschaftlich* (scientific) – *i.e.* not bound together by a rational doctrine, or conducted in the full sense of a need for such a doctrine. . . . If he is an experimentor, he will be thorough in his experiments; if he is a scholar, he will be thorough in his researches. Accordingly no one in this day really studies any subject without having recourse to German books, or else wishing he knew their language that he might have recourse to them; and the footnotes of every good French or English book that appears, whether in scholarship, history, or natural science, are filled with references to German authors. Without them, historical criticism would have been simply nowhere; take away the Germans, with their patience, their thoroughness, their need for a doctrine which refers all transient and material manifestations to subtler and more permanent causes, and all that we most value in our appreciation of early history would have been wanting to us.[8]

Early history, of course, isn't the only thing that would suffer. George Eliot knew well enough that Germans could be pretentious (to an inquiry about protocol she replied, "I don't know specially what is Feuerbach's title, but in case of personal address, you couldn't be wrong in calling him Doctor. Every German calls everybody Doctor 'until further notice'"),[9] but the "largeness of theoretical conception" to which the foregoing passage refers was a welcome contrast to a corresponding narrowness in the English mind. And among other things Feuerbach had large conceptions about a question that was of great personal concern to her at the time of the translation. As Gordon Haight puts it:

> The powerful appeal the book had for her sprang, not from its bold humanism – '*Homo homini deus est*' – for she had long been familiar with that, but from Feuerbach's daring conception of love. . . . She agreed wholeheartedly with Feuerbach's distinction between 'self-interested love' and 'the true human love', which 'impels the sacrifice of self to another'. Such love is, and must always be, particular and limited, finding its expression in the sexual relation, the frankest recognition of the divine in Nature.[10]

George Eliot's idea of translating Feuerbach probably went back to mid-1851, when an earlier book of his, *Das Wesen der Religion*, had come into her hands. But shortly after that time, and well before agreeing actually to do the work, she had been introduced by Herbert Spencer to George Henry Lewes, a married man with four children, one of whom was not his but Thornton Leigh Hunt's; his wife was pregnant with another child by Hunt (it was born two weeks later) and eventually had two more by him. Lewes, a believer in freedom in such matters, had not disowned his wife and, having in effect condoned her adultery, could not apply for divorce. But he was lonely and disillusioned and he and George Eliot rapidly became inseparable. The question was, was it all right for her to join forces with him publicly in what would have been matrimony if circumstances had been different? And working on Feuerbach was just what was needed. According to him (and I borrow once again from Gordon Haight),

> marriage is . . . almost essential to the full development of human personality. But its "religious consecration is not first conferred by the blessing of the priest." It is "the free bond of love," sacred in itself by the very nature of the union; "for a marriage the bond of which is merely an external restriction, not the voluntary, contented self-restriction of love, in short, a marriage which is not spontaneously concluded spontaneously willed self-sufficing, is not a true marriage, and therefore not a truly moral marriage."
>
> George Eliot declared herself wholly in agreement with Feuerbach's doctrine, which was peculiarly relevant to the broken marriage of Lewes, with whom her intimacy grew as she was translating the book. A few days after it was published in 1854 . . . she went to Germany with him as his wife.[11]

They went to Weimar because Lewes was working on his immensely successful life of Goethe, and stayed in Germany, half in Weimar and half in Berlin, for the better part of a year. By the time they returned to England the arrangement seemed to be generally accepted, though never by correct society. As George Eliot gradually became George

Eliot (rather than "the translator of Strauss's *Life of Jesus*") Lewes was the most enthusiastic supporter of her work, which indeed might never have been undertaken in its well-known form without his belief in her genius. One feels that he would not have minded the title of David Williams's biography: *Mr George Eliot.*[12]

George Eliot had always admired the German way with Biblical scholarship; it reinforced what Leslie Stephen called her "anti-theological tendencies."[13] (The origins of these – in her reaction to early Evangelical intensity, in her reading of Scott, in her friendship with Charles and Caroline Bray – lie outside the scope of this essay.) So she had every reason to find Feuerbach attractive. That the book was personally liberating for her there can be no doubt, and this is just as well, since the translation was done for John Chapman, a somewhat hapless publisher chronically short of money, and she made nothing from it. Chapman had contracted for it, along with a book of her own (which seems never to have been written) to be called *The Idea of a Future Life*, for his *Quarterly Series*, and in the course of a letter to Sara Sophia Hennell contrasting Germany and England once again she suggests that he very nearly backed out:

> Your impression of the book exactly corresponds to its effect in Germany. It is considered *the* book of the age there, but Germany and England are *two* countries. People here are as slow to be set on fire as a *stomach*. Then there are the reviewers, who set up a mound of stupidity, and unconscientiousness between every really new book and the public. Still I think the really wise and only dignified course for Mr. Chapman would be to publish it in his *Series* as he has announced it.[14]

The translation reads like a well-turned English work, and would I am sure meet the Rabassa test if anything translated from German possibly could. Wartofsky suggests only one serious modification, recommending "natural scientist of the spirit" as a translation of *geistiger Naturforscher* rather than George Eliot's "natural philosopher in the domain of mind,"[15] which however in its nineteenth-century context sounds appropriate enough. ('Mind' versus 'spirit' as the translation of *Geist* has been a thorny problem ever since Hegel and I'd rather not get into it here, though I lean towards 'mind' myself.)

I have been positing an affinity with Feuerbach and his thought that makes George Eliot's work on *The Essence of Christianity* something more than just another project undertaken in her pre-novelistic career as an editor, essayist and translator. But these relations work both ways, and it has been suggested that she got more back from the enterprise than comfort and resolve in her alliance with Lewes. "Although Feuerbach entered a plea for scientific rationalism," says Valerie Dodd,

> he tempered the over-cerebral implications of empiricism. In *The Essence of Christianity*, he admitted the element of wonder in vision, praised artists, and suggested that a carefully-defined mode of outwardly-directed feeling could aid rational perception. . . . In his argument that, in philosophy and religion, man unconsciously contemplated himself, Feuerbach entered a plea for the abandonment of philosophy, and a turning towards those disciplines which deliberately scrutinised man and the world, such as sociology and science. . . . Novelists, who traditionally contemplated man and the world, also focused upon those areas of concern. . . . The experience of translating Feuerbach's work may well have been one more factor which helped direct Marian Evans's attention away from philosophical speculation as occupation.[16]

Without it she might perhaps have written *The Idea of a Future Life* after all.

What George Eliot's attention turned *to* from philosophical speculation (and she would, I think, have been a remarkably good philosopher – indeed on the evidence she already was) is well known and forms a distinguished chapter in the history of English literature. People are different, times are different: I will not, in keeping with an earlier resolve, ask how this two-way relationship between the scholar studied and the scholar studying, the interpreted and the interpreter, has played itself out in the case of Feuerbach and Marx Wartofsky. But it can safely be said that it has not involved the abandonment of philosophy.

George Washington University

NOTES

[1] What I am calling genuine text has to be at least two-dimensional, having not merely the linearity of the syntagma but also some form of laterality (by which I mean any non-linear intratextual relation, from repetition and reference to complex forms of narrative or argumentative construction).

[2] Gordon S. Haight, ed., *The George Eliot Letters*, New Haven, Yale UP, 9 vols 1954–1978, vol. II, p. 153.

[3] [George Eliot] ed. Thomas Pinney, *Essays of George Eliot*, London, Routledge and Kegan Paul, 1963, pp. 208, 388.

[4] *Essays*, p. 388.

[5] *Essays*, p. 208.

[6] These translations were not published during George Eliot's lifetime, but the *Ethics* has been published as one of the Salzburg Studies in English Literature. [Benedict de Spinoza,

tr. George Eliot, ed. Thomas Deegan, *Ethics*, Salzburg, Institut für Anglistik und Amerikanistik, Universität Salzburg, 1981.]

[7] See note 3 above.

[8] *Essays*, p. 389.

[9] *Letters*, vol. III, p. 131.

[10] Gordon S. Haight, *George Eliot: a Biography*, New York and Oxford, Oxford University Press, 1968, p. 137.

[11] *Letters*, vol. I, p. xlv.

[12] David Williams, *Mr George Eliot: A Biography of George Henry Lewes*, New York, Franklin Watts [London, Hodder and Stoughton], 1983.

[13] Leslie Stephen, *George Eliot*, London, Macmillan & Co. ('English Men of Letters' series), 1940 [1902], p. 43: "Feuerbach had developed Hegelianism into naturalism, and the translation apparently implies an extension of George Eliot's anti-theological tendencies."

[14] *Letters*, vol. II, p. 137.

[15] Marx W. Wartofsky, *Feuerbach*, Cambridge, Cambridge University Press, 1977, pp. 200, 443.

[16] Valerie A. Dodd, *George Eliot: An Intellectual Life*, New York, St. Martin's Press, 1990, pp. 189–190.

WILLIAM JAMES EARLE

IS THE ENLIGHTENMENT OVER?

> The technical history of the automobile in a free market is a rugged rat race of detail modifications and improvements, many of them irrelevant, but any of the essential ones lethal enough to kill off a manufacturer who misses it by more than a couple of years.
>
> – Rayner Banham

> The world is what it is; men who are nothing, who allow themselves to become nothing, have no place in it.
>
> – V. S. Naipaul

> Any nervous system can be subverted if treated in the right way.
>
> – Richard Dawkins

> Let's count the bodies over again.
>
> – Robert Bly

This year, as I assume you know, is philosophy's 2,578th anniversary, a good time to remember Thales who flourished, by predicting an eclipse, approximately that many years ago. Thales is often called 'the father of philosophy' or 'the father of Western philosophy' – though some feel 'Western philosophy' pleonastic, 'philosophy' being a Greek word, the proper name of what people who trace themselves back to Thales and the other Presocratics are up to and not the name of any other things, better or worse, family-resemblant or not, that other people are up to. It is, of course, worth nothing that philosophy has, in a kind of gender-bent immaculate conception, no trace of a mother. For all we really know about the much mythologized and transfigured Thales, *he* might have been a woman, though tradition as well as doxography count against this.

But I don't begin with Thales to worry about his gender (which my brand of feminism teaches should be made as unimportant as possible). What Thales illustrates is that philosophy, from its very beginnings, cannot be separated from inquiries that belong in more mature form to

C.C. Gould and R.S. Cohen (eds.), Artifacts, Representations and Social Practice, 195–203.

the natural sciences. Thales – I am relying on Jonathan Barnes's scholarship – is supposed to have held the following: (1) The magnet has a soul; (2) The earth rests on water; and (3) Everything is from water.[1] A generalization of (1) is also commonly ascribed to Thales: (4) Everything is full of souls, spirits, even gods; but (4) is most likely an adventious agglomeration that stuck to Thales, *qua* famous wise man, by the same mechanism of fabulous elaboration that gave Pythagoras his golden thigh.

'Soul' in (1) probably means 'principle of motion' and records Thales's insight that things other than officially animate animals move without being moved, and can count as a down-to-earth bit of proto-science; (2) and (3) are straightforwardly thoughts about the natural world. Why did Thales say (1), (2), and (3)? The answer can't be 'because they're true' because at least two aren't; nor is 'because Thales thought they were true' likely to be explanatory. There is no such thing, at least among the sane, as an interest in uttering truths. One useful approach, especially with no access to Thales's psychology, is to consider the *objective competitive advantage* Thales could gain through the public utterance of (1), (2), and (3). People generally have a healthy respect for net advantage even if – as Pierre Bourdieu emphasizes – they can gain something from appearing indifferent to competition and from sincerely believing in *ars gratia artis*, the life of the mind, disinterested research, or some other enabling fiction.[2]

But competition isn't – this is the point of 'objective' – primarily a matter of what people believe, but of what is happening in a competitive field. Competitive fields are not stable, 'natural' objects, but themselves objects of competitive definition: if philosophy can limit its students to philosophy courses, those courses compete against each other; if there is no such rule, all courses may be in competition. *Deep Throat* and *Invitation to a Devout Life* compete if the world contains individuals who might choose between – who are not prevented from choosing between – spending their money going to see the first or buying a paperback of the second.

Fields of competitive intellectual endeavour are perhaps the most difficult to define or isolate. Thales's competitors were the retailers of myths, mainly poets like Homer and Hesiod. This is itself subject to argument. Thales didn't compete by providing more beautiful verses, striking images, affecting or resonant stories. He did something else. Doing something else is often *the* competitive move, but it competes only with what it can be seen as better than, a possible (and worthwhile) substitute or replacement for. How can Thales's bald prose statement be seen this way?

The competitive field, at least as philosophy tells it, where mythèmes and philosophèmes compete, can be defined as 'understanding the world.' For Thales to win against his concurrents in this field, his clientele had to learn to prefer a good question like 'What is everything made out of?' to lots of nice stories. It also had to learn to think about the reasons for believing what it heard in terms of demonstrably relevant evidence, plausible argument, something else, in any case, than whatever it was that allowed intelligent, reasonable Greek, in the appropriate modality of belief and truth ('lethargic' is the word Paul Veyne applies to this modality), to believe their myths.[3]

Genuine innovations, unlike the *ad hoc* tinkering with and marginal differentiation of degenerating research programs, create their clienteles, create their customers. Certainly, Thales was a genuine innovator. He was successful. We should not, however, confuse success with triumph and speak of the triumph of *logos* over *mythos*. There was no such triumph. There never is. None of the Presocratics achieved more than a niche-like market share. Indeed, as Walter Burkert writes in *Greek Religion*, "The picture of the religion as practiced changes hardly at all, in spite of the deeds of all the intellectual heroes."[4]

Writers of the big cultural stories – the 'metanarratives' as they are fashionably described – tend to forget the niche-like, nonhomogeneous, and plain unresolved character of even moderately complicated cultures. The natural world is much the same. Wasps, as eaters of spiders, have an objective interest in spider-survival. Doctors are *as if* inventions of parasites designed to keep their host animals alive for as long as possible. The Athenians liked philosophy and killed Socrates. We are against pornography but spend billions of dollars on it. Disaggregation into rival constituencies dissolves these mild paradoxes. The social world, like the natural world, turns out to be a system of semipermeable, imperfectly defensible, pockets.

Thales and the other Presocratics constitute a pocket of enlightenment. 'Enlightenment' is, of course, an evaluative term and in describing the Presocratics as enlightened, we are saying that they had more of something – a kind of vision, clarity, or rationality – than those they outcompeted. We are favorably contrasting their efforts to think about the world with the poets' beautiful, comfortable stories. And we are supposing they satisfied the Kantian criterion for *Aufklärung*: overcoming 'self-incurred tutelage,' learning to think without dictation from the magisterial tradition.[5]

The idea of a plurality of pocket enlightenments – the Presocratic enlightenment being an early instance – is a useful corrective to the grandiose image of a world-historical, across-the-boards-triumphant enlightenment, misconceptualized by its propagandists and salespeople as the Enlightenment and now going down to defeat under the same misconstrual. Here we join an ongoing discussion about Modernity, the end of Modernity, and the possibility of Postmodernity. This is really a discussion of whether the Enlightenment, variously construed, is over and, if so, what we should next expect.

The first point to make about the discussion itself is that the parties to it are all interested, whether they think of themselves as residual legatees defending the gates or gate-crashing outsiders. Some of those anxious to see the Enlightenment end seek a reenchantment of the world for crypto-religious reasons. Some are simply bored professors suffering from neophilia, humanities department types who can discover nothing save the profit, real and symbolic, yielded by restyling essentially unaltered product lines. On the other side are the pedagogues, temperamental or occupational, who uphold the Enlightenment as a fixed stock of values, standards, and excellences threatened by every student to whom they award a 'D' or 'F. (It goes without saying that I am not turfless either or devoid of defensive responses; analytically, anything true in general is true of me.)

In any case, progress in the discussion of the end of Modernity can be made simply by emphasizing the niche/pocket character of all enlightenments. This is basically a methodological individualist point; whatever enlightenment there is is necessarily sustained in the lives of individual humans. A pocket of enlightenment consists of people more enlightened than others in the population at large. Instead of asking, "Is the Enlightenment over?" we should ask questions like "How many Americans are enlightened?" "What are the patterns of net change?" In asking these questions, we are leaving aside the question of how clumped, or enclaval, the enlightened are, whether they cluster in real gardens or disperse in invisible colleges, whether sonship, sisterhood, mentoriality, and the like are powerful in the transmission of enlightenment. A niche in a market is just an abstract statistical aggregate.

But in saying that some people, whether scattered or significantly clumped, of all the people in a larger population are 'enlightened,' we are, of course, making an evaluative judgment as we did in using the same term to describe the Presocratics. Is this sort of evaluative judgment defensible? After all, the people who think the Enlightenment is over aren't announcing the disappearance of the last enlightened person, but claiming that it is senseless and illegitimate to single out some people as enlightened. Isn't who's enlightened relative, a matter of 'says-you'? Isn't any residual appearance of absolute standards a reflection of the residual power, once a tight monopoly, of the standard setters?

There were and are (whether or not there always will be) clubs of privilege, of unearned self-confidence, whose members attempt universalisation of highly contingent, self-flattering standards and, doing so, generate lots of surplus invidiousness. People who speak upper-class boiler plate ('Marvelous party, Darling!') don't speak English better than those who speak one of the 'check it out/and shit' dialects. They only think they do. On the other hand – I note this sadly, not spitefully – the people who speak of marvelous parties have parties to go to which implies places at their disposal and prior accumulation of capital while the other speakers merely party, a magical/alchemical, largely attitudinal state of affairs which requires nothing more than a park bench, a subway car, or a vacant lot. But let me retreat from these minute socio-linguistic particulars and make a few helpful, and more purely conceptual, remarks. This is mainly a matter of basic intellectual housekeeping.

First, we should distinguish the handful of relativities confused in discussions of relativism. Since 'being enlightened' is a matter of degree, enlightenment judgments are *comparative*. The Presocratics were enlightened relative to other Greeks. I call someone enlightened, unless I am making a historical judgment, who is more enlightened than the average member of my own society. Of course, the person it is most important to be more enlightened than is one's old self. That is the way of progress in wisdom.

Reference to 'my own society' introduces another relativity, though not in the way that panicky relativism watchers may suppose. It is not a matter of the *done* and *not done* relative to people *du même monde* (e.g. "In my country, a gentleman uses a butter knife even when dining alone"). My own society for me, your own society for you, is the operational *context* for competition, success, and failure. This is exactly the same relativity that goes with utility, a thoroughly objective, opinion-independent property of habits, tools, strategies, styles, heuristics, etc. It is hard to imagine a context in which it would not be a good idea to avoid the *post hoc* and Monte Carlo fallacies, not a good idea not to be frantic, or frazzled, or self-deceived. Whereas the capacity to make agreeable small talk or appreciate plotless ballets may be, for all I know, of much less general utility. (It is, of course, *really* useful to possess a property or credential *mistakenly* valued within a social world just as – and for the same reason that – it is objectively profitable to cater to a corrupt taste so long as that taste is common. Every arbitrary, or subjective, preference gives rise to one real and exploitable fact.)

Two further relativisms are worth mentioning in this brief review. The first of these, which might be called 'sentimental relativism,' is just a muddle. This kind of relativism holds that there is competition but no winners and losers. The sentimental relativist tells the losers of every contest that they haven't really lost because the decision of the judges, far from final, is indefensible. Down this road lies the discursive transmutation of handicaps into benefits, ignorance into under-appreciated savvy, deadends into promising lines of research, underdogs into top dogs, and everything being beautiful in its own way. Unfortunately, mere discursive transmutation, even if well-intentioned, does not touch the real bases of disadvantage and failure.

Finally, there is what might be called 'logical relativism.' This is just the premise-relativity of every conclusion. Abstractly, every proof is a proof-on-assumptions. Concretely, I can get you to agree with me, *by arguing with you*, only if you share my premises. This is a trivial truth and fairly obvious too, though philosophy beginners sometimes have to be taught it, as when they ask for a *simpliciter* answer to their question: Can God's existence be proved? Were we to argue the question who should count as enlightened we might find ourselves arguing from unshared premises. There is no guarantee that one of us could think of premises, themselves shared, supporting the unshared premises of our initial arguments. We could, in fact, argue about how often interesting human disagreements eventuate in unshared premises. Probably not all that often. Most argumentation, even when it is not logically defective, is uninventive, incomplete. There are also ways of generating premise-acceptance that do not rely on argument. I find, for example, that smacking people around is a good way of getting them to accept the proposition that pain is intrinsically regrettable. (This is supposed to be a kind of joke.) I think it can be argued that some of these extra-argumentative methods for engineering acceptance are not extra-rational. Anyway, the bottom line is that logical relativism, expressing a necessary truth about arguments, isn't something it makes any sense to start worrying about now. Fretting about logical relativism is exactly like fretting about 2+2 equaling 4.

How do these considerations affect discussion of whether the Enlightenment is over? We should give up the idea of the big Enlightenment and correlative global, especially inexorable, progress. Progress is niche-like, scctorial, molecular. And personal enlightenment, the only kind of enlightenment there is, is always fragile. It does make sense, despite relativist worries, to speak of real progress in personal enlightenment understood comparatively and contextually. The case for enlightenment would have to be multiform: a series of arguments for the comparative benefit, the net advantage, of possessing enlightenment-constitutive properties.

This would give us answers to questions (here phrased I admit as rhetorically as possible) like: Why is lucidity better than mental murkiness? Why is imperturbability better than chronic perturbedness? Why is cognitive patience better than rash judgment Why are minds prone to ask *What do you mean?* and *How do you know?*, other things being equal, more serviceable than minds with no such prompting? Answering these questions, I am in effect thinking about the kind of person I want to be or become, and about the good reasons I think I have for trying to become a person of that kind. People placed similarly will (I expect) have similar reasons for similar plans; about others farther away in real or further away in social space, it is pointless to speculate.

We began with Thales who, as a philosopher, did something like science, given there was no science to outcompete him. What happens to philosophy when science kicks in? Aren't the sentences of philosophers "like these beautiful words [I quote Bruno Latour][6] *Scrabble* players love to compose but which they do not know to place on the board because the shape of the board has been modified by other players"? Actually, those who think the Enlightenment is over can finesse the question. Give up Enlightenment faith in science (in scientific progress, objectivity, and rationality) and you give literary intellectuals, including philosophers, a renewed franchise. I want to suggest that everything is wrong with this response. Science is not something you need to have faith in. It pays its way as it goes by generating a mass of interconnected evidence for its detailed – progressively refined – conclusions.

Although phrases, sometimes rhetorically seductive, are formulated suggesting something wrong with science (for example, Baudrillard's "*le cauchmar de l'objectivité perpétrée par la science* [the nightmare of objectivity perpetrated by science]")[7] and although serious work has been done which shows how complicated scientific rationality must be, there is nothing showing science is a scam. It is not the only kind of knowledge (it doesn't tell me where I left my umbrella, or whether X loves me, or whether Y is a better movie than Z), nor does it self-contained cognitive success translate, not automatically anyway, into a better world. Francis Bacon thought science could be done quickly and would fix up everything. We should give up Baconian optimism in favor of a much more discouraged enlightenment.

Finally, philosophy doesn't need, should in fact be loath to accept, a renewed franchise. It can continue its heterogenous activities of interstitial clarification, conceptual commentary, intellectual cartography. It can go hybrid and impure (Katz and Fodor are good local examples). It can go therapeutic and assemble reminders helpful to minds ensnared in conceptual tangles or obsessed by monoconceptual visions. It can also, as Ermanno Bencivenga suggests, invent new language games – on paper only and so relatively cheaply – which may be useful in future conceptual emergencies.[8] What philosophers cannot do (maybe this is like the deck steward telling the seasick passenger he can't get sick here prompting the reply "That's what you think!") is to compete with science over most of the range of the question "What is there?"

This is an application of a more general principle. Philosophers, like any writers, have to be aware of their own belatedness and shape their contributions relative to what is current in the field. There is no use doing what's been done. There is no use doing what your competitors can do better. Someone may try to impose a definition of the competitive field (e.g. deep thinking about being *versus* superficial and technical management of beings) that makes him the only game in town. This isn't likely to work if there are others around – I would I suppose count as one – making their living making fun of *fiat*-ridden strategies of self-protection, of last ditch attempts at disciplinary autonomy.

The City University of New York
Baruch College and The Graduate Center

NOTES

[1] Jonathan Barnes, *The Presocratic Philosophers*. Vol. 1 (London: Routledge & Kegan Pual, 1979), pp. 5–7.

[2] For example, Pierre Bourdieu, *Esquisse d'une théories de la pratique* (Geneva: Librairie Droz, 1972), p. 234: "... la constitution de domaines relativement autonomes de la pratique s'accompagne en effet d'un processus au terme duquel les intérêts symboliques (souvent décrits comme 'spirituels' ou 'culturels') se contituent contre les intérêts proprement économiques tels qu'ils se définissent sur le terrain des transactions économiques par la tautologie orginaire 'les affaires sont les affaires'; l'intérêt proprement 'culturel' ou 'esthétique', comme intérêt désintéressé, est le produit paradoxal du travail idéologique auquel les écrivains et les artistes, premiers intéressés, ont pris une part importante et au terme duquel les interêts symboliques s'autonomisent en s'opposant aux intérêts materiels, c'est-à-dire en s'annulant symboliquement comme intérêts." Translated by Richard Nice as *Outline of a Theory of Practice* (Cambridge: Cambridge University Press, 1977), p. 177.

[3] Paul Veyne, *Les Grecs ont-ils cru á leurs mythes?* (Paris: Seuil 1983), p. 29; translated by Paula Wissing as *Did the Greeks Believe in their Myths?* (Chicago: University of Chicago Press, 1988), p. 18.

[4] Walter Burkert, *Greek Religion*. Translated by John Raffan (Cambridge MA: Harvard University Press, 1985). p. 305.

[5] Immanuel Kant, 'Beantwortung der Frage: Was ist Aufklärung?' in *Schriften zur Anthropologie, Geschichtsphilosophie, Politik und Pädagogik 1*. Edited by Wilhelm Weischedel (Frankfurt am Main: Suhrkamp, 1977), p. 53: "Aufklärung ist der Ausgang des Menschen aus siener selbst verschuldeten Unmündigkeit. Unmündigkeit ist das Unvermögen, sich seines Verstandes ohne Leitung eines anderen zu bedienen." We might well ask – rhetorically, I hope – whether contemporary foes of enlightenment are fans of *Unmündigkeit*?

[6] Bruno Latour, '*The Prince* for Machines as Well as for Machinations' in *Technology and Social Process*. Edited by Brian Elliott (Edinburgh: Edinburgh University Press, 1988), p. 32: "The huge iron and steel plants of Lorraine are rusting away, no matter how many elements they tied together, because the world they were supposed to hold has changed. They are much like these beautiful words *Scrabble* players love to compose but which they do not know how to place on the board because the shape of the board has been modified by other players."

[7] Jean Baudrillard, *Cool memories 1980–1985* (Paris: Editions Galilée, 1987), p. 130.

[8] Ermanno Bencivenga, *Looser Ends. The Practice of Philosophy* (Minneapolis: University of Minnesota Press, 1989). This is presented in a number of places but see, especially, pp. xiv-xv, 116–117, 142–143.

PAUL FEYERABEND

REALISM

1. HOMER AND THE BIBLE

Almost everybody admits that there are dreams, stones, rainbows, fleas, sunrises, murders, planets – and many other things. For the authors of the Homeric epics these entities were real in the sense that thcy occurred, had distinctive properties and affected their surroundings. They formed a rich pattern of interactions of varying nature and strength. The effects of dreams, for example, easily surpassed those of trees and stones (the dreams of kings might lead to war and multiple murder). There was no grand dichotomy such as the dichotomy real/apparent and events did not conceal or hint at a hidden and perhaps inaccessible world.

Biblical stories, on the other hand, are inscrutable in precisely this sense. Homer, says Erich Auerbach (*Mimesis*, Bern and Munich, 1946, 9 and 15) "is all surface . . . nothing is hidden" while a biblical tale is "enigmatic, . . . dark" (17) and "in need of interpretation" (18). There are "many layers, one arranged above the other" (15) and each situation has "subterranean" components that endanger its "clarity" (22).

However, the *Iliad* contains passages where the Homeric world, too, becomes ambiguous and opaque. An example is *Il.* 9, 225ff.

2. ACHILLES' COMPLAINT

Here Odysseus, acting as an emissary of the Greeks, tries to get Achilles back into the battle against the Trojans. Achilles resists. "Equal fate" he says, "befalls the negligent and the valiant fighter; equal honor goes to the worthless and the virtuous" (318f.). He seems to say that honor and the appearance of honor are two different things.

Interpreted as a stable, well-defined and unambiguous entity the Homeric notion of honor (and the geometric notion that may have corresponded to it) did not allow for such a distinction. Like other epic stereotypes, it was an aggregate containing individual and collective actions and events. Some of the elements of the aggregate were: the

C.C. Gould and R.S. Cohen (eds.), Artifacts, Representations and Social Practice, 205–222.

role (of the individual possessing or lacking honor) in battle, in the assembly, during internal dissension; his place at public ceremonies; the spoils and gifts he received when the battle was finished and, naturally, his behavior on all these occasions. Honor was present when (most of) the elements of the aggregate were present, absent otherwise (cf. *Il.*, 12, 310ff. – Sarpedon's speech).

Achilles sees things differently. He was offended by Agamemnon who had taken his gifts. The offence created a conflict between what Achilles received and what he thought was his due. The Greeks who plead with Achilles, Odysseus among them, agree that there is a conflict; their suggestions illustrate its customary resolution: the gifts were returned untouched, more gifts were promised, honor is restored (519, 526, 602f.: "With gifts promised go forth – the Achaeans will honor you as they would an immortal"). So far we are squarely within tradition as expressed in the first part of Sarpedon's speech and reconstructed by our scholars. Achilles is not reassured. Extending the conflict beyond its customary resolution he perceives a lasting clash between honor and its accepted manifestations: honor and the actions that establish and/or acknowledge its presence *always* diverge.

At first sight it seems that we are dealing with a simple case of generalization. Achilles notices a *particular* discrepancy and declares *all* instances of an attribution (or a denial) of honor to be affected by it. There are various difficulties with such an account. To start with, Achilles does not just generalise, he aims at a new *concept*; secondly, the concept is not a summary of familiar actions and events as are many archaic concepts, it *clashes* with any such summary. Some scholars have added that it also clashes with the *rules of the language* Achilles is using. Achilles, says A. Parry "has no language to express his disillusionment. Yet he expresses it, and in a remarkable way. He does it by misusing the language he disposes of. He asks questions that cannot be answered and makes demands that cannot be met".[1]

But if Achilles "has no language to express his disillusionment", if he "can in no sense, including that of language (unlike, say Hamlet) leave the society which has become alien to him", then what does this so-called disillusionment amount to? We have no description of it, not from Achilles who, according to Parry, lacks the terms, not from his visitors who are simply confused, and not even from the poet who, after all, provided the evidence for Parry's account. According to Parry and others who believe in the power of linguistic forms (and corresponding forms of life) they are all victims of a 'conspiracy' from which there is no escape.[2] Yet Achilles acts as if there were no conspiracy and Parry writes as if he knew the (apparently inexpressible) thought behind Achilles' nonsense. Clearly, there is a mistake somewhere in the argument – but what is it?

Achilles makes statements which sound strange to his visitors and to some 20th century commentators. There is nothing unusual about this situation. People often say strange things. But one can ask them and then the matter will either be clarified, or shelved and forgotten until someone else starts asking all over again. Outrage, instant dismissal, lack of interest are other possible reactions.

Nobody can ask Achilles, and his visitors do not pursue the matter. They are puzzled, they argue for a while but they soon give up. This, too, is not unusual. The world is full of garbled messages, unfinished letters, damaged records. Again there are many ways of dealing with the problem, each one having advantages and drawbacks.

The case ceases to be a problem in this familiar, annoying but manageable sense and becomes profound and paradoxical when it is lifted out of its natural habitat and inserted into a model, a sketch, or a theory of it. A theory (model, sketch) of a historical process says both too little and too much. It says too little because it starts from a fragment of what it wants to represent. But it also says too much because the fragment is not just described, it is subdivided into essence and accident, the essence is generalised and used to judge the rest. The distinction between linguistic use and misuse and the associated idea of a 'conspiracy' both arise in this way. But this means that the paradoxes I just mentioned – Achilles crosses boundaries of sense as if they did not exist – don't occur in history. They occur when historical events are first 'reconstructed' and then evaluated on the basis of unhistorical procedures; they are the mirror image of the ideology and the actions of our scholars.

But although there is no paradox, there is still a conflict: Achilles and his visitors have different ideas about the circumstances that establish or restore honor. The ideas of the visitors are well-known and traditional – this, at least, is what the scholars tell us; Achilles' idea is *new*. Not only that; Achilles' new idea seems to *explain* phenomena which, because of his anger and despair, affect his more than others. The new idea, therefore, has a *structure*, namely, the structure of the phenomena to be explained, 'overlaid' by a certain way of relating them to the rest of the world.[3] The structure does not conform to the linguistic habits some scholars, Parry among them, claim to have inferred from the evidence. But stable linguistic habits – if they were indeed stable – are not automatically boundaries of sense – long lasting empirical beliefs can be and often have been corrected by new discoveries. True, Achilles' visitors are disturbed, they do not agree, perhaps they do not even understand what Achilles is up to. But again we have to ask what this attitude amounts to. Does it show that Achilles talked nonsense, or does it only show that what Achilles said was meaningful, though surprising and unfamiliar? Besides, who is going to judge the matter? Achilles, who does not seem to perceive any difficulty, his visitors who don't say enough to enable us to distinguish between the two possibilities, or a 20th century scholar who uses his own parochial distinctions? Considering how much leeway there is, it seems prudent not at once to impose a theoretical frame but to try getting more information from the story itself.

To start with, Achilles' reply does not come out of the blue. It concerns a situation that lies squarely within the common sense of the time – the conflict between custom and Agamemnon's actions. People had violated customs before and had been duly reprimanded. Such a conflict can either be increased, or it can be alleviated. And it can be increased either by further actions, e.g., by a further violation, or by seeing the violation in a different light, for example by regarding it as an instance of a general tendency. Achilles, being driven by his anger, turns the conflict from a particular disturbance, or a series of disturbances, into a cosmic rift.

I already described the two steps a logician can distinguish in his action. The first step leads from Achilles' personal situation to the situation of many – what can happen to me can and does happen to others. It may sound exaggerated and overly dramatic, but it still conforms to what I called, in a simplifying way, the archaic notion of honor. It is perfectly meaningful. The second step turns the implausible accidental generality so obtained into what one might call an absolute difference. It suggests that honor fails not only because of human weakness, but because of the nature of things: real honor always clashes with its alleged manifestations. We know the motor that makes Achilles speak in this way; it is his disappointment, his anger. What we need to identify are the influences that give his utterances structure and the analogies, within archaic Greek, that make the structures seem familiar and in this way give meaning to the second step as well.

Homeric thought (and, we may assume, the common sense of the time) was not entirely unprepared for grand subdivisions. Divine knowledge and human knowledge, divine power and human power, human intention and human speech (an example mentioned by Achilles himself: 312f.) were opposed to each other in ways that resemble the distinction Achilles wants to introduce. This is the analogy. Achilles uses it; for example, he connects his idea with the judgement of Zeus (607ff). But while the judgement of Zeus has a certain amount of arbitrariness and while divine decisions may vary from case to case, the division adumbrated by Achilles has the clarity and power of an objective and person-(god-) independent law. Even this feature has an analogy in the epic: the wilfulness of the gods is never absolute: it is restricted by *moira*, an "irrefrangable order . . . which . . . exists independently of them" and by norms which they may violate but which can be used to criticize and to judge their behavior.[4] Thus Achilles modifies an existing belief by subsuming it under a more general belief that so far had been separated from the first. What is the source that not only favors the subsumption (which to later generations seems entirely natural) but maintains the universal validity of the subsuming principle? Or to express it in terms of a later point of view: what prompts Achilles to speak in a way that suggests an as yet unheard of separation between the social aspect of a property (such as honor) and an independent 'real' nature? Do we have to resort to miracles such as 'creativity', or are there better ways of explaining the transition?

Had Achilles or the poet who wrote his words lived in the 7th or 6th centuries B.C. I could have answered: the source is closely connected with certain social developments. In politics, abstract groups had replaced neighborhoods (and the concrete relationships they embodied) as the units of political action (Cleisthenes); in economics, money had replaced barter with its attention to context and detail; the relations between military leaders and their soldiers became increasingly impersonal; local gods merged in the course of travel, tribal and cultural idiosyncracies were evened out by trade, politics and other types of international exchange, important parts of life became bland and colorless and terms tied to specifics accordingly lost in content, or in importance, or they simply disappeared. Without any help from philosophers, politicians, religious leaders, "words . . . bec[a]me impoverished in content, they . . . bec[a]me one sided . . . empty formulae".[5] One might add that the process was constituted by individual actions, even by conscious decisions which however did not have it as its aim and that it was to that extent unconscious. But it modified the 'conspiracy' of Homeric Greek by means of a second 'conspiracy', actualised some ambiguities of Homeric thought and restructured its content accordingly. Looking at matters 'from the outside' we notice that one area of behavior ('Homeric common sense') is being 'overlaid' by another (the newly emerging structures just described). Seen 'from the inside' we have a discovery: a new important feature of the world is revealed. It is not revealed 'as such', but only with respect to the problem that sensitizes Achilles: the existence of personal qualities detached from the efforts of an individual or the reactions of his peers.

But Achilles did not live in the 7th or 6th centuries. He spoke at a time when the developments I enumerated above were in their infancy. They had started; they had not yet produced the results described. Thus Achilles' observations cannot be explained by referring to these results. But they adumbrate them, which means that Achilles' speech also contains an element of invention. It is still discovery, for it reveals the outlines of a slowly rising structure. It is invention because it contributes to the raising of the structure. It deals with 'objective' facts because it is substantiated by a process that is nourished from many sources; it is 'subjective' because it is part of the process, not independent confirmation of it. Moreover (this again said in favor of 'subjectivity') – the move towards increasing abstractness and the separation of reality and appearance it contained was not the only development that occurred. As becomes clear from funeral inscriptions, passages of comedy, sophistic debates, medical treatises and especially from the Platonic dialogues,[6] the view that things, ideas, actions, processes are aggregates of (relatively independent) parts and that giving an account of them means giving a detailed but open list remained popular right into the classical age of Greece: 'geometric thought' was a seed which grew into many different plants. It is clear that the customary dichotomies (subjective/objective; discovery/invention; etc.) are much too crude to describe complex processes of this kind.

Rather, we have to say that the structures that preceded the 'rise of rationalism' were 'open' in the sense that they could be modified without being destroyed. They contained the paths Achilles was about to enter, though in a vague and infinished way. They were also 'closed', for it needed a stimulus to reveal their ambiguities and alternative structures to reset them. Without the stimulus, words, phrases, rules, patterns of behavior would have seemed clear and unproblematic (clarity is the result of routine, not of special insight); without an (existing, or slowly developing) alternative structure, the possibilities implicit in Achilles' language would have lacked in definition. Thus entities such as 'geometric perception' or 'the archaic form of life' are to a certain extent chimeras; they seem clear when indulged in without much thought; they dissolve when approached from a new direction. The expression 'dissolves', too, is somewhat fictitious – the transition often remains unnoticed and amazes or annoys only a thinker who looks at the process from the safe distance of a library, or a bookstudded office. As always we must be careful not to interpret fault lines in our theories (recent example from physics: the 'fault line' that separates classical terms and quantum terms) as fault lines in the world (molecules do not consist of classical parts and, separated from them, quantum parts). Ambiguity however turns out to be an essential companion of change.

But this means that Homer is 'clear and superficial' only under special and rather restricted circumstances. His stories become 'enigmatic and dark' and even misleading when viewed with a jaundiced eye and approached by way of changing and often unnoticed social tendencies. Was the change effected by Achilles a change for the better? And did Achilles have a choice?

The answer to the second question is simply – no. Achilles participated in a development he was not aware of and could not control. The answer to the first question depends on who is being asked. Achilles himself was not pleased; other participants ran into difficulties. Modern scholars (economic historians, political theoreticians, philosophers) have spoken of progress. Be that as it may – considering the idiosyncratic character of Greek political life there was no way of inhibiting a process that affected all its parts.

3. PHILOSOPHY

Achilles, I said, experienced changes he did not create and enlarged dichotomies he had not conceived. Being swept along by events he lacked the means of opposing the new and, to him, depressing vision of honor and justice. It therefore makes sense to say that his ideas and his actions were 'socially conditioned'.

The explanation loses much of its power when we turn to the early

Greek philosophers. They, too, depended on circumstances they neither noticed nor understood. But these included an element that considerably reduced the impact of ideas, visions and even of the circumstances themselves.

> Far more than their counterparts in most other ancient civilizations, Greek doctors, philosophers, sophists, even mathematicians, were alike faced with an openly competitive situation of great intensity. While the modalities of their rivalries varied, in each the premium – to a greater or lesser degree – was on [novelty], skills of self-justification and self-advertisement and this had far reaching consequences for the way they practiced their investigations. (G.E.R. Lloyd, *The Revolutions of Wisdom*, University of California Press, Berkeley and Los Angeles 1987, 98f.)

Those participating in the competition may have felt that only their ideas deserved attention and that common sense and the notions of their rivals were empty talk. However, to get the attention and to stop the talk they had to consider others, which means that the circumstances of their lives put some distance between their intuitions and their ideas.

"Shepherds of the wilderness" the muses address Hesiod (*Theogony* 26ff), "wretched shameful things, mere bellies, we know how to speak many false things as though they were true; but also know, when we want to, how to utter true things': there is no longer a straight path from impression to truth. Parmenides is guided by a strong vision which occasionally makes him speak with contempt of those who do not share it. Yet he explains, in a series of elementary steps why his vision should be adopted and regarded as a basic truth. The steps themselves are novel to a certain extent;[7] they are also plausible which means that truth now depends on the outcome of a battle between plausibilities. Pythagoras made science part of a comprehensive political-religious movement. Yet, having arrived in Kroton he addressed separately the men, the women and the children of the city and seems to have made a great impression – he knew how to talk to non-philosophers (Pompeius Trogus in Iustinus XX, 4). Plato considered various ways of presenting knowledge and, after a detailed analysis of the possibilities of each used the dialogue, though in a restricted and clearly circumscribed way. Examining a particular topic, Aristotle collected earlier views which then functioned as (modifiable) boundary conditions of his research. The Sophists stayed close to common sense and defended (certain reconstructions of) it against the more abstract systems of their predecessors. "One may assume" write D. E. Gershenson and D. A. Greeenberg ('The Physics of the Eleatics', *The Natural Philosopher, 3* (New York, Blaisdell Publishing Co., 1964) 103, my emphasis)

> that the attraction the Eleatic school held for large numbers of its contemporaries and the notoriety it subsequently achieved were due to the reputation its founders had as masters of dialectic in an age which had recently become fascinated by systematic reasoning. *Logic was the rage of the day*; all over, in the marketplaces, in the streets, in private homes and in public buildings, at all times, sometimes all through the night, people engaged in dialectical disputations and flocked to hear the acknowledged masters of logical argument display their art.

What fascinated the Athenians – for Athens was now the stage on which the different approaches collided – was the multiplicity of the ideas proposed, the strange nature of some of them and the possibility of proving a thesis as well as its opposite (cf. Euthydemus 257 dff). There were many offerings; a person in search of knowledge had to choose not only between results but between methods of argumentation as well. I shall illustrate the problem by discussing some aspects of Parmenides' philosophy.

4. PARMENIDES, ATOMISM, COMMON SENSE AND MEDICINE

The Ionian cosmologists postulated a unity behind diversity and tried to reduce observable events to it. They began with plausible physical hypotheses. Thales, for example, identified the unity with water. He may have argued for his choice; for example, he may have said that water occurs in a solid, a fluid and an airy form and that it is necessary for life. And we must not forget that the Greeks who lived around the Mediterranean "like frogs around a pond" had first-hand experience of the immensity and the many functions of water (mist, air, clouds, etc.). Thales' successors suggested other substances, again with plausible reasons. Then Parmenides pointed out that the physical approach was rather shortsighted. According to Parmenides the most basic entity underlying everything there is, including any hypothetical substance one might propose, is Being. Speaking with hindsight, we can say that this was a shrewd suggestion: Being is the place where logic and existence meet. Every statement involving the word 'is' is also a statement about the essence of the world.

What can we say about Being? That it is (*estin*) and that not-Being is not. What happens on the basic level? Nothing; the only alternative of Being is not-Being, not-Being does not exist, hence there is no change. How is the basic level structured? It is full, continuous, without subdivisions. Any subdivision would be between Being and something else, the only something else on the basic level is again not-Being, not-Being does not exist, hence there are no subdivisions.

But is it not true that we traditionally assume and personally experience change and difference? Yes, we do. Which shows, according to Parmenides, that neither tradition (*ethos poly'peiron* – "habit, born of much experience", Diels-Kranz Fragment 87, 3) nor experience ("the aimless eye, the echoing ear . . ." 87, 5) are reliable guides to knowledge. This was the first, the clearest and the most radical separation of domains which later were called reality and appearance and, with it, the first and most radical defence of a realist position. It was also the first theory of knowledge.

Those who are ready to make fun of Parmenides should consider that large parts of modern science are bowdlerised versions of his result. His premise, *estin* – Being is – is the first explicit conservation law; it states the conservation of Being. Used in the form that nothing comes from nothing (which found its way into poetry: *Lear* i, 1, 90) or, in Latin, *ex nihilo ni[hi]l fit*, it suggested more specific conservation laws such as the conservation of matter (Lavoisier) and the conservation of energy (R. Mayer).[8] The more philosophically inclined practitioners of 19th century physics posited a 'real' world without colors, smells, etc., and a minimum of change; all that happens is that certain configurations move reversibly from one moment to another. In a relativistic world even these events are laid out in advance. Here the world

> simply *is*, it does not *happen*. Only to the gaze of my consciousness, crawling upward along the lifeline of my body, does a section of this world come to life as a fleeting image in space which continuously changes in time (H. Weyl, *Philosophy of Mathematics and Natural Science*, Princeton University Press, Princeton, 1949, 116).

"For us, who are convinced physicists" wrote Einstein

> the distinction between past, present and future has no other meaning than that of an illusion, though a tenacious one. *Correspondance avec Michele Besso* (Paris 1979), 312; cf. also 292).

Irreversibility, accordingly, was ascribed to the observer, not to nature herself. And so on. None of the scientists who supported the dichotomy could offer arguments that were as simple, clear and compelling as those of Parmenides.

And yet Leucippus, who is traditionally associated either with the Eleatics in general or with Parmenides' disciple Zeno,

> thought he had a theory which harmonized with sense perception and would not abolish either coming-to-be and passing-away or motion and the multiplicity of things. He made these concessions to the facts of perception. On the other hand he conceded to the Monists that there could be no motion without a void. The result is a theory which he states as follows: "The void is a 'not-being', and no part of 'what is' is a 'not being'; for what 'is' in the strict sense of the term is an absolute *plenum*. This *plenum*, however, is not 'one'; on the contrary, it is a 'many', infinite in number and invisible owing to the minuteness of their bulk . . ." (Aristotle, *De gen. et. corr.*, 325a23ff, Ross trans.).

In short: Being is many and moves in non-Being. Note the nature of the argument: Leucippus does not try to refute Parmenides by using the 'fact' of motion. Parmenides had been aware of the 'fact' and had declared it to be illusory. Moreover, he had not simply asserted the illusory character of motion, he had presented proofs. He had transcended sense impression on the grounds "that 'one ought to follow the argument'" (Aristotle, *De gen. et corr.*, 325a12f). Leucippus, in contrast, decided to follow perception; one might say that he and those who thought in a similar manner (Democritus, Empedocles, Anaxagoras) wanted to bring physics closer to common sense.

A general principle that supports the move was formulated by Aristotle. Commenting on Platonists who justified the virtues by reference to a supreme Good, he wrote (*EN* 1096b33ff, my emphasis):

> Even if there existed a Good that is one and can be predicated generally or that exists separately and in and for itself, it would be clear that such a Good can neither be produced nor acquired by human beings. *However it is just such a good that we are looking for* . . . One cannot see what use a weaver or a carpenter will have for his own profession from knowing the Good itself or how somebody will become a better physician or a better general once 'he has had a look at the idea of the Good' [apparently an ironical quotation of a formula much used in the Platonic school]. It seems that the physician does not try to find health in itself, but the health of human beings or perhaps even the health of an individual person. For he heals the individual.

Thus we can say that at the time in question (5th to 4th century B.C.) there existed at least three different ways of establishing what is real: one could 'follow the argument'; one could 'follow experience'; and one could choose what played an important role in the kind of life one wanted to lead. Correspondingly there existed *three notions of reality* which differed not so much because research had as yet failed to eliminate falsehoods but because there were different ideas as to what constituted research.

Following his argument Parmenides established a reality that was 'objective' in the sense that it was untouched by human idiosyncracy. Following his different approach Aristotle introduced a reality that depended on the nature, the achievements and, especially, on the *interests* of humans. Leucippus, Democritus and others had an intermediate position; they moved towards common sense but stopped early on the way. Still, their results clashed with established subjects such as medicine.

Empedocles for example used four elements, the Hot, the Cold, the Moist, the Dry. Change is defined as a mixing and unmixing of the elements, health as their balance, illness as their imbalance. Considering these definitions the author of *Ancient Medicine* who seems to have been a practicing physician wrote as follows (chapter 15):

> I am at a loss to understand how those who maintain the other [more theoretical] view and abandon the old method [of direct inspection] in order to rest the *techne* on a postulate [i.e. who introduce abstract principles such as the elements of Empedokles] treat their patients on the lines of this postulate. For they have not discovered, I think, an absolute cold and hot, dry and moist that participates in no other form. On the contrary, they have at their disposal the same foods and the same drinks we all use, and to the one they add the attribute of being hot, to another, cold, to another, dry, to another, moist, since it would be futile to order patients to take something hot, as he would at once ask 'what hot thing?' So they must either talk nonsense [i.e. speak in terms of their theories], or have recourse to one of the known substances [i.e. add their descriptions in an *ad hoc* manner to common practice].

Aristotle himself produced 'intermediate' (see above) subjects and made them a measure of existence. "*We physicists*' he wrote (*Physics* 185a12ff)

> must take for granted that the things that exist by nature are either some or all of them in motion . . . Besides, no man of science is bound to solve all the difficulties that may be raised but only [those that arise from the mistaken use of] the principles of [his] science.

Adding the 'intermediate' notions we obtain not just three conceptions of reality, but an indefinite number. It seems that the 'problem of reality' has many solutions.

5. MODERN SCIENCE

"That is quite understandable" the modern reader will reply. "What you are describing is a period before the rise of modern science. But modern science is (1) based on a uniform approach, has (2) led to a coherent body of results which (3) force us to make science not just *a* measure, but *the* measure of reality." Neither (1), nor (2) nor (3) is correct.

As I have argued (in a talk to be published in the Proceedings of the Erasmus Ascension Symposium of April 1992), scientists from different areas use different procedures and construct their theories in different ways; in other words – they, too, have different conceptions of reality. However, they not only speculate; they also test their conceptions and they often succeed: the different conceptions of reality that occur in the sciences have empirical backing. *This is a historical fact, not a philosophical position* and it can be supported by a closer look at scientific practice. Here we find scientists (Luria in molecular biology, Heber Curtis, Victor Ambarzumian, Halton Arp and Margaret Geller in astrophysics and cosmology, L. Prandtl in hydrodynamics, etc.) who want to tie research to events permitting "strong inferences", "predictions that will be strongly supported and sharply rejected by a clear-cut experimental step" (S. E. Luria, *A Slot Machine, a Broken Test Tube*, New York, 1985, 115) and who show a considerable "lack of enthusiasm in the 'big problems' of the Universe or of the early earth or in the concentration of carbon dioxide in the upper atmosphere", all subjects that are "loaded with weak inferences" (Luria, 119). In a way these scientists are continuing the Aristotelian approach which demands close contact with experience and objects to following a plausible idea to the bitter end.

However this was precisely the procedure adopted by Einstein (Brownian motion, general relativity); by the researchers in celestial mechanics between Newton and Poincaré (stability of the planetary system); by the proponents of the atomic theory in antiquity and later, down to the 19th century; by Heisenberg during the initial stages of matrix mechanics (when it seemed to clash with the existence of well-defined particle tracks) and by almost all cosmologists. "Is it not strange" asks Einstein (letter to Max Born, quoted from *The Born-Einstein Letters*, New York, 1971, 192)

> that human beings are normally deaf to the strongest argument while they are always inclined to over-estimating measuring accuracies?

– but just such an "overestimating of measuring accuracies" is the rule in epidemiology, demography, genetics, spectroscopy and in other subjects.

I repeat that all the subjects I just mentioned have been successful, thus confirming the notions of reality implicit in their theories. Even outlandish conjectures that ran counter to physical common sense were confirmed. An early example is Maxwell's calculation of the viscosity of gases. For Maxwell this was an exercise in theoretical mechanics, an extension of his work on the rings of Saturn. Neither he nor his contemporaries believed the result – that viscosity remains constant over a wide range of density – and there was contrary evidence. Yet more precise measurements turned the apparent failure into a striking success. It pays to "follow the argument".[9]

This is true even for large scale enterprises such as the entire history of modern science. Surrounded by comets, new stars, plagues, strange geological shapes, unknown illnesses that would come out of the blue, linger for a while and disappear again, comets which behaved in exactly the same manner, irrational wars, biological malformations, meteors, oddities of weather, the leaders of Western science asserted the "inexorable and immutable" character of the basic laws of nature (Galileo to Castelli, letter of Dec. 21, 1613). Early Chinese thinkers had taken the empirical variety at face value. They had favored diversification and had collected anomalies instead of trying to explain them away. Aristotelians had emphazised the local character of regularities and insisted on a classification by multiple essences and corresponding accidents. Scientists of an equally empirical bent, Tycho Brahe among them, regarded some anomalous events of their time as miracles or, as we might say today, they took cosmic idiosyncracies seriously; others, like Kepler, ascribed them to the subjective reactions of the telluric soul while the great Newton, for empirical as well as for theological reasons, saw in them the finger of God (*Opticks*, query 31).

Empiricism and early versions of instrumentalism were supported by leading theologians. St. Thomas (*Summa Theol.*, question 8, article 4) had assumed that "inexorable and immutable" universals existed side by side with God, only in a different manner. Duns Scotus and Ockham objected that all we have are the results of God's acts of will. We do not understand why these acts occurred, we do not know what acts will occur in the future, we can only observe their effects, connect them and hope for the best.[10] It needed a tremendous act of faith to continue "following the argument" and not only to *assume* regularities where no regularities had yet been found, but *to work towards them* for centuries, and in the face of numerous failures.

The result that emerged in the 19th century was not a coherent science but a collection of heterogeneous subjects (optics, acoustics, hydrodynamics, elasticity, electricity, heat, etc., in physics; physiology, anatomy and practical knowledge, in medicine; morphology, evolutionism, etc., in biology – and so on). Some of the subjects (hydrodynamics, for example) had only the most tenuous relation to experiment, others were crudely empirical. There were surprising unifications such as Maxwell's unification of magnetism, optics and electricity, but they soon created problems (problem of the reference system) while the allegedly most basic subject of all, point mechanics, ran into conceptual difficulties of the most serious kind (cf. Boltzmann's introduction to his *Vorlesungen ueber die Principe der Mechanik*, Leipzig 1897). Johann Theodore Merz, *A History of European Thought in the 19th Century* (first published 1904–1912, republished by Dover Publications, New York) has described the situation in a precise and detailed manner.

In the 20th century the overall situation is roughly the same. One must not be misled by the fact that scientists, when asked, produce a more or less uniform response. For the ways in which they think and proceed in their own domains (elasticity, ethology, the chemistry of substances, rheology, linguistic, etc., etc. – elementary particle physics is not the only science!) are as varied as they were before. Thus all we can offer today is a *list*, similar to the list given by Merz, enumerating the achievements and the drawbacks of various approaches and we can identify 'science' with this complex and somewhat scattered war on many fronts. Alternatively we can put one view on top and *subordinate* the others to it, either by pseudo-derivations (most so-called 'reductions' dissolve when one looks at them more carefully), or by declaring them to be meaningless. We can also present a *paste job* where each particular view and the results it has achieved is smoothly connected with the rest thus producing the impression of a single and coherent scientific 'reality'. I prefer to think that

> [t]here is no simple 'scientific' map of reality – or if there were, it would be much too complicated and unwieldy to be grasped or used by anyone. But there are many different maps of reality, from a variety of scientific viewpoints (John Ziman, *Teaching and Learning about Science and Society*, Cambridge University Press, 1980, 19)

or, putting reality where the achievements are, there are different kinds of reality defined by different modes of successful research which in turn are backed, more or less consciously, by different ideas about the relation between humans and their surroundings: assumptions (1) and (2) are false. And as regards (3) it suffices to point out that practical results, *taken by themselves*, never forced anyone to accept one view of reality over another. The artisans of ancient Greece, who extracted metals from the ore, prepared them for various purposes, manufactured weapons, designed jewelry, built houses, bridges, ships, theatres, made important contributions to ancient Culture. They had detailed information about materials of the most varied kind. Yet the philosophers did not start from this information, but from some very abstract principles.[11] In book vii of his *Republic*, Plato enumerates the subjects that might be useful for the future leaders of the state: music, arithmetic, geometry, astronomy. Music is useful because it furthers harmony and grace, arithmetic aids the general in distributing his troops, geometry helps him in demarcating a military camp and the barracks within, while astronomy is needed for orientation and the calendar. Plato adds that the first three subjects have also a theoretical side where numbers, lines and sounds are related not to material things, but to each other. According to Plato the resulting structures form an unchanging real world whose perception creates knowledge and stabilises the mind.

This distinction between *abstract theory* which provides understanding and *practical subjects* which are never entirely transparent played an important role in the rise of modern science. It replaced the rich *practical information about materials* by *theories of matter* which were abstract and poor in content,[12] but defined reality and guided research for generations to come: practical results, taken by themselves, do not enforce any particular notion of reality. Note, incidentally, that abstract theory triumphed in the end which shows that an unreflected preference for practical knowledge would have been most impractical.

6. A TASK FOR PHILOSOPHERS

Now if science is indeed a collection of different approaches, some successful, others wildly speculative, then there is no reason why I should disregard what happens outside of it. Many traditions and cultures, some of them wildly 'unscientific' (they address divinities, consult oracles, conduct 'meaningless' rites to improve mind and body) succeed in the sense that they enable their members to live a moderately rich and fulfilling life. Using this extended criterion of success I conclude that non-scientific notions, too, receive a response from Nature, that Nature is more complex than a belief in the uniformity and unique excellence of science would suggest, and that an interesting task for a writer capable of looking beyond the limits of a particular school would be to consider some of its properties. I myself have started from what Pseudo-Dionysius Areopagita said about the names of God. God, he said, is ineffable. But depending on our approach God may respond in a variety of comprehensible ways. God is not identical with any one of these ways and it would be a mistake to identify Him (Her, It) with, say, Nature as conceived by modern cosmology (there are also personal manifestations of divinity). Moreover, describing a response and not Being itself, all knowledge about the world now becomes ambiguous and transparent. It points beyond itself to other types of knowledge and, together with them, to an unknown and forever unknowable Basic Reality. Thus the literary forms used by the composers of the Bible seem far better adapted to our situation than the more lucid but basically superficial stories that have replaced them.

Zürich

NOTES

[1] 'The Language of Achilles', *Transactions and Proceedings of the American Philosophical Association, 87* (1956), 6. The quotation in the next paragraph is from page 7 of the same essay.

[2] "[A]ny language" writes M. Baxandall, *Giotto and the Orators*, Oxford, 1971, 44, "not only humanist Latin [the language Baxandall examines in his discussion of how Giotto's work was perceived by the humanists] is a conspiracy against experience in the sense of being a collective attempt to simplify and arrange experience into manageable parcels."

[3] "To exercise a language regularly on some area of experience or activity, however odd one's motives may be, overlays the field after a time with a certain structure." Baxandall, *op. cit.*, 47.

[4] Walter F. Otto, *The Homeric Gods* (New York, Pantheon, 1954), 276. Cf. also Burkert, *Griechische Religion der Archaischen und Klassischen Epoche* Walter Kohlhammer (Stuttgart 1977), 205f, and F. M. Cornford, *From Religion to Philosophy* (Harper Torchbooks, 1965), 16. Reference to norms occurs implicitly in the *Iliad*, e.g. in *Il.* 24, 33ff, explicitly in tragedy, for example in Sophocles, *Antigone*, 456.

[5] Kurt von Fritz, *Philosophie und Sprachlicher Ausdruck bei Demokrit, Platon und Aristoteles* (Neudruck, Darmstadt, 1966), 11.
[6] Details and literature in my essay *Farewell to Reason* (London, 1987), Chapter 3.
[7] Parmenides' argument employs the form 'If . . . then —'. Such forms were familiar from Near-Eastern law. The Sumeric code of Ur-Nammu (before 2000 B.C.) used the more complicated clause 'if . . . , provided that ///, then —' while simple conditionals occur in the Eshnumma Code of 1800 B.C., the Code of Hammurabi, in *Exodus* 21–23, in the non-apodictic laws of *Deuteronomy* as well as in early Greek law (Gortyn, Draco). Application demanded familiarity with the *modus ponens*. Parmenides inverted the order of reasoning (indirect proof); this procedure, too, had predecessors.
[8] 'Die organische Bewegung in ihrem Zusammenhang mit dem Stoffwechsel', in Robert Mayer, *Die Mechanik der Waerme*, Ostwald's Klassiker der Exakten Wissenschaften Nr. 180 (Leipzig 1911), 9.
[9] Maxwell's calculations are reproduced in *The Scientific Papers of James Clerk Maxwell*, E. D. Niven (ed.) (Dover Publications, New York, 1965) (first published in 1890), 377ff. For more recent examples cf. G. Birkhoff, *Hydrodynamics* (New York 1955), sections 20 and 21.
[10] Ockham, *Scriptum in Librum Primum Sententiarum* (Ordinatio) (Prolog. Distinctio 1), G. Gal and S. E. Brown, eds. (Franciscan Institute, St. Bonaventura University, New York, 1967), 241.
[11] Similarly the historians used legends, not local historical traditions, as the starting points of their research. Cf. Felix Jacoby *Griechische Historiker* (Alfred Druckmueller Verlag in Stuttgart 1956), 220, right column lines 64ff.
[12] Physicians and other early opponents of the excesses of philosophy (cf. the above quotation from *Ancient Medicine*) expressed their objections in writing – they were members of a tradition of written exchange that was soon to dominate Western Civilization. Not all crafts participated in this tradition; we have no written reports from potters, metal workers, architects, miners, painters, etc. We must reconstruct their knowledge from their work and from indirect references to it. Cyril Stanley Smith, a metallurgist from MIT, did this in a book (*A Search for Structure*, Cambridge, Mass, 1981) as well as in an exhibition (photographic displays in *From Art to Science*, MIT Press, 1980).

JAAKKO HINTIKKA

AN ANATOMY OF WITTGENSTEIN'S PICTURE THEORY

1. DECONSTRUCTING WITTGENSTEIN'S "THEORY"

The idea of deconstruction is just fine, but it has been given a bad name by the *soi-disant* deconstructionists. There are plenty of Humpty-Dumpty ideas in philosophy and its history, that is, ideas which not only have several distinguishable ingredients but which are such that they cannot be put together any longer once the difference between the different factors is discovered and recognized. There is an abundance of such concepts and conceptions ripe to be deconstructed which are incomparably more important philosophically than the notions with which Derrida and his ilk have occupied themselves. What I shall do in this paper is to present a case study in the kind of deconstructive method I just indicated.

Indeed, it is the main thesis of this paper that what is usually discussed under the unitarian heading of "Wittgenstein's picture theory" involves several different and largely independent ideas. The inconclusiveness of most of the earlier discussions of Wittgenstein's so-called "theory" is due to a failure to separate these different ingredients from each other. This failure is to some extent shared by Wittgenstein himself, for he apparently began to pay serious attention to differences between the different picture ideas only when he was forced to give some of them up during his middle period. In fact, discussions of whether Wittgenstein "gave up the picture theory" in his later philosophy offer an instructive example of the confusion one inevitably runs into if one does not distinguish the different components of the syndrome that usually goes by the name "Wittgenstein's picture theory". Since Wittgenstein gave up only some of the different picture ideas, it makes no sense to ask whether he gave up "the picture theory" or not.

To express my thesis more explicitly, I shall argue that Wittgenstein's so-called "picture theory" is a combination of at least the following different tenets:

C.C. Gould and R.S. Cohen (eds.), Artifacts, Representations and Social Practice, 223–256.

(1) An elementary proposition represents the (possible) state of affairs that it represents in virtue of being an isomorphic replica of this state of affairs.
(2) The totality of possible combinations of simple objects matches the totality of possible elementary propositions.
(3) Each name (primitive symbol) has the same logical form (logical and categorial type) as the object it represents.
This is the basis of (2) in Wittgenstein's thought.
(4) Elementary propositions are independent of each other.
(5) All non-elementary propositions are (complex or otherwise derived) pictures of facts in the same sense as elementary propositions.
(6) A part of the background of all these different theses is a sixth one. It is the thesis to the effect that in a logically correct language the logical (pictorial) forms of propositions are their *syntactical* forms.

I shall refer in the sequel to these assumptions or claims as follows:
(1) Elementary propositions as pictures.
(2) The mirroring thesis.
(3) The categorial matching thesis.
(4) The atomicity thesis.
(5) Complex propositions as pictures.
(6) Pictorial form as syntactical form, or the syntacticity thesis.

2. PRESUPPOSITIONS OF WITTGENSTEIN'S PICTURE THEORY

Wittgenstein's "picture theory" relies on certain background assumptions. Even though they are not the focus of this paper, acknowledging them will help us to understand the nature of Wittgenstein's ideas about the pictorial character of language.

Wittgenstein's criticisms of Russell's and Whitehead's *Principia* have obscured his indebtedness to Russell's and Frege's logical ideas. The criticisms are in some instances aimed at details, such as the treatment of identity. By and large Wittgenstein's criticisms of Russell's logic are leveled at Russell's ramified hierarchy, to some extent foreshadowing Frank Ramsey's proposal to eliminate this hierarchy altogether.

In fact, the logic envisaged in the *Tractatus* is basically a simple higher-order logic (simple theory of types) with a nonstandard interpretation (in Henkin's sense). In this regard, it is not unlike Russell's and Frege's logical languages.

This type of language is paradigmatic for Wittgenstein in several important respects. First and foremost, Wittgenstein shares with Frege the belief that he has developed a true *Begriffsschrift*, a universal logical language. Hence Wittgenstein assumes that whatever he can say of his higher-order quantificational logic holds of *Sprachlogik* in general.

We can formulate this Wittgenstein assumption as a separate thesis:

(7) The true logic of human language and human thinking is a higher-order quantificational language.

A more specific assumption which Wittgenstein shares with the majority of twentieth-century analysts can likewise be considered a part of the same Frege-Russell syndrome. It is expressed by Wittgenstein as follows (4.221):

> It is obvious that the analysis of propositions must bring us to elementary propositions which consists of names in immediate combination.

This is ambiguous, however. Even if elementary propositions are the only possible stopping-points of analysis, the reality of infinitely deep logics shows that it is not only not obvious but positively wrong to claim that such an analysis will come to an end after a finite number of steps.

Wittgenstein takes the analyzability of all propositions into elementary ones to entail that all the bearers of simple (unanalyzable) names have to be given as a part of the interpretation of the language in question. This assumption is undoubtedly also motivated by Wittgenstein's belief in the universality of language.

The cash value of this primacy of simple objects in Wittgenstein is that they have to be treated as each of them existing necessarily and as collectively being necessarily exhaustive. This enables him in turn to think of quantification as being reducible to truth-functions. In combination with Wittgenstein's thesis (7) this leads him to the most important background assumption of his "picture theory of language":

(8) The whole logic of our conceptual system can be reduced to truth-functional logic.

This thesis is to all practical purposes identical with Proposition 5 of the *Tractatus*:

> A proposition is a truth-function of elementary propositions.

Even though there are interesting issues concerning these assumptions, I shall not discuss them any further here, for they are not the focus of this paper.

3. WITTGENSTEIN'S BACKGROUND IN RUSSELL

As in so many cases, the best way of seeing what the content of Wittgenstein's so-called picture theory is, is to understand the problem situation out of which it grew. In other words, in order to understand Wittgenstein's picture idea, we have to go to its antecedents. Two of them are especially important here: Russell's theory of acquaintance and the Frege-Russell idea of a logically correct language. The plot of this part of our story has been outlined with marvelous clarity by David Pears. To put the main point in briefest possible terms, Wittgenstein's *Tractatus* was an *Aufhebung* of Russell's theory of acquaintance. The fullest exposition of this theory is found in Russell's posthumously published 1913 MS entitled *Theory of Knowledge*. The theory of acquaintance maintained, in a nutshell, that any proposition I can understand must ultimately consist of entities I am acquainted with. But, this apparently cannot be the whole story. For acquaintance with all the simplest ingredients of a proposition will still leave its *logical forms* unexplained and understood.

For instance, it is plausible to maintain that in order to understand simple relational proposition of the form "aRb", we have to be acquainted with a, R, and b. But this cannot be enough, or so it seems, for it does not distinguish understanding the proposition as "aRb" rather than as "bRa". After all, in both cases we are dealing with acquaintance with the same three entities. Russell's solution to this problem was to postulate an additional class of objects of acquaintance, the logical forms. In order to understand "aRb" we have to be acquainted with four, not three, entities. The additional entity is the logical form of the proposition. Russell identifies acquaintances with this form with knowledge of the corresponding completely generalized proposition $(\exists x)(\exists \varphi)(\exists y)\ x\varphi y$. Thus for Russell, even logic (the totality of logical forms) is based on experience (acquaintance), viz. on the experience needed to come to know the truth of the completely generalized propositions.

What Wittgenstein does is to reject totally logical forms as a separate class of objects of acquaintance. As he puts it in the *Prototractatus* (p. 29 of the original), "Common form is not a common constituent."

Logical forms are not for him independent self-sustaining entities. They are not constituents of propositions. This is what Wittgenstein means by such pronouncements as 4.0312:

> My fundamental idea is that the logical constants are not representatives; that there can be no representatives of the logic of facts.

For what the so-called logical constants were supposed to denote are logical forms (or ingredients thereof).

4. LOGICAL FORMS AS FORMS OF SIMPLE OBJECTS

But what is it in Wittgenstein's theory that can do the job that Russell's self-sustaining forms were supposed to do? Here the Frege-Russell paradigm of *Begriffsschrift* again comes into play. For according to it the rock bottom of one's universe – the kind of universe which we can speak of in a Frege-Russell language – is a set of objects, including both individuals (particulars) and higher-order objects. Hence it is tempting to use these basic ingredients of a Russellian universe as the ultimate building-blocks of all logical forms, too. More specifically, it is natural to think that all other logical forms can be assembled from the forms of the basic (simple) objects. This is precisely what Wittgenstein argues for in the *Tractatus*. In this way, Wittgenstein can so to speak have his logical forms and dispense with them, too. In other words, he can have in his theory a class of logical forms all right, viz. the logical forms of simple (i.e., unanalyzable) objects. These forms are no longer self-sustaining, however. They are forms *of* objects; they do not exist independently of the objects whose forms they are; they are given to me together with the objects whose forms they are.

This simple account shows a crucial difference between Russell's and Wittgenstein's simple objects (objects of acquaintance). Both were simple in the sense of being unanalyzable into other, more basic objects. However, Wittgenstein's objects were not simple in the sense of being formless.

5. ELEMENTARY PROPOSITIONS AS PICTURES

This leaves Wittgenstein with a number of problems. First, he has to explain how simple objects can join together so as to form the facts that our propositions are about. If I cannot understand the proposition "aRb" by simply being acquainted with the formula's objects, a, R, and b, how can I know what this proposition is about by being acquainted with these objects which are now supposed to come to us equipped with their respective logical forms?

Wittgenstein's answer is that the proposition, as the corresponding fact, is not simply a set of names or of objects. The fact is a structure formed from its constitutive objects. And the possibility of this structure does not depend on any "logical glue" over and above the objects involved in it and their logical forms. For the combinations of simple objects into states of affairs are precisely what is governed by their logical forms. The formation of a state of affairs is not like gluing two cubes together, it is like fitting two jigsaw pieces together. Whether, and if so in what different ways, two pieces can be fitted together is precisely what their form (in this case, their geometrical form) determines. Or, to use Wittgenstein's own imagery, in a proposition objects hang together like links in a chain, not like plates glued or riveted together by means of some additional ingredient. (See *Tractatus* 2.03.)

We are already approaching the first ingredient in Wittgenstein's syndrome of picture ideas. For how can we express in language the kind of combination of simple objects into a state of affairs? The obvious (though perhaps not inevitable) answer is: by combining their names in an analogous way. This presupposes that our simplest linguistic symbols, (simple) names, must have a logical form, which governs their possibilities of being combined with other simple names. In this sense, names are not simple: they are not formless, Wittgenstein argues.

An elementary proposition, which is a combination of names, says that the objects named are related to each other in the same way as the names in the proposition. For instance, if "a" and "b" stand for certain particular objects and "R" for a certain relation, then "aRb" is true if and only if this relation holds between them (in this order). Of course, strictly speaking it is not the symbol "R" that represents a relation. Rather, the relation is represented by the linguistic *relation* of flanking the symbol "R". Notationally, this relation hence could be represented by a letter plus two empty argument-places next to it, i.e., by something like "–R–". For it is only then that it is literally true that a proposition as a syntactical entity is isomorphic with the state of affairs which makes it true (and with the possible state of affairs which would make it true). It is only when we so to speak build the two empty slots (and their relative order) into the very symbol of a (two-place) relation can we say an elementary proposition like "aRb" represents a fact in virtue of its form (structure).

What easily misleads us here (and what has misled several commentators) is that we apparently do not similarly have to associate any form to a name of a particular. This is an optical illusion however. Not associating any argument-place with a symbol for a particular object is to associate a form to it quite as much as associating an argument-place with it. For what the idea of form here amounts to is a rule governing its possibilitics of combination with other symbols.

Thus this component of the total picture view is essentially a truth condition for elementary propositions. In this respect, it differs conceptually from the other theses, especially from theses (2)–(3), which do not deal with the nature of truth-conditions. This fact already suffices to show the need of distinguishing from each other Wittgenstein's several picture theses.

6. WITTGENSTEIN AND TARSKI

If you step back and look for a moment at the view thus sketched, you are likely to have a *déjà vu* experience. What the rigmarole of objects in a fact being related in the same way as their names in an elementary proposition amounts to is in effect the first clause of a Tarski-type truth-definition, viz. the clause for atomic sentences. This clause presupposes that a valuation of the formal language in question is given. Applied to a sentence like "aRb" it says, roughly speaking, that it is true iff the relation which is the value v(R) of "R" holds between the values v(a) and v(b) of "a" and "b", respectively. At first, this looks precisely the same as Wittgenstein's picture idea as applied to elementary propositions. On both accounts, "aRb" is true if and only if the configuration of the linguistic symbols matches the configuration of the entities in the world they represent.

This comparison between Tarski-type truth-definitions and Wittgenstein's picture view brings out what seems to me the most important idea included in it. As we have seen, this idea is also a direct response to the historical situation in which Wittgenstein's theories were formed.

In the light of hindsight, however, there is one respect in which the Tarski-Wittgenstein analogy is highly misleading. To put the point in a nut-shell, Tarski-type truth-definitions are formulated for one "world" or model only (often taken to be the actual world), whereas Wittgenstein's idea is calculated to apply to different possible situations ("possible worlds").

To put this point in somewhat fuller detail, Tarski's truth-definition clause for atomic sentences presupposes a valuation of the language in question in *one* model (*one* "possible world") only. It does not say anything about the values of our linguistic symbols in other models or "possible worlds". Wittgenstein's idea is much more sweeping. For him a name is the name of its object in any old possible state of affairs. Wittgenstein's meanings specify the references of our symbols for all such possible situations. It is as if the Tarskian valuation function were given to us, not relative to one world, but for all of them in one fell swoop. In any single world, Wittgenstein and Tarski so to speak agree as to what the truth-condition is for elementary propositions, but for Wittgenstein that condition is given globally whereas Tarski's definition applies only locally.

Another way of expressing the same point is to say that Tarski's ideas are (in one sense of this dangerous word) extensional; Wittgenstein's ideas are modal.

This point is easily obscured by other ideas in Wittgenstein. He uses in the *Tractatus* the locution "possible world" only once. However, this is not because he is avoiding the idea of *possible* world, but because he is avoiding the idea of *world* as a totality. In contrast to the locution "possible world" Wittgenstein frequently uses such terms as "possible state of affairs" or "possible combination of objects".

We might say that thesis (1) is already a combination of two theses, viz. (a) the idea codified in the first clause of a Tarski-type truth definition; (b) the extension of this idea to all "possible worlds" or possible situations of language use.

7. THE MIRRORING THESIS

It is obvious that thesis (1) does not do the whole job Wittgenstein needed to be done. He wanted the entire role of Russellian independent logical forms to be played by the logical forms of simple objects. These logical forms spell out how objects can be combined with each other. One particular elementary proposition reflects one such combination. But this is not enough. Obviously it must be the case that the *totality* of all elementary propositions must reflect the *totality* of possible facts.

Otherwise we need an explanation of why certain possible combinations of names do not correspond to any possible facts. This explanation cannot be given by the logical forms of names (or of elementary propositions built up from them). Hence these logical forms would not do the whole job assigned to them. Hence Wittgenstein has to assume the mirroring thesis (2).

What is remarkable here is that Wittgenstein's "picture theory" has to include the mirroring thesis (2), but that it is an assumption different from the thesis (1) of elementary propositions as pictures, an assumption that goes beyond thesis (1). Moreover, it is important to realize that Wittgenstein himself was aware of the difference between the two theses (1) and (2).

This awareness is evinced especially clearly by Wittgenstein's comments on logical form. For it is the logical forms of simple objects that determine the totality of ways in which they can be combined into states of affairs. Hence for Wittgenstein to speak of the totality of possible combinations of objects is to speak of logical forms and ultimately of the logical form of reality.

Once we realize this, we soon realize also that Wittgenstein expressed the distinction between theses (1) and (2) even terminologically. When Wittgenstein is concerned with the way one single proposition represents a state of affairs by means of its form he speaks of a picture and of picturing, *Bild* (or *Abbildung*) and *abbilden*. When he is concerned with the way the totality of possible elementary propositions reflects the totality of possible combinations of objects, he typically speaks of a mirror and of mirroring, *Spiegel* (or *Spiegelbild*) and *gespiegeln* and the same goes (according to the observation just made) for Wittgenstein's remarks on the logical form of reality.

> Propositions cannot represent logical form: it is mirrored (*spiegelt sich*) in them.
>
> What finds its reflection in language (*sich in der Sprache spiegelt*), language cannot represent.
>
> . . .
>
> Propositions show the logical form of reality.
>
> They display it. (4.121)

A closer survey of Wittgenstein's usage in the *Tractatus* shows easily this difference in meaning between the two sets of Wittgensteinian expressions. It is for instance symptomatic that Wittgenstein's "mirroring" terminology is used precisely when we would predict that he would use it, viz. in discussing the idea of logic as consisting of mere tautologies in 6.13 and in discussing all the different forms of propositions in 5.511 and 5.514. How Wittgenstein's notion of tautology and his thesis of the tautological character of logical truth is connected with the mirroring thesis (2) will be indicated below in section 12.

8. LOGICAL FORMS OF NAMES

But, what is the basis of the mirroring thesis in the *Tractatus*? Why does Wittgenstein think that all and only ways of combining names are matched by possible configurations of objects? Sight unseen, one can think of different answers to these questions. For instance, it might be thought that the mirroring thesis is a guideline as to how a truly representative langauge, perhaps an "ideal langauge", ought to be constructed. The rules for combining names might be conventional, but the conventions governing them must be chosen in accordance with the mirroring thesis. This thesis would then be true by an enlightened fiat, not by necessity.

This is not Wittgenstein's view, however. There is a deeper foundation for the mirroring thesis (2) – or perhaps a much closer guideline for choosing our symbolism.

The thesis says that all the possible combinations of simple objects must be matched by possible combinations of names and *vice versa*. Now what determines the totality of combinations of objects? Wittgenstein's answer is contained in propositions like 2.0123 and 3.315. They show that what governs the possibilities of combining an object with others is its logical form. As was emphasized above, this form is given together with the object.

From this it follows that the mirroring thesis is valid as soon as each object is represented by a name which has the same logical form as it. Only then can it be the case that, e.g., "a picture contains the possibility of the situation that it represents".

This, then, is the basis of the mirroring thesis in the *Tractatus*: Wittgenstein is in it requiring that each simple object must be represented by a symbol (name) which shares its logical form. This is precisely the import of the categorial matching thesis (3).

The consequences of this observation depend on Wittgenstein's notion of logical form. He does not say very much about it, but some of the things he says are quite striking. For instance witness 2.0251:

> Space, time and colour (being coloured) are forms of objects.

There is very little that is strange or surprising about 2.0251. Space is a way for objects to be related to, or compared with, each other (in Wittgenstein's language, to be combined with each other into a fact), and likewise for time and color. This is in keeping with the third paragraph of 2.0121, which is to be taken somewhat more literally than one perhaps realizes:

> Just as we are quite unable to imagine spatial objects outside space or temporal objects outside time, so to there is no object we can imagine excluded from the possibility of combining with others?

Hence it follows that Wittgenstein's spatial objects (e.g. spatial relations) are to be represented by spatial objects (e.g. spatial relations), etc. What this means is not entirely clear. What is clear is that Wittgenstein's picture idea, even when applied merely to atomic (elementary) propositions, involves much more than mere isomorphism based on an arbitrary correlation of names and objects. It involves an isomorphism of two structures of the same logical form. And this identity of logical forms extends to the very names of which a proposition consists, in relation to their objects. Thus, in a sense (albeit in a sense that has to be handled cagily), the picture idea does not concern sentences only; it concerns also names.

Perhaps this is only to be expected. We saw above that Wittgenstein was forced by his rejection of logical forms as independent entities to associate forms with simple objects. Hence the requirement of formal identity between language and the world must likewise be extended to names in their relation to objects.

This point is closely related to the observation made earlier to the effect that Wittgenstein's picture idea goes beyond the first clause in a Tarski-type truth-definition even in the case of elementary (atomic) propositions.

The thesis that each simple name shares the same form as the object it represents presupposes a more general idea. It presupposes that a language with its objects is a part of the world. Moreover, each simple name is a member of the same facet of the world as its object. Only through this worldliness of language can Wittgenstein hope to maintain the categorial identity of symbols and objects symbolized.

The idea that language is a part of the world also figured in Wittgenstein's later development. Sometime around 1928, it was sharpened into the idea that language is a part of the physical world. (See, e.g. *Phil. Remarks*, VII, sec. 68: "The language itself belongs to the second [i.e. physicalistic] system".)

9. PICTURES BY SIMILARITY?

Further confirmation of our interpretation is obtained by examining its consequences. From what has been said it follows that the pictures of the *Tractatus* are more than mere isomorphic images of the corresponding facts. They resemble the facts to the extent of sharing their logical form. This makes them in a sense pictures by similarity.

This observation is related in an intriguing way with what Wittgenstein said later. In *The Blue Book*, pp. 35–37 Wittgenstein discusses what he calls "pictures by similarity" and "pictures not by similarity". The nature of the contrast is not perhaps entirely clear, nor is it clear what its relation to Wittgenstein's own earlier ideas is supposed to be. The key passage runs as follows:

> If we keep in mind the possibility of a picture which, though correct, has no similarity with its object, the interpolation of a shadow [meaning entity] between the sentence and reality loses all point. For now the sentence itself can serve as such a shadow. The sentence is just such a picture which hasn't the slightest similarity with what it represents. . . . This shows you the way in which words and things may be connected.

What is unmistakable here is that no identity of form is any longer assumed to obtain between the objects and the words. Their correlation is established by ostension, which does not presuppose any identity of form whatsoever.

A comparison with the *Tractatus* helps us to understand why Wittgenstein can introduce, as he clearly does, the notion of "a picture not by similarity" as a novel idea. A picture is a picture by similarity if identity of logical form is required to hold between its simplest ingredients and the objects they represent. A picture is not a picture by similarity if the relation of the simplest ingredients to their objects is conventional. Since Wittgenstein had held the former view in the *Tractatus*, the idea of a picture "not by similarity" was for him a new idea in *The Blue Book*. We can thus see a major difference between the picture idea Wittgenstein is defending in *The Blue Book* and his views in the *Tractatus*.

That Wittgenstein originally thought of the pictures that propositions are as actual likenesses is also shown among other passages by 4.011:

> At first sight a proposition – one set out on the printed page for instance – does not seem to be a picture of the reality with which it is concerned. But no more does musical notation at first sign seem to be a picture of music, nor our phonetic notation (the alphabet) to be a picture of our speech.
>
> And yet these sign-languages prove to be pictures, *even in the ordinary sense* [emphasis added] of what they represent.

Witness also 4.012.

> It is obvious that a proposition of the form '*aRb*' strikes us as a picture. In this case the sign is obviously a likeness (*Gleichnis*) of what is signified.

Thus the mirroring idea and the categorial matching thesis (thesis (3)) are closely intertwined. The latter is a motivation for the former, but Wittgenstein was led to it via the same considerations by which he was led to the picture idea (thesis (1)), i.e., via a rejection of Russell's view of logical forms as independent objects.

It is nevertheless important to note that the core idea of Wittgenstein's picture view, i.e., the thesis (1), does not entail either of the two theses (2)–(3). The primary application of picture ideas in the *Tractatus* thus is to elementary propositions. It can be extended to names only by courtesy of further assumptions that are independent of thesis (1).

10. ATOMICITY

Neither the idea of elementary propositions as pictures nor the mirroring thesis – nor yet the two combined – yields the conclusions Wittgenstein wanted. In order to see that they do not we may perform a thought-experiment. We can imagine Wittgenstein expounding his theories of picturing and mirroring to a Leibniz *redivivus*. What would Gottfried Wilhelm have to say? An answer is not difficult to imagine. The idea of sentences as pictures, representing what they represent by means of their structure, was embraced in so many words by Leibniz. Hence the pictorial character of elementary sentences would not have posed any difficulties to Leibniz.

Moreover, each simple object, each monad, mirrors according to Leibniz the entire world. If it is to be represented (named) by a similar monad, that monadic name likewise determines all its possible combinations with other objects. And if Wittgenstein's requirement is satisfied and a name of a simple object is a categorically identical object in the world, then Leibniz's other assumptions imply that this is essentially what must be the case. Hence some version of the mirroring idea is clearly in keeping with Leibniz's thinking.

But the ultimate conclusions of Wittgenstein and Leibniz are nevertheless entirely different. For Wittgenstein, the only *a priori* truths (if they can be called truths) are tautologies, whereas for Leibniz we can know *a priori* a number of things about the structure reflected by monads, for instance, that this structure exemplifies the simplest (metaphysically) possible laws.

Where do the two philosophers part company? The answer lies in Wittgenstein's assumption of a radical contingency, or perhaps rather atomicity, of the world. The logical form of each simple object governs the ways it can be combined with other objects. But for two different objects these possibilities are independent of one another. If a and b are combined with each other in a certain way, nothing can be inferred concerning whether c and d are in fact combined in a certain possible (but not necessary) way. This is the atomicity thesis (4) mentioned above. As we have seen, it is independent of theses (1)–(3).

In the jargon of logic, the atomicity thesis asserts that different elementary propositions are independent of each other. From the truth or falsity of one elementary (atomic) proposition one can never validly infer the truth or the falsity of another. Metaphorically speaking, according to the atomicity thesis, the grand mirroring relation, with all possible total states of the world being mirrored by all possible maximally consistent sets of propositions, reduces in the *Tractatus* to a local mirroring with the totality of elementary propositions reflecting the totality of basic facts which can obtain or fail to obtain completely independently of each other. Thus the atomicity thesis (4) is in effect a sharpened form of the mirroring thesis (3).

Why did Wittgenstein believe in the atomicity thesis? No explicit answer can be found in the *Tractatus*, nor is one likely to be deduced from the text. It is not even clear whether Wittgenstein himself fully realized at the time of the writing of the *Tractatus* that the atomicity thesis is really independent of his other assumptions. Some informed guesses concerning its background are nevertheless possible. I have argued (with Merrill B. Hintikka) that the simple objects of the *Tractatus* were phenomenological objects, rather like Russell's "objects of acquaintance". They were not phenomenalistic objects, that is phenomena as distinguished from reality, but they were given to me in my immediate experience.

If so, Wittgenstein's atomicity thesis is a version of the assumption of the radical contingency of all experience. Whether Wittgenstein could find an argument for his thesis based on such Humean principles, will not be investigated in this paper. The phenomenological roots of many of Wittgenstein's ideas are nevertheless clear. For an example, see sec. 18 below.

It may also be that Wittgenstein's tacit adoption of the atomicity thesis reflected his faith in Russell's and his own logical notation. The mirroring thesis asserted that the totality of all possible combinations of names into propositions reflects the totality of possible combinations into states of affairs. But in the truth-functional notation we can in fact negate or not negate an elementary proposition independently of negating or not negating any other one. If this notation is correct (in the sense of mirroring all the *a priori* relationships between propositions), and if the mirroring thesis holds, elementary propositions must be logically independent of each other.

This argument is perhaps not fully persuasive either. It is in any case relevant that the atomicity thesis was the first one to be abandoned by Wittgenstein when he began to revise his ideas in 1928. The precise context of that revision will be indicated below.

Wittgenstein might defend himself here by pointing out, correctly, that he had left open the questions as to what complex interdependencies there might obtain in the world. (Cf. e.g. 5.554.) That question could only be answered by experience. But he is nevertheless excluding the possibility that there might be global interdependencies of the kind Leibniz was envisaging.

11. THE GIVENNESS OF LOGICAL FORM

There remains a puzzling fact about Wittgenstein's notion of the logical form of a simple object. This form has essentially a modal role: it governs the possibilities of the object in question to be combined with all other objects. Yet it is given to me in acquaintance, in one fell swoop, so to speak. How is this possible?

This is an interesting and vexing problem here, but it is Wittgenstein's problem, not mine. For there is no doubt that Wittgenstein did hold that the logical form of an object is given to me together with the object. Perhaps Wittgenstein thought that to think otherwise would threaten to make the form an independent entity, like Russell's logical forms as objects of acquaintance.

As usual, Wittgenstein's later pronouncements, made usually in connection with changes in his philosophical position, throw sharp light on his earlier views. Among other things, they confirm strikingly the depth of Wittgenstein's earlier belief that one experience could provide one with a general rule, governing the logical behavior of an entity. Here we are discussing the laws governing the possibilities of a simple phenomenological object to be combined with others. About the parallel – or perhaps special – case of the rules governing the grammatical use of a word, Wittgenstein wrote in MS 116 (sec. 218 Nyman):

> Earlier I thought at one time that grammatical rules are an explication of what I experience on *one* occasion when I once use the word. They are as it were consequences or expressions of the properties which I momentarily experience when I understand the word.

In somewhat less implausible terms, for a while Wittgenstein held in his middle period that an act of ostension can give me the rule of using the expression to be defined.

Wittgenstein's later change of mind on this point was one of the most important factors that prompted his so-called rule-following discussion. Here it suffices to note the early Wittgenstein's belief that one experience could give me a rule (a logical) law governing the possibilities of a simple object's being combined with others.

The same belief is clearly what underlies the near-identity in the young Wittgenstein's mind between phenomenology and logic (or, as he put it later, "grammar").

12. ATOMICITY THESIS AS A LEADING IDEA OF THE *TRACTATUS*

Logically, and semantically, the gist of Wittgenstein's picture view is the thesis (1) of elementary propositions as pictures. Philosophically, though, much of the action is elsewhere. The mirroring thesis (2) is one of Wittgenstein's most central ideas in the *Tractatus* especially when it is sharpened into the atomicity thesis (4). In order to see the role the atomicity thesis plays there, consider what it means for the theory of truth-functions. Wittgenstein had argued, on independent grounds, that the rest of our language can be reduced to the language of propositional connectives. (See sec. 2 above.) Sentences formed by their means can be considered, as Wittgenstein shows, as truth-functions of elementary propositions. Hence truth-function theory is a part of the true *Sprachlogik*. (This part of the story Wittgenstein never gave up.) But in the *Tractatus* Wittgenstein wanted to argue that truth-function theory is not only a truth and nothing but the truth but also the whole truth about the logic of our language.

What is required for that? Clearly that all elementary propositions are independent of each other. But this is precisely what is asserted by the atomicity thesis (4). This means that the atomicity thesis (8) is closely related to some of the central tenets of the *Tractatus*. If the independence of elementary propositions is combined with the thesis (7) (all complex propositions are truth-functions of elementary ones), it follows that all logical truths are tautologies in the strict truth-functional sense of the word. This is one of the distinctive theses of the *Tractatus*, a thesis which is connected with some of Wittgenstein's most firmly held philosophical doctrines.

Among other things, we can now see why Wittgenstein is doing in the six-propositions what he is in fact doing there. Why, for instance, the preoccupation with the notion of tautology in 6.1–6.121? In this part of the *Tractatus*, Wittgenstein is developing his truth-function theory. (See sec. 13 below.) The answer is obvious: Wittgenstein's completeness (atomicity) thesis is equivalent to claiming that logical truths are precisely the tautologies of propositional logic, that the logic of tautologies *is* the whole logic.

There is another context in the history of twentieth-century philosophy where the same problem of the independence of atomic propositions has come up. Assume that a first-order logic is used as a framework of semantical representation, and assume (for the sake of argument) that the language or language fragment we are interested in can actually be translated into (or otherwise recaptured in) a first-order language. Then first-order logic is all we need in semantics, only if all atomic propositions are logically (conceptually) independent of each other. Otherwise we need additional laws, sometimes called "meaning postulates", to have a full semantical theory. The need – and/or the admissibility – of such meaning postulates was later debated by the likes of Carnap and Quine.

Thesis (4) is also closely related to Wittgenstein's ideas of the totality of language, the limits of language, and of the transcendental. I will not discuss those connections here, however.

13. THE ENIGMA OF PROPOSITION 6

Of all the different components of the picture syndrome of the *Tractatus*, thesis (5) is the one most closely related to the overall argumentative structure of the *Tractatus*. The main facts of the case ought to stare in the face of anyone who is trying to understand Wittgenstein's argumentative strategy in the *Tractatus* but have rarely been noted.

The first massive fact is the strange way in which Wittgenstein expounds his picture idea. In 2.1–2.225 Wittgenstein discusses pictures in general, and in 3.1–3.24 he applies the picture idea to propositions in general.

Elementary propositions do not come into play until 4.21. Yet the picture idea is *a priori* plausible only in the case of elementary propositions. This is tacitly recognized by Wittgenstein in that the only concrete example of a pictorial proposition is an elementary proposition of the Russellian notation.

> It is obvious that a proposition of the form 'aRb' strikes us as a picture. In this case the sign is obviously a likeness of what is signified. (4.012.)

This interpretational problem becomes even more acute against the Russellian background outlined above. How can he extend his pictorial idea from elementary proposition to all others? How can he do so apparently without offering any arguments whatsoever?

We can find an answer by raising another large-scale interpretational question. In this case, it ought to be as obvious as any question concerning the understanding of the *Tractatus*. What is Wittgenstein's overall vision in the *Tractatus*? The integral numbered propositions tell the story.

1. The world is all that is the case.
2. What is the case – a fact – in the existence of states of affairs.
3. A logical picture of facts is a thought.
4. A thought is a proposition with a sense.
5. A proposition is a truth function of elementary propositions.
6. The general form of a truth-function is $[\bar{p}, \xi, N(\bar{\xi})]$
7. What we cannot speak about we must consign to silence.

The odd man out here is number six. If Wittgenstein's organization of his book is to make any sense, Proposition 6 ought to be the culmination of his grand argument. Instead, it seems to be a miserable letdown. What does a cute minor result concerning the details of truth-functional logic that any sophomore can prove have to do with Wittgenstein's overall vision of language, world and their interrelation? Why should Wittgenstein deign to pay any attention to Sheffer's result, which *prima facie* has no deeper theoretical significance, let alone flaunt it as the crowning achievement (the penultimate integral-numbered proposition) of the entire *Tractatus*?

Obviously, the import and role of Proposition 6 need closer attention. What does it say, in plain Queen's English? It says that only one operation is needed to form all truth-functions, viz. simultaneous negation of a number of propositions, that is, in effect, a conjunction of negations. Why is this relevant? Proposition 5 shows that Wittgenstein thinks the entire logic of our language is reducible to the theory of truth-functions. What does that tell about the overall relation of language and the world? The answer many commentators have tacitly adopted is to assume that for Wittgenstein our language consists of truth-functions of pictorially interpreted elementary propositions. The picture theory applies literally taken only to elementary propositions, according to this view. It applies to others only indirectly, in virtue of their being truth-functions of elementary propositions. Thus the picture theory and the truth-function theory are the two separate major ingredients of the *Tractatus*. For if the picture idea were applicable without further ado to all propositions, Wittgenstein would not need the theory of truth-functions and would not need any Proposition 5.

But this view is absurd. It leaves unexplained the plain statements earlier in the *Tractatus* where Wittgenstein applies without any qualifications the picture idea to *all* propositions, and it leaves unaccounted the motivation of Wittgenstein's further excursion into the details of truth-function theory in Proposition 6 and in the related subordinate propositions.

14. TRUTH-FUNCTION THEORY AS A WAY OF EXTENDING THE PICTURE IDEA TO COMPLEX PROPOSITIONS

There is only one way of making sense about this situation. It is to realize that *Proposition 6 is Wittgenstein's way of extending the picture idea from elementary propositions to all others.* It is his oblique way of stating my thesis (5). Since Wittgenstein had in Proposition 5 reduced all complex propositions to truth-functions (to his own satisfaction, at least), what he had to do is to show that, so to speak, truth-functions of pictures are also pictures. But how does Proposition 6 help him? How does it further Wittgenstein's project to show that all truth-functions are conjunctions of negations? It will help him if he can interpret a conjunction and a negation of pictures as also being a picture. Now conjunction is easy: a conjunction of pictures is simply a conjoint complex picture. But what about negation? Wittgenstein's answer turns on his theory of the bipolarity of propositions. In simplest possible terms, he is saying: The negation of a picture is not only a picture, it is the very same picture, but taken with the opposite sense (with the converse polarity). The relation of p and ~p is thus like the positive and negative print of a photograph.

I shall not examine here whether this view of Wittgenstein's is a viable one or not. Instead, it is in order to point out that it is unmistakably Wittgenstein's idea in the *Tractatus*. There are two kinds of reasons for maintaining that it is. First, it makes for the first time satisfactory sense of the overall argumentative structure of the *Tractatus*.

It explains why Proposition 6 is the culmination of Wittgenstein's argument. In Propositions 1–4 Wittgenstein expounds his overall view of the structure of the world (Props. 1–2) and of the language and its relation to the world (Props. 3–4). But he still has to show that this vision represents everything that is going on in our language. This is what he undertakes to do in Propositions 5–6, and this project culminates in the claim that all propositions, not only elementary ones, are pictures of states of affairs.

But why does he not say this in Proposition 6? The reason lies in Wittgenstein's oblique way of marshalling his argument in the *Tractatus*. Wittgenstein had stated his claim that all propositions are pictorial in the beginning of the book. (Cf., e.g., 3 and 3.1.) But this thesis was put forward only as a claim to be defended later. Hence he could not simply repeat the claim, but had to express the basis of holding his view. For this purpose, he first reiterated the claims as applied to elementary propositions, where it can be taken to be more or less obvious. But this left open the problem of extending the picture idea to all propositions, which in the light of Prop. 5 of the *Tractatus* meant extending it to all truth-functions. This extension is made possible by the representability of all truth-functions by means of repeated applications of the Sheffer stroke operation, which is precisely what Proposition 6 states. Thus this proposition does not mean an excursion on Wittgenstein's part into the intricacies of truth-functional logic. It is an absolutely integral part of his overall line of thought in the *Tractatus*, and indeed its last and conclusive step.

Another way of expounding this interpretation is to show that Wittgenstein actually viewed the pictorial character of complex propositions in the way I have suggested. The treatment of conjunction is a relatively trivial matter here. The conjunction of two pictures can be thought of simply as a composite of the two. The crucial question concerns Wittgenstein's treatment of negation.

In this respect the best evidence is constituted by Wittgenstein's theory of the bipolarity of propositions as it is expounded in the materials included in the volume *Notebooks 1914–1916*; see e.g. pp. 94, 97, 101–2, 111–5, 123–5. Witness in particular pronouncements like the following:

> What I mean to say is that we *only* then understand a proposition if we know *both* what would be the case if it was *false and* what if it was *true*.

Speaking more generally, Wittgenstein consistently insisted that the understanding of not-p is implicit in the understanding of p. Of course, both kinds of understanding must according to Wittgenstein be pictorial.

15. THE PRESUPPOSITIONS OF THESIS (5)

Wittgenstein's attempted extension of the picture idea to all propositions (i.e. thesis (5)) results from three assumptions: They are the thesis (8) of the reducibility of all logic to truth-functional logic, the atomicity thesis (4), and Wittgenstein's idea of the pictorial significance of negation and conjunction. The atomicity thesis is an extension of the mirroring thesis (2). Even though the three theses (2), (4), and (5) can – and must – be distinguished from each other, they are at the same time closely related to each other.

At the same time, thesis (5) or, more vividly speaking, the combination of theses (4) and (8) is the cornerstone of Wittgenstein's conception of logic. It is for instance what justifies his idea of logical truths as tautologies. At the same time it governs Wittgenstein's conception of the use of language as a direct comparison of propositions with reality. In view of these implications of thesis (5) and of Proposition 6 of the *Tractatus*, which was seen to be closely related to thesis (5), it is no wonder that Proposition 6 is in Wittgenstein's organization the culmination of the constructive part of the *Tractatus*.

16. THE INDEPENDENCE OF THESIS (5)

It is important to realize that Wittgenstein's thesis (5) of complex propositions as pictures is quite different from the mirroring thesis (2). In principle, one can adhere to the mirroring thesis, in other words, one can believe that the totality of elementary propositions reflects the totality of simple facts, and also adopt the truth-function theory for all other propositions as an entirely independent way of dealing with the rest of our logic. Indeed, it is often assumed that this is what goes on in the *Tractatus*. For instance, one distinguished commentator has spoken of the two cornerstones of the *Tractatus*, the picture theory and the theory of truth-functions. In reality, however, Wittgenstein's discussion of truth-function theory is closely geared to his picture ideas. The thesis (8) of the reduction of the rest of the logic of our language to truth-function theory must admittedly be argued for independently of the other theses. Furthermore, the laws of truth-functional logic are there independently of their possible pictorial interpretation. Wittgenstein was in fact the first to spell out some of the crucial laws of propositional logic. But – and here is one of the most important insights needed to understand Wittgenstein's line of thought – this discussion of truth-functional logic is prompted in the *Tractatus* by an ulterior motive. This is what was argued in the preceding two sections.

Wittgenstein's thesis is closely related to his "main idea" that there are properly speaking no logical constants, no logical forms apart from the logical forms of objects and their combinations – and even these combinations are predetermined by the logical forms of objects.

17. THE INEVITABILITY OF PROP. 6

This last point deserves to be spelled out more fully. It is not only the case that Wittgenstein could (to his own satisfaction) extend the picture idea to all propositions. It is important to realize that he *had* to do so in order to carry out the overall project of the *Tractatus*. Why? Because he wanted the logical forms of the simple objects to do the whole job in the semantics of our language. Each proposition, however complex, had to be capable of being compared with reality directly, in virtue of its very own logical form.

For if it first had to be related to elementary sentences via truth-functional operations, these operations would contribute to the meaning of the complex proposition. Truth-functional connectives, the linguistic expressions of these operations, could not vanish, contrary to what Wittgenstein maintains in such propositions as 5.441. Truth-functional logical constants would be representatives, notwithstanding 4.0312. There would be a logic of facts involving precisely such representatives, viz. truth-function theory. (Cf. 4.0312.) In order to bring out the pictorial meaning of a proposition, we would need a separate propositional calculus. Hence, a complex proposition could not show how things are if it is true; that would have to be figured out by means of truth-functional calculations. An arbitrary proposition would not be a model of reality. (Cf. 4.01.)

In particular, a tautology would not in general show that it is a tautology. (Cf. 6.127.) In brief, Wittgenstein would not have succeeded completely in exorcising Russellian logical forms as independent constituents of our propositions.

18. WITTGENSTEIN'S "REDUCTION TO ACQUAINTANCE"

Here we can also answer a question which may have bothered my readers for a while about the Russell-Wittgenstein link. A major part of Russell's philosophical project in 1910–16 was his epistemological and semantical reduction to acquaintance, or perhaps its mirror image, the logical construction of the rest of the world out of the objects of acquaintance. What is the counterpart to this two-way street between objects of acquaintance and everything else in Wittgenstein? The apparent lack of a counterpart has been adduced as a reason why Wittgenstein's philosophy in the *Tractatus* cannot be considered as an extension and derivation of Russell's theory of acquaintance.

An answer is very simple. Wittgenstein is offering us a linguistic counterpart of the reduction to acquaintance. That is what his extension of the picture theory to the entire language amounts to. What is essential to the language Wittgenstein envisages in the *Tractatus* is not that it is an ideal language or that it is our actual language. What matters is that it is a *Begriffsschrift* in Frege's sense: a codification of our entire conceptual framework. For this reason, Wittgenstein can try to practice what he preaches and to carry out the entire reductions to acquaintance as an exercise in the logical syntax of our language or, strictly speaking, of Wittgenstein's *Begriffsschrift*. The extension of the picture theory to the entire language is Wittgenstein's "reduction to acquaintance".

If this is explanation of the Wittgenstein reduction to acquaintance appears too simple to be true, it is instructive to recall that at one time Russell thought that he could carry out *his* reduction to acquaintance by the sole means of logical analysis. In Russell's case, this analysis was an application of his theory of denoting, including his theory of definite descriptions. Toward the end of his famous paper "On Denoting" Russell writes:

> One interesting result of the above theory of denoting is this: when there is anything with which we do not have immediate acquaintance, but only definition by denoting phrases, then the propositions in which this thing is introduced by means of a denoting phrase do not really contain this thing as a constituent, but contain instead the constituents expressed by the several words of the denoting phrase. Thus *in every proposition we can apprehend* (i.e. not only those whose truth or falsehood we can judge of, but in all that we can think about), all the constituents *are really entities with which we have immediate acquaintance* [emphasis added].

This is a revealing statement. It amounts to saying that every proposition we can apprehend can be expressed in the usual quantificational idiom, they are in the last analysis about objects of acquaintance. This inference is obviously mediated by Russell's tacit assumption that the only possible values of quantified variables are objects of acquaintance.

Likewise, Wittgenstein tacitly based his reduction to acquaintance on his argument which purported to show that all propositions are combinations of names and all facets combinations of simple objects. As in Russell, this implies a reduction to acquaintance only on a further assumption, viz. that the objects named by simple names are objects of acquaintance. In this matter, there hence obtains a remarkable similarity between Wittgenstein and Russell.

19. THE ROLE OF SYNTAX

Of the different theses we are here distinguishing from each other, perhaps the most obvious and *prima facie* least troublesome is thesis (6), the syntacticity thesis. For one thing, it is stated by Wittgenstein in so many words in the *Tractatus*:

> In logical syntax the meaning of a sign should never play a role. It must be possible to establish logical syntax without mentioning the *meaning* of a sign: only the description of expression may be presupposed. (3.33.)

But, although the presence of the syntacticity thesis (6) in the *Tractatus* is obvious, its precise role is not.

At first sight, thesis (6) of syntactical form as pictorial form does not play an especially conspicuous role in the *Tractatus*, but it is nevertheless important to appreciate it. It is closely connected with the question as to whether, and if so in what sense, Wittgenstein was advocating an "ideal language" in the *Tractatus*. As Frank Ramsey emphasized against Russell (who nonetheless has since been followed by a host of interpreters), Wittgenstein's ideas in the *Tractatus* were not by any means restricted so as to apply to an ideal language only. The framework used here enables us to state precisely what the situation is and in what sense Wittgenstein's "theory" was supposed to apply also to ordinary language. For Wittgenstein, most of the picture theses apply both to natural and ideal languages. In particular, in so far as it makes sense to apply theses (1)–(6) to natural languages, they are applicable.

How can we see this? One way of doing so is to recall that Wittgenstein's picture view way calculated to apply also to thoughts and not only to language. Witness 3–3.001, 3.02.

> A logical picture of facts is a thought.
>
> 'A state of affairs is thinkable' – this means that we can picture it to ourselves.
>
> A thought contains the possibility of the situation of which it is the thought.
>
> What is thinkable is possible too.

These theses are *a fortiori* applicable to any meaningful language in which one's thoughts are expressed. In fact, in 3.1 Wittgenstein writes:

> In a proposition a thought finds an expression that can be perceived by the senses.

In 3.02, thesis (2) (the mirroring thesis) is expressed practically in so many words for thoughts instead of propositions. (Cf. here also 3.04.)

The other theses (1)–(5) can likewise be seen to apply to thoughts and through them to natural languages.

The only exception is the syntacticity thesis (6). This can be seen, e.g., from 3.323:

> In everyday language it very frequently happens that the same word has different modes of signification, and so belongs to different symbols – or two words that have different modes of signification are employed in propositions in what is superficially the same way.

It is precisely for the purpose of making the syntactical forms of language reflect its logical forms that we need "an ideal language:"

> In order to avoid such errors we must make use of a sign-language that excludes them by not using the same sign for different symbols and by not using in a superficially similar way signs that have modes of signification: that is to say, a sign-language that is governed by *logical* grammar – by logical syntax.
>
> (The conceptual notation of Frege and Russell is such a language though, it is true, it fails to exclude all mistakes.) (*Tractatus*, 3.325.)

How, then, can a proposition of natural language be a picture of a state of affairs, if its syntactical form does not reveal its logical form? Wittgenstein's answer is: Because the use of the proposition provides the missing ingredients of the form.

> In order to recognize a symbol by its sign we must observe how it is used with a sense. (3.326.)
>
> The *application* of logic decides what elementary propositions there are. (5.557.)

To speak of the application of logic is to speak of the tacit conventions governing the use of language. Of them Wittgenstein has this to say:

> The tacit conventions on which the understanding of everyday language depends are enormously complicated. (4.002.)

Of course, what is needed to understand a language according to Wittgenstein are the elementary propositions. Hence 5.557 and 4.002 concern the same matter.

This explains the apparent discrepancy between such statements as the following:

> Language disguises thought. So much so, that from the outward form of the clothing it is impossible to infer the form of the thought beneath it. (4.002.)

> In fact, all the propositions of our everyday language, just as they stand, are in perfect logical order. (5.5563.)

The simple explanation is that in the former Wittgenstein is talking about the syntactical "clothing" of language, whereas in the latter a proposition is taken to include also its use. This is shown by 5.557 which follows immediately 5.5563.

In other words, as Wittgenstein puts it in 3.262:

> What signs fail to express, their application shows. What signs slur over, their application says clearly.

One reason why it is important to appreciate the thesis (6) of pictorial form as syntactical form is what Wittgenstein later rejected it. This rejection has sometimes been taken to amount to a "rejection of the picture theory" on Wittgenstein's part, even though in reality it does not necessarily affect at all the rest of Wittgenstein's complex of ideas, i.e., does not necessarily affect the other theses (1)–(5). Hence it is a fundamental misunderstanding to take such a rejection of syntacticity thesis (5) to show that Wittgenstein "gave up the picture theory". This is but one example of how the distinction between the different theses (1)–(6) helps to dispel misunderstandings of Wittgenstein's philosophy.

20. CONSEQUENCES FOR INTERPRETING WITTGENSTEIN IN GENERAL

These observations are interesting also because they throw sharp light on other facets of Wittgenstein's philosophy. It is sometimes said that in the transition from his early philosophy to a succession of later positions Wittgenstein changed his philosophical methodology. The suggestion seems to be, in the crudest possible terms, that he changed from a logician into an ordinary-language analyst.

I shall not argue here that Wittgenstein underwent such a transition or that he did not. What is clear is that whatever changes there were in his mode of philosophical methods they were but *consequences of changes in his substantive philosophical views.*

This can be seen from what was said earlier. In his early work Wittgenstein was concerned with logical forms because he thought that the syntactical form of a proposition in a logically correct language contains everything we want to know about it – certainly everything we need to understand it. This explains Wittgenstein's preoccupation in the *Tractatus* with the logical (syntactical) forms of propositions such as they would be in a logically correct language.

But Wittgenstein later came to believe that the true logical form of a proposition can only be shown by the way it is used, no matter how ideal the language is that we are employing. Then the major emphasis will obviously have to be on those facts of use and usage which indirectly show its logical form. And to study them will be to study "ordinary use". Hence, Wittgenstein's alleged change of methodology is nothing but a corollary to certain changes in Wittgenstein's actual substantial views.

Here we also have an illustration of an extremely important point to be kept in mind in trying to understand Wittgenstein. In spite of all the changes, sometimes dramatic ones, in Wittgenstein's philosophical views, almost invariably his later position grew out of seeds implicit in his earlier ideas. In the present case, the idea that "what signs fail to express, their application shows" prefigures his later view of meaning as use.

Similar anticipations of later views can be found elsewhere in Wittgenstein. For instance, his eventual explanation of how "we can refer to sensations and to give them names" is anticipated by his ideas about "the language of gestures" in the early thirties.

21. WITTGENSTEINIAN PICTURES AS JIGSAW PUZZLES

The combined force of the different assumptions that enter into Wittgenstein's so-called picture theory determines the nature of logic according to the *Tractatus*. However, the different theses (1)–(6) affect Wittgenstein's conception of logic differently. It was already pointed out that the mirroring thesis (2) in conjunction with thesis (5) is essentially tantamount to the famous thesis that all logical truths are tautologies.

Theses (1)–(2) and (5) together imply a kind of jigsaw puzzle theory of the nature of logic. All we have in logic is a supply of names, corresponding to a supply of simple objects. Each object comes with a specification of the ways in which it can or cannot be combined with others, just like jigsaw puzzles pieces. Thus all that we have to deal with in logic is how to combine these jigsaw puzzles pieces into actual pictures.

It might seem that this is a hopelessly simplistic idea of logic. In other words, it might seem that Wittgenstein's reduction of entire logic to the logical form of simple objects impoverishes it hopelessly. How can we have a realistic "logical multiplicity" if all we are dealing with is a matter of combining different simple names with each other like jigsaw puzzle pieces, without any logic of propositional connectives let alone quantifiers? Notwithstanding such rhetorical questions, Wittgenstein's concept of logic is in fact not too poor at all. In fact, an interesting partial answer to the critical question is provided by certain types of work in logic. Hao Wang and his collaborators have shown that the decision problem for the entire first-order logic can be reduced to what are known as domino problems. (The crucial papers have been reprinted in Hao Wang, *Computation Logic, Philosophy*, Kluwer Academic, Dordrecht, 1990.) These problems are in effect glorified jigsaw puzzle construction problems.

The general form of such a domino problem is to fill the entire Euclidean plane with square dominoes, of course in such a way that contiguous sides of two adjacent dominoes always match. Moreover, for the construction one has available to oneself an infinite number of dominoes of each of a finite number of different kinds. Finally, it is required that at least one domino of each kind must be used.

Wang's reduction of the decision problem for first-order logic to the totality of such domino problems shows vividly that the Wittgensteinian jigsaw puzzle logic can have the same "logical multiplicity" as the entire first-order logic.

Moreover, it can be argued that Wang's result is not a logician's curiosity but really reflects the essential nature of first-order logic. One way of seeing this is to have a look at the distributive normal forms for first-order logic. (See here my contribution entitled "Von Wright on Logical Truth and Distributive Normal Forms" to Paul A. Schilpp and Lewis B. Hahn, editors, *The Philosophy of G. H. von Wright*, Open Court, La Salle, Illinois, 1989, pp. 517–37.) Their basic ingredients are the formulas known as *constituents*, which are of the form

$$\begin{aligned}&\Lambda_{i\in I}(\exists x)C_i[x] \quad \& \\ &(\forall x)\ V_{i\in I}C_i[x] \quad \& \quad \Lambda_j(+)Aj \qquad (1)\end{aligned}$$

The last term here is a conjunction of negated or unnegated atomic formulas. Obviously, the rest of (1) is in effect a list of all the different kinds of individuals there are.

In them, each $C_i[x]$ has the same form:

$$\Lambda_{j\in J}(\exists y)C_j[y,x] \ \&$$
$$(\forall y)\ V_{j\in J}C_j[y,x] \ \& \ \Lambda_k \pm A_k[x] \qquad (2)$$

When we compare two different conjunctions of the form (2) occurring in the same (1) with each other, we can see two necessary requirements which have to be fulfilled if (1) is to be consistent. First, each x in (1) has to find a slot among the y's in each conjunction (2) occurring in (1). This is very much like saying that any two jigsaw puzzle pieces must be compatible, capable of being fitted into one and the same picture.

Moreover, for each y in (2) there must be a compatible x in (1). This is very much like saying that each gap left by any given jigsaw puzzle piece must be capable of being filled by one of the available pieces.

This is obviously closely related to domino problems and to jigsaw puzzles. Indeed, von Wright has called (in a special case) the two tasks faced by anyone who is trying to show that (1) is consistent "the fitting-in problem" and "the completion problem". Thus the jigsaw puzzle analogy captures in effect beautifully the two basic conditions that a *prima facie* logical "picture" has to satisfy in order to be "coherent" (satisfiable) for the two conditions just sketched in fact suffice to weed out all inconsistent constituents even though for this purpose they have to be applied to constituents of an arbitrary great quantificational depth (number of layers of quantifiers).

22. WITTGENSTEIN'S LATER REJECTION OF THE SYNTACTICITY THESIS (6)

The jigsaw puzzle analogy is misleading in one respect, however. The pieces of the puzzle correspond to names. If the syntacticity thesis (6) is true, however, then in an ideal language a proposition is like an already completed jigsaw puzzle picture.

Wittgenstein himself mentions later an alternative to such a view. In *Philosophical Remarks* II, sec. 10, he writes:

> If you think of propositions as instructions for making models, their pictorial nature becomes even clearer.

By a "model", Wittgenstein here obviously means something tantamount to a picture.

Here the status of a proposition has altered radically. It is no longer like an already completed jigsaw puzzle picture. It now is like a jigsaw puzzle in its original unassembled state. It is no longer a picture; it is a recipe for constructing one.

What this means in that Wittgenstein has in a sense given up the syntacticity thesis (6). For a proposition in its actual syntactical form is no longer considered a picture by Wittgenstein. What is a picture is the result – possibly an unpredictable result – of certain picture constructions.

There are many statements in Wittgenstein's middle writings where he seems to be denying that there is an "agreement" between language and the world. They have mistakenly been taken as evidence that Wittgenstein "gave up the picture theory". This is simply wrong. What Wittgenstein is denying in such passages is merely the syntacticity thesis (6). The quotation given above is instructive in that Wittgenstein there explicitly embraces the picture idea, in spite of having abandoned the idea that propositions are pictures.

Another instructive passage is found in MS 116 published in part in *Philosophical Remarks*, Appendix 4B, pp. 212–14. At first sight, Wittgenstein is here rejecting the picture idea:

> What gives us the idea that there is a kind of agreement between thought and reality? – Instead of 'agreement' here one might say with clear conscience 'pictorial character'. But is this pictorial character an agreement? In the *Tractatus Logico-Philosophicus* I said something like: it is an agreement in form. But that is an error.

A closer look nevertheless quickly shows that the Wittgenstein is rejecting merely the syntacticity thesis (6). For the "agreement in form" Wittgenstein considers as an error is an agreement between the syntactical form of a sentence and the form of the fact it expresses. If we include the use of a sentence in the proposition it expresses, we can restore the "pictorial character". This use is called by Wittgenstein "the method of projection". This is clearly shown by Wittgenstein's words. (He is comparing a proposition to a blueprint drawing which a construction worker is using.)

> We might now express ourselves thus: the method of projection mediates between the drawing and object. – But if the method of projection is a bridge, it is a bridge which isn't built until the application is made. . . . What we may call 'picture' is the blueprint *together with* the method of its application . . . . So I am imagining that the difference between proposition and reality is ironed out by the lines of projection belonging to the picture, the thought, and that no further room is left for a method of application, but only for agreement and disagreement.

Thus Wittgenstein is here rejecting only the syntacticity thesis (6), not the other ingredients of the picture syndrome.

And the idea that "the method of representation" has to be included in the proposition is not even new for Wittgenstein. It is already asserted in the *Tractatus*; witness, e.g., 3.13: "A proposition includes all that the projection includes, but not what is projected". The only essential difference is that this "method of representation" now involved (according to Wittgenstein) human activities. In the *Tractatus*, "The method of projection so to think of the sense of the proposition". Later, it involves according to Wittgenstein calculus-like activities and eventually actual applications of language (Cf. the last displayed quotation from Wittgenstein.).

23. AFTERMATH (AFTERLOGIC?)

We have thus seen how each of the six different thesis (1)–(6) plays a different role in the *Tractatus*. Recognizing the differences between their several contributions to Wittgenstein's line of thought also helps to clarify essentially the argumentative structure of the *Tractatus*.

But does this suffice to show that the several theses are really separate? Perhaps not. However, eloquent further evidence comes from an interesting source: later Wittgenstein. For as I have noted above, in his later development Wittgenstein rejected some of the theses (1)–(6) but left others intact. Furthermore, history has rendered her judgement in other ways. For instance, anyone who believes in a (suitably modalized) version of Tarski-type truth-definitions in effect accepts thesis (1), but few, if any, of such analysts in these days accept any theses like (2) and (3).

To return to Wittgenstein, I have argued elsewhere that the first main change in his views was to abandon, probably sometime in 1928, the atomicity thesis (4). This means, trivially, giving up the categorial matching thesis (3) and the mirroring thesis (2). Less trivially, it means that the thesis (5) of complex propositions as pictures becomes, not invalid, but inoperative. It no longer serves the function of showing that all propositions in a well-formed language are pictures. Moreover, we saw above that Wittgenstein gave up the syntacticity thesis (6). Furthermore, because of Wittgenstein's incipient holism "the concept of an elementary proposition now loses all of its earlier significance", as Wittgenstein puts it, in *Philosophical Remarks* I, sec. 83. This seems to jeopardize thesis (1). Does anything remain about Wittgenstein's earlier theory?

It is perhaps understandable that many interpreters have taken these developments to mean that Wittgenstein "abandoned the picture theory". A closer look shows something quite different, however. It shows Wittgenstein's remarkable persistence in clinging to the picture idea. We saw instances of the persistence earlier in sections 9 and 19.

However, the way in which Wittgenstein was able to salvage some of his picture view has not yet been diagnosed fully. What happened was that by giving up the syntacticity thesis (5) Wittgenstein was able to save some extra ingredients of the picture idea. Even though a proposition as a syntactical entity cannot be construed as a picture, the situation is different if we take it together with the activities that connect it with the world. These activities are what Wittgenstein calls a "method of projection". Then a proposition could still be a picture of a fact, albeit not a "picture by similarity" (cf. sec. 9 above).

Another opening for maintaining the picture idea and even extending it was created by Wittgenstein's abandoning the idea of the logical forms of simple names which is the main presupposition of thesis (3). By giving up this idea, Wittgenstein made it possible for himself to acknowledge pictures which are not pictures by similarity, as we saw in sec. 9 above. In other words, the names could now be related to their objects conventionally, without presupposing any intrinsic identity of logical forms. (Of course, the logical form of propositions could still reflect the structure of the corresponding facts.)

This enables Wittgenstein to extend the picture idea from propositions whose names represent definite objects to propositions where they stand for indefinite objects, e.g. objects of a certain kind. This is what is involved in later Wittgenstein's distinction between "historical pictures" and "genre pictures". This pair of contrasting concepts is one of the recurring themes in Wittgenstein's later thought. (See, e.g., *Philosophical*

Investigations I, sec. 522.) It is found frequently in Wittgenstein's later writings. It is a vivid reminder that, far from giving up the picture idea lock, stock, and barrel, Wittgenstein thus in fact developed it farther in certain directions than he had done in the *Tractatus*.

A general way of describing what Wittgenstein did is to say that by giving up thesis (3) Wittgenstein was able to extend thesis (1) from elementary propositions to a number of others without the benefit of theses (2) and (6).

Boston University

ISAAC LEVI

RATIONALITY AND COMMITMENT*

This paper discusses the function of principles of rationality in inquiry and deliberation rather than the content of such principles. Appealing to the belief-doubt model of inquiry pioneered by C. S. Peirce and J. Dewey, I shall argue that principles of rationality should impose weak constraints on the coherence of the beliefs, values and choices of deliberating and inquiring agents. Efforts to derive substantial moral or theoretical deliverances from such principles are, thereby, ruled out of court.

Weak though these constraints may be, the capacity of human and institutional agents to satisfy them is severely limited. Principles of rationality are ill suited for the prediction and explanation of human behavior. Nor can they be regarded as prescriptions which rational agents are obliged to obey to the letter. The reason is the same in both cases. Persons, institutions and other alleged specimens of rational agency lack the emotional or institutional stability, the memory and computational capacity and the freedom from self-deceit required to satisfy the demands of even weak principles of coherence in belief, value and choice. Our rationality is severely 'bounded'.

In some respects, beliefs, value judgements and other so-called 'propositional' attitudes relevant to deliberation and inquiry resemble religious vows. Just as religious vows often incur obligations only an angel could fulfil, so too, only a rational angel can satisfy the requirements imposed on rational belief, value and choice. Still, so I shall argue, the laws of heaven do have a relevance to rationality on earth. I shall try to sketch an account of what that relevance might be.

I would bet my bottom dollar that Alexander Graham Bell spent some time in Brantford, Ontario. I grant that there is a logical possibility that I am wrong; but I have no living doubt justifying my engagement in inquiry. There is no serious possibility that I am in error in any sense guiding my current conduct. To be sure, very few of the members of this audience will share my conviction. Most of them will not have heard of Brantford, Ontario and those who, perhaps, have may be in doubt as to whether Bell ever lived there or whether he sent a

C.C. Gould and R.S. Cohen (eds.), *Artifacts, Representations and Social Practice*, 257–275.

telephone message from Brantford seven miles to Paris, Ontario. I myself confess uncertainty as to the precise date when this event took place and have some residual doubt as to whether this was the first long distance telephone call as is advertised by the Brantford community. Nonetheless, both those who share my views and those who differ do so by drawing a distinction between what is to be taken for granted as certainly true (or certainly false) and what is doubtful and uncertain.

Dewey thought that Peirce's distinction between what is taken for granted and what is open to question can be extended to attitudes concerning what ought to be done. Almost everyone would agree that murder is wrong. This pleasant unanimity is shattered when we ask whether abortion is wrong. At least some of the participants in the debate remain in honest doubt on the point rather than being firmly committed to either an anti-abortion or pro-choice view.

The distinction between the doubtless and the doubtful can be extended to other propositional attitudes as well. For example, an inquirer can be in doubt as to the appropriate 'prior' probability judgement to use in analysing data. For the present, however, I shall focus on the distinction between what is taken for granted in the sense of fully believed (disbelieved) and what is conjectural. I fully believe that Bell lived for a while in Brantford. That is to say, I am maximally certain that he did so. There is no serious possibility (no real and living doubt) as far as I am concerned that he did not. I remain in doubt, however, as to whether Reagan and Bush were in collusion with the Iranians over the timing of the release of hostages from the American Embassy in Teheran.

Advocates of the Peircean belief-doubt model insist that the inquiring or deliberating agent stands in no need to justify what he or she fully believes. Because I am sure that Bell lived for a while in Brantford, I lack the 'real and living doubt' required to provoke an inquiry into the matter. Justification is only required for changing one's state of full belief – i.e., modifying the distinction between what is judged to be certain and what is judged to be doubtful. Inquirers are sometimes justified in removing doubt and converting conjectures into certainties. On other occasions, inquirers are justified in doubting what they initially took for granted. In the absence of a good reason to change, however, the inquirer should retain the commitments he or she has. The first principle of pragmatism is "where it doesn't itch, don't scratch".

Thus, the fact that most readers do not share my conviction about Bell's sometime residence in Brantford is not sufficient to shake my conviction that he did reside there. However, if someone whom I respected as a competent authority on the history of southern Ontario were to weigh in with a dissent, that would normally qualify as a good reason for my coming to doubt my initial conviction and induce me to inquire further as to whether Bell did or did not live for a brief period in Brantford.

To be sure, if someone in the audience who was not the sort of competent authority I envisage demanded reasons for my convictions, I might try to oblige. But in so doing, I would not be attempting to offer reasons justifying my full belief. I have no real and living doubt which would require me to have such reasons. My attempt to offer reasons would really be an effort to explain to my interrogator why he or she ought to come to full belief that Bell lived in Brantford. I would marshall reasons which would justify the interrogator from the interrogator's point of view coming to full belief that Bell lived there for a while. But justifying a claim to another is different from justifying a claim to oneself.

I do not mean to suggest that when an agent lacks a good reason to question the agent's full beliefs, the agent is justified in suppressing the views of those with whom the agent disagrees. We tolerate the public expression of dissent while expressing our own contempt for it. To be sure, we are not obliged to register contempt either. Teachers are often obliged professionally to pay attention to points of view they judge to be patently false and, indeed, to pretend that their minds are open to such views even when they are not. This insincerity may be justified if it is effectively employed to induce students to critically examine their own views. And even where professional obligation does not support such dissembling, the demands of civility in discourse may. We should, however, distinguish between contemptuous toleration (both when the contempt is overt and when it is disguised for pedagogical purposes or considerations of conversational etiquette) and genuine respect for dissenting views. Respect for a dissenting view is displayed when the agent confronted with dissent recognizes a good reason for genuinely opening the agent's mind by ceasing to be convinced of the view the agent initially endorsed in opposition to the dissenter. Liberals of the sort I admire tolerate dissent but their toleration is often contemptuous. They need to have good reasons for opening their minds up so as to entertain seriously the views of dissenters. The mere presence of disagreement is not such a good reason. If it were, it would equate toleration and respect for the views of dissenters. Since there will be a dissenter for virtually every substantive view, advocates of toleration who conflate it with respect for the views of dissenters must be urging upon us the skepticism of the empty mind.

According to the vision present in the belief-doubt model, the central problem for epistemology is to give an account of intelligently conducted (i.e., rationally conducted) inquiries leading to modifications of states of full belief relative to which a distinction is made between what is settled and what is doubtful.

According to the Peirce-Dewey view of such inquiry, justifying changes of states of full belief (and, hence, of doubt) is attempting to adjust means to ends.[1] The ends of inquiries focused on obtaining information to settle some question of fact or theory may differ from and be irreducible to the goals of political, moral, economic or prudential decision making. Likewise the options faced will differ. But cognitive decision making remains a species of decision making. And like other types of decision making, it should be the outcome of intelligent or rational mean-ends deliberation.

If the principles of rationality regulating scientific inquiry as well as moral, political, economic, religious, aesthetic and self interested deliberation are the same, the principles of rational choice needed would have to be extremely weak. We could not require such principles to be substantial or thick enough to support a morality or a political superstructure or, for that matter, to sustain some strong network of principles for determining how well scientific conjectures are confirmed on the basis of given evidence.

Nonetheless, the principles no matter how 'thin' will be designed to constrain not only the deliberating agent's choices among options but will also presuppose for their applicability that the deliberating agent has certain kinds of attitudes. Thus, it is widely taken for granted that principles of rational choice recommending the maximization of expected utility presuppose that the agent's beliefs can be factored into the agent's full beliefs and credal probability judgements. Those who are skeptical of the propriety of expected utility maximization often question whether agents have credal probability judgements. They may, however, substitute other attitudes as expressions of uncertainty.

I shall follow familiar views according to which proposals for principles of rational choice presuppose that the deliberating agent to whom they are applied is not only in a state of full belief relative to which he or she distinguishes between settled assumptions and conjectures, but is also in a state of credal probability judgement and in a state of value commitment (relative to his or her state of full belief). The state of credal probability judgement furnishes a fine grained discrimination between the conjectures with respect to the agent's judgements of credal probability. The state of value commitments assesses the conjectures with respect to the agent's values, goals and desires. The state of full belief, credal probability judgement and value commitment together with the principles of rationality determine which of the options available to the agent are admissible for him or her to choose.

The trichotomy between states of full belief, states of credal probability judgement and states of value commitment is doubtless far too coarse grained to do justice to our attitudinal life insofar as it impinges on our deliberations. However, even this rough and ready distinction can be used to emphasize another important feature of the approach favored here. An account of rational deliberation needs to impose some constraints not only on criteria for what are admissible options but also on states of full beliefs, credal states and value commitments. Thus, if our theory of rational choice requires that full beliefs be closed under logical consequence, that our credal state obey requirements of the calculus of probabilities and that our preferences should be transitive and obey a so-called 'independence axiom', these requirements are to be enshrined as principles of rational full belief, credal probability judgement and value judgement.

The principles of rationality thereby serve a dual function according to the belief-doubt model as I am reconstructing it:

(1) They specify necessary conditions for a rational agent's being in a state of full belief (probability judgement, value commitment) at a given time. But since they are weak or thin constraints, agents can change from one such state satisfying the requirements of rationality to another. Hence, the principles of rationality also circumscribe the space of potential changes in states of full belief (probability judgement, value commitment).

(2) According to the belief-doubt model, changes in states of full belief (credal probability judgement, value commitment) call for justification. Since changes in states of full belief (credal probability judgement, value commitment) are to be justified by showing that such changes promote the ends of the inquiring or deliberating agent, the principles of rational choice specify minimal conditions such justification should meet.

The principles of rational choice, full belief, credal probability judgement and value judgement are neither explanatory nor predictive. No one is able to recognize as true all the logical consequences of what he or she fully believes, to fully obey the calculus of probabilities or to secure preferences (value comparisons) that are transitive. Nor can anyone even manage to approximate these requirements to any satisfactory degree except in relatively uncomplicated deliberations. Poor memory, limited computational capacity within the time constraints imposed for solving a problem, excessive costs, emotional stress, mental disease or lack of self knowledge can all conspire to prevent fulfilment of even very thin principles of rationality. So we cannot hope to use a belief-desire model of explanation to explain or predict human conduct relying on principles of rationality as covering laws. Nor is it to be expected that belief-desire models relying on qualified or modified versions of the principles of rationality will do any better. The likely outcome of inquiry aimed at identifying the requisite qualifications will be the replacement of presuppositions about the kinds of attitudes in the unmodified principles of rationality with presuppositions concerning other attitudes. The new theory of rationality is not going to do better in prediction and explanation from covering laws than the old. Those who retain a faith in principles of rationality as explanatory and predictive principles may continue to pursue their efforts to vindicate their faith in the face of the daunting obstacles they face. As far as I am concerned, the handwriting is on the wall.[2]

Yet, I do not think that we will or should stop using belief-desire models or the principles of rationality that accompany them for other purposes. Principles of rational choice have prescriptive uses. In particular, they are often intended for use by deliberating agents for the purpose of policing their own deliberations as well as those of others.

In inquiry, such self policing is of crucial importance. The adjustment of means to ends requires the deliberating agent to identify both his or her ends and the potential means (the available options) for realizing these ends and then to determine which of the available options are admissible relative to the ends recognized. That is to say, the deliberating agent will need to identify his or her state of full belief, credal state and value commitments. Without doing so, the agent will not be in a position to apply the principles of rational choice to identify the set of admissible options.

Someone will no doubt object that the same considerations that argue against using belief-desire models for explanation and prediction pose obstacles to their prescriptive use at least by human beings and their institutions. 'Ought' presupposes 'can'. If we cannot fulfil the requirements of thin principles of rationality, it would be foolhardy and irrelevant to impose an obligation upon us to do so.

So some will argue that we should recognize that our rationality is 'bounded' and should tailor our principles of rationality to our capacities to satisfy them. But no matter how we trim our principles of rationality, there will always be predicaments so complex and so stressful as to preclude the applicability of the eviscerated standard. Evisceration will continue until nothing of interest is left to carve out.

Still more importantly, the counsel of those who urge us to trim our principles of rationality is the counsel of complacency. Of course, we cannot be obliged to do at the moment what we cannot at that moment do. We cannot be obliged to recognize all the logical consequences of our full beliefs or even enough of consequences to solve some particular complicated problem. But we can be urged (costs and time permitting) to seek therapy for our distress, to devise prostheses to extend our computational capacities and memories (such as computers, paper and pencil, handbooks of tables, etc.) and to learn logic and mathematics so that we can to some extent overcome our disabilities. Authors who counsel us, instead, to modify the principles of rationality so that we can all be OK, neglect the importance of trying to enhance capacity to fulfil the demands of rationality.

Thus, we need a way of understanding the prescriptive force of the principles of rationality which can recognize our failings as rational agents while avoiding the temptation to convert such recognition into a complacent tolerance of these failings.

The worries just raised concerning the applicability of prescriptions of rational choice are also worries about whether such agents can be in states of full belief, credal probability judgement and value judgement meeting standards of synchronic rationality. This worry is different from the concern about the rational or intelligent conduct of inquiries aimed at improving states of full belief (probability judgement, value judgement) where these states are already assumed to satisfy the requirements of rationality. Nonetheless, the concern about applicability does threaten the belief-doubt model as I am construing it. The belief-doubt model offers an account of conditions under which changes in states of full belief (probability judgement, value judgement) are justified that presupposes that such states satisfy the requirements of synchronic rationality. Consequently, the belief-doubt model may be deemed irrelevant to deliberation and inquiry because no agent is ever in such a state.

We need to determine whether there is an important and useful sense in which flesh and blood human agents and the corporate agents with which they have relations can be said to be in states of full belief (probability judgement and value judgement) satisfying the demands of rationality even though we acknowledge our limited capacities for computation and recovery of information or lack of self-knowledge and our psychological infirmities. And we need show that the attitudinal states so identified are appropriate objects for critical review according to the belief-doubt model.

For the sake of simplicity, let us focus on changes in states of full belief. Much of what I have to say can be transferred *mutatis mutandis* to credal states and states of value commitment:

Consider the following scenario:

Mario took for granted at time t_0 that Albany is north of New York City. He also took for granted at that time that 'is north of' is transitive.

Without any alteration in these beliefs, at time t_1 he also came to fully believe that Montreal is north of Albany – perhaps, as a result of looking up the location of Montreal in an atlas. There is a sense in which Mario simultaneously came to fully believe that Montreal is north of New York. If Mario were offered a bet at t_1 as to whether Montreal is north of New York where he wins \$1 if Montreal is north of New York and loses nothing if it is not, he should be prepared to pay up to \$1 for the opportunity to take the bet. The reason is that it is inconsistent with what he already believes that Montreal is not north of New York and, hence, there is no serious possibility that Montreal is not north of New York. The bet insures him of a sure \$1 and if he thinks more money is better than less, he should be prepared to ante up any sum of money up to \$1 for the opportunity.

Yet, Mario may not consciously or explicitly recognize that Montreal is north of New York and in this sense fails to fully believe that Montreal is north of New York. There are several reasons why this may not be so.

(a) Mario may not have asked himself nor have been confronted by others with the question as to whether Montreal is north of New York.

(b) He may have been asked the question and answered in the negative due to emotional stress, limited computational capacity or the time required to reach an answer. Or he may have been offered the bet mentioned above for an arbitrarily small positive charge and refused it.

(c) He may have been asked the question and pleaded ignorance. He failed to identify the consequences of his full beliefs. Or he may have been offered the bet at various positive prices and refused it for some while accepting it for others. Perhaps again emotional stress, limited computational capacity or the time required to reach an answer prevented his putting two and two together.

Variant (a) of Mario's predicament at t_1 has been commonly handled by distinguishing between belief as a disposition to assent upon interrogation and belief as assent. More generally, when it is acknowledged that belief is not only a disposition to linguistic behavior such as assent but also to various forms of non-linguistic behavior such as accepting bets, distinctions are recognized between belief as disposition and belief as manifestation of the disposition. If Mario were offered the bet at t_2, his disposition to accept the bet for a price up to \$1 is, by hypothesis, already present. The stimulus is present and so is the manifestation. That is to say, he accepts the bet at the appropriate terms. So the difference between the change from t_0 to t_1 and the change from t_1 to t_2 is a difference between acquiring a doxastic disposition and being provoked to display it.

The contrast between belief as doxastic disposition and belief as doxastic manifestation does not accommodate predicaments (b) and (c) so well.

If at t_1, Mario had the requisite doxastic dispositions and had been interrogated, he should have responded by assenting to 'Montreal is north of New York'. In scenario (b), however, he refused the bet even for a small positive charge. Mario does not have the doxastic disposition to behave characteristic of full belief that Montreal is north of New York at t_1 according to the dispositional analysis. Yet given the full beliefs he by hypothesis does have, he should rationally have had the disposition. His failure to have it is a defect in Mario viewed as a rational agent. The remedy for the defect is to be achieved by alleviating Mario's emotional distress (or whatever) and improving his capacity to put two and two together and recognize the logical implications of his full beliefs. The remedy calls for therapy – not inquiry.

Suppose that Mario overcomes the defect at t_2. He acquires the doxastic dispositions he should already have had. There is a change in his beliefs understood as changes in doxastic dispositions. But there is no change in the doxastic dispositions he should, as a rational agent, have. That remains as it was at t_1. This stands in marked contrast to the shift from t_0 to t_1 which involves both a change in Mario's doxastic dispositions and in the doxastic dispositions he should rationally have.

Case (c) is like case (b) except for one point. Even though at t_1 Mario is interrogated, Mario pleads ignorance. This means, of course, that Mario cannot be said to fully believe as doxastic disposition that Montreal is north of New York or to manifest such a disposition. On the other hand, we have no grounds for attributing to him a doxastic disposition to dissent from 'Montreal is north of New York' either. And when offered the bet at varying prices, he displays that he is in a state of uncertainty or doubt as revealed by his accepting the bet at some prices but not at others. As in case (b), Mario has failed to have the doxastic dispositions which a rational agent in his circumstances should have. At t_2, he overcomes the failing.

I do not wish to quarrel with the naturalist view that there is a contrast between belief as a kind of 'multitrack' disposition and belief as a manifestation of such a disposition. But the difference between Mario's failure to believe that New York is north of Montreal at time t_0 and at time t_1 is not a difference in disposition. And it is not the difference between failing to display an already present disposition and manifesting it. At both times he lacked belief in the sense of a doxastic disposition to assent or to behave 'as if' New York is north of Montreal. At both times, he failed to manifest the doxastic disposition he did not have.

Why, however, is this difference between Mario's beliefs at the two times important? Mario's failure at t_1 to believe in the dispositional sense that Montreal is north of New York is a failure to do what he ought to have done. He had no such obligation at t_0.

The question still remains: How can we criticise Mario for failing to live up to a standard he cannot satisfy?

Imagine that Mario had a logically omniscient, perfectly rational *doppelgänger*, Mario's rational angel. Angelic Mario, though logically omniscient, is not omniscient. Hence, at t_0, angelic Mario has all the full beliefs flesh and blood Mario has plus all their logical consequences. At t_1, both angelic and flesh-and-blood Mario add 'Montreal is north of Albany'. Angelic Mario adds all logical consequences including

'Montreal is north of New York' as well. There is no change between angelic Mario's state at t_1 and t_2 except, perhaps, in terms of manifestations of his state as in the type (a) scenario for flesh-and-blood Mario.

We might say that Mario has an obligation as a rational though flesh-and-blood agent to mimic angelic Mario in his beliefs. Insofar as he cannot and costs permit, he should undergo therapy and training for improving his performance. The prescriptions concerning rational full belief (that it be closed under logical consequence and consistent) can be understood as regulating the way obligations are generated while providing for excuses due to human computational and emotional incapacity.

This does not as yet determine whether angelic Mario was justified in shifting from the doxastic dispositions he had at t_0 to those he had at t_1 or whether the flesh-and-blood Mario was justified in altering his obligations to acquire doxastic obligations accordingly. This issue is not a question for therapy or technology to handle. The justification is obtained at the outcome of inquiry. Unlike Angelic Mario, flesh-and-blood Mario stands in need of therapy at t_1. The change in his beliefs from time t_1 to time t_2 is a product of such therapy.

Abandoning the myth of the angelic Mario, from whence comes this doxastic obligation which in scenarios (b) and (c) Mario not only fails to fulfil but is not able to fulfil at t_1?

Clearly, whatever we want to say about the source of Mario's doxastic obligation, we would have to locate it in his adding full belief that Montreal is north of Albany in the dispositional sense to what he already fully believed at t_0. That act of adding new information to the initial belief state generated new doxastic obligations stronger than those he already had.

The idea that doxastic obligations are generated by doxastic dispositions suggests that changes in full belief illustrated by Mario's change from t_0 to t_1 are really changes in commitment (and not merely in disposition) in a sense I shall now try to explain.

Undertakings or commitments resemble contracts and promises in three respects: (1) Making a promise or drawing up a contract requires the occurrence of events describable in physical terms. (2) These events are redescribable as promises or contracts only if their occurrence is understood as generating obligations. (3) Prescriptive principles and their interpretations need to be invoked in order to construe the event described in physical terms as generating obligations. The putting of pen to paper must be interpreted as signing a contract to pay a sum of money by a certain time and such signing of contracts must be seen as generating an obligation to pay the sum of money by the due date. The obligation is generated by the signature in virtue of the laws of contract or moral principles invoked.

Mario's change in state of full belief from t_0 to t_1 resembles signing a contract in these respects precisely. Mario's change in state of full belief is a change not merely in some doxastic dispositions but also and more importantly in doxastic commitment. In acquiring the doxastic disposition to assent to 'Montreal is north of Albany', Mario has undertaken a commitment to maintain that doxastic disposition in addition to the doxastic dispositions he already had at t_0 pending the acquisition of good reasons for altering them. His undertaking generates further commitments to acquiring doxastic dispositions to assent to the logical consequences of 'Montreal is north of Albany' and the doxastic dispositions he had at t_0. In particular, Mario undertook to be disposed to assent to the truth of the claim that Montreal is north of New York. In lieu of the law of contract, the principles of rational belief generate the obligations.

Thus, we have a three way polyguity in our understanding of 'full belief': (A) Full belief as doxastic commitment; (B) Full belief as doxastic disposition – I shall call this full recognition rather than full belief; and (C) Full Belief as manifestation of doxastic disposition – I shall call this manifestation of recognition or doxastic disposition.

Full belief as doxastic disposition and as doxastic manifestation have unavoidably prescriptive and evaluative components. The behavioral or physical dispositions involved are interpreted as fulfilling a doxastic obligation or as generating such obligations. This interpretation cannot be obtained by appeal to information about the natural world alone.

But this ought not be so surprising. That is what Brentano's thesis has already told us about the gap between the intensional and the natural. The only novelty introduced, if it is a novelty, is that the irreducibility of the intensional to the physical is itself equated with the irreducibility of the prescriptive to the physical. More would-be naturalists may be prepared to resign themselves to the intractability of the fact-value dichotomy than to the gulf between the intensional and the natural.

We have not, however, fully addressed the difficulties about lack of logical omniscience which have provoked many commentators to follow

H.A. Simon in trimming the standards of rationality. When one makes a promise one is certain one cannot keep, one's conduct is fraudulent. By comparing belief states to undertakings where the undertakings generate obligations extending well beyond our capacities, belief states are construed as fraudulent commitments.

Doxastic commitments are not contracts or promises. But there are other kinds of undertakings. Religious vows are often seen as generating undertakings well beyond the ability of the person taking the vows to satisfy. Someone vows to be righteous knowing full well that he will never satisfy all the requirements of righteousness. Yet such vows are not fraudulent.

The reason they are not fraudulent is that they do not strictly speaking obligate the one who undertakes the commitment to meet them fully or even approximately but only insofar as the agent has the capacity to do so. However, this rather substantial qualification of the obligation undertaken is accompanied by another substantial obligation. The agent is required to extend his or her ability to fulfil the obligations by training and education, the use of appropriate therapies, and the acquisition of prosthetic devices enhancing his or her capacities. There is no fraud because the person who has undertaken the religious vows is not obliged to do what he or she cannot do although he or she is obliged to seek ways and means to enhance his or her capacities.

Doxastic commitments resemble such religious undertakings in this respect. We are committed doxastically to fully believe the consequences of what we fully believe and to believe that we believe them. This does not mean that Mario at t_1 is obligated to fully recognize that Montreal is north of New York even when, as in scenarios (b) and (c) he cannot do so. However, given that he cannot do so, he should seek ways and means to enhance his capacity to put two and two together.

Thus, the inquiring and deliberating agent is committed to improving the extent to which he or she meets the demands of logical omniscience by studying logic and mathematics, using computers and other prosthetic devices and undergoing forms of psychotherapy.

Suppose then at time t_2, Mario, who at t_1 is in either version (b) or (c) of the predicament, does put 2 and 2 together and successfully recognizes that Montreal is north of New York. The change from t_1 to t_2 is clearly different from the change from t_0 to t_1. The latter change is a change in Mario's doxastic commitment and is best seen as the outcome of inquiry. The former change involves no change in commitment but only in what I shall call 'doxastic performance'. And it is a result of therapy or the use of prostheses.

Version (a) of Mario's predicament at t_1 is slightly different. Here Mario is not tested at t_1. It is left unsettled as to whether Mario has the doxastic dispositions fulfilling his doxastic commitments or does not. If the former is the case, the shift at t_2 is one where there is no change either in doxastic commitment or doxastic performance but only in the manifestation of Mario's doxastic performance. In the latter case, there is a change in performance.

So far all of my examples and discussion have been about full belief. However, the contrasts I am sketching here apply *mutatis mutandis* to probability judgement and value judgement. On the view I am proposing, principles of rational full belief, credal probability judgement and value judgement characterize the way in which states of full belief, credal probability judgement and value judgement understood as states of attitudinal commitment generate obligations to full recognitions, conscious probability and conscious value judgements that only rational angels could fulfil perfectly. We are not obligated so strictly. God is merciful. We are, however, obliged to improve our capacities to fulfil our commitments opportunities and costs permitting.

I first bruited an analogy between belief as commitments and promises or contracts. I admitted the analogy was only partial and introduced a second analogy between belief as commitment and religious vows which overcomes some of the defects in the analogy with promises. But this analogy too may seem only partial. Someone might ask: To whom has Mario made his doxastic vows when undertaking doxastic commitments? And what are the sanctions legitimizing the prescriptions of rational full belief, credal probability and value judgement which generate obligations from performances?

Dewey insisted that one can behave religiously without being committed to any sort of theology (Dewey, 1934). Atheists, agnostics and other secularists can behave religiously. I would go further. We need not suppose that if the vows are not to God they must be to some community or individual. The vows are to no one or no one group. And the ends we take for granted do not require justification as long as they are taken for granted.

I am a secularist. It is enough for me that principles of rationality can be understood as characterizing the commitments undertaken in states of full belief, probability judgement and value judgement. With this understood, we can distinguish between change in belief (probability judgement, value judgement) construed as change in performance and best effectuated by therapy and training and change in belief (probability judgement, value judgement) construed as change in commitment.

I suggest that the province of inquiry according to the belief-doubt model of Peirce and Dewey is to rationalize changes in attitudinal commitments whereas therapy, technology and education address the improvement of performance. Clearly one cannot have the one kind of change without the other. Doxastic performances cannot be recognized to be performances unless they are understood as attempts to realize commitments. Understanding comes when the commitments which the agent is attempting to fulfil are recognized. Commitment without performance is empty. But we should also remember that performance without commitment is blind.

Principles of rationality, therefore have two tasks to perform: They regulate changes in commitment by furnishing the criteria for admissibility that warrant choosing one change in commitment over another. And they regulate changes in performance by indicating the standards to which performance would conform were we rational angels and thereby furnish the ideals of rational health which should guide our psychotherapies, our educational programs and our technologies.

Both Dewey's vision of the intelligent adjustment of means to ends and my view of principles of rationality as thin constraints on rational full belief, probability judgement, value judgement and choice are paradigmatically conceptions of instrumental rationality. We both contend that it is rationality so construed that is central to justifying changes in doxastic and other attitudinal commitments (such as changes in probability and value judgement as the first task requires).

Nonetheless, these principles also constrain our interpretation of the propositional attitudes as dispositions – i.e., as performances and as manifestations fulfilling or failing commitments as the second task demands. In performing this task of characterizing the standards of rational health, the values embodied are themselves not to be rationalized by reference to other goals and values they promote. Although they are secular in the sense they entail no theology, they have the features of religious commitments I have sketched above.

This brings me to the important respect in which the view of rationality I have been outlining may differ from Dewey's. Dewey seems to have held that our full beliefs in the truths of logic are liable to reasoned change. I surmise he would have said the same about commitments to minimal requirements on the coherence of probability judgement and value judgement as well. Nothing is fixed in concrete including the thin principles of rationality.

In one respect, I share Dewey's view. Thin principles of rationality characterizing our doxastic and other attitudinal commitments have changed historically. Even at the present, they are subject to controversy and change.[3] The problem is to give an account of good reasons for making such alterations. Such an account would have to meet two requirements: (a) It would have to be comprehensive enough to address good reasons for changing full beliefs, probability judgements and value judgements and (b) it would need to avoid begging controversial questions about changes in the standards of what constitute good reasons. I do not know how to meet these demands.

Nonetheless, there are certain considerations which are relevant in assessing standards of rationality. I have already invoked them in this discussion. The standards should be strong enough to perform their functions in interpreting the attitudes as commitments and performances and evaluating choices in the light of such interpretations. Yet, they should be weak enough to accommodate a broad spectrum of belief states and value commitments so as to recognize such states and commitments as potentially subject to reasoned review and alteration as is feasible.

In opposition to Kant and other enlightenment figures, we should try to avoid rendering our favorite substantive doxastic and evaluative commitments secure against reasoned change by seeing them as dictates of rationality. Morality and science are one thing. Rationality is quite another. We may expect our morality and our science to conform to the requirements of rationality. But we should resist the pretensions of those who seek to ossify morals or science in the name of reason or, who, while acknowledging the changeability of morals and science, see such variability as manifesting a variability in the standards of rationality which cannot be rationalized.

To the extent that we can meet the two vaguely specified demands mentioned above, we may hope to come close to avoiding these pitfalls and fulfilling the spirit if not the letter of Dewey's view.

Columbia University

NOTES

* Thanks are due to Akeel Bilgrami for his helpful comments and his unsuccessful efforts to save me from error.

1 Consider, for example, Dewey's notion of warranted assertibility. (Dewey, 1938, p. 120, 1977, p. 271.)

The inquiring agent seeks to answer an as yet unsettled question. When an answer is obtained at the conclusion of such an inquiry, it is a "judgement" in Dewey's terminology. The conjectures that constitute potential solutions of the problem as well as the settled background information, techniques and methods taken as noncontroversial resources in the context of the specific inquiry are called "propositions". These are the entertainable means for realizing the given end of solving the problem. For Dewey, propositions are "affirmed" and judgements "asserted". When the inquiry is properly conducted with the appropriate adaptation of means to ends, Dewey called the assertion or judgement which represents the solution to the problem "warranted". The judgement that is the warranted assertion of that inquiry may then become a resource for subsequent deliberation and inquiry. *Qua* means of the new inquiry, it is not a warranted assertion. It is an affirmed proposition. This does not mean that it becomes a conjecture. Unless there is good reason to do so, once the inquirer has become certain at the end of the first inquiry, he or she should not cease being certain subsequently. What it does imply is that the settled assumption is now a means for realizing new ends in new inquiries.

My point in elaborating on Dewey's categories of assertion of judgements and affirmation of propositions is not to endorse their usefulness. I myself have not found them convenient. But they reveal Dewey's conviction that even scientific inquiry is a goal directed activity exhibiting features in common with technological, economic, moral, prudential and political deliberation. In this sense, justifiable change in state of belief is a species of rational decision making.

2 My skepticism about the use of principles of rationality as covering laws in explanation of physical behavior does not derive from doubts about the availability of psychophysical laws covering the causal interactions between mental and physical events as does Donald Davidson's skepticism about psychophysical explanation and prediction (1980, essays 11 and 12). I contend that the principles of rational belief, desire and choice fail to contribute to explanation of behavior as physically described because such principles are false as applied to human beings. It is quite clear that human and institutional agents lack and will always lack the computational capacity and emotional or institutional stability to conform to principles of rationality when the predicaments faced are too stressful or complex. I can at best assent to the truth of finitely many sentential representations of consequences of my beliefs. There are infinitely many others I could not assent to even if asked.

Thus, principles of rationality could not be "constitutive" of what it is to be a rational agent as Davidson (1980, p. 221) appears to say because they are known to be false of agents. Of course, Davidson acknowledges that they are false of agents. But he does say that we cannot coherently regard someone who is subject to "global confusion" as an agent. I am not sure what "freedom from global confusion" means for Davidson. But I suspect that Davidson means that agents free of global confusion behave rationally most of the time. But whether the agent is free of global confusion in this sense would depend upon how frequently he or she faced highly complicated problems which the agent lacked the computational capacity to handle. Someone whom we would regard as a clear headed and highly intelligent agent might be thrust into a position of responsibility where his or her daily fare became complex problem solving transcending the agent's capacity. Would the erstwhile agent then cease to be an agent? Perhaps, Davidson will be able to find a satisfactory explication of the confusing notion of freedom from global confusion. But it is clear that universal conformity with the principles of rationality is not constitutive of agency. Nor are these principles covering laws over the domain of agents. Hence, they cannot be used in explanations of choices or other intentional behavior or of physical behavior. This is so whether or not there are psychophysical laws.

It may, perhaps, be objected that the principles of rationality are covering laws when a *ceteris paribus* clause is added. Those who invoke *ceteris paribus* clauses must be thinking of laws as represented by two components: (a) a domain or scope specification and (b) a formula or principle. It is not required that the distinction between principle and domain of applicability should be context independent. A lawlike universal generalization of the form "All A's are B's" can be parsed into the components (a) and (b) in at least two ways. According to one method, the scope specification is the range of the variable bound by the universal quantifer in "$(x)(Ax \rightarrow Bx)$" and the formula is the open sentence or predicate "$Ax \rightarrow Bx$". According to another, the domain is the set of A's and the formula is "Bx". Often the choice of a way of making the decision depends upon whether in making repairs to a putative law subject to refutation by counterinstance by imposing restrictions on the domain what aspect of the putative law one seeks to save. If one finds an A which is not a B and seeks to identify more accurately the domain in which B holds, "Bx" becomes the formula. But if one seeks a more accurate specification of the domain in which "$Ax \rightarrow Bx$" holds, the formula is "$Ax \rightarrow Bx$". No matter how one proceeds, introducing a *ceteris paribus* clause is to offer a promissory note for restricting the domain of application of the formula or principle.

As defenders of the applicability of psychological principles in explanation and prediction of behavior are fond of pointing out, *ceteris paribus* qualifications are often found in the natural sciences as well as in psychology. However, they are promissory notes which ought to be redeemed by supplying more adequate scope restrictions. In the natural sciences, this responsibility is often undertaken and often yields the result that no adequate replacement of the *ceteris paribus* clause can be found without modifying the formula. Moreover, even the restrictions on the scope and the modification of the principle initially designed to characterize complex systems typically end up being stated within the framework of some more fundamental theory. There is no *a priori* necessity that this be so; but it is quite likely to happen if *ceteris paribus* clauses attached to principles of rationality construed as components of covering laws are redeemed. The restrictions of scope will quite likely be most accurately described in biological or biophysical terms and the principles of rationality will cease being constituents of the formula.

I suppose that one might say that the principles of rationality can be ingredient in covering laws where the domain is the domain of rational, logically omniscient angels. But that set is the empty set and the covering laws for that domain are not of great interest unless other agents closely approximate the behavior of rational angels. Neither corporate nor human agents begin to come close to approximating rational angels.

In sum, the principles of rationality are neither constitutive of nor covering laws for the domain of personal or corporate agents.

[3] I myself have championed substantial changes in ideals of rational choice envisaged by strict Bayesians and other students of rational choice. (Levi, 1974, 1980, 1986.)

REFERENCES

Davidson, D. (1980), *Essays on Actions and Events*, Oxford: Clarendon Press.
Dewey, J. (1934), *A Common Faith*, New Haven: Yale University Press.
Dewey, J. (1938), *Logic: The Theory of Inquiry*, New York: Holt.
Dewey, J. (1977), *Dewey and His Critics*, ed. by S. Morgenbesser, New York: *Journal of Philosophy*.
Levi, I. (1974), 'On Indeterminate Probabilities,' *Journal of Philosophy 71*, 391–418.
Levi, I. (1980), *The Enterprise of Knowledge*, Cambridge, Mass.: MIT.
Levi, I. (1986), *Hard Choices*, Cambridge: Cambridge U. Press.

ALASDAIR MacINTYRE

THE *THESES ON FEUERBACH*: A ROAD NOT TAKEN

When we reread Marx nowadays, that reading has to address two salient and related features of our recent experience. The internal collapse of the Communist state apparatus in so many countries has left behind a variety of groups in those countries struggling to attain or rather to reattain the standpoint of civil society. At the same time the distinctively contemporary social theorizing of our own political culture, theorizing which gives a voice to the now dominant forms of power, either asserts or presupposes that the standpoint of civil society cannot be transcended. What then was and is the standpoint of civil society?

The expression 'civil society' and its cognates in other European languages had first been used to translate Aristotle's '*koinōnia politikē*'. But by the early nineteenth century it had come to be used in a variety of very different ways and Hegel, who learned it from Adam Ferguson, adopted it to name those social, economic and legal relationships into which individuals enter in order to satisfy their needs, forming by so doing "a system of complete interdependence wherein the livelihood, happiness and legal status of one human being is interwoven with the livelihood, happiness and rights of all" (*Rechtsphilosophie* 183). The individual from the standpoint of civil society is to be distinguished from and contrasted with the set of social relationships into which she or he has chosen to enter. Those relationships, often understood as contractual, are on the one hand a means to the attainment of each individual's ends and on the other a system so constructed that by entering it each individual becomes a means for the attainment by other individuals of their ends. Among the needs generated by such a system therefore is one for the protection of individuals from being so used by others as a means that their pursuit of their own ends becomes frustrating rather than fulfilling. Hence appeals to moral and legal norms affording such protection have an important function within civil society. The central conceptions informing thought within civil society about human relationships are therefore those of utility, of contract and of individual rights. And the moral philosophy which gives expression to the standpoint of

C.C. Gould and R.S. Cohen (eds.), Artifacts, Representations and Social Practice, 277–290.

civil society consists of a continuing debate about those concepts and how they are to be applied.

Up to 1844 Marx had engaged in a philosophical debate with Hegel, with the Left Hegelians and with Feuerbach about the nature of civil society, its relationship to the state and to religion, and the inadequacy of the criticism of civil society by its critics so far. The subtitle of *The Holy Family* summarizes his enterprise: 'a Critique of Critical Critique'. In 1845 Marx and Engels together embarked on a new project in the course of first writing and then abandoning the manuscript of *The German Ideology*, that of offering an historical and analytical account of the genesis and dynamic of modern capitalist economies. What were Marx's reasons for thus turning away from philosophical enquiry? When I speak of a turning away from philosophical enquiry I am not of course denying that Marx's later historical and economic analyses are themselves informed by philosophical presuppositions. That would be absurd. But philosophy is no longer the object of his enquiries and the questions which he poses are generally not philosophical questions.

Lucio Colletti has remarked on how few of Marx's later writings advert to this change by giving us "the reasons, philosophical as well as practical, which had induced Marx to give up philosophy after his break with Hegel and Feuerbach" ('Introduction' to *Karl Marx: Early Writings*, ed. Q. Hoare, New York, 1975, p. 8), citing the short text to which Engels later gave the title 'Theses on Feuerbach', the *Preface* to *A Contribution to the Critique of Political Economy* published in 1859 and the *Postface* to the second edition of the first volume of *Capital*. But the latter two were written a very considerable time later from what had become Marx's mature standpoint. Only in the first of these do we have a genuinely transitional text.

Colletti's own diagnosis of the change focuses upon Marx's critique of Hegel, in which Marx had worked through Hegel's conceptions of dialectic and of the state to the point at which their empty abstractions and incoherences had led him to reject them. And Colletti understands Marx's next stage as one that had progressed beyond the limitations of those conceptions. I want instead to suggest that the important question is not so much why Marx rejected Hegel and Feuerbach as why, in rejecting them, he rejected philosophy, and moreover that, by rejecting philosophy, at a stage at which his philosophical enquires were still incomplete and were still informed by mistakes inherited from his philosophical predecessors, Marx allowed his later work to be distorted by presuppositions which were in key respects infected by philosophical error.

Marx's *Theses on Feuerbach*, on this view, are in part a successful, but in part an unsuccessful attempt to identify what is involved in transcending the standpoint of civil society. In distinguishing the success from the failure I cannot but presuppose some particular interpretation of the text and there is no interpretation which is not contentious. But I shall bypass the scholarly disputes, because within the space of this paper any adequate treatment of them would be impossible. And in so doing I shall also fail to acknowledge scholarly debts. Informed readers will notice, for example, that I take for granted George L. Kline's thesis ('The Myth of Marx's Materialism', *Annals of Scholarship* 3, 2, 1984) that Marx did not have a materialist ontology and that the word '*Materiell*' has, in *The Theses on Feuerbach* as elsewhere, to be construed with care; and that I presuppose the truth of Carol Gould's account of Marx's ontology of individuals-in-relation (*Marx's Social Ontology*, Cambridge, Massachussetts, 1978, chapter I, especially pp. 30–39) and its Aristotelian antecedents. But they and a number of unacknowledged others cannot be held responsible for an interpretation that I assert rather than argue.

My approach will be to identify and to comment upon six central assertions expressed in Marx's eleven theses. The first of these is that the standpoint of civil society cannot be transcended, and its limitations adequately understood and criticized, by theory alone, that is, by theory divorced from practice, but only by a particular kind of practice, practice informed by a particular kind of theory rooted in that same practice. The philosophers have hitherto tried to understand, but their understanding was not guided by the aim of transforming the social and natural world in the requisite way. The eleventh thesis does not tell philosophers to abandon the attempt to understand; it tells them to direct their tasks of understanding towards the achievement of a particular *telos*. What *telos*?

It is the *telos* of some form of what Marx in the first thesis calls *objective* activity, taking over this expression from Fichte and Hegel. Objective activity is activity in which the end or aim of the activity is such that by making that end their own individuals are able to achieve something of universal worth embodied in some particular form of practice through cooperation with other such individuals. The relationships required by this type of end are such that each individual's achievement is both of *the* end and of what has become her or his own end. Practices whose activity can be thus characterized stand in sharp contrast to the practical life of civil society. It is a contrast which is best expressed in Aristotelian rather than in Hegelian terms.

In activities governed by the norms of civil society there are no ends except those which are understood to be the goals of some particular individual or individuals, dictated by the desires of those individuals, and no goods are recognized except those involved in the satisfaction of the wants and needs of individuals. Because there are many goods which individuals can achieve only by cooperative attention to the goods of others, civil society recognizes as common goods those goods which are pursued in common by individuals. But the only available conception of a common good is one constructed from and reducible to conceptions of the goods pursued by various individuals in their attempts to satisfy their desires.

By contrast the ends of any type of practice involving what Marx calls objective activity are characterizable antecedently to and independently of any characterization of the desires of the particular individuals who happen to engage in it. Individuals discover in the ends of any such practice goods common to all who engage in it, goods internal to and specific to that particular type of practice, which they can make their own only by allowing their participation in the activity to effect a transformation in the desires which they initially brought with them to the activity. Thus in the course of doing whatever has to be done to achieve those goods, they also transform themselves through what is at once a change in their desires and an acquisition of those intellectual and moral virtues and those intellectual, physical and imaginative skills necessary to achieve the goods of that particular practice. So, as Marx puts it in the third thesis, there comes about a "coincidence of the changing of circumstances and of human activity of self-changing".

Yet at once it is plain that there are at least two objections to construing Marx in this way. First and most obviously on this construal Marx is presented as if he had made a distinction which is expressed in an Aristotelian vocabulary, a vocabulary which he did not in fact use and some of whose presuppositions he had rejected. The conception of a type of practice teleologically ordered to the achievement of a or the common good may, it will be said, be at home in an Aristotelian or Thomistic perspective, but it is alien to Marx's. To this I respond by agreeing in part: what I have ascribed to Marx is indeed not what Marx said. Nonetheless I am contending that, if Marx had done the work of spelling out in detail the key distinction which the argument of *The Theses on Feuerbach* needs, he would have been compelled to articulate it in something very like Aristotelian terms. Hegel's idiom is just not adequate to the task.

To this a second objection must be that what Marx says is far too compressed and elliptical to support this kind of interpretation. Those interpreters who have elucidated the theses by drawing on Marx's other writings have at least had the evidence of those other writings to offer. To this I reply that, if we understand the theses as on the one hand marking a significant break with what Marx had done hitherto, and on the other pointing in a direction which Marx did not in fact take, then reliance on his other writings may itself be misleading. What we should be looking for is an attempt to articulate, in terms that are not deformed either by the errors of Hegel or by those of Feuerbach, an effective rejection of civil society. So our first question should be: what was it that, on Marx's view, rendered any rejection of civil society in either Hegelian or Feuerbachian terms ineffective?

Marx's rejection of all Hegelianism, and more especially of that of the Left Hegelians, was in important part a rejection of purely theoretical enquiry as an instrument of social change. What the Left Hegelians had characteristically supposed was that to exhibit the incoherences of the principles embodied in the social and political *status quo*, thereby exposing its irrationality, was by itself to have made an important and effective contribution to bringing about its downfall. We should not be harsh in condemning their error. After all we have over a century and a half of experience which they lacked, an experience of modern social orders not merely surviving the exposure of the incoherence of their governing principles, but even in some cases seeming to flourish because of that incoherence. The modern state, for example, behaves part of the time towards those subjected to it as if it were no more than a giant, monopolistic utility company and part of the time as if it were the sacred guardian of all that is most to be valued. In the one capacity it requires us to fill in the appropriate forms in triplicate. In the other it periodically demands that we die for it. This deep incoherence in the modern state is not a secret, but the fact that it is plain for everyone to see of itself does nothing at all to undermine the modern state. And Marx was perhaps the first to recognize how very little the exposure of incoherence generally achieves.

Other aspects of Marx's critique of Hegelian philosophy he shared with and had indeed learned from Feuerbach. By the time that he wrote the theses, he needed to take great care to distinguish his own positions from some of those of the later Feuerbach (on the complexity and subtlety of Feuerbach's development see Marx W. Wartofsky, *Feuerbach*, Cambridge, 1977). Feuerbach after all had already written in *Principles of the Philosophy of the Future*, published in 1843, that "The road taken so far by speculative philosophy from the abstract to the concrete, from the ideal to the real . . . will never arrive at true objective reality", but only at a reification of philosophy's abstractions, and that "The transition from the ideal to the real takes place only in practical philosophy" (II, 252). This seems to anticipate Marx. How then is Marx's position in the theses to be differentiated?

Marx in the theses makes one of the main heads of his criticism of Feuerbach a charge that Feuerbach's critique of religion had been inadequate. Feuerbach had understood religion as a distorted expression of human sentiment, one in which truths about love were expressed in a disguised form, one in which the true relationships of subject and predicate were inverted. Philosophy was to pierce through this disguise and by setting out the relevant truths in rational form dispel supernaturalist illusions. But Feuerbach then went on to announce what he took to be his discovery that philosophy also by its abstractions generates illusions. Marx's criticism of Feuerbach on religion is therefore perhaps best read as a prologue to Marx's criticism of Feuerbach on philosophy.

Marx's criticism of Feuerbach on religion had two parts. First he complained that Feuerbach, while understanding, in Marx's view correctly, that religion has to be wholly explained in terms of its secular basis, does not then ask what it was in "the cleavages and self-contradictions" (the fourth thesis) in that basis which had engendered illusion, and how that secular basis would have to be transformed so that it was no longer liable to engender illusion. Secondly Marx contended that Feuerbach had not analyzed "the religious sentiment" adequately as a social product, since Feuerbach's explanation of it terminates at the point at which he analyses it psychologically as a sentiment of individuals, rather than as a mode of expression characteristic of a particular type of social order (the eighth thesis). What matters to Marx here is the particular conception of the individual upon which Feuerbach relies. Feuerbach does not understand about that conception, first that it too is among those conceptions which are defective because abstract, and secondly that it belongs to the conceptual scheme of one particular type of social order.

If we apply Marx's criticisms of Feuerbach's account of religion as distortion to Feuerbach's later account of philosophy as distortion, what then is it that Marx is saying? He is asserting, first that philosophy has its secular basis in a particular type of social order, that informed by the standpoint of civil society, and secondly that, if we suppose that in understanding philosophical enquiry and argument as the activity of individuals, and that by giving an account of the secular basis of that activity as the activity of individuals, we have successfully moved from the abstract to the concrete, then we shall be deceiving ourselves. Marx's use of the notion of abstraction has of course often been criticized. Surely, it has been said, all concepts, all uses of language, involve abstraction, and therefore it cannot be a criticism of any particular concept or conception that it is abstract. But this criticism misses the point, something that has also been often said, but which still bears repeating. In Marx's semitechnical Hegelian usage, to abstract is always to frame a concept in a way which deprives it of the contextual connections in which alone it is at home and therefore to present it as having application independently of the relevant set of contexts. It always in consequence involves conceptual error and misunderstanding. We should perhaps note in passing that Marx's use of this notion of abstraction is often Wittgensteinian rather than Hegelian.

What then is it about the concept of the individual, as deployed by Feuerbach, which renders it abstract, and what is it about that abstraction which enables it to play its part in the thinking and acting characteristic of civil society? An answer to this latter question itself supplies an answer to the former. We already noticed that all transactions in civil society are understood to be between individuals and sets of individuals and that those individuals are only contingently related through their own acts of will to the social circumstances and the social relationships which they happen at any particular moment to inhabit. It may be of course that an individual finds her or himself entangled in certain social relationships without having willed this to be the case. But that she or he continues in them is, except when force or fraud are at work (the force and fraud which the legal and moral protections of civil society are designed to prevent), her or his own doing. Everything that comes about is understood to be either an intended or an unintended consequence of the actions of one or more individuals. The human individual must therefore be viewed in abstraction from her or his social relationships and the human essence must be specifiable by reference only to properties possessed by individuals apart from and in independence of their social relationships.

What is important is to recognize that it is this conception of the individual which is actually embodied not just in the thought, but also in the activities characteristic of civil society. To regard individuals as distinct and apart from their social relationships is a mistake of theory, but not only a theoretical mistake. It is a mistake embodied in institutionalized social life. And it is therefore a mistake which cannot be corrected merely by better theoretical analysis. Better theoretical analysis is of course necessary and in the sixth thesis Marx indicates what kind of theoretical account is required. The human essence is not given by considering the properties of individuals in isolation. "In its reality it is the ensemble of social relations". And of course what Marx means by "*the* ensemble" in this aphoristic utterance is not entirely clear. What is clear is that human beings who genuinely understand what they essentially are will have to understand themselves in terms of their actual and potential social relationships and embody that understanding in their actions as well as in their theories.

In civil society however there has to be a contradiction, a cleavage between how human beings really and essentially are and how they understand themselves to be. This "cleavage and self-contradiction", whereby civil society is a social order in which human beings are generally deprived of a true understanding of themselves and their relationships, is the source of the illusions diagnosed by Feuerbach. Even though these are philosophical illusions, whether about religion or about philosophy itself, they can have no philosophical cure. Here Marx becomes anti-Wittgensteinian. The only remedy for such illusions is an alternative form of practice of just that kind which we have already seen to be incompatible with the standpoint of civil society. Why so?

Civil society is characterized not only by its abstract individualism, but by a particular way of envisaging the relationship between all theory, including social theory, and practice. The adequacy of a theory to its objects is conceived of as a matter of the conformity of "thought objects" to "sensuous objects" (the first thesis). On this view we are to correct "abstract thinking" by our contemplation of the sense-experience afforded by the physical and social world. From the point of view of what Marx calls Feuerbach's "contemplative materialism" what contemplation of the social world reveals are "single individuals" and their agglomeration in civil society (the ninth thesis). Theoretical investigation leads to the materialist conclusion that these individuals are what they are because of their circumstances and their upbringing. Human beings are then taken to be a product of causal agencies over which they have had no control.

The social theorist who has arrived at this conclusion has by so doing completed the work of elaborating an adequate theory. Now she or he has to understand her or himself as about to embark on a second distinct task, that of applying this theory in order to effect change. But in so envisaging their task theorists have, characteristically without recognizing it, made the sharpest of distinctions between how they understand themselves and how they understand those who are the subjects of their enquiries. They understand those whose actions and experiences are to be explained by their theory as the wholly determined products of circumstance and upbringing. Their biological and social inheritance makes them what they are, independently of and antecedently to their own reasoning and willing, which are no more than products of that inheritance. By contrast such theorists understand themselves as rational agents, able to and aspiring to embody their intentions in the natural and social world. They understand others in terms of a determinist theory. They understood themselves in terms of a rational voluntarism. Marx put this in the third thesis by saying that "The materialist doctrine concerning the changing of circumstances and upbringing forgets that circumstances are changed by men and that it is essential to educate the educator himself. This doctrine must, therefore, divide society into two parts, one of which is superior to society." What do we need to learn from these remarks?

A first lesson concerns the relationship between the autonomy of theory and the social order of civil society. Marx had already identified the autonomy of theoretical enquiry as characteristic of civil society. Now he adds to that as also distinctive of civil society the self-conception of the theorist as an autonomous agent and as therefore in her or his own understanding always a potential legislator for and on behalf of others. Marx here suggests what was misleading in an antithesis which was later to dominate a good deal of discussion of Marxism, that between the view that theoretical enquiries inescapably function within and as part of economic and social formations and the rival view that theoretical enquiries can be autonomous and independent of the social contexts inhabited by theorists. Marx however identifies one particular kind of autonomy and independence of theory as themselves characteristic of and inseparable from a particular type of social order.

Secondly Marx, in asserting that such autonomous social theory cannot but envisage human beings in two incompatible ways, on the one hand as products of objective social and natural circumstances and on the other as rational agents, identified a continuing *aporia* for all modern social theory. It has not been uncommon in this century for social theorists to announce that they have solved the problem thus posed, among them Parsons, Sartre, Habermas and most recently Bourdieu. But various as their solutions are, they have all been solutions from *within* theory. Even Bourdieu, whom Loïc J. D. Wacquant has recently congratulated on "the inclusion, at the heart of a theory of practice, of a theory of theoretical practice" (P. Bourdieu and L.J.D. Wacquant, *An Invitation to Reflexive Sociology*, Chicago, 1992, p. 43) relates theory primarily to the practice of the scientific enquirer, that is, of the theorist. But Marx's point, I take it, was that no solution from within theory or even from within the practice of theorists is possible. It is only from the standpoint of social practice of a very different kind, one prior to both enquiry and theory, that a solution will be possible. What kind of practice might that be? It cannot be the kind of practice envisaged by those concerned to reform the institutions of civil society, without however abandoning its basic beliefs. For Marx had identified and found reason to reject the hierarchical structure of that type of reformist theory and practice. Those who without abandoning the standpoint of civil society take themselves to know in advance what needs to be done to effect needed change are those who take themselves to be therefore entitled to manage that change. Others are to be the passive recipients of what they as managers effect. This hierarchical division between managers and managed is thus legitimated by the superior knowledge imputed to themselves by the managing reformers, who have cast themselves in the role of educator. Marx almost certainly had foremost in mind Robert Owen, whom he had described in the Paris manuscripts as the author of "an abstract philosophical philanthropy". Owen was to have numerous successors in the subsequent history of socialism, among them both Lenin (at least on occasion) and Beatrice and Sidney Webb.

Notice that in the sixth thesis we once again confront the problem of interpreting what Marx expressed only in compressed and elliptical terms. What is it about the social educator's possession of theory that is taken to legitimate the educator's superior role? It must surely be that the educator takes her or himself not only to know more, but also to know best, that the educator takes her or himself to know what is genuinely good for others, something that they do not themselves know. Hence educators suppose themselves to be entitled to impose upon others *their* conception of the good. Marx contrasts the activity of this type of educator in respect of knowledge of the good with the activity involved in quite another kind of practice, one such that those engaged in it transform themselves and educate themselves through their own self-transformative activity, coming to understand their good as the good internal to that activity. Here again the elucidation of Marx's anti-Hegelian and anti-Feuerbachian thesis has had to be in Aristotelian terms. But this elucidation could scarcely be justified, if it proved impossible to cite any relevant example of just such a form of practice, one which would *both* be entitled to be called 'revolutionary' (the first and third theses) *and* be adequately characterizable only by an Aristotelian reference to the goods internal to it.

We find just such an example in the account given by Edward Thompson in *The Making of the English Working Class* (London, 1963) of the communal life of the hand-loom weavers of Lancashire and Yorkshire before and during the greatest prosperity of those weaving communities at the end of the eighteenth and the beginning of the nineteenth century. At its best the hand-loom weaver's way of life sustained his family's independence and his own self-reliance. Honesty and integrity were highly valued and what Thompson calls the "rhythm of work and leisure" allowed the cultivation of gardens, the learning of arithmetic and geometry, the reading and the composition of poetry. What the hand-loom weavers hoped to, but failed to sustain was "a community of independent small producers, exchanging their products without the distortions of masters and middlemen" (p. 295). At their best they embodied in their practice a particular conception of human good, of virtues, of duties to each other and of the subordinate place of technical skills in human life, but one which they themselves had no theory to articulate. By so doing they had, to the extent that it was possible, placed themselves outside civil society. And a theory which had successfully articulated their practice and which had been formulated so that its dependance on that practice was evident would have supplied just the kind of example of the relationship of theory to practice which the argument expressed in the theses on Feuerbach so badly needs.

What made the practice of the hand-loom weavers revolutionary? It was the degree to which, in order to sustain their mode of life, they had to reject what those who spoke and acted from the standpoint of civil society regarded as the economic and technological triumphs of the age. So Thompson relates how capitalist progress in the end "transformed the weavers into confirmed 'physical force' Chartists . . ." (p. 302). Marx himself had experience of the militancy of weavers in the insurrection of the Silesian weavers of the Eulengebirge in 1844. But he seems not to have understood the form of life from which that militancy arose, and so later failed to understand that while proletarianization makes it necessary for workers to resist, it also tends to deprive workers of those forms of practice through which they can discover conceptions of a good and of virtues adequate to the moral needs of resistance. Yet in the theses on Feuerbach Marx came very close to formulating just the distinctions which might have enabled him to understand this. But to have expressed those distinctions clearly and to have developed their implications would perhaps have left Marx unable to define his relationship to the large-scale revolutionary changes which he had identified as imminent, tied instead to what he took to be already defeated forms of past life. Marx therefore may have had the alternatives either of rejecting philosophy or else of depriving himself of the possibility of immediate effective participation in great events. And perhaps this is why he rejected philosophy.

Some of Marx's thoughts in the theses do of course reappear in his later writings. But with his rejection of philosophy in 1845 he lost the opportunity to develop those thoughts systematically and to understand their implications for the relationship of theory to practice. In so doing he left behind him unfinished philosophical business and, when, later on, philosophy was revived within Marxism, it was typically either as the dialectical and historical materialism of Plekhanov, which emerged from Engels' misunderstanding of Marx's relationship to Feuerbach, or as the rational voluntarism of the young Lukács, in which Lukács revived strains in Marx's thought whose fullest expression had been in the Paris manuscripts of 1844. But this opposition, between on the one hand the philosophy of Engels and Plekhanov and on the other that of the young Lukács, revived in a new version, or rather in a series of new versions, one of the antitheses already put seriously in question by Marx himself in the theses on Feuerbach. Each party in these subsequent debates had an excellent diagnosis of the errors of its opponents. The partisans of the younger Lukács understood very well that if human beings were the products of circumstance and upbringing, in the terms propounded by Engels and Plekhanov, then the kind of revolutionary agency through which the limitations of circumstance and upbringing could be transcended became unintelligible. The partisans of Engels and Plekhanov understood equally well that if the possibilities of revolutionary agency were what the Lukács of *Geschichte und Klassenbewusstsein* and of *Lenin* took them to be, then the nature of the historical determination of social and economic orders became quite unclear. In an early review of the former book Ernst Bloch predicted of the Bolsheviks that "Some of them will say that Marx had not placed Hegel on his feet so that Lukács can put Marx back on his head" (*Der Neue Merkur*, October 1923 – March 1926). Both the Bolsheviks who did so and Lukács himself were in different ways victims of this same misleading metaphor and with it of a philosophical inheritance which prevented both parties from understanding the significance of Marx's theses on Feuerbach.

This failure is peculiarly evident in two aspects of the history of moral philosophy within Marxism. One is the degree to which, when Marxists have been forced into moral debate, they have had to fall back upon the resources already provided by the moral philosophy of civil society, so that the contending parties merely repeat what had been already said earlier and better by the protagonists of that moral philosophy. So Kautsky in one way and Trotsky in another repeated theses of Benthamite utilitarianism, while theorists as different as Bernstein and Guevara echoed Kant. But even more significant is a second aspect of Marxist moral thinking.

From Marx and Engels onwards Marxists have generally supposed that an historical and sociological understanding of moral concepts and precepts as articulated within practices was incompatible with an appeal to objective standards of goodness, rightness and virtue, standards independent of the interests and attitudes of those engaged in such practices. And Marxists of course have not been alone in so supposing. But here once again there is a false antithesis. What the objectivity of moral and other evaluative standards amounts to is to be understood only from within the context of and in terms of the structure of certain types of historically developed practice, in which the initial interests of those engaged in such practices are transformed through their activities into an interest in conforming to the standards of excellence required by those practices, so that the goods internal to them may be achieved. These are types of practice socially marginalized by the self-aggrandizing and self-protective attitudes and activities characteristic of developing capitalism, types of practice alien to the standpoint of civil society. But they are the types of practice within which moral thinking is put to the relevant practical tests and achieves objectivity. It is only in such contexts that the question of "whether objective truth can be attributed to human thinking" can be answered in respect of moral thought, and this question is not one "of theory, but is a practical question" (the second thesis), answered by members of fishing crews and farming cooperatives and string quartets, just as by eighteenth century handloom weavers and their medieval and ancient predecessors, only through and by reference to forms of practice which precede the theory that they so badly need. Only in and through such practices can the standpoint of civil society be transcended.

We still therefore need to take serious account of the insights of the theses on Feuerbach, if we are to be able to take a road forward which Marx himself did not take. I have noted how Marx was unable to develop his own insights in the theses. But the important thing now about the errors that resulted is not so much that they were Marx's errors, as that for so many of us they were *our* errors and the defeat of Marxism has been *our* defeat. But Marxism was not defeated, and we were not defeated, by the protagonists of the standpoint of civil society, who now mistakenly congratulate themselves on the collapse of Communist rule in so many states. Marxism was self-defeated and we too, Marxists and ex-Marxists and post-Marxists of various kinds, were the agents of our own defeats, in key part through our inability to learn in time some of the lessons of the theses on Feuerbach. The point is, however, first to understand this and then to start out all over again.

University of Notre Dame

JOSEPH MARGOLIS

DONALD DAVIDSON'S PHILOSOPHICAL STRATEGIES

I

The single theme that is most nearly the essential thread of Donald Davidson's entire philosophical output is this: to bring into coherent and plausible order all the categories by which we understand the functional relationship between language and the world – primarily with regard to the effective success of practical life in cognition and action – without relying on the mediating role of theories, conceptual schemes, privileged philosophical doctrines, or other such *tertia*. Davidson himself says as much, for instance in the John Dewey lectures of 1989 (presented at Columbia University). There, though he chides his friend Richard Rorty for presuming to assimilate his view of truth to his own (Rorty's) notorious conviction, he tempers the objection, observing that "he [Rorty] sees clearly that for me this is related to the rejection of a representational picture of language and the idea that truth consists in the accurate mirroring of facts."[1] Rorty made his judgment about Davidson's having abandoned philosophy in tendering the well-known charge that "truth is not the sort of thing one should expect to have an interesting theory about."[2] The mistake was understandable (but not excusable), both because it is not easy to see *what* remains philosophically eligible if *all* intermediary conceptual devices for orienting our understanding of truth and meaning are to be avoided and because, like Rorty, Davidson is not entirely opposed to identifying himself as a kind of pragmatist. What is at stake is *what* we should construe as a philosophical *tertium quid* and what sort of philosophy can proceed without *tertia* altogether. Rorty condemns the whole of professional philosophy in this regard; his is a verdict he and his critics agree to label 'postmodernist.' Davidson holds instead that there *are* genuine options that successfully meet the constraint – his own in fact; he is quite content, therefore, to rely on what his critics call his 'pragmatism.' What needs to be appreciated is how exceptional this general theme is in contemporary analytic philosophy. It catches up a certain presumption that may well be widespread, but I do not believe there is another philosopher of Davidson's stature in the

C.C. Gould and R.S. Cohen (eds.), Artifacts, Representations and Social Practice, 291–322.

Anglo-American analytic tradition who subscribes to the thesis. For example, it cannot be found in the work of W. V. Quine, Nelson Goodman, Hilary Putnam, Michael Dummett, C. G. Hempel, Thomas Kuhn, or Karl Popper. In fact, it supplies a measure of the profound division between the views Quine espouses and those of Davidson, which are usually thought to be very closely linked. They are, but the difference, which I shall return to, is of the most profound importance. I find it generally neglected in the analytic literature. It is also a baffling doctrine, as we shall see.

The pragmatist theme is nicely grounded in the opening sentence of the Dewey lectures: "Nothing in the world, no object or event, would be true or false if there were not thinking creatures."[3] Quite so. And yet, that very pronouncement is surely common property for all philosophies, regardless of their different theories of truth. It is not an altogether vacuous utterance, since it does draw attention to truth's implicating a relationship of some sort between truth-claiming creatures and the world about which their claims are made: but what that relationship is cannot be fathomed from the single utterance itself.

Davidson provides the required clue: "I propose," he says, "an approach . . . that makes the concept of truth an essential part of the scheme we all necessarily employ for understanding, criticizing, explaining, and predicting thought and action."[4] He means this very seriously and he takes it to signify that, among other things, "epistemic views" of truth are "untenable' and "realist views [are] ultimately unintelligible," since "both invite skepticism."[5] At one stroke, then, he signifies that the views mentioned cannot fail to be defeated as potentially valid philosophies and that answering skepticism *is* a decisive criterion of philosophical pertinence and strength. Of course, Davidson's critics might counter that no theory of these sorts is worth considering that does not "invite" skepticism; that risking skepticism is not the same as succumbing to it; that the theories in question do mean to meet the skeptical challenge; and that his own proposal is either unsatisfactory because it does not "invite" the skeptical response in any serious way or, if it does, it does not actually meet its challenge.

In any case, the remark explains Rorty's error at the same time it distances Davidson from other well-known analytic philosophers who have examined the question of truth – namely, all those who have favored correspondence and coherence theories of truth and all those who have engaged in the recent realist/anti-realist disputes: notably, Hilary Putnam and Michael Dummett. That is, Davidson not only insists that the notion of truth is not philosophically eliminable – for instance, he rejects the so-called disquotational theory of truth and denies that Tarski's formal semantics, which he accepts, is, effectively, a disquotational theory – but he also believes that he himself has gone beyond Tarski's account in a philosophically important way that eschews *tertia*.

This is still not enough to capture the nerve of Davidson's philosophy, for it is impossible to formulate a suitable overview of his thesis without featuring his commitment to a thoroughgoing extensionalism and physicalism, both of which he regards (surprisingly) as benignly neutral to the conceptual thread I have just drawn out. In a word, he plainly believes he had succeeded in showing that his own versions of extensionalism and physicalism do not implicate *tertia*. I take this to be the boldest and most baffling feature of his entire philosophy. Of course, those themes, which he shares with the dominant philosophies of the Anglo-American world (though they are certainly not entirely of domestic origin), are the very same themes his own mentors have favored in consolidating what they take to be the best of the work of Frege, the logical positivists, the unity of science theorists, the logical atomists, and similarminded figures. I mean, by "mentors," W. V. Quine and C. G. Hempel specifically. And yet, of course, in the philosophical papers of those two theorists, physicalism and extensionalism have always been pursued in a strong, doctrinally partisan spirit that could never claim to be part "of the scheme we all necessarily employ." In Rorty's terminology, which Davidson does not seem disposed to avoid, their doctrines – or, *those* doctrines in *their* hands – amount to palpable *tertia*.[6]

This marks the distinction of Davidson's undertaking; for his entire career has been devoted, in effect, to forming a plausible picture of those (much-disputed) doctrines in full accord with the irenic vision I mentioned a moment ago, namely: (i) that *they* – or something very much like them – *are* "an essential part of the scheme we all necessarily employ"; and (ii) that what satisfies (i) may be shown to escape philosophical *tertia*.

Let me insert here, for the sake of avoiding confusion, the following caveat: theorists like Carnap, Russell, Hempel, and Quine would hardly deny that the extensionalism and physicalism they favor are essential to any philosophy that has pretensions of universal adequacy. That is not what is at issue. The point is rather that they never argue for those doctrines in terms of what *may* be drawn from the practical philosophical life and behavior of ordinary human creatures not yet infected with the virus that invokes *tertia*. Davidson means to recover the best work of these and similarminded philosophers but by way of a strategy none of them ever pursued – or would ever have considered philosophically responsive. In this respect, Davidson is rightly viewed, currently, as the principal champion of a comprehensive extensionalism and physicalism preferred in English-language philosophy, formulated in terms of the most salient issues of the analytic tradition (language, knowledge, truth, reference, meaning, action, belief), but in a unique way that appears to stalemate familiar objections to the more 'partisan' versions of those same doctrines. In fact, there is some reason to believe that Davidson might resist admitting that he subscribes to philosophical 'doctrines,' as being a phrasing already too concessive to the mediating and representational notions he opposes. In fact, his master contribution lies more with formulating the most promising general program of analysis in accord with these objectives than with detailed arguments regarding the most strenuous parts of it.

If valid, Davidson's feat would be a remarkable one: it would supply the pragmatist's maximally lean alternative to Aristotle's and Kant's classic accounts of the same question. Put another way: *if* extensionalism and physicalism proved not to be "an essential part" of the scheme "we all necessarily *employ*," then, if I understand Davidson correctly, they would be no more than philosophical intrusions of the interfering sort he opposes. That single concern identifies correctly Davidson's particular brand of pragmatism. I emphasize the expression "we all necessarily employ," because Davidson finds his philosophical bearings in the *praxis* of ordinary life. He is suspicious of high-flying philosophers who claim to *describe* what is 'necessary' to ordinary human discourse and action – which ordinary discourse and action would never in any pertinent non-philosophical sense confirm: Davidson claims to eschew all *tertia* and to theorize about what, without such philosophical mediation, *is* arguably "an essential part" of our ordinary practice. That is the complex thesis of the opening line of the Dewey lectures.

The thesis entails, I believe, a project that cannot succeed. But I also think it is characteristic of Davidson's uncanny knack (for isolating in a fresh light the most strategic issues of contemporary analytic philosophy) that he should be able to improvise several apparently unencumbered ('neutral') ways of entering the most volatile disputes. This, then, is the essential key to his particular mode of inquiry: search for what, without invoking any privileged philosophy, *we must concede* is "essential" to what "*we all* necessarily employ" in ordinary cognitive, linguistic, and active life. The pressure point for any responsible resistance to Davidson's theories, therefore, lies in this: exposing the partisan philosophy (in effect, the *tertia*) that Davidson would have us believe is (or are) no such thing. Rorty had (wrongly) believed that Davidson's commitment to what is ubiquitous in the viable lives of intelligent creatures like ourselves signifies that conventional philosophy is at an end; whereas what Davidson interprets his own remarks to mean to that philosophies that do not hew to the essential, the invariant, the universal, the necessary *in* the common-place *praxis* of our world have surely gone astray. In that sense, Putnam and Dummett *have* gone astray: after all, they 'invite' skepticism. Tarski and Hempel and Quine may have been attracted to philosophical strategies that also entail *tertia*; but, unlike Putnam and Dummett (it appears), their particular claims may be fully recovered without the offending intrusions: presumably, their claims do not ultimately 'invite' skepticism.

II

Why should we join Davidson in picturing the course of philosophy thus? one wonders. The answer is a dialectical one: it lies partly with the salient quarrels of the history of philosophy – in particular, with the local history of American philosophy – and it lies partly with an anticipated advantage we may not at first perceive. Its charm rests with its would-be economy at the same time it reconciles the most powerful, most technically developed, most dominant strain of professional thinking and what it supposes to be already latent (as "an essential part" of what "we all necessarily employ") in ordinary life. Thus, for instance, Davidson staunchly advocates applying Tarski's model in explicating the formal semantics of truth *in natural languages* (despite Tarski's reservation – even dismay – at the enlarged use of his important notion).[7] There, Davidson's maneuver is meant to demonstrate the philosophical *neutrality* of Tarski's model thus applied, while at the same time conceding that Tarski did not himself pursue the full (philosophical) account of truth required.[8] This is how Davidson means to redeem extensionalism.

He argues in a similar way for what has come to be called a 'nonreductive physicalism'[9] – in effect, for the thesis he identifies as a 'supervenience theory'[10] – which he treats as benignly respectful of our ordinary recognition of the beliefs and desires that dominate our practical life. He therefore opposes eliminative and reductive physicalisms with the same zeal with which he opposes epistemic realist and anti-realist theories of truth. But, in doing that, he nevertheless means to ensure the philosophical victory *of* extensionalist treatments of truth (along the lines favored by Tarski and Frank Ramsey) and physicalist treatments of mental life (along lines considerably softened from, yet still in accord with, Hempel's and Quine's views). The undertaking is ingenious and arresting. The question remains: whether it can possibly be successful. My own view is this: (i) that physicalism and extensionalism are not an essential part of the scheme we all necessarily employ; and (ii) that, as a consequence, Davidson's own thesis *is* a *tertium* of the kind he would escape. Furthermore, I should add: (iii) that, at the present stage of Western philosophy, there is no plausible way of sustaining Davidson's (i) and (ii); and, as a consequence, his project may be seriously misconstrued.

I have started too many hares at once here. I shall try to catch up with some of these shortly. But before doing that, it will pay to dwell a little longer on a certain puzzling feature of Davidson's comprehensive effort. If I am not mistaken, Davidson is disinclined to treat his own philosophical interventions as constituting the separate parts of a systematic argument in favor of a single encompassing theory. He does indeed have a unified vision, but he does not regard it as a partisan claim to be *tested* against the supposed advantages of other dialectically engaged doctrines. That, apparently, would be the way of employing philosophical *tertia*. Davidson is content to remind us of what "we all necessarily employ." This is surely his reason for opposing Putnam's and Dummett's argumentative strategies. (It also helps to explain Putnam's bafflement on the same score.) He believes his own commitment is *not* a doctrinal choice in the way in which philosophy is ordinarily pursued. For example, Putnam and Dummett think of the validity of their views primarily in terms of the artifactual contingencies of their particular dispute.[11] This is what Rorty misread, what Davidson's insistence on the 'essential' and 'necessary' was meant to convey.

Viewing matters thus, Davidson construes his own papers primarily as occasional discussions intended to meet local and transient provocations as they arise, all the while he leads us back to the self-effacing recovery of "the scheme we all necessarily employ" – which *does not* depend for its own force on the skill of any such argumentative contingencies. Frankly, the idea is that the recovery of *that is* not the advocacy of a partisan conception – in the way, say, of theorizing about language, knowledge, reality, truth by postulating a link between cognizer and cognized. No, it is a form of philosophical therapy – or sanity (partly at least Wittgensteinian, partly Aristotelian, partly pragmatist) – to recover (presumably, in a theory-free way) what *is* necessary, essential, universal, invariant, ubiquitously entailed in the great central mass of our ordinary practices.

What puzzles those who read Davidson's papers with care is: first, that what *he* takes to be distinctly neutral *as* a philosophy, but not neutral to the fortunes of philosophy, others are unwilling to characterize as neutral (physicalism and extensionalism, for instance); secondly, that what the compelling grounds for this general optimism are are nowhere very clearly laid out (for instance, regarding truth and meaning); thirdly, that what might reasonably persuade those who still practice philosophy in the familiar way to abandon *their* intervening *tertia* is nowhere supplied (for instance, with respect to correspondence or coherence theories of truth, realism or anti-realism); and fourthly, that why we should not judge Davidson to have deceived himself into thinking that he is not also championing certain distinctly partisan views is never satisfactorily explained. But if the general project fails in this way, then all of its subaltern doctrines do as well.

I shall elucidate Davidson's various ingenious claims in a moment. But I must emphasize the beguiling simplicity of Davidson's overall 'position' and the puzzling absence of a certain contentiousness in it that one associates with the history of the doctrines he actually favors. One cannot read Carnap for instance on the prospects for translating 'mental' terms by physicalist terms without noticing how readily that project may be opposed (Carnap eventually abandoned it) – and how keenly Carnap was aware of its contentiousness.[12] Of course, Davidson does not espouse Carnap's original translational thesis. But can it possibly be true that a mere shift from a reductive physicalism to a nonreductive physicalism signifies an escape from the dialectical strategies of invoking philosophical *tertia*? I cannot see the evidence for thinking that it does – or the reasonableness of supposing that we can bypass altogether 'ordinary' philosophical quarrels by recovering what "we all necessarily employ."

You may think this insistence on my part, regarding Davidson's wording, ungenerous or unreasonable. But I assure you, I intend it as an essential clue and, as it happens, as the most vulnerable part of Davidson's philosophy. More than that, Davidson's program is meant to be an improvement on the principal strategies of the kind of philosophical analysis Quine and Tarski and Hempel have represented so ably for the past forty years. If I am right in this, then, on the argument I intend to supply, that entire strategy – surely the best-known and most influential in English-language philosophy – may be called into mortal question. At any rate, there is no point in exploring Davidson's detailed philosophical claims without a proper sense of this executive conviction. I suggest therefore: (a) that, at the present time, Davidson may be thought to offer the most commanding and original overview of English-language analytic philosophy; (b) that his own inquiries proceed in a somewhat pragmatic manner that has proved difficult to characterize correctly; and (c) that the radical economies intended bear in a decisive way on the prospects of philosophy in the next century.

III

I am certain I have taxed your patience with what will seem a kind of blunderbuss complaint. I move at once, therefore, to provide supporting evidence drawn from Davidson's account of truth. No one will deny that the question of how we should treat truth philosophically is the single most persistent theme in Davidson's papers. I suggest that there are two quite different concerns that Davidson pursues and means to reconcile: the first, the earlier and perhaps better known, bears on his intended use of Tarski's formal semantics – what, applying Tarski's 'satisfaction' condition to natural language, he terms Convention T; the second, the later (later, at least in terms of a full formulation), the one clearly influenced by pragmatist and Wittgensteinian convictions, possibly also colored in some way by exchanges with Rorty – what I shall term Davidson's 'epistemic holism.'

I take Davidson's account of truth to join in a most original way Tarski's formal semantics with a pragmatist view of the truth of the massive core of our ordinary beliefs. Neither of these ingredient doctrines can be convincingly sustained without the other; each would be deformed if not interpreted in accord with their conceptual linkage.

Look now at Davidson's remarks about truth. (I must warn you that Davidson uses familiar terms – quite carefully – in unfamiliar ways. Frankly, he often expressly denies what he appears at first to affirm.) Consider the following important remark from a fairly recent paper:

> It should be clear that I do not hope to define truth in terms of coherence and belief. Truth is beautifully transparent compared to belief and coherence, and I take it as primitive. Truth, as applied to utterances of sentences, shows the disquotational feature enshrined in Tarski's Convention T, and that is enough to fix its domain of application. Relative to a language or a speaker, of course, so there is more to truth than Convention T; there is whatever carries over from language to language or speaker to speaker. What Convention T, and the trite sentences it declares true, like '"Grass is green" spoken by an English speaker, is true if and only if grass is green,' reveal is that the truth of an utterance depends on just two things: what the words as spoken mean, and how the world is arranged. There is no further relativism to a conceptual scheme, a way of viewing things, a perspective. Two interpreters, as unlike in culture, language and point of view as you please, can disagree over whether an utterance is true, but only if they differ on how things are in the world they share, or what the utterance means.[13]

Davidson draws two conclusions: "First, truth is correspondence with the way things are. . . . Second, a theory of knowledge that allows that we can know the truth must be a non-relativized, non-internal form of realism."[14]

You will notice that Davidson apparently espouses both a 'coherence' and a 'correspondence' theory of truth and, indeed, a 'realism' and a 'disquotational' theory. But he does not do so in any way that violates his executive constraint. He explicitly says that he does not *define* truth "in terms of coherence," and he recommends correspondence only in the "trite" sense he illustrates. The same is true of his realism, which he explicitly opposes to Putnam's ('internal realism') and his disquotational reading of Tarski's model, which, elsewhere, he characterizes (correctly) as *not* a mere disquotational theory (one, for instance, that would retire the theory of truth).[15] He rejects theories of these sorts that implicate *tertia* and he espouses others, seemingly similar, that preclude *tertia*. And yet, there are difficulties with this apparently innocuous policy.

For example, there is no clear discussion, in Davidson, of the philosophical connection between theories that invoke and theories that eschew *tertia*, or of what clearly marks the difference between them. The last sentence of the passage I cited just now is, for instance, either utterly trivial or false in terms of the role of *tertia*. On the first possibility, nothing follows of any philosophical interest at all; on the second, 'interpreters' *may* disagree "over whether an utterance is true" *if they differ in their philosophies*! Davidson objects to 'internal realism' because, as he says, it "makes truth relative to a scheme, and this is an idea [he adds] I do not think is intelligible. A major reason," he continues, "for accepting a coherence theory is the unintelligibility of the dualism of a conceptual scheme and a 'world' waiting to be coped with."[16] He construes the point of his remark to be the same as that pursued in his well-known paper, 'On the Very Idea of a Conceptual Scheme.'[17] But, for one thing, the doctrine of that paper is not easy to accept; and, for another, there is a perfectly straightforward (and important) sense in which (whether valid or not) internal realism need not mean what Davidson claims it means. It may say, *not* that the *meaning* of 'true' is relativized to a conceptual scheme, but only that the *ascription* of truth-values is relativized to the world interpreted in accord with a conceptual scheme. There is (then) no neutral world for us to know.

There is every reason to believe that Putnam intends the latter reading and not the former, since he is so openly opposed to any 'relativism' of the first sort – on the grounds that any such reading is incoherent or self-defeating – and because, in the very same context, he explicitly reconciles 'internal realism' and 'conceptual relativity' (*not* relativism).[18]

If I understand the matter rightly, then Davidson *cannot* disallow Putnam's philosophical option as a possibly viable alternative to his own position – *and on his own terms*. But if that is so, then Davidson has failed to draw an effective demarcation line between those philosophies that invoke *tertia* and those that eschew them. This is the point of Putnam's attack on 'metaphysical realism' (which Putnam may believe Davidson is attracted to), when he (Putnam) specifies "the mistake of supposing that [the question.] 'Which are the real objects?' is a question that makes sense *independently of our choice of concepts*."[19] Discourse about "how things are in the world," Putnam says, is "version-relative": there *is* no "neutral" way of viewing the world.[20]

Does Davidson dispute this? If he does, then it may be that he rejects *tertia* in the more quarrelsome sense the *he* is willing to invoke some form of cognitive privilege (which remains a matter for further dispute); and, if he does not, then the two conclusions he draws (which I have cited) must be vacuous and, therefore, not in the least unfriendly to Putnam's conjecture. I find no sustained discussion of the issue in Davidson's papers, apart from the charge of skepticism just aired.

But look again at the passage cited, this time in terms of Davidson's comment about Tarski's model and the nature of truth. "Truth is beautifully transparent," he says. Does he mean that there is an abundance of 'transparent' truths, truths that we couldn't possibly be mistaken about; or does he mean that the *concept* of truth is transparently obvious, whether or not what *is* true is? The first 'invites' both the specter of cognitive privilege and the threat of skepticism (which, you remember, Davidson turns against the partisans of philosophical *tertia*); and the second is trivial in just the sense in which the last sentence of the long citation (given above) is trivial: that is, the sense in which, first, ascriptions of truth-value are not yet at stake, and the sense in which, second, its admission does not entail the exclusion of *tertium*-ridden philosophy.

We know on independent grounds that Davidson means to say that truth is "beautifully transparent" when construed in Tarski's way, because *Tarski's* model *is* philosophically 'neutral.' Certainly, at times, Davidson has suggested that the neutrality of Tarski's model is linked to its disquotational feature and to its yielding 'trivial' sentences like the one mentioned in the extended citation.[21] This is a vexed matter and requires a measure of care. (A great deal hangs on whether the specimen sentences *are* genuinely 'trite' or 'trivial'.)

There are, I suggest, two quite different (irreconcilable) readings of Tarski's model that Davidson collects as "Convention T": one, I think, *is* simply disquotational (whatever Davidson says to the contrary); the other he takes to be 'empirical' and 'open to test.'[22] What I say is this: *if* Convention T is merely disquotational, then it cannot (and need not) be empirically testable. Certainly, it could not support Davidson's late discovery that the second reading commits him to admitting that "since I was treating theories of truth as empirical theories, the axioms and theorems [of those theories] had to be viewed as laws."[23]

There is at least one decisive claim of Davidson's about Tarski's theory (which has puzzled many readers) that argues, at least provisionally, in favor of disquotation. (It is not compelling, I admit, but the reason has more to do with interpreting Tarski correctly than with Davidson's intention.) "If we characterize T-sentences by their [syntactic] form alone, as Tarski did, [Davidson insists, then] it is possible, using Tarski's methods, to define truth using no semantical concepts. If we treat T-sentences as verifiable, then a theory of truth shows how we can go from truth to something like meaning – enough like meaning so that if someone had a theory for a language verified in the way I propose, he would be able to use that language in communication."[24] There are two serious mistakes here: for one, Tarski does not proceed in a purely syntactical way, "using no semantical concepts"; and, for another, there is in principle no way of disjoining questions of truth and questions of meaning – anywhere. These objections are not merely local complaints against Davidson's theory: they bear on the fortunes of philosophy in general and on the leading role analytic philosophy has assumed.

Consider, then, that Davidson's version of Tarski's theory, *applied to natural languages*, can only be disquotational; any other reading would be profoundly problematic. *Why* should the theory of truth regarding natural languages be confined to 'formal' considerations alone? What could possibly assure us that the syntactic and semantic features of a natural language were in any way separable from one another, so that, following Davidson (and, on his say-so, Tarski), we *could* "characterize T-sentences [sentences matched, remember, to the sentences of a natural language] by their form alone"?

On any familiar reading, the syntactic structure of a natural-language sentence is a function of what the sentence means. For example – and of course pertinently – it is problematic whether intentional sentences *can* be assigned extensionally structured T-sentences. As far as I know, Davidson never offers a sustained *empirical* argument (the point of the second interpretation of Tarski) that shows that the extensionalism of Tarski's account *is* convincingly 'neutral.'

Davidson is attracted here to the seeming analogy between a possible 'deep' extensional logic and Noam Chomsky's deep grammar for natural languages. He says that he wrote 'Semantics for Natural Languages' under the suggestion that "the deep structure of a sentence should correspond to the logical form a theory of truth assigned to that sentence." He sought to "provide a formal semantics for natural languages to match the sort of formal syntax linguists from Chomsky on have favored."[25] This, perhaps, is the sense in which Davidson views Convention T as suitably neutral (that is, as empirically confirmed) regarding the 'transparency' of truth. But, of course, Chomsky's theory is open to a related objection: namely, that Chomsky should *first* show that natural languages in actual use (as opposed to sentences artifactually fitted to his own ideal conditions) exhibit this or that grammar, and *then* show that his universal grammar fits all such empirical grammars perspicuously. On the argument, the first grammar *cannot* be detached from the semantics of the sentences in question.[26]

To my mind, Tarski's theory is *not* neutral, since extensionalism is not neutral. Or, if it *is* neutral, it is only because it is disquotational; but, then, the disquotational theory has nothing to do with extensionality and *does not characterize T-sentences "by their form alone."* It has nothing to do with either their form or their content: that's what makes the theory 'neutral' or (if valid) trivial. More important, the disquotational theory is *not* a theory *about* truth (as Davidson realizes); it is a theory only about the dispensability of truth-predicates in the use of natural-language sentences.

Tarski never viewed his own theory as philosophically neutral. In the Postscript to his deservedly famous paper, 'The Concept of Truth in Formalized Languages,' he admits to greater reservations about his thesis than appeared in the body of the original paper. He had originally worked with "formalized languages possessing a structure which is in harmony with the theory of semantical categories and especially with its basic principles." But in the Postscript he says: "Today I can no longer defend decisively the view I then took of this question," and he specifically wonders what "the consequences would be for the basic problems of the present work if we included in the field under consideration formalized languages for which the fundamental principles of the theory of semantical categories no longer hold." He then considers "only those languages in which occur, in addition to the universal and existential quantifiers and the constants of the sentential calculus, only individual names and the variables representing them, as well as constant and variable sentence-forming functors with arbitrary numbers of arguments."[27] In the original paper, Tarski clearly found that "[in colloquial language] it seems to be impossible to define the notion of truth or even to use this notion in a consistent manner and in agreement with the laws of logic."[28] (There is no discussion, in Davidson, of this important caveat.)

The upshot of these remarks is simply that Davidson is mistaken in reporting Tarski's view and wrong about its empirical fit vis-à-vis natural languages. Given the prominence of his adherence to Tarski's view, this is surprising. It leaves only the disquotational reading; but the disquotational reading has nothing to do with Tarski's substantive views. Davidson clearly says that "the main, if not the only, ultimate concern of philosophy of language is the understanding of natural languages."[29] But then, Tarski's sense of the impossibility of applying his model to natural languages should have been Davidson's conviction as well. Clearly it is not, but there is no explanation in Davidson as to *why* we can (afford to) be more sanguine about the extensional syntax of natural languages. It is true, as he says, that "Convention T, in the skeletal form I have given it, makes no mention of extensionality, truth functionality, or first-order logic. It invites us to use whatever devices we can contrive appropriately to bridge the gap between sentence mentioned and sentence used." It is in this sense that he wants "to defend . . . the Convention as a criterion of theories [of truth], not any particular theories that have been shown to satisfy the Convention in particular cases, or the resources to which they may have been limited."[30] Given these constraints, Davidson's version of Convention T *cannot* be Tarski's. Also, if it is merely disquotational (which is the easiest reading), it cannot seriously serve as a 'criterion' of any interesting sort. It certainly could not help in confirming the extensionality thesis.

The explanation lies already in Tarski; for, on Tarski's view, the *semantic* features of the formalized languages Tarski considers *are already known* to be hospitable to a purely syntactic characterization of the structure of the sentences in question. Tarski never attempts to treat truth *in a purely formal way*, except in the Pickwickian sense in which the semantics of the sentences in question *are* known not to affect adversely the syntactic conditions drawn up. Davidson supposes, remember, that "a theory of truth shows how we can go from truth to something like meaning." That is, the *syntactic* characterization of truth may (in Davidson's view) be supplied first, and then the *semantic* complication of meaning may be added. That now threatens to be either radically false or incoherent.

We cannot go *from* truth to meaning *if* the theory of truth holds that the truth of sentences is inseparable from their meaning – everywhere. Davidson had hoped to follow Tarski by distinguishing between 'absolute' and 'relative' truth: roughly, by distinguishing between conditions of truth formulated first and separately in syntactic terms ('absolute') and then *adding* conditions that reflect the use of semantic concepts ('relative'). This is the meaning of that pregnant pronouncement that closes the discussion of Convention T: "'absolute' truth goes relative when applied to a natural language."[31]

What is mistaken or misleading about this summary is the suggestion that there *is*, first, a largely correct theory of ('absolute': *non*-semantically infected conditions of) truth (Tarski's supposed contribution) that can *then* be *supplemented* by further attention to the contingent vagaries of particular natural languages. That is why Davidson claims that such an 'enrichment' does not affect the fundamental theory: "for a truth theory satisfying Convention T," he says, "is extensional."[32] That cannot be true on the disquotational view (which is indifferent to the matter) *or* assumed on Tarski's view (which is risked on the evidence). Davidson nowhere shows that it holds *empirically*, and he certainly has not shown that it is part of what "we all necessarily employ." I cannot see how it can be denied that, on Davidson's theory, *Tarski's* demurrer regarding natural languages should strike *Davidson* as a concession in the dircction of *tertia*.

IV

I said there were two foci to Davidson's theory of truth: one, Tarskian (which I have now reviewed); the other, Wittgensteinian and 'pragmatist,' which I shall address more briefly,. The second is developed most fully in 'A Coherence Theory of Truth and Knowledge' and in 'Radical Interpretation.' You may see at once from the following citation how Davidson means to supplement Tarski and how, at the same time, he brings his account of truth into line with the rejection of *tertia*:

> What makes interpretation possible . . . is the fact that we can dismiss *a priori* the chance of massive error [in belief]. A theory of interpretation cannot be correct that makes a man assent to very many false sentences: it must generally be the case that a sentence is true when a speaker holds it to be. So far as it goes, it is in favor of a method of interpretation that it counts a sentence true just when speakers hold it to be true. But of course, the speaker may be wrong; and so may the interpreter. So in the end what must be counted in favor of a method of interpretation is that it puts the interpreter in general agreement with the speaker: according to the method, the speaker holds a sentence true under specified conditions, and these conditions obtain, in the opinion of the interpreter, just when the speaker holds the sentence to be true.[33]

The use of the term 'interpretation' is bound to mislead. Davidson does not mean to invite intervening interpretive schemes in order to overcome skeptical challenges about other minds and other speakers. That *was* the point of Quine's 'analytical hypotheses,' which, at one and the same time, introduced conceptual *tertia* in accord with his well-known doctrine of the 'indeterminacy of translation' and met the threat of skepticism there by favoring speakers and their home language.[34] (The theme has obviously influenced Putnam's notion of 'version-relative' accounts of the world and is matched, more eccentrically, by Nelson

Goodman's doctrine of 'world-versions,' which recommends instead that we speak not of "multiple possible alternatives to a single actual world but of multiple actual worlds."[35] Davidson restricts the use of 'interpretation' to the problem of other minds.

Davidson is unique among the theorists mentioned (whom one might have thought held somewhat similar views) in that he alone regards the following two admissions as completely unacceptable and as entailing *tertia*: (a) that we must concede the eligibility of alternative conceptual schemes in characterizing what we take 'the' world to be; and (b) that there is no way of talking about 'the' world except *via* the indissolubly symbiotized complex that, in the form of conceptual schemes, is at once world and language. (The world, we may say, is 'languaged' and language, 'worlded'; and there is no unique or neutral way of characterizing either world or language.[36])

In this respect, Davidson appears to be 'pre-Kantian' – or perhaps Kantian in that (Humean) sense in which Kant wondered whether Hume had anticipated him. The other theorists mentioned – Quine, Goodman, Putnam – are all, I believe, influenced by a historicizing or contextualizing of Kant's transcendental argument, at the same time, clearly, they suppress the transcendental, history, and context in the interest of favoring extensionalism and (somewhat less reliably) physicalism. (Putnam has become increasingly skeptical of his former adherence to these programs.[37])

Davidson precludes the 'invitation' to skepticism by the pronouncement cited. (It is, in fact, the master theme of the important paper, 'The Very Idea of a Conceptual Scheme.') At the same time, he completes, by doing that, a sketch of what he takes to be needed for an adequate theory of truth. Both elements preclude *tertia*. By the first, he offers a universal account of the 'deep' formal syntax of natural-language sentences that (he believes) conforms with Tarski's Convention T; and, by the latter, he ensures a massive space of reliable truths that admit all the contingencies of culturally diverse semantic practices. What I wish to emphasize is that, first, Davidson *needs* the second thesis in order to supplement his own disjunction between syntactic and semantic issues; and, second, he supplies what he needs, ingeniously, by invoking a doctrine that eschews *tertia*. I have already indicated why I think the first leg of the program fails: there is no convincing account (certainly not in Tarski, not in Chomsky either, and not, because the matter is nowhere discussed, in Davidson) that there are *any* pertinent conditions of truth (short of trivial disquotation) that can be formulated in formal or syntactic terms alone, 'first' or without reference to the semantic content of sentences in actual use. The second fails simply because it then has nothing *to* supplement: it cannot be supposed to contribute semantic distinctions *within* the terms of a comprehensive extensionalism (of Tarski's sort). Alternatively put, the validity of this thesis must be considered without regard to the Tarskian side of Davidson's undertaking.

In the interest of economy, I shall confine myself to one counter-consideration and its import. In the passage cited just above, Davidson says very plainly that "we can dismiss *a priori* the chance of massive error" in the ordinary beliefs of quotidian speakers. He draws from this the conclusion (with some reservation, it is true) that, for speaker and interpreter, a particular sentence is true "just when the speaker holds the sentence to be true." This is the thesis I had in mind in speaking earlier of Davidson's 'epistemic holism.' It is somewhat like Wittgensteinian holism, in that we are able to detect a false belief (for example, the belief that the world is flat, which Davidson mentions) only because, in doing that, we "must depend on a background of largely unmentioned and unquestioned true beliefs." Hence, no single mistaken belief "necessarily destroys our ability to identify further [presumptively true] beliefs."[38] (That is what the holism makes possible, without the threat of skepticism.)

This last remark of Davidson's is, of course, quite reasonable. But it remains uncomfortably true as well that Davidson is in danger of uttering a fallacy (the so-called 'fallacy of division') in inferring – from the holistic truth that there must be a massive core of true beliefs, if humans are to function at all – the further truth that this or that particular belief is likely to be true or may be reasonably taken to be true "when the speaker holds the sentence to be true." In the Wittgensteinian literature, a similar difficulty has been noted and turned aside.[39]

What is provocative about Davidson's proposal is that, if valid, it would appear to ensure a large body of truths: (i) with some significant measure of certainty or reliability, yet without invoking any specific source of cognitive privilege; and (ii) in a way that functions transculturally and transhistorically, without invoking any intervening interpretive devices by which to ensure relevance and understanding. In that sense, eschewing *tertia*, it counts as Davidson's version of pragmatism. It fails, if the charge of fallacy is sustained. But it fails for a much less quarrelsome reason. In the important paper, 'Radical Interpretation,'

Davidson offers an altogether different account, one that seems more nearly true or at least less doubtful:

> The method of devising a theory of truth for an unknown native tongue is intended to solve the problem of the interdependence of belief and meaning by holding belief constant as far as possible while solving for meaning. This is accomplished by assigning truth conditions to alien sentences that make native speakers right when plausibly possible, according, of course, to our own view of what is right. What justifies the procedure is the fact that disagreement and agreement alike are intelligible only against a background of massive agreement. Applied to language, this principle reads: the more sentences we conspire to accept or reject (whether or not through a medium of interpretation) the better we understand the rest, whether or not we agree about them.[40]

This second account is *not* the equivalent of the first; it may well undermine it. For, in the first account, Davidson claims that "massive error" in belief must be impossible; whereas, in the second, all that is needed is "massive agreement" about what to take as true, "according, of course, to our own view of what is right." The second account actually admits an interpretive *tertium* – in fact, one that is very close to Quine's view of 'analytical hypotheses' – whereas the first rejects *tertia* by retreating to what look suspiciously like a form of cognitive privilege. If this line of objection is valid, then Davidson has failed to provide a viable theory of truth: first, because he has sought to apply Tarski's 'satisfaction' condition to natural languages under conditions that either reduce, vacuously, to disquotation or, more relevantly, fail to address the empirical adequacy of Tarski's strong extensionalism; and, second, because he has sought to supplement the first condition in a way that either, illicitly, implicates a form of privilege or, more relevantly, sensibly stops short of the requirements of a theory of truth.

A convenient way of putting the point is this: if the intended supplement were sufficient, then, in accord with Quine's well-known argument, epistemology could be 'naturalized,' that is, treated as a part of empirical psychology, or causally explained, or something of the sort.[41] It is clear that Davidson would be hospitable to such a program. (It is already implicit in the analogy he finds with Chomsky's analysis of language.) But the fact remains that, although naturalistic epistemologies – which American analytic philosophy has very much favored – act to retire transcendental arguments and interpret normative questions naturalistically – they never address the legitimative question that appears not to yield to the same strategy: namely, the question as to why we *should* regard as justified the first-order resources by which we normally confirm and disconfirm our truth-claims.

The point is explicitly acknowedged in Quine:[42] the validity of construing knowledge as subsumable within psychology is not decided by doing that; it cannot be. In fact, ultimately, it poses the same question as the question Davidson fails to address regarding the defining conditions of truth. To admit this is to admit the profound sense in which analytic strategies in general cannot succeed. For, on the gathering argument, to admit the eligibility of the legitimative (second-order) question – *not*, let us admit, a transcendental question in anything like Kant's or Husserl's sense – *is tantamount to admitting the ineliminability of tertia.*

There you have the completion of the argumentative circle regarding the first of Davidson's two principal themes. If I merely add, now, more or less as a reminder of what is obvious (rather than as an argument in its own right), that Davidson has not taken into account the profoundly disquieting possibility that truth-claims *and* their validation may be (and may be taken to be) artifacts of the tacit, contingent, diverse, shifting, historically evolving, possibly moderately incommensurable forms of life of the societies in which they obtain, then the difficulty of his own project affords an important instruction about philosophy's prospects.

V

I come now to the second of Davidson's themes: the physicalist reading of the mental or psychological. It is entirely fair to say that Davidson would not characterize himself as a 'physicalist.' But the reason is that the use of the term in analytic circles favors either eliminationism or reductive physicalism, and Davidson is clearly not an advocate of either of these positions. I judge them to be incompatible with his executive concern with what "we all necessarily employ."

In any case, Davidson advances two very large claims that confirm the sense in which he is a physicalist of sorts and, joined to his extensionalist commitment, a linear descendent of positivism and the unity of science program – though by way of the 'Humean' pragmatism already sketched.[43] Here, he seems closer to Hempel's projects than to Tarski's. It may be that a pragmatist's middle ground had already been prepared by Quine: *Word and Object* is, of course, hospitable to both extensionalism and physicalism, but Quine's book does not engage the first in epistemic terms nor the second dialectically. The evidence may be found in Quine's notorious dismissal of Brentano's insistence on the puzzles of intentionality and his entirely formal solution for retiring referential resources in favor of predicative ones.[44]

In a curious way, the same is true of Davidson, though he gives the impression of having come to terms with what Quine neglects.[45] Let me add a few sentences here, from another of Davidson's papers, to confirm the fairness of the charge with regard to the topic of truth. Obviously, I cannot afford to pursue the issue in depth, though it is too important to ignore altogether. In 'Reality without Reference,' we find the following remarks, the import of which, in the light of what I have already argued, should be self-evident:

> Translation is a purely syntactic notion. Questions of reference do not arise in syntax, much less get settled. . . . I propose to defend a version of the holist approach [to meaning], and urge that we must give up the concept of reference as basic to an empirical theory of language. . . . The argument against giving up reference was that it was needed to complete an account of truth. I have granted that a Tarski-style theory of truth does not analyze or explain either the pre-analytic concept of truth or the pre-analytic concept of reference: at best it gives the extension of the concept of truth for one or another language with a fixed primitive vocabulary[46]

The short refutation of this proposal requires no more than recalling that: first, *no* 'meaning holism' could possibly confirm the eliminability of reference; no natural language has a 'fixed primitive vocabulary'; and reference is inseparable from the meaning of sentences uttered and the intentions of speakers. Davidson's proposal is not necessarily to eliminate reference altogether, 'only' to eliminate reference as 'basic' – as basic as truth: "words, meanings of words, reference, and satisfaction [semantically construed] are posits we need to implement a theory of truth. They serve this purpose without needing independent confirmation or empirical basis."[47] But this fails to address the profound Leibnizian question about the identity of indiscernibles (which Quine has encouraged us all to disregard); and its initial plausibility depends entirely on the plausibility (now disputed) of a purely syntactic reading of Tarski's account of truth. Certainly, the claim about reference (or extensionality) cannot fail to be seen to be a thesis that depends on intervening *tertia* rather than on an appeal to what "we all necessarily employ."

I turn without further ceremony to the promised second line of inquiry: Davidson champions what he calls 'supervenience' and 'anomalous monism.' These are surely the central doctrines of his treatment of the mind/body problem, though there are other relevant papers of his to consider (particularly with regard to the analysis of human actions).[48] The two doctrines, in fact, offer two different faces of what appears to be the same theory. (I do not believe they can be that, however.) Davidson provides rather little in the way of clarifying supervenience, though it has become a very attractive option in recent interpretations of physicalism.[49] Here is what be says:

> supervenience might be taken to mean that there cannot be two events alike in all physical respects but differing in some mental respect, or that an object cannot alter in some mental respect without altering in some physical respect. Dependence or supervenience of this kind does not entail reducibility through law or definition: if it did, we could reduce moral properties to descriptive, and this there is good reason to *believe* cannot be done; and we might be able to reduce truth in a formal system to syntactical properties, and this we *know* cannot in general be done.[50]

As far as I can see, there is no actual argument offered in favor of supervenience, except by way of analogy with the treatment of Tarski's theory and whatever may favor anomalous monism. Davidson says as much.[51]

The point that is common to the doctrines is the claim of an extensional equivalence between sentences about mental events and sentences about (corresponding) physical events, without attempting any reduction and without insisting on the elimination of the mental. But no independent argument of any kind is given, empirical or *a priori*, to show that supervenience is true or must be true. In 'The Material Mind,' which appeared three years after 'Mental Events,' Davidson declares, in the very context in which he opposes reductive physicalism, that "the physical characteristics of an event (or object or state) *determine* the psychological characteristics." He borrows the term for this dependence ('supervenience') from G. E. Moore, and he then affirms: "it is impossible for two events (objects, states) to agree in all their physical characteristics . . . and to differ in their psychological characteristics. . . ."[52]

He favors the modal thesis, therefore. But surely that is indemonstrable, in the plain sense that its denial is not self-contradictory or empirically implausible in any obvious way. I should say the modal thesis is at least unlikely. First of all, even in terms of the mind/body question, Herbert Feigl had for many years worried about the coherent possibility (which Davidson does not address) of what Feigl called the 'many-many' problem: namely, that, for any mental (or psychologically informed) event (say, deciding on a chess move), there are indefinitely many possible physical realizations; that, for any physical event, indefinitely many mental events may be realized by the same event; and that there is no formulable algorithm for matching events from the two series.[53] But, beyond that, *if* what is mental or psychological among humans includes, prominently, what is linguistically or culturally informed (say, along Wittgensteinian or pragmatist lines), then supervenience is implausible *if* linguistic and cultural predicates cannot be extensionally matched with physical predicates and cannot be reduced to the mental. This of course raises the specter of ineliminable *tertia.*

I know of no one who supposes that either move can be supported on empirical or *a priori* grounds. Certainly, Davidson does not address the question. Hence, if the culture/nature problem cannot be reduced to the mind/body problem, and if the mind/body problem cannot be solved (for humans) without solving the culture/nature problem, then supervenience is either false or extremely implausible. Furthermore, the idea that the physical 'determines' the mental may be cognitively inoperative: the mental may be complex enough to 'include' or 'incorporate' the physical in some sense; but neither the mental nor the cultural need depend in any way on *inferring*, from some determinate physical event, what is pertinently 'mental' or 'cultural.' If supervenience is false or unreliable, then, trivially, the physical cannot be said to 'determine' the mental. The 'mental' and the 'cultural' may be 'emergent' (in some sense) with respect to the physical – without admitting dualism – and yet not be 'supervenient.' Which is simply to say that (nonreductive) physicalism may fail.

Turn, now, to anomalous monism. This is Davidson's most famous thesis on the mind / body problem. It appears completely formed in the paper 'Mental Events.' It is actually a triad of "assumptions or principles," which Davidson attempts to show is self-consistent.[54] Two preliminary considerations are in order. First of all, Davidson treats two of the principles (the first two in the citation given below) as "undeniable facts," which is to say, he sees his project – that of explaining how they are compossible – as akin to Kant's question regarding the compatibility of freedom and causal necessity.[55] And, secondly, the first admission counts as reminding us (I judge) of what "we all necessarily employ," in the sense already indicated. But the solution is itself a garden-variety philosophical thesis (the third principle): one that under other circumstances (for instance, circumstances akin to what Davidson objects to in Putnam and Dummett) might have been thought to invoke some *tertium quid.*

My own view is that the triad is inconsistent – on Davidson's own grounds. Furthermore, anomalous monism is a stronger thesis than supervenience, in that it offers a form of mind / body identity.[56] If I am not mistaken, the thesis is original with Davidson. It is not altogether clear what the connection between the two doctrines is, unless supervenience is already to be construed in terms of the other (which is nowhere affirmed): if supervenience were not so constrained, then anomalous monism might well be false or not known to be true. That is, a strong identity thesis might yet be confirmed. On the other hand, supervenience countenances an extensional equivalence between the mental and the physical but opposes any version of the identity theory. Oddly, the two doctrines are presented in the same paper without resolving the issue of their relationship.

Here are the principles that form the triad:

1. "at least some mental events interact causally with physical events." ("the Principle of Causal Interaction")
2. "where there is causality, there must be a law: events related as cause and effect fall under strict deterministic laws." ("the Principle of the Nomological Character of Causality")
3. "there are no strict deterministic laws on the basis of which mental events can be predicted and explained" ("the [Principle of] the Anomalism of the Mental.")[57]

The terms 'mental' and 'physical' are used with care but somewhat loosely: "an event is physical if it is describable in a purely physical vocabulary, mental if describable in mental terms."[58] Clearly, the two idioms are not treated in precisely the same way. Still, Davidson does say that he means to treat the mental, at least in terms of the argument, along the lines of "what Brentano called intentionality."[59]

The difficulty with Davidson's argument lies elsewhere. Davidson says that he wants "to argue for a version of the identity theory that denies that there can be strict laws connecting the mental and the physical."[60] That is what anomalous monism means. The curious thing is that if supervenience is true, either empirically or *a priori*, then anomalous monism is either false or successful in ruling out a nomological reading of extensional equivalence: the formula just given shows that the third principle (cited above) must be read in a modal way. But no satisfactory argument is ever given. In fact, Davidson expressly says: "Although the position I describe [anomalous monism] denies there are psychophysical laws, it is consistent with the view that mental characteristics are in some sense dependent, or supervenient, on physical characteristics."[61] I do not say that supervenience entails psychological laws, but I cannot see how it precludes them either.

Both the required clue and the difficulty in applying it effectively lie in this: on Davidson's view, (a) causality, identity, and individuation all behave extensionally; but (b) the mental behaves non-extensionally. From (a), it follows that supervenience and the Principle of Causal Interaction are both possible; but from (b), it follows that supervenience and the Principle of Causal Interaction are impossible. I cannot see how it can be denied that anomalous monism is therefore inconsistent.

I have, I admit, suppressed Davidson's key argument – that regarding the 'holism' of the mental. He says, for example, on the question of 'definitional reduction' or synonymy:

> If we were to find an open sentence couched in behavioral terms and exactly coextensive with some mental predicate, nothing could reasonably persuade us that we had found it [that is, synonymy]. We know too much about thought and behavior to trust exact and universal statements linking them. Beliefs and desires issue in behavior only as modified and mediated by further beliefs and desires, attitudes and attendings, without limit. Clearly this holism of the mental realm is a clue both to the autonomy and to the anomalous character of the mental.[62]

The trouble is this: *if*, despite this holism – which, in effect, is the holism of every model of rationality – the mental can be fixed tokenwise sufficiently well to confirm causal interaction, then it cannot in principle fail to allow for nomologicality; and, *per contra*, if it cannot be fixed well enough to confirm psychophysical laws, then it cannot in principle be fixed well enough to confirm causal interaction:

> There is no assigning beliefs to a person one by one," says Davidson, "on the basis of his verbal behavior, his choices, or other local signs no matter how plain and evident, for we make sense of particular beliefs only as they cohere with other beliefs, with preferences, with intentions, hopes, fears, expectations, and the rest[63]

Davidson cannot have it both ways. He means to admit 'token-identity' but not 'type-identity,' that is, nomological identities, identities ranging over natural kinds. He says that, on his own view, "mental events are identical with physical events."[64] This signifies (if I understand it rightly) that mental events behave in the same way, logically, as do physical events: they may be identified and reidentified extensionally; but if that is so, then they *may* enter into nomological relationships (psychophysical laws), no matter how weak those laws may be and no matter how often we identify individual mental events in 'cross-category' (non-nomological) ways.

Elsewhere, Davidson says, speaking of actions:

> We must conclude . . . that our primitive actions, the ones we do not do by doing something else, mere movements of the body – these are all the actions there are. . . . If an event is an action, then, under some description(s), it is intentional.

He adds: "the concept of being primitive, like the concept of being intentional, is intensional, and so cannot mark out a class of actions."[65] But if the intensional is as ubiquitous as this, then even the physical is anomalous; and if it is not enough to preclude nomologicality in general, then it is hard to see why it precludes the mental necessarily. I find myself forced, therefore, to the conclusion that anomalous monism is an inconsistent triad.

VI

I shall bring this account to a close with two brief observations. First, Richard Rorty has succeeded, in his extremely useful essay, 'Pragmatism, Davidson and Truth,' in isolating, however inadvertently, just the version of 'pragmatism' he (Rorty) espouses and Davidson means to avoid. He does this, I suggest, by offering the following judgment: "On my interpretation . . . Davidson joins the pragmatist in saying that 'true' has no explanatory use. His contribution to pragmatism consists in pointing out that it has a disquotational use in addition to the normative use seized upon by [William] James."[66]

Now, it is true, as we have seen, that Davidson opposes explanations of the correspondence and coherence sorts, though he also subscribes to (what he calls) 'mild' versions of these doctrines. Moreover, he explicitly opposes what Rorty offers as the disquotational theory: which, effectively, is to say that truth *does* have an 'explanatory' function but only if it escapes *tertia*. Davidson is not a pragmatist (on his own view) *if* that means "that language is . . . a screen or filter [a *tertium quid*] through which our knowledge of the world must pass."[67] This is his argument against empiricism, pragmatism, transcendental idealism, 'internal' realism (and incommensurabilism). They are all, Davidson claims, "forms of relativism that [he says] I find as hard to understand as the empiricism I attack."[68] I have already shown that the objection is seriously problematic. Nevertheless, it helps to fix our sense of Davidson's objection to *tertia*.

The point of Rorty's essay is to press this admission in the direction of claiming Davidson as an ally of his own program and as a strong pragmatist. I believe he is right to say that "the function of the *tertia* which Davidson wishes to banish was precisely to provide a mechanism outside the causal order of the physical world, a mechanism which could have or lack a quasi-causal property with which one might identify truth."[69] What Rorty means is that truth *sans phrase* cannot have an explanatory function in that sense, but that Davidson need not preclude a causal role for true beliefs. For, if we are in touch with the world in a way that is not mediated by language and the like, *some* sort of causal account will be needed. This, I think, *is* the point of Davidson's distinction between 'absolute' and 'relative' truth: the semantic content of true beliefs must be causally linked to the human condition in a way that the massive core of true beliefs can be reasonably assured. I take this claim to be untenable, simply because it ignores too sweepingly the entire question of the cultural and historical diversity of belief-formation. But I admit, at the same time – or rather, I insist – that to accommodate the point is *to admit the ineliminability of tertia*. That, I cannot dissemble, is the fatal weakness of Davidson's entire project and the philosophical programs that have been attracted to his ingenious recommendations.

Where Rorty goes wrong is in supposing that his reading of Davidson precludes truth's having an 'explanatory' role of any sort. That is indeed his own view (and apparently, also, his view of William James's position[70]). But it cannot be Davidson's. *For Davidson is committed to Tarski's extensionalism*: that is, he is committed to extending Tarski's account, over Tarski's own demurrer, to natural languages. He cannot fail to have an explanation of the validity of that doctrine, if his theory is to have any plausibility at all. I must report, however, that he nowhere offers an explanation.

The second observation concerns a rather late paper of Davidson's, 'A Nice Derangement of Epitaphs' (1986) – the title of which of course catches up a malaprop from Sheridan's *The Rivals*. It is a strenuous paper which deserves a sustained analysis of its own and which has generated a good deal of concern as to whether Davidson has remained consistent with his earlier views on truth and meaning.[71] My own judgment is that he *has* remained consistent – more than consistent – with his program against *tertia*; but that he has failed to perceive or admit that his own versions of extensionalism and physicalism are ruled out by the new adjustment – or, ruled out insofar as he precludes the need to rely on philosophical *tertia*. If the conclusion holds, then, at the present time, there is no reliable and coherent overview of the whole of Davidson's philosophy.

The central thesis is suggested by the candid admission that linguistic communicants do remarkably well in understanding one another even when linguistic errors, slips, malapropisms, and the like intrude. "We want [says Davidson, reflecting on the meaning of the problem] a deeper notion of what words, when spoken in context, mean; and like the shallow notion of correct usage, we want the deep concept to distinguish between what a speaker, on a given occasion, means, and what his words mean. The widespread existence of malapropisms and their kin threatens the distinction, since here the intended meaning seems to take over from the standard meaning."[72] His new thesis is that, to understand this distinction correctly, "we must . . . modify certain commonly accepted views about what it is to 'know a language', or about what a natural language is. In particular, we must pry apart what is literal in language from what is conventional or established"[73]

I am persuaded that Davidson's principal objective here *is* to confirm his opposition to *tertia*; but if I am right in my summary judgment in broaching the matter at the start of this section, then Convention T, supervenience, and anomalous monism cannot be secured by the same strategy. Generally, Davidson (now) opposes our understanding one another by way of speakers and interpreters possessing *any* "correct theory in advance" (for instance, Convention T). There is and can be no such theory: "there is no word or construction that cannot be converted to a new use by an ingenious or ignorant speaker."[74] The startling conclusion Davidson draws is this: "there is no such thing as a language, not if a language is anything like what many philosophers and linguists have supposed. There is therefore no such thing to be learned, mastered, or born with. We must give up the idea of a clearly defined shared structure which language-users acquire and then apply to cases."[75]

I find this a congenial thesis, one that brings Davidson closer to Quine but leads him to the point of making concessions to the profound intentionality, contextedness, historicity, and cultural divergence of human language and behavior. Just the themes, in short, that characteristically divide Anglo-American and continental European currents.

But let me give the issue the briefest of analyses, for the sake of rounding out this account. One thing is clear. The 'Nice Derangement' paper is centered on linguistic communication, on speakers and listeners understanding one another in spontaneous exchange. It cannot, therefore, recognize – in the spirit of avoiding *tertia* – Davidson's rather two-storey account of the 'absolute' and 'relative' aspects of the syntax and semantics of natural languages. Convention T surely has to go, for it *is* a theory that, on an earlier view, was regarded as universally applicable (syntactically) and neutral to every episode of natural-language discourse. It now finds no basis for vindication. Secondly, though speakers and hearers do possess *theories* by which they approach one another's utterances, no theories of such sorts can be said to capture the *linguistic competence* they share. Here, Davidson conveniently distinguishes between a "prior theory" and a "passing theory": "For the hearer, the prior theory expresses how he is prepared in advance to interpret an utterance of the speaker, while the passing theory is how he *does* interpret the utterance. For the speaker, the prior theory is what he *believes* the interpreter's prior theory to be, while his passing theory is the theory he *intends* the interpreter to use."[76] (There is a somewhat less than direct sense in which Davidson is here addressing H. P. Grice's well-known puzzles of meaning.[77])

Davidson's essential point is that "no one now has explicit knowledge of a fully satisfactory theory for interpreting the speakers of any natural language," and that "communication does not demand that any two people speak the same language" in any sense in which "prior theories" and "passing theories," learned theories and innate "theories," can be said to be *theories of the shared linguistic competence of native speakers.*[78]

I break off at this point to add only this: *if* this general condition holds, and if all that is required for successful communication is that "what must be shared is the interpreter's and the speaker's understanding of the speaker's words,"[79] then it is at once clear that the fate of Davidson's entire program depends on the validity of his earlier theory regarding the reliable truth of the massive corpus of our ordinary beliefs. But it is extremely difficult to see, now, *how*, if speakers and hearers do not share the same theories, or if we cannot assure ourselves that they do share the same theories, speakers and interpreters can be known to share an "understanding of the speaker's words." There is no doubt that linguistic communication is successful. But how can we be sure that that means that speakers and hearers *share an understanding of the speaker's words*? Either, now, that seems vacuous because impossible to test, or trivial because the question never arises, or incipiently skeptical because the question is thought profound but beyond our resources. In short, on the argument, it seems impossible to reconcile the 'Nice Derangement' paper and 'The Very Idea of a Conceptual Scheme.'

* * *

I hope I may now say, having demonstrated the profound arbitrariness of Davidson's philosophical program, that we cannot, at this late date, avoid admitting that all the standard questions of epistemology implicate interpretive *tertia* ineliminably. That is a conceptual bet, of course, hardly an *a priori* truth. But if that were conceded, I should be heartened to go on to say that, in drafting the foregoing argument, I was motivated by the need to provide a space for accounts of truth and knowledge that suppose: (a) that human thinking has an inherently historical structure, and (b) that questions of knowledge and questions of reality are indissolubly one. I have not argued for this. I have only cleared a space for such a view to thrive. It is my conviction, however, that Marx Wartofsky has been a pioneer in the development of a viable 'historical epistemology.'[80] I am pleased to be able to join him in pursuing the possibility. I am convinced it is the master theme of the next century.

Temple University

NOTES

[1] Donald Davidson, 'The Structure and Content of Truth,' *Journal of Philosophy*, 87 (1990), p. 281.
[2] Richard Rorty, *The Consequences of Pragmatism* (Minneapolis: University of Minnesota Press, 1982), pp. xiii-iv; cited by Davidson.
[3] 'The Structure and Content of Truth,' p. 279.
[4] 'The Structure and Content of Truth,' p. 282.
[5] 'The Structure and Content of Truth,' p. 298.
[6] See Richard Rorty, 'Pragmatism, Davidson and Truth,' in Ernest LePore (ed.), *Truth*

and Interpretation: Perspectives on the Philosophy of Donald Davidson (Oxford: Basil Blackwell, 1986).

[7] See Donald Davidson, 'Semantics for Natural Languages' and 'In Defence of Convention T,' *Inquiries into Truth and Interpretation* (Oxford: Clarendon, 1984). These are the principal sites of Davidson's advocacy of Tarski's model.

[8] See for instance 'The Structure and Content of Truth,' p. 315.

[9] See John F. Post, *The Faces of Existence* (Ithaca: Cornell University Press, 1989).

[10] See Donald Davidson, 'Mental Events,' *Inquiries into Truth and Interpretation.*

[11] See, for instance, Hilary Putnam, 'Realism and Reason,' *Meaning and the Moral Sciences* (London: Routledge and Kegan Paul, 1978).

[12] See Rudolf Carnap, 'Die Physikalische Sprache als Universalsprache der Wissenschaft,' *Erkenntnis*, II (1931); trans. as 'Psychology in Physical Languages," by George Schick, in A. J. Ayer (ed.), *Logical Positivism* (Glencoe: Free Press, 1959).

[13] Donald Davidson, 'A Coherence Theory of Truth and Knowledge,' in LePore, *Truth and Interpretation*, pp. 308–309.

[14] 'A Coherence Theory of Truth and Knowledge,' p. 309.

[15] See, for instance, 'The Structure and Content of Truth,' pp. 183–295; see, also, Paul Horwich, 'Three Forms of Realism,' *Synthese, 2* (1982), cited by Davidson.

[16] 'A Coherence Theory of Truth and Knowledge,' p. 309.

[17] Donald Davidson, 'On the Very Idea of a Conceptual Scheme,' *Inquiries into Truth and Interpretation.*

[18] See, for instance, Hilary Putnam, *The Many Faces of Realism* (La Salle: Open Court, 1987), Lecture I, particularly pp. 16–21.

[19] Putnam, *The Many Faces of Realism*, p. 20.

[20] Putnam, *The Many Faces of Realism*, p. 19.

[21] See Donald Davidson, 'Introduction,' *Inquiries into Truth and Interpretation*, p. xv.

[22] See 'In Defense of Convention T,' p. 73.

[23] 'Introduction,' p. xiv.

[24] 'In Defense of Convention T,' p. 74.

[25] 'Introduction,' p. xv.

[26] See Noam Chomsky, *Knowledge of Language: Its Nature, Origin, and Use* (New York: Praeger, 1986); and 'Semantics for Natural Languages.'

[27] Alfred Tarski, 'The Concept of Truth in Formalized Languages,' *Logic, Semantics, Metamathematics*, trans. J. H. Woodger, 2nd ed. John Corcoran (Indianapolis: Hackett, 1983) p. 268.

[28] Tarski, 'The Concept of Truth in Formalized Languages,' p. 153.

[29] 'In Defense of Convention T,' p. 71.

[30] 'In Defense of Convention T,' p. 68.

[31] 'In Defense of Convention T,' p. 75.

[32] 'In Defense of Convention T,' p. 73.

[33] Donald Davidson, 'Thought and Talk,' *Inquiries into Truth and Interpretation*, pp. 168–169.

[34] See W. V. Quine, *Word and Object* (Cambridge: MIT Press, 1960), §§ 15–16.

[35] See Nelson Goodman, *Ways of Worldmaking* (Indianapolis: Hackett, 1978), Ch. 1, particularly p. 2.

[36] This *is* the thesis opposed in 'On the Very Idea of a Conceptual Scheme.'

[37] See Hilary Putnam, *Representation and Reality* (Cambridge: MIT Press, 1988).

[38] 'Thought and Talk,' p. 168. The pertinent passage in Wittgenstein appears in Ludwig Wittgenstein, *Philosophical Investigations*, trans. G.E.M. Anscombe (New York: Macmillan, 1953), Pt. I, § 242.
[39] See, for example, the relatively early discussion in Sydney Shoemaker, *Self-Knowledge and Self-Identity* (Ithaca: Cornell University Press, 1963), Ch. 6.
[40] Donald Davidson, 'Radical Interpretation,' *Inquiries into Truth and Interpretation*, p. 137.
[41] See W. V. Quine, 'Epistemology Naturalized,' *Ontological Relativity and Other Essays* (New York: Columbia University Press, 1969).
[42] See Quine, 'Epistemology Naturalized,' pp. 75–76; also, Barry Stroud, 'The Significance of Naturalized Epistemology,' *Midwest Studies in Philosophy*, 6 (Minneapolis: University of Minnesota Press, 1981).
[43] The 'Humean' theme is itself criticized by Davidson in terms of his advocacy of 'epistemic holism' (the holism of linguistic meaning) and against what he calls 'the Building-Block theory' of meaning of the empiricists. See Donald Davidson, 'Reality without Reference,' *Inquiries into Truth and Interpretation*, pp. 219–220.
[44] See Quine, *Word and Object*, pp. 176–190, 219–221.
[45] See 'Mental Events'; 'Reality without Reference'; and Donald Davidson, 'Actions, Reasons, and Causes,' *Essays on Actions and Events*.
[46] 'Reality without Reference,' p. 221. The point about holism is explained at p.220: "Words have no function save as they play a role in sentences; their semantic features are abstracted from the semantic features of sentences, just as the semantic features of sentences are abstracted from *their* part in helping people achieve goals or realize intentions." You would be right to see in this both Davidson's thin insistence on human *praxis* and its contributing to our sense of what "we all necessarily employ" – hence, the rejection of *tertia*.
[47] 'Reality without Reference,' p. 222.
[48] The most important paper associated with both supervenience and anomalous monism is 'Mental Events.'
[49] See for instance Colin McGinn, *The Problem of Consciousness* (Oxford: Basil Blackwell, 1991), pp. 179–181.
[50] 'Mental Events,' p. 214.
[51] 'Mental Events,' p. 215.
[52] Donald Davidson, 'The Material Mind', *Essays on Actions and Events*, p. 253.
[53] I have heard Feigl mention the matter a number of times, but I have not found the expression in his published papers.
[54] 'Mental Events,' pp. 208–209.
[55] 'Mental Events,' p. 207.
[56] 'Mental Events,' p. 209.
[57] 'Mental Events,' p. 208. Davidson is prepared to weaken the determinism.
[58] 'Mental Events,' p. 210.
[59] 'Mental Events,' p. 211.
[60] 'Mental Events,' p. 212.
[61] 'Mental Events,' p. 214.
[62] 'Mental Events,' p. 217; see, also, pp. 221–222.
[63] 'Mental Events,' p. 221.
[64] 'Mental Events,' p. 209.

[65] Donald Davidson, 'Agency' *Essays on Actions and Events*, pp. 59–61.
[66] Rorty, 'Pragmatism, Davidson and Truth,' p. 342.
[67] 'Introduction,' p. xviii.
[68] 'Introduction,' p. xviii.
[69] Rorty, 'Pragmatism, Davidson and Truth,' p. 346.
[70] See Rorty, 'Pragmatism, Davidson and Truth,' pp. 333–334.
[71] See, for instance, Ian Hacking, 'The Parody of Conversation' and Michael Dummett, 'A Nice Derangement of Epitaphs: Some Comments on Davidson and Hacking,' in LePore, *Truth and Interpretation: Perspectives on the Philosophy of Donald Davidson*.
[72] Donald Davidson, 'A Nice Derangement of Epitaphs,' in LePore, *Truth and Interpretation*, p. 434.
[73] 'A Nice Derangement of Epitaphs,' p. 434.
[74] 'A Nice Derangement of Epitaphs,' p. 440.
[75] 'A Nice Derangement of Epitaphs' p. 446.
[76] 'A Nice Derangement of Epitaphs,' p. 442.
[77] See H. P. Grice, 'Meaning,' *Philosophical Review, 66* (1957).
[78] 'A Nice Derangement of Epitaphs,' p. 438.
[79] 'A Nice Derangement of Epitaphs,' p. 438.
[80] See, for instance, Marx W. Wartofsky, 'Epistemology, Historicized,' in Abner Shimony and Debra Nails (eds.), *Naturalistic Epistemology* [*Boston Studies in the Philosophy of Science*, vol. 100] (Dordrecht: D. Reidel, 1987).

JOËLLE PROUST

TIME AND CONSCIOUS EXPERIENCE*

> "Let the jury consider their verdict," the King said, for about the twentieth time that day.
>
> "No, no!" said the Queen. "Sentence first – verdict afterwards."
>
> Lewis Carroll, *Alice in Wonderland*

In a recent work, Dennett examines a series of empirical findings in neurology and psychology of perception and maintains that "in certain time spans, the distinction between pre-experiential and post-experiential content revisions cannot be maintained" (Dennett & Kinsbourne, 1992, 192). He claims instead that the "multiple draft theory", which he advocates, makes the identification of a particular "moment of processing in the brain as the moment of consciousness" "arbitrary" (194). My aim here is reviewing the conflicting hypotheses in order to assess Dennett & Kinsbourne's skeptical conclusion. While doing so, I hope to clarify the notions of experience and of consciousness, and more generally the relationship between neural and mental properties. I will first summarize some of the startling results of neurosurgeon Benjamin Libet which triggered a philosophical controversy in the late seventies and early eighties. I will then discuss two of the conflicting views, Ted Honderich's and Patricia Churchland's, which will allow me to suggest a reformulation of Libet's hypothesis. A final discussion will deal with Dennett and Kinsbourne's solution.

The kind of common-sense view the contemporary layman and the early neurophysiologists share is that there should be a one-one correspondence between a chain of physical stimuli directed at the sensorial receptory areas of a nervous system and a sequence of resulting sensations. According to that view, if physical events having the role of distal stimuli E1 and E 2 are such that E1 is prior to E 2, then the corresponding proximal stimuli keep the same temporal relation, and so at the end of the processing sequence, the physical sensations S1 and S 2 should do. If a stimulus is sent at time t, it should be 'felt' or 'experienced' *before* a stimulus sent at time t+1.

C.C. Gould and R.S. Cohen (eds.), Artifacts, Representations and Social Practice, 323–341.

Now it is well known that stimuli belonging to *different* sensorial modalities do routinely violate this ideal picture of a smooth temporal isomorphism between physical events, proximal stimuli and sensations per se. The example of a stormy night when the flash of lightning precedes the sound of the thunderbolt illustrates the potential disruption of the temporal order from the distal to the proximal stimuli. But in this example, what explains this difference is only a matter of diversity in speed of propagation between light and sound; the specific properties of the receptor do not come into play, over and above the generic features which allow it to be responsive to the particular properties of the given signals. Only generic properties such as the relative position of the receptor and the source of the signals, as well as the nature and characteristics of the channel, such as sound wave or light wave, are thus relevant for the explanation.

Why don't we take into account the mental *processing speed* required in dealing with this kind of phenomena? Obviously because the physical differences in the channels of the signals *suffice* to explain the discrepancy between two simultaneous physical events and two temporally distinct proximal stimuli. This does not preclude the possibility that indeed the two kinds of signals are also processed *at different speeds*, in the sense that it takes them different durations to achieve what Libet (1973) calls 'neuronal adequacy', i.e., to be submitted to the cerebral activities which, upon reception of a given stimulus, elicit the resulting conscious sensory experience. No doubt this additional source of temporal discrepancy looks negligible, when it is compared with the previous one. For it takes 'strong' stimuli from 50 to 100 msec to achieve neuronal adequacy, and approximately 500 msec for stimuli close to threshold level, and the difference between auditory and visual stimuli ranges between 30 and 40 msecs[1]. Nevertheless, there is indeed the question of what makes it possible to achieve a synthesis from the activities of different sets of neurons dealing with disparate features of the objects of perception. This is part of the *binding problem*, which Damasio (1989) addresses by invoking 'convergence zones' in which feedback projections from the various cortical areas help coordinate synchronous activity[2].

But the kind of discrepancy we will be mostly concerned with here is related to the timing of a conscious experience limited to a single sensory modality and even to a single signal type, of a very elementary nature, i.e., close to the kinds of psychophysical stimuli usually applied in attentional studies. For in this very restricted domain, we already encounter a major theoretical fork concerning the nature of subjective timing and its relationship to physical time. It is precisely in exploiting these kinds of data that Dennett & Kinsbourne try to show the empirical equivalence of two very different kinds of hypotheses.

For lack of space, I concentrate here on one type of findings from neurophysiology explored by these authors. In experiments related in Libet (1973), (1978), (1979), (1981), (1985), (1992), Benjamin Libet has shown that there is a significant discrepancy between the subjective timing of an experience and the time necessary to achieve the "neuronal adequacy" of that same experience; in other words, although a delay of approximately 500 msecs is necessary for a subject to feel a close to threshold tactile stimulus applied on the back of his hand, he uses a different temporal marker than this one to date the stimulus. In other words, "the *subjective timing* of the experience *is (automatically) referred backwards in time*" (Libet, 1981, 48). To show this, Libet uses neurosurgical evidence (experiments made during brain surgery in awake patients with local anesthetics, as well as (intracranially) electrode-implanted ambulatory patients).

It is an established fact that a somatic sensory experience, such as the feeling of a skin stimulus of the back of the hand, can be elicited in two ways. Either by a direct stimulation of the relevant part of the body, or by an electrical stimulation of the contralateral relevant part of the somatic cortex (postcentral gyrus). Any stimulation effected somewhere on the neural pathway between the sensory receptors in the skin and the sensory cortex has the property of producing "the same sensation", i.e., say, the same tactile impression on the back of the hand. But there will be some important differences, which need to be spelled out.

The first and major difference has to do with the production of a "primary evoked potential"; this particular electric response in neurons in the sensory cortex to a sensory stimulation is elicited by all the stimulations effected on the neural pathway from skin to thalamus, but *not* by electrical stimulations effected in the somatic cortex itself. This difference in the electric activity recorded in the brain seems to be causally correlated to the subjective timing of the sensory experience and is of course the dependent variable explored in the various experiments reported by Libet and Libet *et al.*

A second difference has to do with the duration of the stimulus at a threshold intensity level necessary to elicit a sensory experience. It was found that if liminal stimuli were applied to the surface of sensory cortex or to any other (subcortical) part of the neural ascending pathway in the brain (such as the section included in the medial lemniscus or in the thalamus), those electric pulses had to last about 500 msecs to produce a sensory response. On the other hand, a single pulse electrical stimulus may achieve threshold intensity when applied on the back of the hand.

But, and this is also a finding of Libet's, there is no significant difference in the delay of the so-called 'induced sensation', for the cortical stimulus and the skin stimulus both need approximately 0.5 secs to achieve neuronal adequacy.

Now what will happen if a subject is given two successive stimuli, the first on the relevant part of the somatosensory cortex for the left hand and the second one on the skin of the back of the right hand? Everything else being equal, one could predict that the stimulus on the left hand would be felt before the stimulus on the right one. In fact, the sensation on the right hand is consistently reported by the subjects as appearing *before* the cortically-induced sensation on the left hand. We have here an illustration of an *inversion of subjective timing relative to the objective stimulus onset*. How then is it possible that there could be such a blatant temporal inversion between the objective timing of the distal stimulus and the subjective timing reported by the conscious subject, though both stimuli take approximately the same time to achieve neuronal adequacy?

As we shall see, the problem is not only to understand the cerebral mechanism(s) which underly this peculiarity of the cerebral processing in real time of successive stimuli, but to examine the consequences for what *sensation* and *consciousness* mean, once those two concepts are severed from the pretheoretic intuition of 'presence' and 'simultaneity'. Our task will be first to examine the available empirical hypotheses which could help solve the problem at hand, and second to draw some philosophical corollaries about the meanings of concepts such as consciousness and sensation.

To account for the difference in temporal reception of cortex vs. skin induced stimuli, Libet explores the ways in which the cerebral responses develop in each of those conditions. Two differences emerge. (1) Whereas a peripheral stimulus (or a stimulus applied anywhere on the relevant neuron between skin and thalamus) always produces a specific electric response, named 'primary evoked potential', no such effect appears in the case of a stimulation applied anywhere between thalamus and cortex. (2) Another difference is the duration of a stimulus at a constant, near threshold intensity needed for a conscious experience. We saw earlier that, if the stimulation is effected in the cortex or in a subcortical area, it must last at least 500 msecs to produce a conscious experience. But if it is effected at the periphery, a much shorter stimulation can produce a cortical response. Of course, this is not to say that neuronal adequacy requires less time in the latter case: the duration of a peripheral stimulation and the duration of the central response to this input (cf. Libet, 1992, p. 260) do not have to coincide. (Fig. 1)

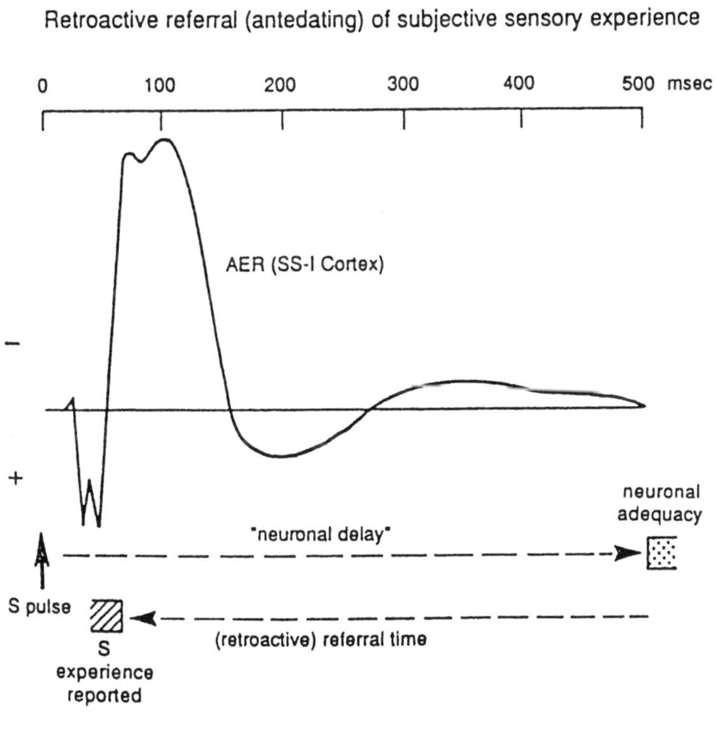

Figure 1. The retroactive referral of subjective sensory experience. Reproduced with permission of Libet (1992).

Therefore a brief near to threshold stimulation on the back of the hand will bring about first after 10 to 20 msec a primary evoked potential in the cortex, second a general neuronal activity in the cortex during about 500 msec.

Putting together those two clues – presence of a specific electric signal in the skin case only, same mean duration to reach neuronal adequacy – Libet comes up with a revolutionary theory of the relationship between cerebral and subjective temporality: the time of the skin induced stimulation is referred back to the moment when primary evoked potential occurs. In other words, the experience at the output of the full processing of the stimulus is "subjectively antedated" (Libet, 1992, 260) to the time of the primary evoked potential. What thus explains the difference between the time of the skin-induced and the time of the cortex-induced experiences is that in the latter case there is no backwards referring in time for lack of any primary evoked potential telling, so to speak, when the stimulation is supposed to have begun. (Fig. 2)

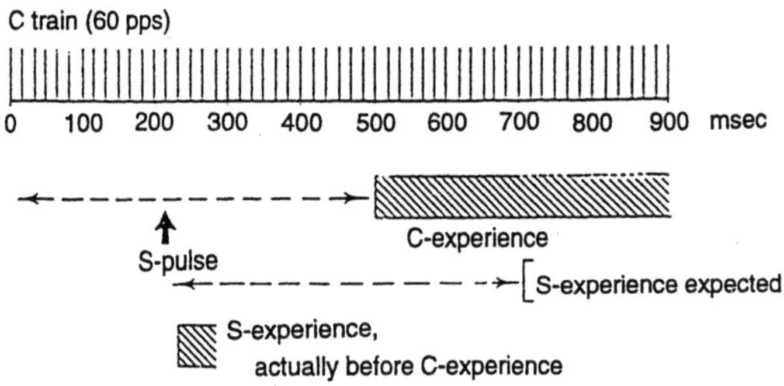

Figure 2. Subjective time order of a cerebrally-induced vs. a skin-induced sensation. Reproduced with permission from *Revue de Métaphysique et de morale*, Avril-Juin 1992.

This hypothesis is further confirmed by a comparison between a peripheral stimulation and a central (subcortical) one in which both characteristics of the preceding stimuli are combined. It requires a stimulus as long as the cortical one, but it triggers a primary evoked potential as the skin one. Libet reports that, in that particular "central" case, it is experienced as simultaneous with the peripheral stimulus (1992, 262). This is an important fact, which played a decisive role in the controversy with Patricia Churchland.

Now Libet's hypothesis of a subjective backwards referring in time of conscious experience allows us to raise the question of the identity between a physiological process and a subjective, qualitative, experience.

What does it mean to speak of "the time of the conscious experience of a stimulus"?

(1) Does it mean the "subjective" time, i.e, the time which is allocated to the stimulus on the basis of the corresponding primary evoked potential? In this case, one will have to say that the subjective experience is antedated in relation to the "true" moment when the information processing is being completed. In other words, subjective experience will first have had to take place, with a certain duration, and then will be so to speak artificially antedated.

(2) Or does it mean the time necessary to reach neuronal adequacy for the experience in question? In the latter case, experience only occurs at the end of the processing line, and *not* at the moment it has been "referred (back) to".

Honderich's (1984) comments on Libet's results outline the difficulty inherent to the first of these hypotheses:

> Despite what is said of the experience's involving "subjective referral backwards in time", the experience itself occurs only about 0.5 sec after the beginning of stimulation. The experience does not occur at the time to which it is "referred". (1984, 118)

To be able to claim that what is antedated is experience itself, one has to allow that the experiential content can be severed from the temporal awareness which is relative to it. As Honderich puts it,

> The central idea appears to be that a subject has a conscious sensory experience including or involving or accompanied by the belief or impression that it is not *then* happening. (1984, 124)

But insofar as the actual having of a sensory experience includes the impression that it is *present*, Libet's hypothesis leads one to attribute contradictory beliefs to subjects, which they could describe with the words "present sensation past" or "now sensation then", or perhaps "later sensation earlier" (1984, 124). We are thus facing a full-blown paradox: each subject would be at every moment entertaining contradictory beliefs by reporting his own experiences as being both present and past; not the familiar impossibility of having present experience coincide with its verbal report, rather the impossibility of having a present experience as such.

Honderich's point has the following interest: it reveals what is wrong with the notion of *experience*, or of *being aware of a sensation*, as they are being used in this debate. Honderich's discussion relies on the idea that the *present of consciousness coincides in an absolute way with an objective clock indicating the very moment when neuronal adequacy is reached.*

The subject says – or thinks, or feels – "now sensation X". But what does the term "now" refer to? Is it some point of his stream of consciousness? Or, so to speak, does he go beyond the mirror of his sensations to reach some objective temporal correlate, which would be the time on the objective clock? Even if we concede to Honderich that "we cannot choose to regard a conscious sensory experience as something of which the subject is unaware" (1984, 125), we do not have to suppose that perceived time is nothing else than the time of the corresponding perception. Being conscious does not imply having a stream of consciousness isomorphic to the underlying cerebral processes or taking as a temporal rule the succession of neuronal events.

If we return to Figure 2, we see a large area corresponding to the time it takes for experience to be constructed through a series of processing steps. One can imagine some Laplace demon having access to the developing experience every ten milliseconds, i.e., to "phase Pi of the experience in progress". One can finally imagine that once experience has been fully constructed in all its sensory and associative aspects, the final stamp is added: the one indicating at which point in time to "deliver" the experience, i.e., to add the information contained in the primary evoked potential.

But of course what is being here confused is experience and the kind of information processing which makes it possible. Experience can be said to be "antedated" only if it is considered as already synthesized before the final temporal marker is applied to it. If on the other hand one takes experience to be what is immediately given to a subject, one should not take an intermediary processing phase as being *already* his or her experience; and it is not appropriate to confer a qualitative content to what is neither conscious nor susceptible to becoming conscious.

What does it mean exactly to say that a subject is presently aware of some sensory experience? It is a familiar problem in philosophy of language to make explicit the meaning of indexical terms such as 'now'. Whenever it is attempted to spell out their meaning, such as 'the moment at which I speak', other terms such as 'I' and the present tense reinject again contextual elements in the purported definition. Libet's idea is that, whenever a subject feels some stimulus as simultaneous or prior to another stimulus, he will use the information which is cerebrally carried by the evoked potential associated to that stimulus. It is thus tempting to try to construct a one-one mapping between a phenomenological event, an informational event and a neurophysiological event. In particular, the large, sudden oscillation characteristic of the primary evoked potential is taken as constituting both a functional and phenomenological distinct phase.

In philosophical terms, it would be tempting to say that this signal is the rigid designator on which everything else depends, the very notion of 'I' becoming dependent on the so-tagged mind/brain state.

But one sees immediately why, however tempting, it is not so easy to transfer the physical-neuronal signal into the mental. The 'now' of the clock registering the cerebral state of 'neuronal adequacy' is different from the 'now' of the mind sensing the stimulus. The first one is anchored in the simultaneity of two physical mechanisms, one of which is assumed to elicit (upon completion, i.e., after some delay) the state of awareness; the second one – the qualitative, mental 'now' – is anchored in the sensation reported by the subject. As remarked by William James, an isomorphism between brain states and states of awareness does not help us to understand the specific sensation of a present, as contrasted to a past stimulus.[3] As Grünbaum (1973) comments:

> The hypothesis of isomorphism of traces and states of awareness renders only the succession of states of awareness, but not the *instantaneous awareness of succession*, which is an essential ingredient of the meaning of 'now': the now content, when viewed as such in awareness, includes an awareness of the order of succession of events in which the occurrence of the awareness constitutes a *distinguished element*. (p. 325)

Even supposing that the subjective experience of 'now' is made possible by the early cortical response possibly completed by a global synchronization of neuronal activity across the brain, it is clear that we should not *identify the reconstructed 'now' with the mechanism effecting the portraying*.

The fact that subjective awareness of 'skin stimulus now' is achieved over time does not mean that the subjective experience is itself isomorphically spreading over time. In the last sentence, it becomes clear that 'time' has been understood in two different ways: the first one refers to cerebral (processing) time, the second to subjective (feeling) time.

In fact, Libet's experiments do not tell us anything about primary evoked potential and the sense of [present experience]. What the evoked potential carries information on is more likely to be intrinsically ordinal rather than indexical. In other words, the information it carries is not [present sensation] (or: [sensation now]) – opposed to [past sensation], but rather the position of the sensation on a before-after complete ordering. Had he emphasized this purely serial function of the primary evoked potential, Libet would not have had to face Honderich's objections.

This purely ordinal interpretation of the informative function of the designated electrical signal allows one to understand what may be the vital role of the signal within a global oscillatory neuronal activity, i.e., something like a synthetic principle of object construction and of temporal succession of external events in time. While he underlines this kind of potential role, Libet understands it as due to its absolute, versus relative temporal value. But the ordinal property of the mechanism suffices to explain the synchronous character of information processing pertaining to different sensory modalities. The duration of the processes varies from one modality to the next. If the subjective timing of experience onset depended on the *end* of the processing, a subjective asynchrony would ensue unless some other mechanism implements Kantian synthesis of experience. As we already saw, the time necessary to reach 'neuronal adequacy' depends among other factors on the intensity of the stimulus, which does not have to be the same on the different qualitative dimensions. Thus the only useful way of clustering sensations seems to be to tag them on their common distal origin. Whenever an external event triggers the activations of several sensory modes of stimulation, the common origin of those proximal stimuli will only be retained by the system if it keeps track of the beginning of the processing. If it was the *end of the processing* which tagged the temporal order of each sensation, another principle would be required to bring back distal order among internal data.

Despite all the difficulties raised by this solution, it seems to be precisely the hypothesis which Patricia Churchland favours in (Churchland, 1981)[4]. Libet's data, according to Churchland, are equally compatible with a different, 'less spectacular', account:

> Rather than suppose one sensation is referred backwards in time, explain the phenomenon by saying that one stimulus is put on hold, so to speak, until it can be 'admitted' to consciousness. (Churchland, 1981, 177)

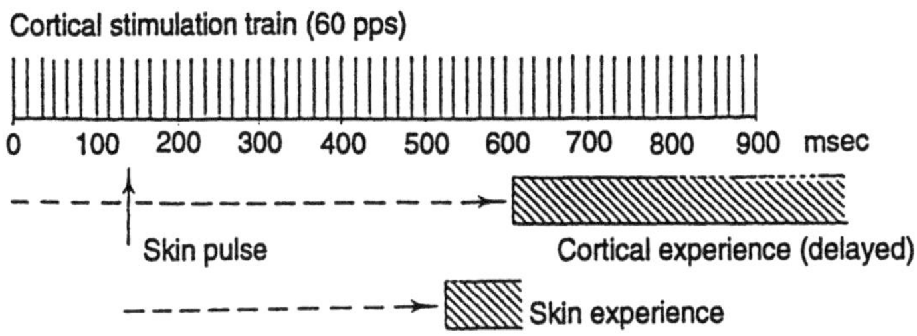

Ordering Test:
Stimuli: skin pulse administered 200 msec after pulse train to cortex is begun.
Reports: after trial end, some subjects report that the skin sensation was felt before the cortical sensation.

Figure 3. The postponement hypothesis. Reproduced with permission from Churchland (1981).

The alternative hypothesis maintains that instead of having the peripherally induced sensation antedated, the cortically induced sensation is postponed in the double stimulations experiment [Fig. 3]. In this new case as in the former, experience can be had (with its distinctive present quality) only once neuronal adequacy is reached. Therefore the *ordinal* relations described by Libet still hold. But

(1) Instead of conscious experience being made 'on line', only the temporal tagging which intervenes at the end of the processing is acknowledged as plausible.

(2) The difference in the cortically induced sensation case is explained via a mechanism which has nothing to do with the primary evoked potential: it is supposed that some property of the cortical stimulation allows it to be 'put on hold' – put in a buffer memory? – "until it can be 'admitted' to consciousness".

The arguments Patricia Churchland gives in support of her hypothesis have to do with the results of backward masking presented by Libet's previous papers (Libet 1973, 1978). Masking is the well-known phenomenon in learning which shows that, for certain sufficiently short temporal intervals between stimuli, the perception of the first stimulus is entirely masked by perception of the second. Such a masking seems to affect mainly conscious perception while not entirely suppressing the processing of the information which it contains. Masking in Libet's earlier works was of the backwards operating kind. When a near-threshold skin stimulus was followed, 200 msec later, by a powerful cortical stimulus, the skin stimulus was either non-consciously perceived, or inversely, amplified (a phenomenon known as "retroactive enhancement"), according to the location of the stimulation. These results allow Libet and his coworkers to compute the time taken to reach "neuronal adequacy", the maximal duration of a retroactive effect of about 600 msec.

To this line of reasoning, Churchland objected that, in the case of an interference between a skin stimulus and a cortical one, it is possible that what happens is not masking "in the usual sense", but "blanking in short term memory". She argues that the subject reports on his experience at the end of the trial (1981, 172), and that "there are believed to be stepwise connections from the primary and secondary cortex to the hippocampus", the hippocampus itself being involved in memory and attention tasks. Churchland states that the latter argument does not "add up to anything like a proof", but should at least be tested by comparing retroactive masking skin-cortex with central "skin-skin" masking. Nevertheless the hypothesis finally favoured suggests that information is processed, stored in memory, and then later readmitted into consciousness.

While Libet admits that a subject could "forget" his primary sensation once subjected to the action of the cortical stimulus, finally he believes the suggestion has to be given up. For such a cortical stimulus is proved to have two types of effects on a skin stimulus: either the former masks the latter, or it enhances it retroactively (Libet 1981, 190). In such a case, it is somewhat difficult to invoke the causal role of "forgetting".

Libet reviews the evidence against the postponement hypothesis. The evidence consists in the comparison between the various durations required by the various kinds of stimulations and the judgments offered by the subjects about the subjective order of succession of their corresponding experiences. Let us remember here the counter-experiment consisting in directly stimulating subcortical areas. If one directly stimulates the median lemniscus, the resulting primary evoked potential is comparable to one resulting from skin-induced stimulation, and both stimulations require a duration of about 500 msec to become conscious. As Libet is able to predict, a stimulus is perceived as simultaneous with the skin stimulus if the onsets of both stimulations are. This result seems unexplainable unless one gives to primary evoked potential a distinctive informational role about order in time.

Another difficulty of the postponement hypothesis, already mentioned, appears with the problem of synthesizing experience in the unity of subjective time. How could we avoid 'temporal jitter' between the various features of a perceived event if the subjective construction of succession is controlled by the end of each processing line, given the fact that the time of processing is known to be different in different sensory modalities, and that inside one and the same modality, stimulation time necessary to produce a sensation varies with the intensity?

Aside from these facts which seem to speak against its empirical adequacy, Churchland's hypothesis relies on the assumption that a methodological principle of simplicity should have us reject the retroactive hypothesis:

> The retroactive hypothesis invites circumspection because it needlessly ushers in quite daunting complexity. Consider that the primary evoked response is taken to signal the unconscious receipt of information from the periphery, and precedes the neuronal activation time necessary for awareness. Hence the evoked potential wave precedes the awareness, but additionally it is alleged to provide the 'time marker' for the sensation to be referred back to. Thus *the awareness is simultaneous with what precedes it;* an event A both precedes and is simultaneous with an event B. (Churchland, 1981, 177)

The invoked economy principle could apply if the temporal information used in a cerebral process had to be nothing else than the temporal characteristics of the process itself. But although the brain likes to code spatial properties with spatial features of the representations encoding them, it is not likely that temporal properties have to be coded as a function of the objective succession of outputs. There is not the slightest evidence to show that signal processing has to be temporally isomorphic to conscious sensation. Libet's results on the other hand lead us to think that temporal information is in fact linked to early processing: it is a matter of cerebral functional architecture; nothing belonging to the intentional characterization of the processes can conflict with the retroactive hypothesis.

Dennett (1991) and Dennett & Kinsbourne (1992) attack the concept of central mental headquarters where all informations come together. Tracing this idea back to Descartes' pineal gland, which operates as a gate towards the conscious mind, they show that it still is at work in many theories of cognitive science, even though the (ontological) dualist tradition has been largely abandoned. For example, Geschwind (1970) still locates conscious overview of input and conscious formulation of intentions in the left angular gyrus (Kinsbourne, 1988, 240). As Kinsbourne remarks, no selective damage up to date has ever been found supporting such a claim.

Available neuropsychological evidence shows on the contrary that some lesions may affect selectively, although not simultaneously, the capacity to entertain certain sensations and the capacity to become aware of having them. A cortically blind patient

> can see little or nothing, but acts and speaks as if he could see as well as ever. The victim of cortical blindness not only sees nothing; he does not experience the fact that he is not seeing anything. His conversation is rich with allusions to visual experiences which conflict with physical reality. His confabulations about visual percepts represent his attempts to use his previously acquired knowledge about visual events in the face of the sudden loss of the ability even to imagine what it is like to have a visual experience. (Kinsbourne, 1988, 241)

In the same way, a patient with a frontal lesion cannot choose to organise any activity on his left side. But he is not aware of this incapacity. If one tries to tell him, he may either deny it, or verbally acknowledge it, while at the same time keeping all of his behavior unchanged (*Ibid.* 241). Lobotomized patients have their cerebral hemispheres linked to the cerebral stem, but not to the other hemisphere; many losses in control ensue. But the patient is not aware of these losses. If he is asked to tell (using his left hemisphere, which usually interprets perception and action) what he saw (with his right hemisphere), his answers take into account only whatever information is available on the left side, although with an emotional influence coming from scenes watched by the right hemisphere (Gazzaniga, 1988, 234).

All these clinical observations suggest that there could be no clear-cut distinction between a sensation and a conscious sensory experience. Clinical evidence appears then to sever the intuitive link between a subjective experience and its objective correlate. A cortical blind, for example, is not deprived of qualitative experience; he only ceases to interpret in the usual way his retinal stimulations. The same evidence also severs the intuitive link between having an experience and knowing that one has it. For example, an heminegligent has no qualitative experience corresponding to half of his visual field, but is not aware of that deficit. He is nevertheless able to process part of the information available in that part of the visual field which he does not perceive consciously. Are we prepared to say in that case that he has a sensation, insofar as he is capable of being causally influenced by what he unconsciously perceives? Or should we say that he has no sensation, since he cannot have any access to it?

Reflecting on these neuropsychological data leads Dennett and Kinsbourne to a global skepticism concerning the scope of consciousness in normal perception and action. The new theory which they advocate, the 'multiple draft theory' is meant to oppose the Cartesian theater model.

Three major claims in this model need to be outlined. First, it denies that a representation has to be constructed out of a final assemblage from multiple finer detections, which should first be made, then synthesized in *re-presentation.* Mental functioning should be conceived of as *holistic.* Each discrimination is made in the context of a largely distributed and cyclic processing, and occurs only once.

Second, the highly parallel character of neural processes, the high connectivity of the brain, tend to blur the line between a purely bottom-up and a purely top-down way of information processing. In other words, there is no output which could be singled out as being the stream of consciousness 'narrative'.

> 'The stream of consciousness' is *not* a single, definitive narrative. It is a parallel stream of conflicting and continuously revised contents, no one narrative thread of which can be singled out as canonical – as the true version of conscious experience. (Dennett & Kinsbourne, 1992, 185)

Finally, the temporal properties of discriminative operatives may well be objectively described; they do not determine the temporal properties of subjective experience. In other words, an operation occurring before or after another does not make the corresponding contents before or after one another.

Dennett & Kinsbourne contrast two approaches of temporal facts involved in consciousness which they respectively name 'Orwellian' and 'Stalinesque'. The difference between these two approaches depends on whether "the editing or adjustment or tampering occurs before or after a presumed *moment of onset of consciousness* for the contents in question" (1992, 236). When an element is supposed first to reach consciousness, then to be forgotten (erased, replaced), the account is Orwellian. If an element does not appear in consciousness, even if it leaves traces in memory, the account is Stalinesque.

Clearly, Libet's and Churchland's hypotheses can be read with this contrast in mind. While Libet's is Stalinesque in suggesting that primary evoked potential is the crucial, although unconscious, element organizing conscious perceptions temporally, Churchland's is Orwellian in postulating that the subject first has a sensation (cortically induced), then forgets it and stores it, and finally recovers it. In the first approach, editing takes place early on, on the shooting set. In the second, editing occurs later, after the shooting is completed. The question is obviously to know whether empirical evidence affords us a choice among these two different hypotheses. We examined earlier data supporting Libet's 'Stalinesque' account. Dennett and Kinsbourne's claim is that no fact of the matter allows us to prefer one hypothesis over the other.

> Because perception turns imperceptibly into memory, and 'immediate' interpretation turns imperceptibly into rational reconstruction, there is no single, all-context summit on which to direct one's probes. Any probe may elicit a narrative (or a narrative fragment), and any such elicited narrative determines a 'time line', a subjective sequence of events. (Dennett & Kinsbourne, 1992, 195–6)

This skeptical move can be understood in two ways, none of which seems to be compatible with the constraints which are given to us by the notions of consciousness and of conscious experience. The crucial question is to know whether there can be an 'optimal time of probing'. Here though, two different interpretations of the course of the underlying argument seem possible.

First version: Skepticism is directed at probing in absolute time. Although the authors repeatedly claim that no confusion should be made between constructing conscious experience in time and time represented in consciousness, they do rely on this identification to dismiss the possibility of having distinctive evidence in favour of one of the hypotheses. Their argument consists in concluding from the fact that cerebral processes are holistic, cyclic, highly connected, that one can "probe" at any point of the process and get totally different "narratives".

Second, they seem to conclude from the fact that the distributed character of cerebral activities make *content itself* "spatially and locally distributed" (185). But what kind of data is probing supposed to bring to the fore? Are we to collect activations on output units in a connectionist neural network, mathematical vectors at the output of a filter at a given level of representation? In fact, none of these probes will give us any conscious *content*. Only from a certain level of processing on will the possibility of a conscious experience appear. In other words, one has to acknowledge that the notion of a conscious content only has relevance at a certain level of representation and processing (cf. Crick & Koch 1990). Saying that there is no isomorphism between brain operations in time and temporal awareness should not lead us to believe, and indeed seems to conflict with the idea, that the parallel activity of the brain makes the probing perfectly arbitrary. For even though we don't know much about the nature of the informational processes involved, what we do know is that phenomenological consciousness (access to mental content) has to result from a certain type of informational sequence of events and processes, which themselves supervene on some sequence of neurophysiological events and processes. As we saw earlier, Libet's results can be interpreted as showing that the brain has specific informational means to convey the subjective order of experiences. Therefore, although we have to refrain from any 'absolute' interpretation of temporal sequencing, there seems to be no methodologically principled rule forbidding us to look for temporally ordered informational processes, i.e., for causal relations between these processes. Why could we not use standard neuropsychological or attentional data to study the relationship between memory and awareness to know whether a Stalinesque or an Orwellian hypothesis has more empirical validity[5]?

Second version: Skepticism is about consciousness itself. The point of the authors is, as suggested earlier, to track down and dismantle the Cartesian fallacy, which leads many cognitive scientists and neurologists to believe that there is some central "finish line" for information processing which opens up on the theater of consciousness. But as is suggested by some of their critics, they may overdo it, and "throw the conscious baby out with the Cartesian bath water".[6] Should the arguments given against Cartesian materialism lead us to consider consciousness perfectly indeterminate? Some of Dennett & Kinsbourne's claims may lead one to believe that they indeed accept this conclusion. This idea of a multiplicity of narratives seems to imply that indeed people are mistaken to believe that they have just one – even if complex – present experience. Furthermore, if no kind of empirical evidence can speak in favour of an Orwellian versus a Stalinesque hypothesis, it might seem that being once conscious (as in the Orwellian case) makes no difference. But phenomenologically it is surely a fact that consciousness has certain qualitative properties, such as being or not directed at something, and having unity. When a subject experiences a break in the unity of the stream of her consciousness, for example in certain cases of schizophrenia, it is generally a devastating and distressful experience. Dennett & Kinsbourne insist that they only want to defend the infalsifiability of hypotheses which aim at answering inappropriate types of questions (1992, p. 235). But if they want to appear as *bona fide* realists, they should be more explicit about the reasons why there is no optimal time of probing, no particular moment when some informational level becomes endowed with the property of becoming conscious. Looking at a chair as is done in physical science enables one to understand what properties of the atoms account for the apparent solidity of the object. Looking at consciousness using neurophysiological tools should no more change the phenomenological notion of consciousness than the physical stance affects the ordinary concept of what a chair is. But it should clearly allow us to understand the macrolevel from the microtakings, and not *vice versa*.

C.N.R.S., Paris

NOTES

* I express my gratitude to Sam Guttenplan, Elisabeth Pacherie, Philip Pettit, and Adriano Palma for reading a former version of the present paper. Special thanks to Adriano Palma for his help with the American translation.

1 Dennett & Kinsbourne (1992) report Pöppel's finding (Pöppel 1985, 1988) that the horizon of simultaneity is about 10 meters.

2 See also Crick and Koch (1990).

3 I owe the reference to James (1950), 628–9, to Grünbaum (1973).

4 This paper exposes methodological worries concerning Libet's work which I will not comment on here.

5 A similar point is made by Block in his response to Dennett & Kinsbourne (1992); see Open Peer Commentary, [same reference] p. 206.

6 An expression from Aronson, Dietrich and Way's contribution in Open Peer Commentary, [previous reference] p. 202.

REFERENCES

Castiello U., Paulignan Y. & Jeannerod M. (1991) 'Temporal dissociation of motor responses and subjective awareness. A study of normal subjects', *Brain 114*.

Churchland, P. (1981), 'On the alleged backwards referral of experiences and its relevance to the mind-body problem', *Philosophy of Science 48*, 165–181.

Churchland P. (1981), 'Discussion : the timing of sensations : reply to Libet', *Philosophy of Science 48*, 492–497.

Crick F. & Koch C. (1990), 'Towards a neurobiological theory of consciousness', *Seminars in the Neurosciences 2*, 263–275.

Damasio A. R. (1989), 'The brain binds entities and events by multiregional activation from convergence zones', *Neural Computation 1*, 123–132.

Dennett D. (1991), *Consciousness Explained*, Boston, Toronto, London, Little, Brown and Company.

Dennett D. & Kinsbourne M. (1992) 'Time and the observer: the where and when of consciousness in the brain', *Behavioral and Brain Sciences 15*, 183–247.

Gazzaniga, M. S. (1988), 'Brain modularity : towards a philosophy of conscious experience', in *Consciousness in Contemporary Science*. A. J. Marcel and E. Bisiach eds., Oxford, Clarendon Press.

Geschwind N. (1970), 'The organization of language and the brain'. *Science 170*, 940–44. Also in Geschwind, *Selected Papers on Language and the Brain* [*Boston Studies in the Philosophy of Science*, vol. 16] ch. 21, pp. 452–466 (1974).

Glynn I. M. 'Consciousness and Time', *Nature 348*, 477–479.

Goodman N. (1978), *Ways of Worldmaking*, Hassocks, Sussex, The Harvester Press.

Grünbaum A. (1973) *Philosophical Problems of Space and Time*, Dordrecht, Boston, Reidel. [*Boston Studies in the Philosophy of Science*, vol. 12]

Honderich T. (1984), 'The Time of a Conscious Sensory Experience and Mind-Brain Theories', *Journal of Theoretical Biology 110*, 115–129.

James W. (1950), *The Principles of Psychology*, New York, reprint Dover Pub.

Jeannerod M. (1992), 'The where in the brain determines the when in the mind', *Behavioral and Brain Sciences 15*, 212–213.

Kinsbourne, M. (1988) 'Integrated field theory of consciousness', in *Consciousness in Contemporary Science*, A. J. Marcel and E. Bisiach eds., Oxford, Clarendon Press.

Kolers P. A. & von Grünau M. (1976), 'Shape and Color in Apparent Motion' *Vision Research 16*, 329–335.

Libet B. (1973), 'Electrical stimulation of cortex in human subjects and conscious sensory aspect' in *Handbook of Sensory Physiology*, vol. II, A. Iggo (ed.), Berlin, Springer-Verlag, pp. 743–790.

Libet B. (1978), 'Neuronal vs. subjective timing, for a conscious sensory experience', in *Cerebral Correlates of Conscious Experience*, P. A. Buser and A. Rougeul-Buser (eds.), Amsterdam, Elsevier/North Holland Biomedical Press, pp. 69–82.

Libet B. (1981), 'The experimental evidence for subjective referral of a sensory experience backwards in time : reply to P. S. Churchland', *Philosophy of Science 48*, 182–197.

Libet B. (1985), 'Subjective antedating of a sensory experience and mind-brain theories: reply to Honderich (1984)', *Journal of Theoretical Biology 114*, 563–570.

Libet B. (1992), 'The neural time-factor in perception, volition and free-will', *Revue de Métaphysique et de Morale 97*, 255–272.

Libet B., Wright E. W. Jr., Feinstein B., and Pearl, D. K. (1979), 'Subjective Referral of the timing for a conscious sensory experience: a functional role for the somato-sensory specific projection system in man' *Brain 102*, 191–222.

Pöppel E. (1985), *Grenzen des Bewusstseins*, Deutsche Verlags-Anstal., translated as *Mindworks: Time and Conscious Experience*, Harcourt, Brace Jovanovich 1988.

Popper K. R. & Eccles J. C. (1977), *The Self and its Brain*, Berlin, Springer Verlag.

Pylyshyn Z. W. (1979) 'Do mental events have durations?' *The Behavioral and Brain Sciences 2*, 277–278.

van der Waals H. G. & Roelofs C. O. (1930) 'Optische Scheinbewegung', *Zeitschrift für Psychologie und Physiologie des Sinnesorgane 114*, 241-288 (also (1931) *115*, 91–190).

ABNER SHIMONY

TEN PHILOSOPHICAL POEMS

EPITAPH

My logic was a Delphic coracle,
Tossing on uncertainty.

ETHICS

The *is* that flouts the good
Troubles childhood.
The *ought* that should not be
Rends maturity.

HISTORY

Artist Nero mocks his epigones,
Whose instruments sound only cries and groans,
Imperial Caesars look with scorn on czars.
Dark archetypes, and darker exemplars.

C.C. Gould and R.S. Cohen (eds.), Artifacts, Representations and Social Practice, 343–348.

TO ATHENA

In spite of cultural inhibitions against praying to pagan deities
And philosophical inhibitions against praying at all,
I humbly present a petition,
Because our need is great.
Turn a deaf ear to those who would constitute you
Guardian of the Military-Industrial Complex,
And reveal to those who worship you
As Goddess of War and Goddess of Wisdom
That they can no longer have it both ways.
Let the quiet voice of your servant Thucydides
Be audible through the din of fanatics.
In the bleak landscape of our century
Restore the sacred olive groves.

GHOSTS OF PROPHETS

Do ghosts of prophets anxiously
Observe the course of history?
Does the demographic chart
Make Malthus proud, or sick at heart?
Does Marx cry out, "Revise, revise"?
Does Jefferson avert his eyes?
Does Wagner angrily deplore
How time interpreted his score?
Is Herzl's song of exultation
Distempered by new agitation?
Does Henry Adams darkly smile?
Do some plead on, "Abide a while,
Eternity can have no last –
The future must redeem the past"?

LEVIATHAN

Leviathan is will and might,
And bulk and boundless appetite;
His wake commands the northern sea.
But where resides his sovereignty?

In the fortress of his sides,
Where the serried army hides?
Or in his weapons of attack,
Which terrorize and crush and wrack?

In the engine of his heart,
Which drives reserves to every part?
Or in the hidden factory,
Where armorers toil ceaselessly?

In the endless skein of nerves,
Which regulate how each cell serves?
In his jaws' triumphal arch,
Through which his helpless captives march?

Or in the cruel digestive juice,
Which transmutes selves to his sole use?
Oh, stark disparity – a man,
To bowels of Leviathan.

LOOKING INTO THE ABYSS

Beware contemplative delights
Inspired by fishes of the sea:
The wonder of their darting forms
And of their subtle modes of sense,
Their steady meritocracy,
Where niches are not gained or lost
Except upon sufficient cause.

Serene detachment soon must lapse.
From viper fish's spiky jaws
And swallower's dark bloated sack
Revulsion cannot be contained.
The diodonts, the synodonts,
The chaetodonts, the polyodonts –
All pack their essence in their teeth,
Each is a swimming abattoir.
Ten million centuries of life
Conceived and staged this Grand Guignol.
Theodicy is stunned and mute
And natural piety appalled.

Then come the misanthropic hours,
When Bacon's portraits say the truth:
That eyes are apertures of fright,
And vice is chronicled on skin,
That hands are instruments of force,
And teeth are lacerating blades.

O hasten from the gallery
And bend the mind to changeless things.
Look not too deep in the abyss,
Lest it look into you.

THE IRON LAW

Paralysis, the dreadful oracle,
Engraves the iron law on bone –
"Matter obeys itself alone."

Mobility, the daily miracle,
Proclaims what humblest life foresaw –
"Emancipation from the iron law."

TORTOISE

Spring comes,
And the plod of the tortoise is heard in the land,
Whose eros is thin-lipped, hard-crusted, wrinkle-skinned,
Deep-seated, lumbering, and persistent.
Praise to the Life Force that creates forms as it will,
And makes each species beautiful to itself.

A PLEDGE

If a spirit meet a spirit
On the further shores of Psi,
Should a spirit greet a spirit
With a rattle or a sigh?

Are souls so thin of substance
Or so bound by etiquette,
That Sappho could not whisper
An impassioned epithet?

Elysium receives its guests
To hear the sages teach;
Could Socrates among the dead
Have suffered loss of speech?

Music rules the planets
And the elementary fire;
Are human voices silent
In the universal choir?

If you and I should chance to kiss
Upon the shores of Psi,
Our discourse shall be sweeter
Than a rattle or a sigh.

– *Boston University*

JOSEPH AGASSI

THE PHILOSOPHY OF OPTIMISM AND PESSIMISM

OPTIMISM AND PESSIMISM AS MYTHS

Intellectual fashions never cease to puzzle me: only yesterday Claude Lévi-Strauss was all the rage, and today he is all but forgotten. Yet he did make an important observation: myths think for us, he said, and they usually come in sets of polar couples, such as tall-short, hot-cold, raw-cooked, honey-ashes. The absence of gradation makes us speak crudely as if instances of the two polar extremes exist and no instances of any intermediate case, whereas usually (almost) only the intermediate cases exist; and we mix the extremes to create the intermediate: we call the more-hot-than-cold more often hot than cold and, symmetrically, the more-cold-than-hot we call more often cold. At times, Lévi-Strauss says, we find this insufficient, and then we invent a mediating pole. Finally, he observes, polar couples come in bunches, so that we can say, the two poles a and b are like the two poles c and d, or that a goes with c as b goes with d, or more generally and more vaguely,

a : b :: c : d.

All this fits our present discourse very well, as we can say,

rationalism : irrationalism :: optimism : pessimism,

to mean, the rationalist persuasion created optimism, and the reaction to rationalism was pessimistic. The *locus classicus* for this is the once so famous and then so neglected lecture 'The Two Cultures and the Industrial Revolution' by C.P. Snow, later Sir Charles and then Lord Snow, where the equation is to be found more concretely, in line with Lévi-Strauss's observation of myth as "the science of the concrete":

scientists : artists :: future-oriented : past-oriented.

To these he adds an explanation, by adding more poles,

:: rationalism : irrationalism

C.C. Gould and R.S. Cohen (eds.), Artifacts, Representations and Social Practice, 349–359.

and

:: objectivist : subjectivist,

to mean, scientists are rationalists, future-oriented, and objectivists, whereas the opposite is true of the artists, who are, he said, "natural Luddites".[1]

(This is right to the extent that artists of the irrationalist persuasion were usually traditionalists, as they belonged to the Romantic reaction to rationalism; it is not true of the artists of the rationalist persuasion, as even Snow admits but proceeds in a mythical fashion undisturbed by counter-examples, though he is scientifically-minded enough to observe them.) To the pairs thus-far mentioned, more pairs can easily be added, such as,

:: *Gesellschaft* : *Gemeinschaft*,

namely,

:: the abstract society : the face-to-face society,

which is, more-or-less,

the open-society : the closed-society,

always remembering that the central idea about poles is mythical, namely, that poles are scarcely ever to be found in pure state, and that the poles only approximate the real, and at times so poorly that a mediating pole is created.

As the situation in current philosophy is very sad, we may take recourse to Lévi-Strauss's idea that when mythical thinking gets too crude, the way to remedy it is to seek a mediating pole; indeed, many philosophers are seeking these days the *via media* that could serve as a *modus vivendi*, between rationalism and irrationalism, between the Enlightenment movement and the Romantic movement. This turned out to be quite a tall order.

The key to the solution is in the added poles,

rationalist : irrationalist :: hypercritical : uncritical.

This invites a short explanation before its usefulness as a key to the solution can be seen. That irrationalism is uncritical need not surprise anyone: irrationalists do not accept the tyranny of reason, and so they are at liberty to be critical or not, and it is easy for them to be critical towards the theories and activities of the rationalists, but not towards their own. The leading irrationalist philosopher of the century, Martin Heidegger, expressed his hostility to both science and criticism by his observation that poetic truth – whatever this may mean – is higher than scientific truth. His follower Hans Georg Gadamer explicitly recommended that we follow our prejudices unashamed in the pursuit of the understanding of any text we happen to study. The fact that the [classical, Enlightenment-style] rationalists are extremely hyper-critical is very important: it was hyper-criticism that made the scientific tradition dismiss as worthless any theory that is not absolutely true, no matter how good an approximation to the truth it may be, and this created the crisis in the philosophy of science when Albert Einstein offered a theory that was better than Newton's so that the hyper-critical had to reject Newton's theory despite its great merit and its having served as the instance of a theory that was scientific *par excellence*.[2] Today this move is not endorsed by most of the followers of classical rationalism, and they are therefore obviously inconsistent. The famous physicist-philosopher P.W. Bridgman first stuck to the classical position with extreme tenacity, and then gave it up completely.[3]

That the medium pole, between the hyper-critical and the uncritical, is the critical, hardly needs saying. So my proposal is that in the philosophy of rationality and of science the desired *via media* that may serve as a *modus vivendi* is the mediating pole of the critical philosophy.

It is easy to construe the mediating pole to each pair thus far mentioned, except for optimism and pessimism. How is this related to optimism and pessimism? Briefly, the rationalists offered a scientific utopia, whereas the irrationalist reaction, in its habit of reversing every rationalist idea and in the despair that (reasonably) followed their despair of reason, offered a negative utopia, an Armageddon, a *Götterdämmerung*. The critical philosopher Karl Popper was the first to disassociate the scientific ethos from all utopianism in his celebrated *The Open Society and Its Enemies*.

SCIENTIFIC OPTIMISM

Optimism and pessimism are crude categories; since the world is a set of lost opportunities, as any side-glance at the past indicates, it is really impossible to say which of the two is nearer the mark. Clearly, then, they are mythical categories. The word 'myth' was traditionally used to mean error and prejudice, thereby expressing the hyper-critical attitude of the [classical, Enlightenment] rationalists. Here the word means ideas without clear-cut reference, ideas that one can hardly ever declare refuted even in the face of much seemingly contrary evidence. This accords not only with the current attitude of anthropologists to myths; it accords also with the views of Claude Lévi-Strauss and with the views of Ernest Gellner who wondered what could ever induce people to give up the myth world for the world of science. It also accords with the important fact, noted by Nathan Sivin, that in Chinese astrology predictions could be verified or not but not refuted.[4]

The specific to the myth system discussed here, is not so much that it deals with science, as science can easily be viewed as a myth system once the word 'myth' is stripped of its derisive overtones, especially since the metaphysical foundations of science are certainly irrefutable. What is specific about the myth system discussed here is the possibility to examine it diachronically, as Lévi-Strauss himself never did, as he was ambivalent about the view of science as a myth.[5] The reason that the two scenarios, the optimist utopia and the pessimist Armageddon, were picked as the most significant, as two myth poles, is merely historical. Traditionally, rational action was divided into the immediate and the distant, the short-term and the long-term, the short-run and the long-run, even though clearly, the center of the stage is well in between. In Nevile Shute's *On the Beach* there is a description of life without prospects for the slightly distant future; the assessment of Shute is that life, even if it is as satisfactory as possible, loses all meaning if there is no next generation. Somehow, all social science has ignored the next generation, in favor of either the immediate future or the end of days.

Why? The answer is historical: the idea that utopia is essential goes back to Plato. Tom Paine, the beloved optimist, has put it as clearly as possible: if one wants to act rationally, one has to determine one's goals. And the theory of rational action, presenting action as the logical consequence of one's circumstances and choice of goals, has led social scientists, from Adam Smith to date, to consider only immediate goals – simply because no one can work out how action contributes to more long-terms goals. Admittedly, economists do occasionally discuss the long range; at times even sociologists do; but in the analysis of their discussion of the long-range action, of long-term considerations, no efforts are required to find that the long-term factors enter the picture only as stop-gaps, only when all short-term explanations fail. (Indeed, the theory of deferred gratification deftly avoids stating for how long the deferment is; all it really means is that its proponents cling to utilitarianism despite evidence to the contrary, evidence that is explained away as mere deferment of gratification, not the wish to act for some motive other than gratification. This is an instance of the mythical mode of thinking in sociology.) The long-range gratification being but an excuse for the oversight of the short-range one is behind John Maynard Keynes' cutting observation: in the long run we are all dead, he said; the only considerations that are open to empirical examination, that is, belong to the short terms. (His observation did not stop him from outlining in the conclusion of his celebrated *General Theory* a plan for a socialist utopia, be it long-term or short-term; yet he intended his long-term end to avoid clash with the short-term ones, and this is the novelty of his plan. It is an amazing fact that this innovation was not noted in the diverse reviews and commentaries on his book.)

The temptation is to ask, how can there be an empirically testable theory of the middle range? But the better question is, what is the significance of the medium-range considerations? Even the long-range considerations are not as important as Tom Paine has suggested: the question is not so much whether the future is bright or dark, as no one can possibly predict that anyway and as it is doubtful that we can influence even the not so distant future, let along the very distant future. What is important is to find the opportunities and seize the day. In other words, the optimism that is rational is not so much a matter of prediction but a matter of obligation – the obligation to do the best under bad conditions.

There is no doubt that the classical scientific optimism was long-range. The matter is very complex, as it involves considerations of all sorts of matters vital to our future, including the matter of the stability of the solar system. Today, when we have got used to the possibility that we will destroy life on our planet – by pollution or by nuclear war – we are less concerned with the stability of the solar system over the centuries; but not quite, as our bias towards the relevant literature shows: The current literature still regularly refers to Laplace's difficult, invalid proofs of the stability of the solar system and overlooks the obvious and simple proof to the contrary by Kant from the law of conservation of energy [the first law of thermodynamics] and the thermodynamic dissipation of energy [the second law of thermodynamics], including the loss of kinetic energy due to friction (the tides). (The collapse of the terrestrial orbit is described in detail in H.G. Wells' *The Time Machine.*)[6]

The classical optimism was nevertheless also somehow based on medium-range considerations, even though this fact was not recognized. It was smuggled into the assessment that science is progressing rapidly, and that on the whole science renders all medium-range forecasts bright and long-term forecasts really medium-range and even short-range. (Karl Marx, for example, waited for the revolution every day.) Not that it was ever forgotten that the fruit of science are technologies that can be used for good or ill; rather it was suggested that scientific technology will find ways to overcome any obstacles that science and scientific technology might produce. The optimism of the classical rationalists was rooted in the faith that science is progressive because everyone with a modicum of good will can contribute to scientific progress, that all one has to do is to express readiness to forego prejudice and pursue the truth; the idea was that the advantages of science are so obvious that all sacrifice in its interest will be highly justified and willingly made.

This was the ideology of the Enlightenment movement, which led to the French Revolution, which was a great failure, as it led to the French terror and to the Napoleonic Wars. The retreat from this ideology was made in two steps, one called the Reaction and one called the Cold War.

THE MYTHICAL ROOTS OF TECHNOLOGICAL PESSIMISM

The Reaction divided human knowledge into the natural and the historical sciences, or the sciences and the humanities, or the two cultures, or the I-it and I-thou languages: the natural sciences retained their optimism, but humanity was beyond scientific method, as its study was historical. What this meant, put in simple unadorned language, is that whereas the natural sciences can be evolved by unprejudiced, objective investigators, human studies are conducted from the viewpoint of one given tradition or another. This made all natural science universal but not human studies, where the national interest reigned supreme. The result was appalling: the intellectual leaders of the European nations were ready to defend what they considered the national prejudices even when they themselves did not share them. In particular their endorsement of the natural sciences made them lose their religion, but their politics enhanced its endorsement. The endorsement was uncritical, and thus it enabled them to advocate the worst aspects of their religious tradition, particularly their unchecked anti-semitism, which was the supreme expression of their rejection of universalism.

The best study of this situation that I know is that of the late Tel-Aviv historian, Uriel Tal, who sought the roots of Nazism in the intellectual climate of nineteenth-century Germany.[7] He argued that the best scientific minds there and then were Universalists in science but not in politics. The result was that German scientists of Christian origins advocated that German scientists of Jewish origins convert to Christianity – not out of a religious conviction, of course, but out of political necessity. The result of this tension between intellectual or scientific optimism and political pessimism was, of course, unbearable, yet it also offered a ray of hope. Thus, Max Planck, though a reactionary, retained his optimism through his faith in science and in its rationality, and remained an optimist, albeit a bitter one, despite the rise of Nazism. Yet also, his reactionary optimism allowed him to view Nazism at first as merely an exaggeration and he did express the hope that if and when they attain political power they would learn to check themselves. The refutation of his expectations did not persuade him to make a radical change. Also, scientific optimism allowed Werner Heisenberg to attack Nazi ideology as soon as the war was over by offering a defence of scientific internationalisism. This attack was not quite convincing, as the Nazis themselves were willing to grant Jewish scientists the status of honorary Aryans, and this was never a secret. Nevertheless, the significance of the contributions of Jews to science may be viewed as the weakest link in the logic of the better educated Nazi sympathizers like Heisenberg himself: despite the Nazi official hostility to Jewish science, its faith in science and in its positive contribution was unshaken, and the defence of the Nazi ideology by biological arguments, as offered by some of the best German biologists, clearly testifies to that.[8]

The picture changed after World War II, even before the Cold War started, and the heralds of the change were people who prepared the ground for the Cold War, such as General Patton and the newly instituted Central Intelligence Agency of the United States of America, who, it is very well known, offered shelter to German war criminals on the alleged ground of national security.

I do not think the atrocities committed after the war could take place so easily, despite all arguments from national security, but for the fact that the Enlightenment optimism suffered the *coup de grace* when the results of Auschwitz and Hiroshima became known. It was not so much the contribution of science to the horrors of the time that matter as the fact that the ideology of science was altered quietly and surreptitiously by the representatives of the scientific community, the leading scientists, the philosophers of science, and the public-relations spokespeople of science. All of a sudden, they all said the atrocities were no more the results of scientific progress than the gun is responsible for the murder it helped commit. This is not so much to argue that this idea is false, as to argue that there was a radical change in the ideology of science, and its surreptitious implementation caused much confusion.

Let me mention two or three small examples of the great confusion caused by the surreptitious change, by the mythical, uncritical move from the Enlightenment optimist pole to the Reactionary pessimist and/or the mediating pole of conditional optimism/pessimism.

One famous debate that has raged in post-war Germany, and it was and still is of supreme political importance, was the question, was the Holocaust unique in any significant way? The genocide involved was surely not new: genocide is mentioned a few times in the Bible and is even advocated there *à propos* the Amalekite tribe. The magnitude of the mass murder, it was alleged, is compared to that of other events in history, when relative numbers rather than absolute numbers are concerned, and even in absolute numbers, perhaps, when the Stalinist atrocities are taken into consideration as a whole (except that they were spread over a somewhat longer period). Yet the obvious fact is that the shocking aspect of the Holocaust was that it took place in a civilized country and with the aid of science which was mobilized by the death-machine in ever so many ways, material and organizational and ideological.

Another famous debate was the already mentioned one, regarding the two cultures, which began when Snow claimed that scientists are good guys as they are optimists – future-oriented, he called them – and artists who are bad guys as their art made them introspective and their introspecting made them natural Luddites and thus reactionaries. The idea of the two cultures, which is the foundation of the Reactionary philosophy, was here turned around against the arts and in a fake defense of the sciences. The very fact that such a confusion was taken seriously when it first appeared in 1960 or thereabout, however, is most remarkable.

The greatest confusion that the Cold War effected was the confusion of science and technology under the umbrella of research and development. The umbrella was endorsed at once not only because the military used it to intervene most rudely and openly in the lives of the academy by bribing the academic administration everywhere; it was endorsed because the very idea that scientific technology is inherently good was thrown out without proper discussion, without any discussion of the consequences of the intellectual change.

Taking science and technology to be one, and taking technology as mere instrument, made science itself nothing but a tool; there was nothing left to guide choice in broad outlines except politics, and this lent tremendous and specious credit to murky, scarcely articulated arguments from national security, based on data shrouded in secrecy. The pessimism rooted in the loss of the hope to be able to control our environment through scientific progress was the natural conclusion of the story.

THE OPEN FUTURE

The question is, what can be done to restore the readiness to use reason for the sake of saving humanity from itself? This question has many facets. One of them was already mentioned: how can we make empirically testable medium-range theories? Another, no less significant, is, how can we educate the public to think scientifically, to act rationally, to participate in politics in a rational manner? These are tall orders. I have tried to address them in my book *Technology*, where I outlined some strategies as far as I could.[9] I had hoped that someone would take up the challenge, but the book remained unnoticed. So let me only add now some general remarks that might perhaps whet the appetite of some of you. I think we do make medium-range forecasts all the time, and that we have to make them, that we can also analyze them and make explicit the blanket assumptions that go into the forecasts that we usually make, and that we can subject these assumptions to critical scrutiny and at times improve upon them. The most obvious and rational blanket assumption is that the sun will rise every morning in the foreseeable future. It is rational because if true we should act upon it and if false whatever we do makes little difference anyway. The most obvious and false and irrational blanket assumption is that we can get good results by improving rational physical and biological technology while letting social and political technology take care of themselves. The most harmful blanket assumption is that the western educational system is sufficiently efficient and functional to be better left more-or-less to continue in its present manner with but little tinkering, with little increased emphasis on mathematics and on natural science. All this will not do. We must develop better, more honest, more democratic, and more critical education from the earliest stages or else we will not be able to recruit the human capital needed for the heavy tasks that lie ahead.

Despite the presence of varieties of factors contributing to the evolution of an optimistic or a pessimistic outlook, the major factor that deserves special attention is this. Political frustration leads to pessimistic desperation and helplessness whereas the opportunity for political action that has any chance for bringing about some improvement contributes to healthy optimistic conduct. What this suggests is that the need is for the creation of opportunities for meaningful actions. For this the awareness is required of the fact that any reform that creates this kind of opportunity is a political reform in the direction of a pluralist participatory grass-roots democracy. I have elaborated elsewhere on this matter, and on its connection with the need for educational reform. Here I only wish to say that those who are sincerely involved in the effort to save humanity should recognize that the failure of the peace and the environmentalist movements thus far was rooted in their paternalism and deep distrust of common people and of democracy. Democracy is not any specific regime, and certainly pluralist participatory democracy is a new kind of democracy; what all democratic movements share is the idea of the siblinghood of humanity and of the dignity of all humans.

Tel-Aviv University
York University, Toronto

NOTES

[1] C.P. Snow, *The Two Cultures and a Second Look*, Mentor, New York, 1964 (the original lecture was published in 1959). This work is so very famous that the bibliography on it is too extensive to present here. For a detailed critique of the work see my *Science and Society: Studies in the Sociology of Science*, [*Boston Studies in the Philosophy of Science*, Vol 65, 1981], Chapter 1, 'Introduction: Science In Its Social Setting', pp. 1–17. Snow's chief proposal was to emulate the varied educational system of the Soviet Union, which

is now known to be too silly for words, but at the time it did a lot of harm, like much of the apologetics for the Communist regimes that was so popular at the time. And so today I think my analysis, though correct, is as defective as shooting sparrows with a cannon; I think the chief motive for Snow's analysis is to explain why the apologetics in question was at the time more popular among physicists than among members of the faculties of arts and letters. My paper was written when this was no longer true, as in the days of the Vietnam war the 'leftist' vociferous spokespeople were from the arts.

2 See my 'Newtonianism Before and After the Einsteinian Revolution', in Frank Durham and Robert D. Purrington, editors, *Some Truer Method: Reflections on the Heritage of Newton*, Columbia University Press, New York, 1990, pp. 145–174, and the references there.

3 P.W. Bridgman, *The Logic of Modern Physics*, 1927, and many later editions, Preface, Chapter 1 and *passim*. Bridgman's tenacity is best seen in his contribution to Paul Arthur Schilpp, editor, *Albert Einstein, Philosopher-Scientist*, 1948, Harper Torchbook, New York, where he rejects general relativity on the ground that it enables one to find out the proper metric only by approximation so that metric is not defined operationally in it. His change of mind is expressed in his posthumous contribution to Mario Bunge, editor, *The Critical Approach to Science and Philosophy: In Honor of Karl R. Popper*, Free Press, London, 1964, 'The Mach Principle', where local Galilean coordinates are freely used to the very end.

4 For the standard use of the word 'myth' today, and for a discussion of the views of Lévi-Strauss, see my *Towards a Rational Philosophical Anthropology*, Kluwer, Dordrecht, 1977. For the views of Gellner see my 'Deconstructing Post-Modernism: Gellner and Crocodile Dundee' in John A. Hall and I.C. Jarvie, editors, *Transition to Modernity: Essays on Power, Wealth and Belief*, Cambridge U.P., 1991, 213–30 and references there. For Nathan Sivin see his *Chinese Science: Explorations of an Ancient Tradition*. Edited by Shigeru Nakayama and Nathan Sivin, Cambridge MA, MIT Press, 1973.

5 See previous note.

6 See my 'The Consolations of Science', *Am. Phil. Q.*, 23, 1986, 129–41 and references there.

7 See Uriel Tal, *Christians and Jews in Germany: Religion, Politics and Ideology in the Second Reich, 1870–1914*. Ithaca, NY, Cornell UP, 1975.

8 For Heisenberg's belated critique of the Nazi ideology see my 'Unity and Diversity in Science', in my *Science in Flux*, [*Boston Studies in the Philosophy of Science*, Vol 27] and references there.

9 See my *Technology: Philosophical and Social Aspects*, Reidel, Dordrecht, 1985.

BERNARD ELEVITCH

LIFE IS NOT A POEM?

Posing the extremes of Kantian autonomy (a "true self") on the one hand, and a Sartrean "hole in being" on the other, Richard Rorty turns for mediation to Freud:

> . . . [B]y helping us see ourselves as centerless, as random assemblages of contingent and idiosyncratic needs rather than as . . . exemplifications of a common human essence, [Freud] opened up new possibilities for the aesthetic life. He helped us become increasingly ironic, playful, free, and inventive in our choice of self-descriptions.[1]

The theme of one's life as aesthetic self-creation is developed later through constant association with poetry: "He de-universalizes the moral sense, making it as idiosyncratic as the poet's inventions."[2] "What ties Nietzsche and Freud together is . . . the attempt to see a blind impress as not unworthy of programming our lives or our poems."[3] "For Freud's account of unconscious fantasy shows us how to see every human life as a poem. . . ."[4] Here aesthetics replaces metaphysics: life, in familiar romantic terms, is a work of art.

The key to Rorty's interpretation seems at first to be "ourselves . . . as random assemblages." Approaching Freud at this level of generality, I should say on the contrary that Freud has taught us to see ourselves as centered, as anything but random collections. And this is because as a theorist he offers coherent accounts of fantasies and behaviors that had only *seemed* bizarre, disconnected, unintelligible. Whether or not his explanations are warranted – or how in principle they could possibly be – is not at issue; one does not have to defend a system as *well*-ordered to see that it is able to organize data in accordance with theoretical principles and criteria. Let us at least give Freud credit for insisting on lawfulness, for seeking patterns or regularities of behavior. While the center may not hold, since it is not a fixed point, it is nevertheless the object of an intellectual pursuit.

It is a commonplace, moreover, that Freud's general psychology occurs in a direct line of descent from the neurological model originally proposed in letters to Fliess and published as the *Project for a Scientific Psychology* more than half a century later.[5] Thus such concepts as "mental energy,"

C.C. Gould and R.S. Cohen (eds.), Artifacts, Representations and Social Practice, 361–364.

"sums of excitation" and the like become intelligible in light of the neurological descriptions and speculations of the *Project*.[6] Even more generally, the Newtonian model applicable to the accumulation and release of energy may be seen as linked to the general theory's pervasive determinism. In Freud's view, there are no "occurrences, however small, which drop out of the universal concatenation of events. . . . If anyone makes a breach of this kind in the determinism of natural events at a single point, it means that he has thrown overboard the whole *Weltanschauung* of science."[7]

That Freud speaks of slips of the tongue and similar "mistakes" from the standpoint of antecedents that are meaningful rather than causal does not weaken his point. "Chance events," mere contingencies, add up to meaningful representations of the whole person. And the achievement of the case histories, acceptable or not, is exactly to unify disparate behaviors, memories, fantasies, dreams – to find the center that Rorty disclaims. That metaphor plays a significant part in all of this is undeniable, but it cannot lead the way to a self which is somehow both "a tissue of contingencies" and the focus of poetic self-creation.

Circumventing the unconscious, which might well be set in opposition to a "central faculty . . . called 'reason,'" Rorty finds an integrative function in the particularity or specificity of individual analyses:

> Anything from the sound of a word through the color of a leaf to the feel of a piece of skin can, as Freud showed us, serve to dramatize and crystallize a human being's sense of self-identity.[8]

This is yet another variation on "life as a poem," since a poet might very well pay attention to the color and feel of a particular experience. For anything like a defense of the life-poetry association Rorty in fact defers to Lionel Trilling, "who said Freud 'showed us that poetry is indigenous to the very constitution of the mind; he saw the mind as . . . exactly a poetry-making faculty.'"[9] But Rorty fails to take note of Trilling's important qualification:

> This puts the case too strongly, no doubt, for it seems to make the working of the unconscious mind equivalent to poetry itself, forgetting that between the unconscious mind and the finished poem there supervene the social intention and the formal control of the conscious mind.[10]

Whatever our common dependency on fantasy, the poet differs from the rest of us in that he or she is able to discipline what might seem to be spontaneous: a poem is of course the result of deliberate art.

Freud aside, an aesthetic model of selfhood is itself too metaphorical to bear much analysis. Why not "Life is a journey," "Life is a dream," or any other familiar expression in the same vein? In point of fact the poem metaphor fails in its task because we already know what is special about poets, how they differ from the rest of us. Or to turn over the coin, the metaphor works well for poets themselves: *their* lives are poems. They are the ones who, in Rorty's expression, clothe their lives in their own respective metaphors.

An extreme and poignant case would be that of John Berryman, author of *The Dream Songs*, who suffered throughout his life from guilt associated with his having accepted his stepfather's name:

> . . . [W]hat I cannot forgive myself for not having done, was to take the name John Smith [his father, a suicide, was John Allyn Smith]. This act of disloyalty I will never, never be able to repair. To "make a name" for myself . . . Can you see how ambivalent my feelings are about this ambition?[11]

The memoir from which this confession is taken makes it clear that Berryman lived his metaphor in all of its ambiguity. The circumstances of his father's death, his mother's variations on the theme, the poet's obsessive preoccupation with every detail of the story – all of this constituted a personal and familial myth that informed Berryman's daily choices and responses no less than his poetry. To get his life straight, and thus to realize his poetic ambition, it was necessary to get his story straight. But that of course was primarily a question of its form and significance, not an autobiographical pursuit of factual data. Finally Berryman ended his life as his father had ended his: a final sentence, and, as in the poet's craft, self-imposed.

Now I do not cite Berryman as representative of poets in general or even of poets (Robert Lowell also comes to mind) who are quite obviously autobiographical. On the contrary, so long as our subject is the nature of selfhood, I see no purpose in generalizing about poets at all. A model that depends on the identification of unconscious fantasy with poetic imagination is, it seems to me, an intellectual evasion; it succeeds only in diminishing our understanding of poetic craft, our respect for the singularity of poets, and – perhaps most seriously – the very possibility of self-understanding. It is just possible that life is not a poem after all, but only a journey. Or a dream.

Boston University

NOTES

[1] Richard Rorty, 'Freud and Moral Reflection,' in *Pragmatism's Freud*, ed. J. Smith and W. Kerrigan (Baltimore: Johns Hopkins University Press, 1986), p. 12.
[2] Rorty, *Contingency, Irony, and Solidarity* (Cambridge: Cambridge University Press, 1989), p. 30. For a succinct commentary on this work, see Thomas McCarthy, *Ideals and Illusions* (Cambridge, MA: MIT Press, 1991), pp. 35–42.
[3] Rorty, *Contingency*, p. 35.
[4] *Ibid.*
[5] In *The Origins of Psychoanalysis*, ed. M. Bonaparte, A. Freud and E. Kris (New York: Basic Books, 1954).
[6] See James Strachey's introduction to *The Interpretation of Dreams* (New York: Basic Books, 1955). K. H. Pribram and M. M. Gill quote approvingly from Strachey in their *Freud's 'Project' Re-assessed* (New York: Basic Books, 1976).
[7] Freud, *Introductory Lectures on Psychoanalysis* (New York: Norton & Co., 1966), p. 28.
[8] Rorty, *Continency*, p. 37.
[9] *Ibid.*, p. 36. Rorty's reference is to Trilling, *Beyond Culture* (New York, 1965), from a talk delivered in 1955. A nearly identical passage occurs in an earlier work (see following note).
[10] Lionel Trilling, 'Freud and Literature,' in *The Liberal Imagination* (New York: Viking Press, 1950), p. 52.
[11] Eileen Simpson, *Poets in Their Youth* (New York: Noonday Press, 1990), p. 61 (ellipsis in text).

ROGER S. GOTTLIEB

LEVINAS, FEMINISM, HOLOCAUST, ECOCIDE

These remarks focus on what might be termed the transcendental but empirical conditions of ethics – ethics as indistinguishable and contrasting sets of presuppositions about what makes possible other than instrumental relations among human beings. The discussion is 'transcendental' in a loose Kantian sense of asking how or why certain beliefs are possible, especially when this possibility is itself not directly addressed in the beliefs themselves. By 'empirical' I mean concrete historical realities whose dark mysteries ethics are designed – consciously or unconsciously – to solve.

1. LEVINAS

I do not pretend to be a Levinas 'expert', whatever such a being might be. And the reader should be aware that what follows here is hardly a scholarly examination of the man's work. In fact, my intellectual background[1] often makes reading his work a sort of extended theoretical hallucination. I have a real distaste for his hyper-inflated language and bald statements of vast proportions, underdetermined (often untouched) by argument or reasons. Yet I also sense his situated moral seriousness and his urgent relation to the Holocaust. While I have not fully decided if he is a philosophical prophet or an intellectual fraud, I am sure that his thought is important in the way it signals what may be the ultimate limit of patriarchy. *Ultimate*, that is, because in Levinas, patriarchy is trying its best – though with inevitable lack of success – to undermine its own foundations and negate its ruinous consequences.

What is for me at once strange and wonderful about Levinas is that he can assert, with so few reasons, a very strange claim about human relationships.[2] This claim, as I see it, defines his work, and can be summarized as follows: Other philosophies of human existence have tended to describe our ethical obligations as consequences of historically, conceptually, or developmentally prior structures of social life, rational thought, or experience. These philosophies generate the need for ethics

C.C. Gould and R.S. Cohen (eds.), Artifacts, Representations and Social Practice, 365–376.

out of the contradictions of a life without ethics (as in contract theory or, to some extent, Hegel); or they generate ethics out of the dialectical development of self-consciousness; or they generate ethics out of ontological assumptions about the nature of humanity, nature, Reason or God. Traditionally, in short, ethics is secondary to knowledge of 'things' (with that term construed as broadly as possible), including the knowledge of, or concerns about, oneself.

It is this sense of knowledge of things that Levinas tries to capture under various rubrics – most importantly, from his two major philosophical works, as 'totality', 'essence', and 'being'. (As for the attempt to generate ethics out of self-knowledge or interest, Levinas believes that this is simply a form of war.) He believes that knowledge is necessarily aimed at or inevitably leads to domination, objectification, and alienation. Therefore knowledge cannot be the basis of ethical life – that is, of a kind of transcending concern for other people, a concern untouched by our own needs, desires or attempts to control. As Hume could not get an 'ought' from an 'is', Levinas finds an unbridgeable gap between knowledge and ethics. If we begin with knowledge – in the guise of science or philosophy, technique or ontology, rational reflection or psychoanalysis – we will never get to the Other person as irreducibly Other. Knowledge is something acquired, dispensed and ultimately instrumentally used by us. Consequently, knowledge of others necessarily reduces the Other to something we possess, something we have acquired, and something – ultimately – we will use.[3] If the foundation of our relation to Others is knowledge, then the Other will be reduced to the same. Otherness will not be allowed to coexist with the agent of sameness.

What Levinas poses as an alternative is the irreducibility or underivability of our concern for the Other. This concern does not stem from an empirically or conceptually based sense of the 'facts', or the ultimate ontological structures of the universe. It does not come from an expansion of self-interest through identification with the other, either practically (as in contract theory from Locke to Rawls), or transcendentally (as in Kant). Nor does it come from the discovery of common interests in the realm of historical struggle (as in Marxism, feminism, or anti-racist movements). Levinas leaves little doubt that the terrain of history, in the sense of political struggle, is too implicated in the wars of self-interest to be a site for ethics.

Where exactly our obligations for the other come from is not clear.

Like a negative theologian, Levinas is most effective in characterizing what the grounds of our concern for the other are *not*. Not a consequence of our knowledge of things (totality) or of the ultimately knowable character of things themselves (essence). *Not* how the things appear to us or exist in their truth (being). Working through, behind and beyond essence and totality, being and knowledge, and leaving a subtle trace of itself in our capacity to speak with care and openness of these things, the call of the Other simply breaks through and across the barriers of science and philosophy, and the greedy attempt to satisfy my desires by 'knowing' about the world and others. This call leaves us infinitely concerned with the Other, with the way our very existence on Earth takes up her space, with our unlimited responsibility which constitutes or makes possible – rather than follows from – ego-bound interests, communication, or subjective freedom.

Thus, if we are to have ethics at all, it must be, in Levinas's phrase, our "first philosophy". The capacity to speak and to know, to haggle over questions of truth and evidence, are *signified by* rather than *signify*, this ultimate responsibility for that which cannot be knowingly reduced to myself: the Other for whom I must act and be concerned; the Other in answering whose call I receive the distinctive imprint of my humanity.

As I read Levinas his strange conclusions about ethics really follow from 'nothing'; or, rather, they come from the sheer fact of 'otherness' – that which somehow penetrates my psychic and ethical space from outside.

> The freedom of another could never begin in my freedom, that is abide in the same present, be contemporary, be representable to me. *The responsibility for the other can not have begun in my commitment, in my decision.* The unlimited responsibility in which I find myself comes from the hither side of my freedom, from a "prior to every memory, . . . "
>
> The knot tied in subjectivity, which when subjectivity becomes a consciousness of being is still attested to in questioning, signifies an allegiance of the same to the other, *imposed before any exhibition of the other, preliminary to all consciousness* – or a being affected by the other whom I do not know and who could not justify himself with any identity, who as other will not identify himself with anything.[4]

While there is much to be said concerning Levinas's exposition of this position (the relation to Husserl and Heidegger, the importance of his views about language and time, etc.), what has always sprung to my mind is the question of *motivation. Why* would anyone want to believe such a thing, want it to be true? What do we receive and what do we escape from, if this strange and wonderful view – that my obligation to the other, especially to those most in need does not follow from who I am or who they are, but precedes those contingent identities? What are, as I briefly defined above, the empirical and transcendental conditions of Levinas's ethic?

2. FEMINISM

There are many problems posed by Levinas's categorical assertions of human responsibility without knowledge, emotional connection, or self-interest. Generally most of these problems are not directly addressed by him, nor by his commentators. We can ask first: why *must* all knowledge be objectifying, tied to domination and the eradication of difference? Why is Levinas compelled to accept the instrumental view of knowledge and leave our ethical connections, therefore, to the realm beyond essence and outside of knowledge? What would he make for instance, of Habermas's attempt to situate forms of knowledge in relation to distinguishable human interests in the control of nature, communicating with others, and social emancipation? Of these three, certainly the first, possibly the second, but it would seem *not* the third, serve the tasks of domination or control. Similarly, there is the Western Marxist concept of *praxis*, in which self-knowledge, knowledge of the world and emancipating political action reinforce each other. In Lukacs, for example, the proletariat's knowledge about its own social status as a commodity helps undermine that status, enabling it to make the transition to different beliefs about, and practices in, society. This process of self-knowledge-social-position-progress-towards-emancipation can be expanded to apply to the development of women and ethnic/racial minorities. (What is consciousness-raising, for example, if it does not include women's transcending their objectification by patriarchy in the process of coming to see that objectification?) From a rather different source, we might ask Levinas why self-interest cannot lead, by a Kierkegaardian existential dialectic, towards the choice of ethical life, principles, and commitments to the good.

Reading Levinas prompts all these queries, and it is part of the frustration of reading him that he seems not to know that attention to them is necessary. But of all such questions what struck me most directly is the fact that Levinas seemed completely oblivious to the feminist counterview to his essential premise: i.e., a view that would argue that his radical disjuncture of self and other is not the basis for a metaphysics of ethics, but simply the consummation of a culturally male perspective on human relationships. The idea that our ethical connection to others is possible because the Other to whom I am ethical hostage leaves a 'trace' in the objective realm is simply at odds with what a host of cultural feminist writers have, in a variety of ways, described. This feminist view is not monolithic (all the more reason it should have been seen by Levinas and his commentators), but within a range of views there is some distinct commonality.[5]

The position can be summarized in this way:

The culturally male ego is predominantly formed through a process of separation, towards an ideal of autonomy, and results in a bounded, competitive and dominating self. By contrast, the female ego is shaped through affiliation, towards an ideal of 'self-in-relation', and results in an empathic, nurturing, and connected self. Women's selfhood stems from women's role as primary caretakers of infants and their responsibility for emotionality and nurturing in adult relationships. Consigned by patriarchy to the 'labor of relatedness', to the production of sexuality, emotional intimacy and affection, women approach the moral realm from a radically different sense of themselves and others than men. Partly as a consequence of their distinctive ego structures, men and women reason differently about moral problems: men favor abstract principles of justice, while women think in terms of concrete relatedness, and reason via empathy rather than abstraction. Feminist ethicists have developed the concept of an ethics of care, of 'maternal thinking' to refer to moral perspectives *based in* a sense of emotional kinship between self and other, as distinguished from those stemming from abstract principles, self (as opposed to other)-interest, or Levinas's own infinite obligation across an irreducible gap. Cultural feminists have further argued that social domination and hierarchy express highly individuated and competitive egos, and thus political injustice and economic exploitation are male forms of relationships. These evils cannot be overcome by the application of abstract principles of liberal democratic/rights theory, or by Marxist-oriented strategies of class struggle, since both these perspectives reproduce the individualism, abstraction, and aggression endemic to the male styles embedded in the evils themselves. Neither, clearly, are they addressed by Levinas's attempt to bridge the irreducible gap between persons by a 'trace' of presence left in the objective world. (Nor by his rejection of history as somehow outside the realm of ethics.) Rather, a social order based in the cooperative, nurturing, non-competitive style of female identity might overcome the antagonisms and oppressions of male-dominated society. The 'feminine virtues' of relationality, empathy and cooperation could make possible a social order that escapes from the domination, exploitation, and violence endemic to both capitalist and bureaucratic state societies. It is further suggested that the image of rigid ego boundaries between people is itself, to an extent, a masculine prejudice inflicted on psychological theory. Certainly women, and perhaps men as well, develop not as Self and Other but as 'Selves-in-Relation' – so that even theoretically we must speak of persons in the contexts of their relations, unknowable outside of those relations.[6]

In other words, Levinas is lost in a world in which we know the Other and answer the Other only through this imponderable call to a responsibility divorced from every other facet of my being – knowledge, emotionality, self-interest, direct relationships – *just because he accepts the basic premises of masculine theoretical culture.* These premises include the notions that human identities are formed in rigid isolation and opposition to one another, and that bridging the gap between self and other always requires some extended process of reflection, self-development, or transformation. In this culture we start as isolated owner/producers (Locke) or isolated minds (Descartes) or aesthetic enjoyers of amoral experience (Kierkegaard) or isolated ego-id-superego complexes, whose struggle for mastery and sex can lead, at best, to the autonomous ego of bourgeois adult masculinity (Freud). In patriarchal thought we never *start*, from time out of mind, in connection to others. We are not seen as beginning, as we in fact do, as babies at our mother's breast, after having come out of her body. Or if the beginning is there, that image of connection is not carried into the heart of the theoretical representation of adult ethical life. Men have obliterated the memory traces of their own relations to their mothers.[7]

Levinas's self – cruelly abandoned in a world of objectifying knowledge and self-interested war – can only hope to remember that trace of the Other, and know that she or he is obligated to answer the call of the other: to put oneself in the place of the Other and seek the Other's good. In a theoretical world of purely masculine possibilities, Levinas may have found his solution as the best alternative possible. But would the whole edifice come crashing down if he realized that it is possible to see human identity as based in a relation to the other *from the start*? That the other is not a trace, not an uninteriorizable 'outside', not something that can only get flattened into sameness if it is brought 'in'; but that the other has been known, connected to and made part of ourselves *from the beginning*, always already, as inter-identification, empathy, need, and intuition. If he could have conceived of a Self so implicated in the other that the Face we see *is in some sense our own*, because the boundaries of self and other – far from being obliterated by a reductionist or instrumental Knowledge – are fluidly constructed by a reality of a shared relationship in which both find their selves.[8] If he could have read (and not as if these perspectives do not carry their own set of problems) Nancy Chodorow, Carol Gilligan, Sara Ruddick, Jean Baker Miller, Janet Surrey, Miriam Greenspan, Ann Ferguson, Nel Noddings . . .

3. HOLOCAUST

But while this is all true, the matter as a whole is not that simple. For Levinas's work does not simply move in a realm of ethical theory. His problems are not simply those posed by phenomenology and post-modernism, Kant and Heidegger. His ethic is at once an intellectual edifice *and* and extended prayer. He can – and this he must know – 'prove' nothing. He can only beg that it be so. And this returns us to the question of motivation. Why does he so want us to feel, or if not to feel to have it true about ourselves, that beyond knowledge and history, we are ethical hostages for the Other whom we do not know?

In my view this motivation, this condition at once transcendental and empirical, can only be the Holocaust.

In his world the destruction of the Jewish people is the basic fact to be known: more than theory, more than self-interest, more than the conventional forms of ego development or psychic, emotional, familial or even communal inter-identification.

Ethically, what were the Jews of Europe during the Holocaust but the irreducibly Other? What were the Jews but the Other for whom the gentile world could find so little to identify with, so few interests in common, so little to have knowledge about? The Fascist world did not know the Jews, or did not know them as Jews, but only as the enemy, as vermin, traitors, insects, germs. The only 'knowledge' that was possible was the knowledge of genocide, to reduce the Other to the Same . . . by murder.

Even more particularly: why does Levinas violate the conventions of symmetry that are virtually ubiquitous in philosophical ethics – i.e., why does he insist that our obligation to the Other is in some sense *greater* than the Other's obligation to us? Consider, quite simply, what it meant in Nazi-dominated Europe to side with the Jews. To shelter or protect, to stand against the regime of evil meant – with few exceptions – torture and death, possibly for one's family, friends and village as well as for oneself. To be responsible, as many were, was to reach out to those who could not reach back to you.

In such a world one might well pray with Levinas that we feel a kinship, a bond, to the Other we do not know; or that we feel an infinite obligation of care, holding us hostage before any choice on our part, to that Other of whom we know nothing. We are bound solely to the fact of Otherness.

For Levinas, *only this prayer will do*, for only this prayer really speaks to that terrible Jewish loneliness. The Jew who is not known by the gentile world; or the knowledge of whom reduces to the technology of the death camps. Levinas is not arguing for a new philosophical system, he is praying or dreaming or simply hoping against hope that what he says might be true: that out of the sheer fact of otherness, there is hope of ethical life. Nothing else, as he has seen, can protect the "widow, the stranger, the orphan" – the Jew. In a different setting, yet another group will be the Otherness that is grist for the mill of power and murder.

In other words, it is just because Levinas finds himself in a world of cultural masculinity – of violence and domination towards the Other, of the use of instrumental knowledge to reduce the Other to the Same – that he *must* create a vision of moral responsibility across an unbridgeable gap. It is just because he is stuck not simply in the *theory* of cultural masculinity but in its *reality* that he is compelled to theorize an unrationalizable moral connection based simply in the fact of Otherness. In a culturally masculine world every Other is, *a priori*, a kind of enemy. Having seen how such a world operates, Levinas is praying that the opposite might somehow come to be – that is, be true against all the appearances to the contrary; that the Other, far from being the object of hostility, is the unknown and infinitely deserving subject of our ethical devotion.

4. ECOCIDE

Is there then no way out? Are Levinas's failings simply those of patriarchy? Is the feminist critique hopeless against the historical reality? Is it perhaps not Levinas who is the dreamer but the feminist, for in a world made by men, empathy, connection and interidentification have so little chance or place? Must the transcendental and empirical conditions of feminism remain with the privatized and domesticated realm of the family or the intimate relationship? Is the feminist dream of an ethical cosmos of care and compassion related to the real world family of patriarchal power, exploited wives and child sexual abuse as Levinas's dream of infinite obligation to the stranger, the widow and the orphan is to the real world of the Holocaust?

Perhaps.

But perhaps not. Perhaps there is another, even more encompassing Death Event, actually in progress now, which poses a counterweight to the dialectic of savagery, despair and hope which defines the moral and intellectual world of Levinas. An event that can serve as a new set of transcendental/empirical conditions for a feminist oriented ethic of compassion and care.

I speak of the specter of ecocide, the continuing destruction of species and ecosystems, and the growing threat to the basic conditions essential to human life. What kind of ethic is adequate to this brutally new and potentially most unforgiving of crises?

Here I will at least begin in agreement with Levinas: the anthropocentric, human-interest oriented perspectives of conservation or liberal environmentalism cannot take us far enough. Our relations with non-human nature are poisoned not just in that we have set up feedback loops leading *already* to mass starvations, skyrocketing environmental disease rates, and devastation of natural resources. Our uncaring violence also violates the very ground of our being, our natural body, our home. Such violence is done not simply to the Other – as if the rain forest, the river, the atmosphere, the species made extinct are totally different from ourselves. Rather, we have crucified ourselves-in-relation-to-the-other; fractured a mode of being in which self and other can no more be conceived in isolation from each than can a mother and a nursing child.

We are that child; and non-human nature is that mother.

What kind of ethic meets *this* transcendental condition? My thought is that here we have a barrier that sets the limits to Levinas's perspective. The Other which is non-human nature is not simply known by a trace, nor is it something of which all knowledge is necessarily instrumental. This Other is inside us as well as outside us. We 'know' it with every breath we take, every bit of food we eat, every time we drink a glass of water. We do not have to find traces on or in the faces of trees or lakes, topsoil or air: for we are made from them.

It is this realization that leads us to an ethics that is appropriate to our time, that is a blend of cultural feminism, eco-feminism and deep ecology.[9] For this ethics, our *identity* is shaped by intimate connection and interidentification not only with other people, as in the cultural feminist vision, but with all of nature. Our *knowledge* does not assimilate the Other to the Same, but reveals and furthers the continuing dance of interdependence. And our *ethical motivation* is neither ontological system nor individualistic self-interest, but a sense of connection to all of life.

> The deep ecology sense of self-realization goes beyond the modern Western sense of 'self' as an isolated ego striving for hedonistic gratification . . . Self, in this sense, is experienced as integrated with the whole of nature.[10]

> Having gained distance and sophistication of perception [from the development of science and political freedoms] we can turn and recognize who we have been all along . . . we are our world knowing itself. We can relinquish our separateness. We can come home again – and participate in our world in a richer, more responsible and poignantly beautiful way.[11]

> Ecological ways of knowing nature are necessarily participatory. [This] knowledge is ecological and plural, reflecting both the diversity of natural ecosystems and the diversity in cultures that nature-based living gives rise to.
> The recovery of the feminine principle is based on inclusiveness. It is a recovery in nature, woman and man of creative forms of being and perceiving. In nature it implies seeing nature as a live organism. In woman it implies seeing women as productive and active. Finally, in men the recovery of the feminine principle implies a relocation of action and activity to create life-enhancing, not life-reducing and life-threatening societies.[12]

Here the noble and terrible burden of Levinas's purely individuated responsibility for sheer existence gives way to a different dream, a different prayer, one which serves as the very possibility that the ecological nightmare of the present might not end with an irrevocable finality.

> Being rock, being gas, being mist, being Mind,
> Being the mesons traveling among the galaxies with the speed of light,
> You have come here, my beloved one . . .

> You have manifested yourself as trees, as grass, as butterflies, as single-celled beings, and as chrysanthemums;
> but the eyes with which you looked at me this morning tell me you have never died.[13]

In this prayer, we are, quite simply, all in it together. And while we might say that this new ecological Holocaust, this creation of planet Auschwitz, is under way, it is not yet final. We have time to step back from the brink, to repair our world. But only if we see that world not as an Other across an irreducible gap of loneliness and unchosen obligation, but as a part of ourselves as we are part of it, to be redeemed not out of duty, but out of love.

Worcester Polytechnic Institute

NOTES

1 Perhaps best summarized as analytical philosophy in the pragmatic tradition, Marxism and critical theory, feminism, reflections on the Holocaust, and influences from spiritual dimensions of Judaism, Buddhism, and Kierkegaard.

2 Along with works referred to in later footnotes, see Emmanuel Levinas, *Totality and Infinity* (Pittsburgh: Duquesne University Press, 1969); *Nine Talmudic Readings* (Bloomington: Indiana University Press, 1990).

3 It seems that Levinas is relatively untouched by environmental critique of instrumental knowledge applied to nature, and that he rejects the Heideggerean concern with Being, except insofar as our concern with that abstraction signals our own moral dimension. For an interesting discussion of Levinas and animals see John Llewelyn, 'Am I Obsessed by Bobby? (Humanism of the Other Animal)' in Robert Bernasconi and Simon Critchley, eds., *Re-Reading Levinas* (Bloomington: Indiana University Press, 1989).

4 *Otherwise than Being Or Beyond Essence* (Boston: Kluwer Academic Publishers, 1974), pp. 10, 25, my emphasis in both quotations.

5 The proliferation of feminist viewpoints makes it hard to speak of any 'general' feminist position. My focus is on the trend often called 'cultural' feminism, which stresses how women's interests, experiences and capacities have been ignored or misrepresented by male-biased theory; and how women's social role in advanced societies confers on them a distinctively 'feminine' form of thought and action, and enables them to be the potential bearers of virtues that will overcome current forms of male based domination and exploitation.

6 Some central texts on the relation between female psychology, women's social role, feminist ethics, and social theory are Jean Baker Miller, *Toward a New Psychology of Women* (Boston: Beacon, 1976); Miriam Greenspan, *A New Approach to Women and Therapy* (New York: McGraw-Hill, 1983); Nancy Chodorow, *The Reproduction of Mothering: Psychoanalysis and the Sociology of Gender* (Berkeley: University of California Press, 1978); Carol Gilligan, *In A Different Voice* (Cambridge: Harvard

University Press, 1983); Susan Griffin, *Women and Nature: The Roaring Inside Her* (New York: Harper and Row, 1979); Carolyn Merchant, *The Death of Nature* (New York: Harper and Row, 1980); Eva Feder Kittay and Diana T. Meyers, eds., *Women and Moral Theory* (Totowa, N.J.: Rowman and Littlefield, 1987); B. Andelson, C. Gudorf, and M. Pellauer, eds., *Women's Consciousness, Women's Conscience* (New York: Harper and Row, 1987); Nancy Hartsock, *Money, Sex and Power: Toward a Feminist Historical Materialism* (Boston: Northeastern University Press, 1983). I have raised critical questions about this perspective in 'Broken Relations: Some Barriers to the Triumph of Feminine Virtue', in Roger S. Gottlieb, ed., *Radical Philosophy: Tradition, Counter-Tradition, Politics* (Philadelphia: Temple University Press, 1993).

[7] This last phrase comes from Miriam Greenspan.

[8] Even when Levinas seems to be hinting that we have some kind of direct relation with others, that relation – at least as I read him – is always abstract, constructed, distant, formal; in a word, metaphysical rather than emotional or psychological. We may, as he insists in the crucial chapter in *Otherwise Than Being*, necessarily "substitute" ourselves for the Other, in fact, for the whole world. But in that substitution there is no real connection to the other, just (once again) that limitless responsibility for the whole universe of suffering and vulnerable Others.

> Responsibility for my neighbor dates from before my freedom in an immemorial past, an unrepresentable past that was never present and is more ancient than consciousness of . . . .
> ['Ethics as First Philosophy', in Sean Hand, ed., *The Levinas Reader* (London: Blackwell, 1989), p. 84.]

Just because our responsibility is so absolute, so pre-existing to everything in our personal life or social world, it never seems to shine with any direct connection to another real human being with whom I, as (in Kierkegaard's words) an actually existing human being, am in an actually existing relationship.

[9] The literature is very large here. For a beginning on eco-feminism, see: Irene Diamond and Gloria Orenstein, eds., *Reweaving the World: The Emergence of Ecofeminism* (San Francisco: Sierra Club Books, 1990). For deep ecology, see Christopher Manes, *Green Rage: Radical Environmentalism and the Unmaking of Civilization* (Boston: Little, Brown) 1990, as well as the works by Vandana Shiva and Joanna Macy referred to in later footnotes.

[10] Bill Devall and George Sessions, 'The Development of Natural Resources and the Integrity of Nature', *Environmental Ethics* 6 (Winter 1984), pp. 302–03.

[11] Joanna Macy, *World as Lover, World as Self* (Berkeley: Parallax Press, 1991), p. 14.

[12] Vandana Shiva, *Staying Alive: Women, Ecology and Development* (London: Zed Books: 1989), pp. 41, 53.

[13] 'The Old Mendicant', by Thich Nhat Hanh quoted in Joanna Macy, *World as Lover, World as Self.*

CAROL C. GOULD

MARX AFTER MARXISM

Contrary to the way it has usually been taken, the most enduring legacy of Marx's thought is philosophical, not practical. Marx's account of political economy, of class and of revolution have occasioned legitimate criticisms, and the practical outcomes of the social movements based on Marxism have largely been negative. Without denying the power of his critique of capitalism, I will argue here that the most important and viable aspect of Marx's thought resides in some of his basic philosophical ideas – particularly the normative concepts – and the turn that they led to in political philosophy. This is somewhat ironic, of course, since on the usual view – whether Left or Right – Marx's normative, philosophical ideas are the least central and certainly the least developed part of his work, compared with the central role of his political economy and revolutionary theory.

If, in the spirit of Marx or at least Hegel, one were to speak dialectically, one could say that the future of Marxism lies in its negation; not in the sense of abstract negation, i.e., simple rejection – say as anti-Marxism in neoconservative or neoliberal forms, but rather so-called determinate negation (*Aufhebung*), replacing what is corrupt or mistaken, and retaining (at a higher level of course) what is viable or promising. Thus the future of Marxism cannot mean going back to some allegedly pure original text, that is, returning to Marx *before* Marxism. It means instead determining what is left of Marx *after* Marxism that can serve for the development of fresh thinking.

However, even such a renewed core of what is living in Marx's thought will not be enough. To my mind, it will need to be combined with elements from other theoretical frameworks, and especially from liberalism and feminism. Therefore, even the determinate negation of Marxism is not enough of a next step if what we want and need is a philosophically adequate social and political theory which could also serve to guide contemporary practice.

In this paper, I will begin with a brief account – no more than a summary – of the inadequacies and limitations of Marxism and of Marx's theory. Most of these criticisms in one form or another are familiar

C.C. Gould and R.S. Cohen (eds.), Artifacts, Representations and Social Practice, 377–396.

from the literature and from current discussion but will help to provide a framework for my later considerations. I will then give an account of what I take to be the strengths of Marx's view and will propose how certain of his basic philosophical conceptions need to be developed beyond his framework, drawing on feminist and liberal theories. Finally, I want to briefly address a problem that arises from Marx's view and confronts any social theory that takes history and culture seriously: namely, the idea of an end of history and the problem of a final stage of human development.

Starting with the criticisms, then, perhaps the simplest and most obvious applies to Marxism as a set of social and political movements. This critique could have been gleaned from the morning headlines of the past several years, with datelines in Moscow, Warsaw, Belgrade, Beijing and points East, West, North and South. The key words of critique, to put it impressionistically, have been authoritarian Marxism, totalitarianism, cultural and intellectual repression, central planning, economic breakdown. The once vaunted crisis of world capitalism has now been joined by the actual economic and political demise of world socialism. Yet, whatever the connections are between Marx's thought and the Marxist movements of the last century, it is wrong I think to lay all of this at Marx's feet, as some commentators have done. However, Marx's own view may be criticized with respect to at least five major dimensions.

The first concerns Marx's ambivalence, lack of clarity, and sometimes plain wrongness about the rights and liberties stressed by classical liberalism. Though he sometimes implies that these are necessary preconditions for any further development,[1] at other times he dismisses rights as bourgeois abstractions, and most egregiously advocates at one point the dictatorship of the proletariat as an appropriate form for the transition to socialism.[2] He is also ambivalent about the market. On the one hand, he sees it in its revolutionizing role in the development of worldwide economy, universal culture, and in the articulation and satisfaction of human needs;[3] moreover, his critique of capitalism is primarily directed against the sphere of production rather than against exchange or the market itself. On the other hand, he criticizes the market as reducing human relations to a cash nexus, and as exemplifying the anarchy of capitalist economy, which is to be replaced by rational direction in socialism.[4]

A second area of criticism concerns some of Marx's general systemic and methodological concepts. One of these concerns the pervasiveness of the economic and of the activity of production in Marx's account of social reality. Though Marx is explicitly critical of capitalism's reduction of all human and social categories to the economic and thus is not crudely economistic, there is a residual overemphasis on the primacy of production and the sphere of material needs. Thus there is an ambivalence here that has opened the way to reductionist economistic interpretations, which disregard his projection of a future in which the activities of economic production are relatively subordinated to other forms of activity. Marx's account of the base/superstructure relation similarly expresses this view of the dominance of the economic, and in fact exacerbates it by suggesting an economic determinism where the superstructural elements – whether political, cultural, or philosophical – are merely epiphenomenal.[5]

Another systemic and methodological problem, related to the first, is the holism that Marx sometimes falls into, whether in his account of the relation of individuals to society or class, or in his occasional reference to overarching laws of history or of economic development. Although I myself have argued in my book on Marx[6] that he basically holds to a view of the ontological primacy of individuals, especially in his more philosophical writings, e.g., in the *Grundrisse*, nevertheless he sometimes speaks holistically of individuals as "ensembles of relations", or of laws that operate "behind the backs" of the agents. Here, individuals seem to disappear as the active agents of change that Marx elsewhere takes them to be and therefore seem to become the mere vehicles of supervening laws of development, a view which Marx had himself criticized in Hegel.[7]

Third, Marx's view may be criticized from a feminist standpoint for failing to take into account male domination both within the economy and institutional forms, as well as in personal relations. Indeed, one may suggest that because of his emphasis on the economic sphere and on the activity of production, and because of his focus on institutional analysis, Marx fails to give an adequate account of reproduction and childraising, the family, and the private sphere in general. Although Marx focuses on social relations within production, he rarely attends to social relations of domination or reciprocity outside of production, which play a fundamental role in social life. Moreover, though Marx and Engels (in the *Origins of the Family, Private Property, and the State*) make some relatively enlightened remarks about women's equality, there is no recognition of shared responsibility for childraising as a central condition for the achievement of such equality.

Fourth, though Marx was perhaps right in criticizing utopian socialism and abstract morality, he was wrong not to offer a more developed projection of new forms of social organization, as well as theoretically developed normative concepts of freedom and justice, which could serve as a guide to practice. Though he does discuss future forms of social life and the concepts of freedom and justice, too much remains implicit or vague.

There is one additional point of criticism which may be added here, but which may seem paradoxical because it appears to go directly against the fourth criticism just presented. This concerns the failure of those forecasts or predictions which Marx did in fact make or which were almost universally imputed to him by the movements that bore his name. Whereas the previous criticism asserts that Marx failed to project a normatively desirable future with sufficient clarity, here he is criticized for having projected too much, in terms of more empirically oriented predictions of large historical trends. But many of these predictions have been factually falsified and therefore their premises would need to be critically reexamined or rejected.

A central one is his empirical and historical prediction of the increasing and absolute impoverishment of the working class, which would function as the motive and guarantee of social revolution. Now one may perhaps speak of increasing immiseration in a global context and especially in the context of the north-south disparity between industrialized and dependent nations. But Marx's account concerned the immiseration of a very classically conceived industrial working class and here two things have happened to confound his predictions. First, the very character of the working class has changed historically and has become much more complex and diverse even in its relations to the means of production and to ownership.[8] And second, this transformed working class in the advanced industrial countries – the ones in which the overthrow of capitalism was initially supposed to take place – has in fact become better rather than worse off, even if at the same time an impoverished underclass has emerged alongside it.

In passing, one may note that some of Marx's predictions have fared rather better, and in particular his uncanny anticipation of what he called "automated" and "self-regulating system of production" and his characterization of the unceasingly dominant role of science and technology in the economy.[9]

Despite the serious criticisms one may make of Marx's thought, there remains a core of significant ideas that have had a formative effect on subsequent social and political theory and that continue to offer valuable insights. I begin with a brief overview of some of the obvious and sometimes less than obvious strengths of Marx's view. These I will simply mention here without any elaboration. They are worth mentioning, however, precisely because some of them have become so commonplace and even generally accepted that it is easy to forget Marx's contribution. I will then go on to focus on five major ideas which bear especially on normative issues of politics and society, the formulation of which owes much to Marx's theory, even if he leaves these ideas vague and undeveloped. These are: the *critique of domination* as a central theme for a theory of justice; *reciprocity* as a model of nondominating and nonexploitative social relations; the concept of *freedom* as requiring access to the means or conditions of human activity; *social individuality*; and the idea of *democracy* as involving participation in decision making in economic and social life. In each case, I will suggest what Marx's contribution is and also how the concept needs to be enlarged or revised if it is to be viable. Specifically, I will point to the ways in which liberal democratic and feminist ideas should be combined with Marx's contribution to yield more adequate conceptions. Indeed, we may observe that it is precisely the vagueness and lack of development of these ideas in Marx that allowed the misinterpretation of the thrust of Marx's normative views by later Marxism, where such misinterpretation also contributed to the degenerate practices of authoritarian and undemocratic socialism.

Among the major strengths of Marx's thought which I want only to note here before proceeding are the following: The most obvious, perhaps, is the fundamental core of his critique of capitalism (though not necessarily the details of its elaboration). By this core I refer to his idea of the exploitation and alienation of labor which results from the control by capital over means of production and the lack of control by workers over these means. Another significant feature of Marx's theoretical approach which has been widely appropriated is his emphasis on the basic conditions of existence, i.e., on means of subsistence, and on practical needs. Related to this is Marx's insistence on the importance of history and of understanding social phenomena in terms of process or change. Yet, another important feature of his approach is the connection he draws between philosophy and social life, and between theoretical or analytical critique and its application in social practice. Other strengths, perhaps less obvious ones, concern more explicitly philosophical issues. Here we may note Marx's refusal to accept a sharp dichotomy between fact and value in the domain of social theory. Further, as a systematic thinker, Marx introduced a significant revision of Hegelian philosophy by combining it with Aristotelian, Kantian, and proto-Heideggerian insights.[10]

I turn now to a discussion of what I take to be five basic normative conceptions to which Marx made a distinctive and important contribution, but which need to be developed beyond his thought. These ideas are not always fully explicit in Marx's work and need to be reconstructed from his text, where they are sometimes intermixed with quite different and even contrary notions. As I have suggested, these normative conceptions will need to be elaborated in new ways, drawing on insights from liberalism and feminism, if they are to be more adequate theoretically and if they are to serve as guides to a renewed practice.

The first of these normative ideas is the critique of domination. Marx's signal contribution here consists first of all in placing the concept of domination at the center of his critique of society. In this way, Marx adds an essential element lacking in more abstract conceptions of distributive justice, specifically an emphasis on the concrete, particular and differentiated social relations of domination and subordination. Furthermore, he reveals the forms of domination to be historically changing. A second contribution is the particular analysis that Marx offers of domination in which it is understood not simply as a matter of control over a person's actions by coercion or force but rather as control over the conditions of agency, that is, of those things that are required in order for persons to carry out their activities. Of course, Marx understood this primarily in terms of conditions for productive activity, i.e., as means of production, and therefore, as ownership and control of these means. Domination is thus seen by Marx at least in its capitalist forms in modern society as economic exploitation of labor or "the extraction of surplus value".

Here, the feminist critique becomes relevant. Marx's emphasis on production and the correlative understanding of domination in these terms is regarded by feminist critics as failing to recognize the pervasive form of social domination represented by sexism or male domination in its various historical modes. Moreover, the feminist critique appropriately points to the existence of domination in the realm of personal relations and to the use of subjective and psychological means of domination in both the personal and the public domains.

An adequate conception of domination beyond Marx's would also draw on one aspect of liberal theory. Despite the fact that liberalism does not include a critique of domination, relying instead on a notion of negative liberty as noninterference with an individual's actions, it does introduce the emphasis in freedom of choice, which may be seen to contribute to an adequate analysis of the concept of domination in the following way: If agents do have such fundamental freedom of choice, then domination cannot be analyzed as direct or causal control of one person's actions by another, in which we might say that their will is obliterated. Rather, we came to understand domination more adequately as a matter of indirect control over the activities of agents by means of control over the conditions which are necessary for the exercise of their agency. We may add that these conditions may be broadly understood as including not only the material or social requisites for the activity of agents in realizing their purposes but also the subjective and psychological conditions necessary for carrying out such activity.

Closely related to the critique of domination is the positive normative concept of reciprocity. Here Marx's contribution involves a distinction between the simple reciprocity among equals by which he, like Hegel, characterizes the relation of market exchange or the exchange of commodity values, on the one hand, and on the other, reciprocity as a relation in which each acts to enhance the free development of the other in a community of social individuals.[11] This second notion of reciprocity is left as a vaguely and somewhat rhetorical notion in Marx, though it may be inferred from his dialectical account of stages of historical development.[12] It is also implicit in his critique of the nonreciprocal relations of alienated labor and exploitation under capitalism, where the normative thrust of the critique is that such nonreciprocal relations are unjust and will be replaced in a future society by "individuals in mutual relationships".[13] The first form of reciprocity by contrast, is clearly described as a relation of economic exchange in which each agent treats the other as both free and equal and in which each enters into the relation in order to satisfy his own need or interest only. This relation may be called formal and instrumental reciprocity. Marx says that this reciprocal relation among equals in exchange masks the nonreciprocal relations of exploitation and alienation at the level of production.

The critique that Marx makes here of merely formal reciprocity as limited to the recognition of abstract equality between agents is paralleled by the recent feminist critique of such abstract equality both for masking concrete inequalities in personal and social relations between men and women – specifically women's experience of powerlessness – and for failing to recognize the noninstrumental aspects of mutuality. These latter include the recognition of the other as a concretely differentiated individual and not simply as an abstractly equal other, the concern for the other out of motives of care and love rather than self-benefit, and relatedly, acting in order to enhance the development of the other.[14] Some of these emphases on noninstrumental and concrete reciprocity are already present in Marx though they are not very developed and certainly not in the contexts of personal and gender relations. But we may note that he makes an especially important contribution in stressing the recognition of differentiated needs as a requirement for economic justice.[15] Feminism, on the other hand, has also not as yet clearly articulated an adequate conception of reciprocity.

What is needed, therefore, is a more fully worked out account of reciprocity as a normative concept and of the various forms it takes.[16] Granting the critiques of the limitations of merely instrumental reciprocity, this does not mean that formal reciprocity – the reciprocal recognition by each agent of the other's equality and freedom – can be dispensed with. Here liberal theory, especially in its Kantian version, is relevant to this understanding of the reciprocity among equals, within the framework of the Kantian dictum of treating persons not as means only but always as ends. But I would add that this recognition of other persons as ends entails a recognition of the conditions of personhood and of the rights to these conditions. Where these conditions concern the basic requirements for activity – i.e., life, liberty, and means of subsistence – we may speak of basic rights.

This noninstrumental recognition of other persons as ends goes beyond formal reciprocity when it involves the recognition by each of the other as a concretely differentiated individual, as this person with a unique identity rather than simply as an anonymous other, or as *any* other or *every* other. When there is this kind of reciprocal respect among individuals, which goes beyond formal and instrumental reciprocity, we have what I would call a mode of social reciprocity, social in the sense of *Gemeinschaft* rather than *Gesellschaft* or merely a societal relation.[17]

Another mode of social reciprocity is the relation among individuals engaged in common or joint activity towards a common goal. I would call this cooperative reciprocity, meaning by this not simply the contingent or aggregate activity of a group of individuals which happens to be coordinated towards some end, but rather the deliberate cooperation of individuals in a commonly understood task. In this case, each recognizes the other as participating in this common activity.

Finally, there is a sense of reciprocity, which one may call mutuality, in which each agent not only recognizes the other's freedom and equality and concrete individuality, but also chooses to act to enhance the other's development. Feminists have frequently suggested that such a relation of deliberate mutual enhancement is normatively required in relations among individuals. This has recently been formulated in terms of replacing a perspective of justice and universality with a perspective of care and particularity. The central model for such a relation of care or nurturance is that of mothering.[18] Some feminist theorists have proposed that such a care perspective and the requirement for mutual enhancement ought to be extended to the domain of politics, either to replace or to complement the justice perspective.[19]

However, I would argue that while such mutuality has an important place in ethics as a normative ideal for a range of personal and social relations, it is too strong a requirement to be posited as a political norm. It is unreasonable to expect that political relations, which are institutional in nature, should require individuals to enhance each other. By contrast, what I have called social reciprocity in its two modes is relevant in political contexts. This is not to say that a care perspective is not relevant to politics and to social life. But we have to make some distinctions here which are not generally made: (1) Care as involving a reciprocal recognition of the individuality and specific needs of the other can be distinguished from (2) care as involving mutual enhancement; and both of these can be distinguished from (3) care as a nonreciprocal relation of nurturance or concern for the vulnerable, that is, for those who are unable to care for themselves and are dependent for their well-being and development on another or others. Here too we may distinguish among the vulnerable between (3a) children as dependent on parenting and (3b) others who are dependent by virtue of their weakness, illness, or deprivation. I would propose that it is care in sense (1) (as social reciprocity) and sense (3b) which are directly relevant to the political domain. Mutual enhancement and mothering fall outside this realm.

A more general implication of the foregoing analysis for the feminist theory of care is the need for a distinction between the relation of care for the vulnerable which may not be reciprocal and is not necessarily a matter of mutuality, and care as the relation of mutuality which essentially presupposes reciprocity. These two senses are sometimes confused in the feminist literature. They tend to be drawn from two different models: The first from the mold of mothering (or more properly parenting), and the second from relations of love or close friendship. But they are clearly not the same although both involve care and share some elements in common.

In a related way, I think we should distinguish between two kinds of nonreciprocal relations, both of which have been dealt with earlier. The first kind is the relation of domination and subordination, including both economic exploitation of which Marx provides an analysis, and male domination, as feminist theory has developed it. This kind we may designate normatively as malignant nonreciprocity. The second kind of nonreciprocal relations is that of care as care for the vulnerable where the relation involves some fundamental dependence which is nonreciprocal. (There may of course be some reciprocal elements in such relations as well.) This second kind of nonreciprocal relation may be characterized, in contrast with the first, as benign nonreciprocity.

The third normative concept to which Marx has made an important contribution – one which we might add has been severely compromised by Marxism after Marx – is that of freedom. This is perhaps the most fundamental of all normative concepts and has the widest scope. I cannot do more here than to point to certain new emphases that Marx brought to the concept. Some of this is adumbrated in his critique of domination, a critique that presupposes the norm of agency and control over the conditions of one's activity, as noted earlier. On reconstruction, one may say that, for Marx, freedom as agency requires the availability of means or conditions in order for choices to be effective. Marx thus has a conception of what later came to be called positive freedom in distinction from negative freedom as simply the absence of constraint or interference with one's actions. The form of activity on which Marx focuses is productive activity. This requires in the first place access to material conditions, that is, the means of production – raw materials, tools, and techniques. But according to Marx, this productive activity is typically an activity of social production which requires social interaction of the agents and the development of the institutional forms of such relations. Although such productive activity involves an element of creativity at all stages of historical development, since work entails the creation of value, it only emerges as free creative activity on Marx's view in a society of the future in which the sphere of material necessity takes a subordinate place in human activities and in which institutional forms of domination, most especially, class, have been eliminated. In this context, Marx speaks of human activity as an activity of the free development of individuality, and sees productive activity as transformed into an activity of self-development.

Although freedom in this sense emerges fully for Marx only at the end of the historical stages of forms of production, it nevertheless appears to be present at every stage in that process in which human beings constitute themselves by their activity. Thus Marx conceives of human activity as self-constituting or self-transforming, that is, he conceives of human beings as creating and transforming their own natures through their activities. But Marx sees this self-transformation not as an immediate relation of the self to itself but rather as proceeding through relations to nature and to others.

There are some serious problems however with Marx's formulation of the concept of freedom, since it sometimes seems to be contradicted by other things he says. For example, as noted earlier, Marx occasionally talks in deterministic terms of laws of history or of "iron laws" of economic development, which "operate behind the backs of the subjects", that is, as objective determinants of their actions. This has led to interpretations of Marx as an economic determinist which would directly contravene any coherent notion of agency. Marx's view about freedom thus remains somewhat unclear and at times deeply ambiguous.

A related conceptual difficulty stems from another tension in Marx's thought between the view that (1) freedom is only fully realized in a future communal society though it is present in germ earlier, as self-constitution or as creative activity; and the alternative view (2) that all previous history is a history of unfreedom as domination or enslavement and that freedom as self-development arises only in the society of the future. This second view seems to me obscure in that it remains mysterious or magical how freedom comes to appear out of a situation of constraint or determination; or how even transformative, revolutionary activity can be generated if people do not have the capacity for such free, transformative society.

I would argue instead that if freedom as self-development is to emerge, whether in the society of the future or in the life of an individual, then it presupposes agency or free choice as a capacity for such self-development. Freedom in this sense remains abstract without access to the conditions necessary for the realization of choices, but such realization must be based on the intentional activity or choices which only agents can make and which, further, I would argue, is what characterizes human beings as human. This emphasis on freedom of choice is what liberal theory has insisted on, though it does not sufficiently recognize the indispensability of access to the conditions of activity to make choices effective. Thus liberalism has offered a clearer defense of this freedom of choice and of the commensurate importance of negative freedom – including civil liberties and political rights – needed to protect it. I would argue that these two emphases – on choice on the one hand and conditions on the other – need to be combined, and I have shown elsewhere how they are compatible.[20]

As is well known, Marx was a sharp critic of the liberal conception of the individual as a separate, self-interested ego, standing in what we might call external relations to other such individuals. He saw this as a particular historical product of alienated forms of activity characteristic of the social relations of capitalism. By contrast, he advanced a normative conception of what he called social individuality, which constitutes the fourth of the concepts I am examining here. This type of social relation Marx took to be immanent in forms of social or cooperative work which exist in all stages of historical development but which burgeon with the development of industrial capitalism and large scale production which involves the coordination of large numbers of workers in a common task. However, this social individuality comes to fruition only in the communal society of the future where it takes the form of the free association among individuals engaged in common activities. Here, on his view, it would have the form of direct relations among persons who mutually recognize each other.

Marx's account of social individuality suffers from an ambiguity however: On the one hand, Marx conceives of the socially related individuals as concretely existing agents who constitute themselves through the relations they enter into. On the other hand, in a shift in the direction of holism, Marx sometimes talks about individuals as being constituted by their relations, and the individual is seen as no more than an ensemble of relations. The individual is sometimes subsumed in a class or as a member of a social whole. There is enough evidence in Marx's work to support an interpretation of the first sort, emphasizing the fundamental role of individual agency.[21] Nonetheless, this ambiguity persists across the range of his works.

Plainly, I think that the first view is the only viable one. Social relations have no independent existence apart from the concrete individuals who stand in these relations and who constitute them by their choices. Yet these individuals come to be who they are and transform themselves in and through their social relations. In this way, then, they are social individuals or individuals-in-relation.[22] Thus this view again may be seen to draw on that part of liberal thought which stresses the agency of individuals. However, unlike classical liberalism, this view does not take these relations to be merely external relations but rather conceives them as internal relations. This is clearly a long story which I will not go into here.

The account of social individuality presented here has significant elements in common with feminist theory as well. Feminists have emphasized the interpersonal and extrainstitutional social relations among individuals which exist beyond the economic and political spheres. This is an essentially relational view like the one proposed here, but has had alternative emphases, sometimes in the direction of preserving the individuality of those who stand in the relations but sometimes blurring this individuality in the direction of a notion of total interdependence.

The fifth of the normative concepts to be considered here is that of democracy. Here too, Marx makes a distinctive contribution, particularly concerning what we would call economic democracy and concerning the relation of economic factors to political democracy. Implicit in Marx's critique of the lack of control that workers have over the conditions of their productive activity under capitalism is the correlative norm of control over their activity and the conditions for it. Such control would involve the workers having a say in determining the work process and its outcome. This would in effect introduce democratic decision procedures into economic life. This is made more explicit in Marx's reference to a future in which what he calls the associated producers would control their own productive process, though it remains unclear as to whether this model is intended to apply firm by firm in the form of worker control (as I think it should) or whether it is to apply on a national scale to economic production as a whole. Marx also reveals critically how political democracy can be subverted by the economic power of a dominant class, even though its forms may be maintained. The implication is that political democracy needs to be complemented by economic and social democracy if it is to realize its ends.

Even though on reconstruction of his views we can say that Marx is a radical democrat in these ways, yet the concept of democracy is not developed very far by him and is sometimes compromised by the admixture of what must be called antidemocratic themes. This is most obvious in connection with his (and Engels's) notorious proposal for a dictatorship of the proletariat as a necessary form for transferring state power in the revolutionary transition from capitalism to socialism. It is also apparent in Marx's ambivalence towards parliamentary democracy which he sometimes dismisses as a mere instrumentality for masking class rule and facilitating its power. At other times, however, he characterizes such political democracy as a necessary precondition for the further development of a fully human society.[23]

Though Marx's critique of the limitations of liberal political democracy has been widely recognized as sound in many respects, yet his sometime disdain for the language of rights and abstract justice and for the forms of political democracy is mistaken. It fails to recognize, as liberal theory does, the central importance of the guarantees of personal freedom and negative liberties and for equal political rights and representative democracy which afford protection against intrusive state power and the worst forms of political oppression. Even where we want to argue for the extension of democracy to the economy and for more ramified opportunities for direct participation in all contexts of democratic decision-making, I would propose that a conception of rights to democratic participation must play an important role. (And in fact, rather than giving up on or eschewing talk about rights for some other desiderata as has been the case in some recent political theory, I think that a somewhat more extensive formulation of rights is required.)

Feminist thought also contributes to the project of more extensive and enlightened democratic theory. Its distinctive addition is to relate democracy as an institutional form to personal relations of reciprocity and mutuality which typically exist outside this institutional context. The normative justifications for democracy on the one hand and for reciprocity on the other are analogous in that both turn on the recognition of the freedom and equality of agents, and we may suggest that each stands a better chance of implementation if the other is present as well. Reciprocity in interpersonal contexts enhances the possibilities of democratic decision making and conversely equal rights to participation in economic social and political decision making would strengthen tendencies to reciprocity outside these spheres.

There is also a way in which the feminist concern with parenting and the relation of care contributes to democratic theory. It qualifies the model of democracy which takes it to be a process of conflict mediation among opposing interests which relies on compromise, by emphasizing the development of cooperation and mutual respect as values which need to be introduced also into politics and other contexts of democratic deliberation and decision making, i.e., into the discourse of democracy. The feminist concern with parenting and with the moral and social development of the child also has import for what I have described elsewhere as the democratic personality[24] and for what used to be called education for citizenship, for example, the cultivation of a disposition to reciprocity and the readiness to take the position of the other. More prosaically, feminists have stressed the importance of childcare and shared parenting as preconditions for the possibility of genuinely equal participation in democratic procedures.

Through this review of the five central normative concepts, I wanted to show what Marx's positive contribution is and also what the weaknesses and inadequacies of his view are. I have suggested some new directions for the rethinking of each of the concepts, drawing in part upon insights from both liberal and feminist theory, though what I have been able to present in this paper is only a sketch. It seems to me to be clear that it is only when these normative concepts are developed fully and their interrelations worked out that a more acceptable social and political practice can emerge. This need is clearly as great in Eastern Europe and the countries of the former Soviet Union – in the countries of what one may now call Formerly Actually Existing Socialism – as it is in the capitalist democracies of the U.S. and Western Europe. In brief, in Eastern Europe and the CIS – what we may call the countries of ex-Marxism – what is required is full political democracy and the introduction of a market framework for what, in my view, ought to be primarily self-managed and worker-owned firms. And what is needed in the capitalist West (or North) is an extension of democracy to economic and social life beyond the political sphere and a realization of the full potential of political democracy.

I want finally to bring up a major problem, both methodological and substantive, that arises with respect to the foregoing discussion of normative concepts. It is a problem that has confronted Marxism in a particular way and also faces any reformulation of normative social theory which includes as it should a conception of these norms as coming to realization through a process of historical development. The problem is this: If the norms are not empty then they must be achievable, for why should people aim at what can't be achieved? Yet, if these norms are in fact achieved, does this then mean that history, conceived in terms of a process of the historical realization of these norms, has come to an end? In short, does the premise of normative concepts entail the conclusion of an end of history? It would seem to me to be important to avoid this end of history idea for it implies a closure of development where human ingenuity has run out of steam or is no longer needed and where human perversity has run its full course. In other words, it implies the achievement of a utopia where no further progress can be conceived. Even worse, if one applies this to one's own time, it would be a simple case of hubris and one could expect the gods to strike one down.

These questions have been posed as a problem for Hegel's theory of history and certainly also for Marx's. Marx foresaw a future society arising on the basis of previous history in which the constraints of necessity which marked class societies were overcome and a realm of freedom ensued in a classless society. Does a similar problem arise for the reformulation of norms that I have suggested here? Clearly, this problem is too large and complex to be dealt with adequately at the end of this paper. For one thing, it would entail a consideration of how norms may be said to emerge, develop, or be realized in history. For another thing, it concerns the question of whether one can reflexively define history exclusively in terms of a process of the realization of such norms. I set both of these questions completely aside here. What I want to do is delineate a few main elements of an approach to the problem and the direction of its resolution.

Assuming we do not want to accept the end of history idea, there are several alternative ways to conceive of history in relation to norms. The first is to move to the opposite extreme from the teleology of an end of history to a purely descriptive and nondevelopmental view of history as mere chronicle – just facts or events strung out in their temporal sequence – in which norms remain ethical desiderata which might or might not be achieved and whose place in history is like that of raisins in a cake, that is, located here and there, changing the local flavor, but unlike yeast, not affecting the growth and development of the cake as a whole. (So much for feminist or at least feminine metaphors.) An alternative view sees history as a process of increasing approximation to the realization of norms, which themselves stand outside history in some problematic Platonic relation, where, however, these norms could never be actually achieved but only more or less instantiated. But what seems to be inadequate in this view is that a norm which, whether in principle or in practice, is unachievable, functions only as an abstract ethical ideal, which moreover would fail to motivate the activity necessary to realize it since it is not a practical goal. The advantage of this view over the first is that one can at least talk about, and make some sense of, the notion of progress in history.

What seems to me the most viable alternative is a view of norms, immanent in historical practices, which can be achieved, but in a certain sense that needs to be made clear. Such an achievement would not mark an end of history but rather a continuation of history in a different way.

The central norm by which I believe one should measure historical progress is freedom.[25] By this I mean something quite open-ended, as I suggested earlier and which I have discussed at some length in my recent book on democratic theory, namely, freedom as self-development, therefore, a characteristic of the activity of individual agents, understood as social individuals. It seems to me a matter of fact that people are widely if not universally engaged to some degree in such self-developing activity; and moreover, that they seek the further realization of freedom in this sense and the conditions necessary for it. In this way, as characteristic of the activity of individuals, one may say that this norm is implicit in history. In fact, the twentieth century, and perhaps most sharply the recent period, has exhibited an extraordinary efflorescence of movements for the removal of barriers to freedom and for the provision of its positive conditions (accompanied at the same time by some barbarous attacks on this development).

I would propose that what it makes sense to talk about is the eventual removal of institutional barriers to the achievement of freedom as self-development and the provision of at least the basic institutional means required for it. That is, one can envision the increasing elimination of domination and exploitation in economic, social and political life to the degree that they flow from the structure of institutions. Likewise, one can envision the spread and further realization of democracy, as a positive institutional condition for freedom. There are also other social and material conditions which we imagine can be provided to facilitate the self-developing activity of individuals.

What it is less easy or impossible to guarantee is that the noninstitutional and personal conditions for freedom, whether negative or positive, can be fully provided. For example, one cannot eliminate all personal or social forms of domination and one cannot guarantee reciprocity in interpersonal relations. However, I think it is reasonable to say that to the degree that people aim at freedom in their activities and since they are social individuals who engage in common activity towards shared ends, we can expect considerable further progress in these areas as well. Of course, even such progress does not entail the elimination of all personal and social conflict, which will undoubtedly continue. What can be entertained as a real possibility is the elimination of structural or systemic constraints on free self-development and the securing of the basic conditions for it. Beyond that, it is up to the actions and choices of the individuals themselves, whether separately or jointly with others.

In discussing the sense in which this norm of freedom can be achieved in history, it is important to distinguish freedom as a general norm of self-development from the particular ways in which individuals or groups will choose to realize this freedom for themselves, including their choice of particular norms, aims, and projects. In this sense we may observe that whatever the regional or cultural differences are among the various ways in which individuals, groups, or whole peoples choose to develop themselves, the general norm of freedom defined in terms of the human activity self-development is not Eurocentric as some critics have argued. It is an openended concept precisely because it is open to the widest variety of instantiations "with no predetermined yardstick" as Marx says in the *Grundrisse*, in one of his most enlightened moments.[26]

I may conclude by observing that if there is a future for Marx after Marxism, it lies, as I suggested at the outset, in the development of the normative philosophical concepts which lie at the heart of his work in directions which go beyond it. It is only such new formulations of these normative concepts that can inform a new democratic practice.

Stevens Institute of Technology

NOTES

[1] Cf. Karl Marx, 'On the Jewish Question,' in Robert C. Tucker, ed., *The Marx-Engels Reader* (New York: W. W. Norton & Co., 1972), pp. 26–52, esp. pp. 35 ff.
[2] Cf. K. Marx, 'Critique of the Gotha Program,' in Tucker, *Ibid.*, p. 538.
[3] Most notably, in K. Marx and F. Engels, 'Manifesto of the Communist Party,' in Tucker, *Ibid.*, esp. pp. 475–77; and K. Marx, *Grundrisse* (Hammondsworth: Penguin Books, 1973), pp. 409–10.
[4] Marx's critique of the market, and of alienated labor and commodity fetishism is elaborated in detail and is well-known. He has very little to say on economic planning. But see *The German Ideology*, p. 191 and *Critique of the Gotha Program*, p. 529 ff., both in Tucker, *Ibid.*
[5] For a somewhat more nuanced view of the relations between economic and other factors in history, see K. Marx, *The Eighteenth Brumaire of Louis Bonaparte*, (Selections) in Tucker, *Ibid.*, pp. 594–617; and F. Engels, 'Letter to Joseph Bloch,' in L. S. Feuer, *Marx and Engels, Basic Writings on Politics and Philosophy* (New York: Anchor Books, 1959), pp. 397–400.
[6] Carol C. Gould, *Marx's Social Ontology, Individuality and Community in Marx's Theory of Social Reality* (Cambridge, Mass.: MIT Press, 1978), esp. pp. 30 ff.
[7] Cf. K. Marx, 'Contribution to the Critique of Hegel's Philosophy of Right,' p. 18; and *The German Ideology*, pp. 148–49, 164–66, both in Tucker, *Ibid.*
[8] Cf. Harry Braverman, *Labor and Monopoly Capital* (New York: Monthly Review Press, 1974).
[9] Cf. K. Marx, *Grundrisse*, pp. 692–95.
[10] Cf. C. Gould, *Marx's Social Ontology*.
[11] K. Marx, *Grundrisse*, pp. 243–44.
[12] See my discussion of this in C. Gould, *Marx's Social Ontology*, pp. 3–5.
[13] K. Marx, *Grundrisse*, p. 712.
[14] Cf. C. Gould, 'The Woman Question: Philosophy of Liberation and the Liberation of Philosophy,' in C. Gould and M. Wartofsky, eds., *Women and Philosophy* (New York: G. Putnams's Sons & Capricorn Books, 1976), pp. 5–44; and 'Philosophical Dichotomies and Feminist Thought,' in H. Nagl, ed., *Feministische Philosophie* (Vienna: R. Oldenbourg Verlag, 1990), pp. 184–90; S. Benhabib, 'The Generalized and the Concrete Other,' in S. Benhabib and D. Cornell, eds., *Feminism as Critique* (Minneapolis: University of Minnesota Press, 1987), pp. 77–95; C. Gilligan, 'Moral Orientation and Moral Development,' in E. Kittay and D. T. Meyers, *Women and Moral Theory* (Totowa, N.J.: Rowman and Littlefield, 1987), pp. 19–33; M. Friedman, 'Care and Context in Moral Reasoning,' in Kittay and Meyers, *Women and Moral Theory*, pp. 190–204.
[15] K. Marx, *German Ideology*, p. 160; and *Critique of the Gotha Program*, pp. 530–31, both in Tucker, *Ibid.*
[16] Cf. C. Gould, 'Beyond Causality in the Social Sciences: Reciprocity as a Model of non-Exploitative Social Relations,' in R.S. Cohen and M. Wartofsky, *Epistemology and Method in the Social Sciences* (Boston & Dordrecht: D. Reidel, 1983), pp. 53–88.
[17] Cf. C. Gould, 'Feminism and Democratic Community Revisited,' in J. W. Chapman and I. Shapiro, eds., *Democratic Community: NOMOS XXXV* (New York: New York University Press, forthcoming).

[18] Cf. S. Ruddick, *Maternal Thinking* (Boston: Beacon Press, 1989); V. Held, 'Non-Contractual Society: A Feminist View,' in M. Hanen and K. Nielsen, eds., *Science, Morality and Feminist Theory* (Calgary: University of Calgary Press, 1987), pp. 111–137.
[19] Jane Mansbridge, 'Feminism and Democratic Community,' in Chapman and Shapiro, *Democratic Community: NOMOS XXXV*, forthcoming.
[20] Carol C. Gould, *Rethinking Democracy* (Cambridge: Cambridge University Press, 1988), Chapts. 1 and 7.
[21] Carol C. Gould, *Marx's Social Ontology*, pp. 30 ff.
[22] *Ibid.*, pp. 25–26, 30–38.
[23] Cf. K. Marx, 'On the Jewish Question,' in Tucker, *Ibid.*, pp. 35 ff; and 'Critique of Hegel's Philosophy of the State', in L. Easton and K. H. Guddat, *Writings of the Young Marx on Philosophy and Society* (New York: Doubleday, 1967), pp. 173–175.
[24] Carol C. Gould, *Rethinking Democracy*, Chapt. 11.
[25] This is, of course, also Hegel's norm, and the *telos* of historical development in his philosophy of history. But I mean something different by 'freedom' than Hegel does, and I see its mode of development differently as well. Cf. G. W. F. Hegel, *Philosophy of History* (New York: Dover Publications, 1956), esp. Introduction, pp. 18–20.
[26] K. Marx, *Grundrisse*, p. 488.

ERAZIM KOHÁK

THE GOOD AND THE RATIONAL

In this paper, I wish to raise a question which I may not be able to answer but which I hope at least to clarify. For reasons which we shall examine, I believe that the project of technical rationality which has been the hallmark of modernity is today ethically, sociologically and ecologically bankrupt. Does that mean, though, that we must abandon our commitment to reason and turn to some form of post-modern a-rationality? I do not believe so. It does, however, call for a recognition of the primacy of practical reason or, less obscurely, the recognition that the universe does not turn from a random aggregate of discrete entities into a meaningfully ordered whole in virtue of what we believe to be true about it, but in virtue of what we recognise as good within it.

That, to be sure, is a major shift. The underlying assumption of Western philosophy at least since Descartes has been that of the primacy of theoretical reason. We have assumed that the fundamental question about reality and our place within it is not what is good, but rather what is true – and truth, we have assumed, is the quality of that proposition which faithfully mirrors reality. The task of knowledge, that is, is to compile a faithful catalogue of the furniture of reality; all else is secondary.

For the most part, we still assume as much, even though it is a highly problematic assumption for a discipline which claims to be a love of wisdom. Wisdom, after all, is not a matter of simply knowing what there is but one of knowing how to comport ourselves with respect to it. It is a matter of knowing what is good, not simply what is true.

The putative primacy of theoretical reason is problematic on systematic grounds as well. Meaning – including value, which is one mode of meaning – is a relational reality. Something can be meaningful /or valuable/ only in relation to an agent. Propositions pertaining to meaning /and value/ cannot, by definition, meet the fundamental criterion of theoretical reason, that of objectivity. Assuming the primacy of theoretical reason then leaves us two rather undesirable choices. One is to exclude from philosophical discourse all propositions pertaining to meaning or value, as Hans Reichenbach and his positivist kin sought

C.C. Gould and R.S. Cohen (eds.), Artifacts, Representations and Social Practice, 397–417.

to do. The other is to disqualify critical reason from dealing with such propositions and rule that philosophical discourse dealing with meaning or value belongs to the realm of the arational or the irrational, exempt from critical reflection.

The inherent undesirability of either option is, I believe, sufficient reason for examining the other option, the primacy of practical reason or, in other words, the assumption that the basic question is not what is true but what is good. That is the assumption embedded in the Socratic definition of philosophy as a love of wisdom rather than simply of knowledge, in Plato's insistence that the Form of Forms is the Form of the Good, or in Augustine's conviction that the ultimate goal of knowledge is salvation. It is also the recognition, elaborated by Edmund Husserl in *Die Krisis der europäischen Wissenschaften*, that reality is primordially a *Lebens Welt*, constituted as a meaningful context in relation to purposeful activity. The universe is intelligible because it is a context of action: in Goethe's metaphor, "Im Anfang war die That." Yet because being is a value for itself – whatever is, wishes to remain in being – that action is not arbitrary. It can be critically evaluated in terms of practical reason; whether it enhances being in its harmonious totality or whether it detracts from it or destroys it. What is true in the context of theoretical reason may be observer relative. What is good in the context of practical reason – in *Lebenswelt*, life's world – is not.

For a historian of philosophy, this raises an intriguing question: was Husserl aware of his shift from theoretical to practical reason when he asserted the primacy of the life-world in the order of reality? The life-world is by definition a context constituted by purposeful activity and so, as Jan Patočka notes, a world of good and evil as it aids or hinders that activity. Husserl in effect noted as much already in *Ideen II* when he recognised motivation rather than causality as the fundamental order of the world of persons. Yet Husserl also clung to the descriptive terminology of theoretical reason. The crisis of modernity in his description is the result of flawed theoretical knowledge, based on constructs rather than on lived experience, not a failure of the practical vision of the good. Historians of Husserl's thought have yet to render their verdict.

For us, though, the question is a systematic one. Assuming the primacy of theoretical reason leads to scepticism and irrationalism. If, though, we assume the primacy of practical reason, then the crisis of modernity is not a matter of inadequate or flawed theoretical knowledge but a flawed vision of the good. Because I am convinced of the sterility of theoretical reason but am not prepared to abandon philosophy to irrationality, I want to explore that suggestion and explore the crisis of modernity not as a problem of theoretical reason but as a problem of practical reason, of our conception of what is good.

Here, though, a word of caution is in order. I do not propose to undertake an investigation in ethics in the strict sense of the word, as the summary or the study of the rules of interaction among humans and between humans and the rest of reality. In periods of civilizational breakdown, as the decline of the Greek city-state or of the Roman empire, ethics inevitably comes to the forefront of attention as time-honored customs cease to operate and elders bemoan the manners of the young. Our time is no exception.

Yet I think the preoccupation with ethics is a false clue. At its core, ethics is really a rather unproblematic discipline, its basic outlines given by the requirements of human coexistence. The Decalogue, the Code of Hammurabi, the ethical codes of other civilizations all testify that, very early in their history, humans discovered that certain practices are conducive to successful social functioning while others are destructive of it. For all the local variations, they have learned that they ought not to kill one another, steal from each other, lie to each other, break sexual taboos, neglect the elderly, envy each other or simply let life sink to a mindless routine of feeding, procreation and elimination. The Ten Commandments are just one, rather specific expression of what is really an all-human consensus, shared even by other higher primates. Disagreement in ethics, however widespread, is not a matter of basic principles.

Traditionally, disagreement in philosophical ethics has had to do less with what we ought to do than with why we ought to do it. From the earliest, humans have claimed the sanction of God himself for their ethical codes. Alternately, we can, with Kant, justify them in the name of reason: undesirable courses of action lead to a contradiction destructive for a rational being. Or again, we can justify them in terms of utility, as Bentham or Mill did. The Decalogue – which both God and reason demand – turns out to be a good prudential rule as well. This is so, we can in turn claim, because its rules are "natural," codifications of the rules of primate interaction adapted to the needs of a predator mode of feeding, as Desmond Morris would have it. Yet when we ask how, concretely, an ethical person ought to behave, there is little difference between the kind of everyday behavior deemed decorous by Kant, Mill, Dewey or the average parson. They may disagree as to what is true, but they largely agree as to what is good. Ethics really is not a very controversial discipline.

There is, to be sure, another area of disagreement in which we might even speak of a "progress" in ethics. That has to do with the relations to which ethical considerations are relevant. At their most depraved, humans extend ethical considerations only to the members of the most direct in-group and exclude harm done to a stranger from ethical significance. One of the great advances in ethics is the Hebrew extension of ethical consideration to "the stranger that dwells in your midst for ye yourself were strangers in the land of Egypt." Another great forward step was the Stoic extension of moral considerations to all humans, bearers of *logos spermaticos*. In our time we are seeking to extend ethical considerations to all sentient beings: perhaps we shall see the day when the vivisectionist will appear as abhorrent as his colleague, the police torturer. The great ethical triumph would be the recognition that no relations are exempt from ethical considerations. Still, such disagreement, however basic, is not a disagreement over principle, what is good.

Actually, most of our disagreement in ethics has to do not with principles but with the way we ought apply them: not universalization, but particularization is the problem. There are few rules for applying rules. We can readily agree that we should not steal, lie, kill . . . but what does it mean in each specific situation in an incredibly complex society? How shall we apply stereotypical rules to nonstereotypical situations, be it the ethics of electronic privacy or of life-support machinery? There will always be a need for ethical discussion to clarify the issue – and for a personal decision to resolve it. Still, such discussion, however passionate and needed, is philosophically uninteresting. The basic rules of respect among humans are just not particularly controversial. Though from the perspective of practical reason the basic question is one of value, it is not a question of ethics.

I would like to suggest that the crisis of value basic to the crisis of our civilization is of a different order, pertaining to what we once designated as "moral philosophy." The term itself is hopelessly archaic, but the problem to which it points is exceedingly timely: what is the human *summum bonum*? What is the good? Or, less archaically, what is the point of it all? Wherein does human life find its fulfilment? What ought I to strive for, what ought I avoid, that I may look back on my life in peace? In the century just past we were wont to speak of "finding happiness," though the Westminster Catechism may have phrased it more accurately: What is the chief end of humans?

From the perspective of practical reason, that, I believe, is the most basic philosophical question. There is significant reason for believing that the great European civilization preceding ours, that of ancient Rome, died because it could not answer that question and give its people a reason for the effort of preserving it. Our situation may be somewhat different. We think we know what the chief end of humans is and how to go about achieving it. We have even developed an incredibly effective machinery to motivate the multitudes to the effort. Our problem is that our vision of the good has become counterproductive. Striving for it produces moral decay and vicious conflict; achieving it would destroy us. It may be vain to ask whether our conception of the Good is true or false, but it is not at all vain to ask whether it is good or ill, sustaining or destroying life – and the answer is not at all ambiguous.

What is our conception of our *summum bonum*? The question is not a theoretical one but eminently practical: what is it that we strive for, wherein do we hope to find life's fulfilment, as attested by what we do and by the value choices we in fact make? Or, as we ourselves are likely to put it, how can we have fun? *Was macht Spaß*?

A casual by-stander in a Prague stand-up buffet summed it up for me succinctly: "Barák, bourák, pivo v plechu" – a house, a car and beer in cans, something of a Czech version of the notorious "American dream." Ghanaians of a generation ago chanted, "Jagwah, fridgeful, been-to, those are the people." The happy few are those who have been to England, have a sports car and a refrigerator. Reflecting the ethos of the Reagan presidency, Americans of the last decade would say that "Whoever has the most toys when he dies, wins" – a perspective advertisers do their utmost to preserve. Less crassly, American public figures speak of an "expanding economy" – and mean by it ever rising individual material consumption, new loans for new gadgets. The vision of the *summum bonum*, the Form of Forms with which the West conquered the world, is the vision of ever rising affluence, ever greater consumption for consumption's sake – literally sacrificing all else not even for the sake of acquiring something specific but for the sake of ever ongoing acquiring. How can we have fun? Let's go shopping. . . .

Seen from within, the consumer mentality seems almost natural. Of course people want more. . . . Yet if anything breaks the spell, it becomes utterly in-comprehensible. Why in the world would anyone expend such heroic effort, accept so much hardship, choose to breathe polluted air amid wasteland, just to keep acquiring toys s/he does not particularly want? Only if s/he were convinced, on some deep level, that affluence really can bring happiness, can bring fulfillment. The reasoning, such as it is, is really quite primitive. A hungry person is an unhappy person. Give that person a heel of bread and s/he will be happier. Give her/him a strap of smoked meat and s/he will be happier still. Given a Lucullan feast, s/he would presumably be ecstatic. So stated, it may not sound overly convincing. Still, to humans disoriented by the collapse of familiar certitudes, exhausted by war and dulled by the failure of fanaticism, it may seem to hold out a hope. Life does have a meaning, after all. For the affluent, it is to acquire a house in the suburbs, a luxury automobile and a prestige wife. As for the rest of the world, Nikita Sergeevich said it all: "To catch up to and to surpass capitalism." The all-overriding race for affluence is on.

It is a powerful motivating vision, a vision of heaven on earth, offering fulfilment and requiring perhaps a lot of work but not really much effort: science and technology will do it for us, aided by The Invisible Hand, as we pant to gratify our greed. The only problem is that it has become as bankrupt as the flat earth hypothesis on Vasco da Gamma's return. Or worse: it has become self-destructive.

That the consumerist conception of the *summum bonum* is morally destructive is something moralists have been pointing out and the affluent demonstrating over the millenia. Acquisition can create a sense of happiness only when it meets an experienced need, and human needs are very quickly met. Only the need for meaning is infinite and possessions are rather poorly geared to meeting that need. Perhaps the moment of acquisition can bring a thrill – the thrill of hunting, the thrill of shopping – but infinite acquisition also becomes meaningless. Perhaps the most expressive metaphor are Roman vomitoria, designed to permit banquetters to regurgitate the previous course so they could go on consuming. America does not have vomitoria, but it does have its dumps and its yard sales, serving an analogous purpose. Here the quest becomes as pointless as it is endless.

Nor pointless alone: it is also destructive. A cherished world can enrich a human life, a possessed, dominated world is reduced to soulless, meaningless raw material and, as Heidegger among others has pointed out, amid a meaningless wasteland, human lives also lose meaning. Reducing the world from a cherished Other to raw material, humans themselves become raw material. The project of affluence through domination and exploitation, which modernity perceived as "progress," is as morally bankrupt and personally destructive as it was when Jesus spoke of the camel passing through the eye of a needle.

Such moral arguments are wearily familiar, true enough but oh, so boring – and just a bit sanctimonious. However, as populations explode all over the globe, a second problem with the consumerist motivating vision stands out: that it is not only personally, but also socially destructive. As a civilizational goal, the quest for individual affluence for a few inevitably generates impoverishment for the many. Old Testament prophets – as Amos, railing against the "cows of Bashan" – recognised that while minor differences can be the result of difference in diligence and ability, major differences are the product of exploitation. This is glaringly evident in Central and Latin America. There are, to be sure, exceptions, as Costa Rica, whose government has long undertaken a major effort to build up a social infrastructure and to prevent a polarization between the super rich and the desperately poor. Those countries, though, which have set simple affluence as their goal, have produced polarized societies prone to conflict. Interestingly enough, during the years when it followed such policy, even the affluent United States showed that tendency.

More viciously still, the quest for affluence, elevated to the role of a civilizational goal, has the same effect globally. The incredible explosion of individual consumption in Europe and North America since the second world war has created a viciously polarized globe. The demands of northern affluence increasingly draw both natural resources and investment capital out of the "third world," exporting waste and poverty in return. The ever more desperate misery produces civilizational breakdowns and becomes a breeding ground for fanaticism. Most of all, it becomes a breeding ground of hungry and desperate billions. Older scenaria envisioned third world countries, then called "the global proletariat," waging war against their affluent exploiters. The prospect is far less dramatic: starving millions and billions swamping the isle of privilege by their sheer numbers, much as the migrating masses overran the Roman empire, as migrant workers long before they came as warriors. A strategy of massive redistribution might save the globe, a strategy of ever expanding affluence dooms it.

In the affluent North, few people note the building up mass of poverty in the third world, just an in affluent suburbs few people note the buildup of despair in the ghetto until it explodes. The few who do note have traditionally proposed a universalization of affluence as the solution. The more callous invoked the putative "trickle-down" effect: first world affluence would spin off third world progress, much as sparrows get fed when a horse gets oats. In practice, that failed disastrously. Without a conscious effort at redistribution, affluence generates more affluence for the affluent and more poverty for the poor. What, though, of a conscious effort? The Marshall plan did raise Europe to the American level of affluence. Why not a Marshall plan for the whole world? Why not, in effect, save the consumerist conception of the *summum bonum* by universalizing it?

The question was raised at the world ecology conference in Rio de Janeiro in 1992 – and the data presented there provided an answer which no one was quire ready to confront openly: if extended beyond the narrow circle of the affluent, such affluence would destroy the earth. At present, less than a fifth of humankind consumes – and generates waste – on the American level. Yet even that fragment consuming at that level is destroying the globe – consuming resources and generating waste – faster than nature and human effort can compensate. Globally, ecological efforts have slowed down the rate of destruction; they have not reversed it. If all humans were to consume and waste at the first world level, the earth would become uninhabitable in a matter of days, not months, its resources exhausted, its surface buried in litter. Yet if we do not share our affluence with the third world, we are likely to perish in a cataclysmic conflict of the rich and the poor.

Altogether, for moral, sociological and ecological reason, the vision of the *summum bonum* as ever increasing affluence has become untenable. The point is not that it is not "true" or that it is "false." Those are categories of theoretical reason and simple inapplicable. The point is that it is *not good*, that it is destructive. If the affluent world, which functions as the role model for the other 80% of the globe, sets the example of continued mindless greed – euphemistically, of "expanding development" – we shall all perish, and we shall in all likelihood perish morally even before we perish physically. If we hope to survive and bequeath this earth to our descendants, we need to go back and ask anew the question of practical reason: what is the good? what is worth striving for, what is beside the point, and what is destructive? What is the chief end of humans, their *summum bonum*?

What is the chief end of humans? The Westminster catechism provides an authoritative answer: "To love God and enjoy him forever." Contemporary believers may substitute inclusive terminology but might well consider that still an adequate answer. For the culture as a whole, that answer has become effectively unavailable. Yet since theoretical reason demands authoritative, "true" answers and the habit of theoretical reason is very much with us, we continue to search for some putative human nature that would enable us to provide an authoritative answer and avoid the difficult question, what is the good? Is there a "natural teleology" to our being of which we could simply say that that is the way it is, that is "true"?

Depth psychology, especially of the Jungian variety, and analogous "deep" reflection in philosophy and the human sciences offer one such approach. They represent the assumption that there is a teleology built into the human psyché simply as a matter of fact, there to be discovered. Specifically, they assume that the deep and archaic contents of our consciousness are not simply an affective mass but, rather, represent a definite orientation to a definable goal, so that the researcher could determine, authoritatively, our *summum bonum*. The imagery may be Jungian, mythical, religious or philosophical with a mystical bent. The basic paradigm, though, tends to be constant, describing the conscious ego as alienated from its sustaining /unconscious/ ground. The vision of fulfillment – the *summum bonum* – then takes the form of "connecting" with that sustaining ground, whether by communing with nature, listening to the voice of Being /or of the inner self/ or by accepting God's grace through faith. The goal – which, more or less plausibly, is taken to promise an overcoming of the world's ills as well – is the overcoming of the alienation between actual humans and the depth of their being.

Sociobiology – again, together with all its variations – provides a variant of this approach, seeking an authoritative answer in our genetic inheritance rather than in the depth of our psyche, but seeking it no less. Thinkers like Konrad Lorenz or Desmond Morris also search for our true humanity, though for the polarity of conscious/unconscious they tend to substitute a polarity of artificial/natural. It is then in our nature as higher primates who adopted a predatory feeding mode – in Mr Morris's delightful if not altogether convincing metaphor – they seek both an explanation of our modes of comportment and an authoritative answer to what would constitute desirable behavior.

Both the psychodynamic and the sociobiological approaches have a great deal to contribute to our self-understanding. We really are incarnate beings and the minds and bodies wherein we are incarnate do have a dynamic of their own. Much of the grief of our lives can indeed be traced to our insensitivity in riding rough-shod over the grain and rhythm of our psyché and of our genetic endowment. Yes, for optimal functioning, it is crucial to be aware of our animal nature as well as of our psychic endowment. There is a "natural" and a "violent" way of being human.

And yet, being human is more than being a psyché, more than being a kind of a higher primate. It is also a matter of our freedom, the *vollkommene Freiheit* of which Husserl speaks. That is not reducible to our theoretical reason or to our affective sensibilities. It is a matter of our practical reason – of our ability to direct our lives to a goal, not as a matter of cognitive necessity but freely, not as a matter of affective spontaneity but for good and sufficient reason, not because it is "true" but because it is good. Because that is a matter of practical reason, we cannot expect to find authoritative answers encoded in our genes or our unconscious. We can, however, find our several possibilities acted out in history, writ large, together with their consequences. Rather than asking what the "true" telos of being human is, we might turn to history and ask what conception of the good humans have chosen and ask not which is true but, more modestly, which is good.

What have humans considered worth striving for over the centuries? It seems not unreasonable to assume that for our earliest ancestors at the dawn of humanity simply survival might have appeared as a sufficient goal, since even bare survival constituted a major achievement. Yet decorations preserved on the walls of cave dwellings testify that it was not a sufficient goal for long. As soon as our ancestors placed even a modest buffer between themselves and cold, hunger and death, even while they still subsisted marginally as hunters and gatherers, they found other concerns.

To be sure, it is rather unlikely that those cave drawings were idle doodles, or *ars gratia artis*. It is quite reasonable to suppose they had a religious significance – and that that religion in turn related quite directly to the pressing task of procuring food. What it is not reasonable to suppose is that religion at that stage related only to that task, as Auguste Comte assumed when, somewhat naively, he projected a 19th century mentality into prehistory. All evidence suggests that religion does not start with a task but in the moment when humans look up from their preoccupation and suddenly become aware how much more there is to the world than the little world of their concern, become aware of the vastness of the sky, of the depth of the earth and of the entire transcending dimension of reality. "Deus, sive natura," says Spinoza while Rudolf Otto describes that transcending dimension as *mysterium tremendum et fascinans*. It is infinity breaking into the finite.

That basic experience can evoke a dual response. One is utilitarian, the practical question of how to harness that vastness to human purposes. It is the posture of negotiation and manipulation expressed by the Roman dictum, *do ut des*, I give that you may give. The trade can be crass, a candle for a favor. It can be lofty, a noble deed for God's favor. Its perspective, though, remains ego-centric and in time will become the source of first magic, then science, ways of manipulating the world for human purposes.

Religion, though, has an other dimension as well, one which responds to the wonder of the vastness with awe and worship. That, too, leads to action, but the motivation is no longer ego-centric. It is wonder and gratitude going out toward the Other. I give because you have given. Ultimately, all moral philosophy reflects the basic opposition between these two postures, that of self-gratification and that of self-transcendence.

That, to be sure, is far too schematic a presentation. The conceptions of the ultimate good which we encounter on the projection screen of history are rather more concrete. One of the oldest is the Egyptian longing for permanence which had been a part of Western consciousness long before the West became aware that there had been an Egyptian civilization. Ancient Greeks did frequently attribute their greatest wisdom to the Egyptians, and an old tale. since shown to be spurious, even had Plato travelling to Egypt in quest of knowledge. In the last century, archeological research bore out the parallel between Plato's basic motives and the lasting theme of three millenia of Egyptian culture, the sorrow of perishing and the longing for what endures. Throughout their long history, the Egyptians appear to have remained convinced that this life and world of passing and perishing is less than real, a mere preparation for the true life to come. They lavished care on the homes they built and furnished not for this world but for the true world, the world ever-lasting – their tombs. For the real, that which is worth longing and striving for, is that which is perfect, unchanging, lasting, eternal, not of this world.

In that respect, Plato was truly an heir of his Egyptian predecessors: he, too, was convinced that the real is the unchanging, Being, not of this world of Becoming. Most of his countrymen, though, did not share that conviction. The excellence ancient Greeks sought, from Homeric times to the Golden Age, was an excellence of this world. The Greek conception of an afterlife, such as it was, bore no resemblance to the Egyptian idea of eternal perfection. Hades was at best a world of shadows, pale and anemic, at most a postscript to this life. This life's fulfilment could not come in that land of shadows. It had to come here, in the land of the living. The Greeks sought it in a perfection of this life, perfection of both the mind and the body. To achieve a harmony of the good and the beautiful, of a perfectly atuned bronzed athletic body and of an agile, no less athletic mind, that was the vision of life fulfilled. Or not quite: the true goal was not simply goodness and beauty but rather the esteem of the *polis* which they bring. Fame, recognition by the community, was what mattered, and Greek heroes were willing to lay down their lives for it. Not wealth: though the Greeks appreciated the advantages of wealth – a laboring man, they agreed, could not aspire to excellence – they had little regard for the pursuit of wealth. Recognition by the *polis* was the one goal worth striving for – and the qualities which the polis appreciated, physical beauty and mental excellence.

In an odd way, their Hebraic counterparts were also people of this world. The *sheol* of the Old Testament is no more of an ideal reality than the Greek *Hades*. "They that go down to the Pit" are cut off from the land of the living, and it was in the land of the living that ancient Hebrews sought the fulfilment of their lives. There was another parallel as well. In their own way, the ancient Hebrews also played to an audience. That audience, though, was not the *polis* but God, the Lord of the cosmos who is not impressed by affluence and ostentation, be it of body or of the mind. "Your new moons and your appointed feasts my soul hateth . . . I am weary to bear them," God says through Isaiah /1:14,16/, "Learn to do well; . . . relieve the oppressed, judge the fatherless, plead for the widow." The sacrifice of the Lord, we learn, is a contrite heart, purity of spirit, mercy and righteousness. Nor is the reward the esteem of the community or some putative reward "in heaven." In pre-exilic times, we encounter the promise of prosperity on earth, though already the book of Ecclesiastes stresses the vanity of earthly goods and of the world's empty praise. The blessings are of a different order: "Blessed are the peacemakers for they shall be called the children of God" /Mat. 5:9/. The reward is being set right with God: life's fulfilment is God's service – and in a real sense, that is the ultimate reward: to serve the Lord. In a much later idiom we could say, anachronistically but not inaccurately, that here self-realization becomes self-transcendence.

The idea of service as life's fulfilment emerges again, in a very different context, among the late Roman Stoics. When Marcus Aurelius bids his readers act as "a man and a Roman," upholding the stout virtues of Roman antiquity, he holds out no hope of a reward. Amid the decadence of late imperial Rome, earthly rewards had reached the nadir of absurdity, banquets punctuated by regurgitation, and there was no television advertising to lend them an air of spurious desirability. As for heavenly rewards, the Stoics, though not actively atheist as their Epicurean counterparts, put little stock in them. There are no rewards, nor are rewards called for. To live as a man and a Roman, living up to the full stature of noble *humanitas*, that is life's point and fulfilment. To think of rewards would be unworthy of a Roman.

Christianity may well have triumphed over Stoicism in great part because it did promise rewards, a fulfilment beyond the limits of this transient life and this flawed world. "Rejoice and be exceedingly glad for great is your reward in heaven." Actually, consistent with the pharisaic Judaism on which Paul drew heavily in shaping Christian doctrine, the early church spoke of resurrection. The Lord has risen, we are told, as a kind of a first fruit, and will return – within the lifetime of believers – to raise the dead and establish His kingdom on this earth. That is what the Jehovah's Witnesses still believe. Most of the church, though, grew weary of waiting. When the second coming did not materialize, it reinterpreted the promise. The Kingdom of God was to be found on a second plane, in a true reality which parallels but does not intersect the line of earthly time. Life's fulfilment is not of this earth: it belongs to the eternal kingdom where there is no change, no waxing and waning, nor passing and perishing. Somewhat ironically, Christianity, fusing Hebrew piety with Greek philosophy, revived an Egyptian theological conception. As the mediaeval church would later present it, there are two parallel realms, this world of the earth in time, and the eternal world of heaven and hell. This world is but preparation, true life will begin with death, now understood as a translation from the earthly to the heavenly realm. To strive for heaven, that now appeared as the *summum bonum.*

The reformation was born of the fear that the sacraments administered by an unworthy priest could not guarantee heaven, that *summum bonum.* Its early strategy, represented by Wycliffe and Jan Hus, was to reform and purify the church, so assuring the efficacy of its sacraments. The mediaeval schema still held: heaven is the fulfilment of life, the sacraments of the Church are the means of heaven, only the church needs to be purified to make those means efficacious. A century later, a second generation of reformers, most notably Luther, introduced a variation in the mediaeval schema. While heaven is still life's fulfilment, the means to it is no longer the Church and its sacraments but God's infinite grace received through faith / Luther / or perhaps destined from all creation for some / Calvin./

Finally, a third generation of reformers, represented by the Puritans and the pietists, unwittingly brought about another change. Though the Puritans definitely did continue to use the rhetoric of heaven, heaven ceased to be a goal of human striving. Predestination placed it outside human control. God alone decides whether I shall be saved, I need not worry about it. What remains for human concern is the works of charity and mercy which are the outward sign of salvation. Perhaps I have been saved, perhaps not . . . but ultimately that is God's concern, not mine. What matters for the human is to do faithfully that to which God calls him/her, to be a faithful servant. It was with that step that protestant Christianity returned from the mediaeval detour into Egyptian eschatology to the basic insight of the early church, echoed by the Stoics, that self-realization is not to be found in self-gratification but in self-transcendence. "Seek ye first the Kingdom . . ." – for "whosoever shall seek to save his life shall lose it."

While the masses of the faithful looked to the Christian beginnings for guidance, the educated discovered another forgotten source, the antiquity, now rediscovered or reborn. Some of the noblest thinkers of the Renaissance, Erasmus of Rotterdam or Thomas More, were no less Christians and heirs of the Middle Ages than their contemporaries, the Reformers. Yet while the Reformers looked to special grace – God's Word in the scripture – for guidance, the men of the Renaissance looked to general grace, God's presence in His creation or, in later terminology, to the order of nature. As their attention shifted increasingly from revelation to science as the means of understanding that grace, the meaning of salvation shifted as well, from heaven to earth. For families like the Medicis or the Borgias, earthly success came to be the meaning of salvation, and knowledge and power its means. After all, Bacon tells us, knowledge *is* power. As Europe approached the cataclysmic wars that would put an end to the middle ages once and for all, it did so with a dual image of the good, as service and as domination. Or, in the same anachronistic terminology, we could say that it sought self-realization either in self-transcendence or in self-gratification.

The thirty year trauma of the wars of religion utterly shattered Europe's confidence in God's guiding hand. Even Comenius, though he would remain a pious believer throughout his life, became suspect to all warring factions because he could not accept any of the traditional schemata. Certainly in Europe at large the profound conviction of God's presence and guidance which Erasmus once shared with Luther and Augustine was irretrievably gone. It was not a matter of an intellectual change as much as of an overall sense of reality. Much later, Nietzsche would employ the metaphor of God being dead. Laplace simply says that he has no need for that hypothesis. For centuries, God had been the basic reality, making the world and the place of humans therein intelligible. Now, amid the horrors of war, the world feels alien, an impersonal mechanism to be manipulated, not a home to be cherished. For centuries, the ultimate responsibility had been God's. Now humans had inherited it.

In retrospect, we could describe the profound change that marks the early 17th century as a shift in the experience of nature. As humans had for centuries experienced it, Nature really did have all the properties traditionally ascribed to God. Humans surviving precariously in its margins encountered it as deep and dark, mysterious and infinitely self-renewing, providing humans with all the wherewithal of life, sometimes severe, sometimes merciful. Given that experience of nature, Spinoza's equation, *Deus sive natura*, might well seem self-evident.

That is what changed with the rise of modern technology. Nature no longer appeared as the mysterious Other. It became an object of manipulation, no longer a mystery, simply a problem. Schiller will later speak of an "Entzäuberung", a dis-enchantment of nature, Hegel of an "Entgötterung der Welt," a secularization of the world. The world became, as Heidegger would say much later still, a reservoir of raw material on order, to be used and abused at our will. Now we are reaching a double recognition. One is that nature is really finite. For some two hundred years, we no longer approach nature with respect and awe, yet we continued to assume it is as infinite as we assumed when we encountered it as *Natura sive Deus*. Now we are discovering that it is very finite, fragile and vulnerable: Nature is an endangered species. Our second discovery is that not only is nature finite, but that also we are a part of that nature. If we transform it into meaningless raw material, we shall ourselves become raw material, on order for the anonymous demands of the process of "conquest of nature" we had once regarded as "progress."

Today, as we survey the devastation all around us and look a probable ecological catastrophe in the face, many of us might be tempted to see the Enlightenment, in retrospect, as a time of unmitigated *hubris*, of raw will to power mascarading behind a smokescreen of "reason," and turn bitterly away. The project of modernity – "conquest of nature" in the name of ever expanding individual consumption for the privileged – has failed drastically, reducing our world to a barren wasteland of raw materials and ourselves to the role of mindless consumers whose sole task in life is to consume, vomit and consume more. It is understandable that writers whose familiarity with the Enlightenment is limited to a selection from Descartes and Hume might consider Enlightenment rationalism the source of all evil and turn to romantic nature mysticism for relief. It is even understandable that thinkers sensitive to the fragile beauty of nature might consider technology and even conscious human effort as much the root of all our ills. We have indeed caused so much damage with our best intentions and our unyielding determination to have our way: the Gabčíkovo water works is just one recent instance. Why not just let things happen for once? Why not write off the entire project of the Enlightenment with a resolute "Ohne mich!", "Count me out!"?

Given the desperate plight of nature and the naked greed posing as "progress," it may be tempting. We certainly, unquestionably do need to become more sensitive to the intrinsic value of our non-human kin, we need to overcome our hopeless anthropocentrism and become sensitive to the autonomy of all being. We do need to introduce ethical considerations into all our interactions with the world, we need to learn humility. The arrogance of technical reason is staggering and destructive. We need to recognise how little we know and be sensitive, not overbearing, when the world resists our efforts. All that – and far more of all that the more sensitively tuned environmentalists try to tell us – is all too true and urgently needs hearing.

Still, however tempting and rewarding it may be to retreat to a forest clearing, beyond the powerline and the paved road, as Thoreau in his day, we cannot simply give up responsibility and surrender our fortunes to the spontaneity of being, of nature, or of the Invisible Hand. We have irretrievably disrupted the autonomy of nature. For better or for worse, the responsibility has become ours and we can no longer give it back.

Let one example serve. A great many American environmentalists look with admiration to the native Americans, and not without reason. They do seem to have created, in their time, what today we would call "sustainable development," though "sustainable harmony" would be more accurate. They managed both to control and to sustain populations, avoiding the vicious cycle of rise and fall of urban civilizations. They managed to live in harmony with nature, with the deer and the buffalo, making no greater demand on it than it could meet without permanent damage. It is not surprising that to some writers, as H.W.F. Saggs, regard the transition from a hunter-gatherer to an agricultural life style as the original sin.

Yet however enviable we may think that life, there is no returning to it. The life style of native Americans required a low population density: in their time, the North American continent could support at best some 3.5 million humans. Today, there are some 350 million in the same area, and the numbers are rising. Short of a 99% decline in population, we cannot return to it. Certainly, we need to learn from the native Americans, we need to accept gratefully the gift of their presence, but even that is something we must do consciously. It is not something that can just happen.

That basic paradigm applies throughout. Somaliland provides a graphic example of what happens when, after a cycle of colonization, modernization, decolonization, people simply let go. Or again, were we today to abolish all constraints on fishery, the oceans would be fished dead long before starvation had time to reduce the demand for fish. For that matter, the results of the deregulation of the S & Ls in the United States provide another example. The spontaneous regulatory mechanisms are no longer there. We have disrupted the "natural" patterns of interaction too deeply just to let things happen. Having once arrogated responsibility for the world, we can now hope to exercise it wisely and gently, as for long we have not, but we cannot hope to abrogate that responsibility. Or, metaphorically, we cannot undo the Enlightenment.

However, we can and have to rethink just what it is that we understand under that term. The Enlightenment, for one, was not yet the "age of progress," of dramatic transformation – and devastation – of the earth. That came later: in the 150 years of "progress," between 1840 and 1990, the world changed utterly and irreversibly. Yet during the 150 years of the Enlightenment, between the Peace of Westphalia in 1648 and the French revolution, the world remained recognisably the same. The Enlightenment was a perspective, it was not yet a project.

What the Enlightenment was, was in the first place the assumption of responsibility, less as an act of arrogance than as a result of default. The phenomenon of the Baroque suggests that Europeans after the Thirty Years' War would have far rather returned to an age of faith than launch upon an age of responsibility. They had no option: the structure that once bore that responsibility – *Natura, sive Deus* – was no longer there. It finally vastly does not matter whether we explain that as a product of increasing population densities, of social complexity or of technologization. The fact remains that things would no longer simply happen: the responsibility had shifted to humans. Nor does it matter that it would take two more centuries before they would recognise it. It did: still during second World War no one worried about the environmental impact of sunk tankers. During the Gulf war, at least some people did worry about it. That new environmental consciousness is also a part of our Enlightenment heritage, of the acceptance of responsibility.

That is the crucial trait of the Enlightenment, the conviction that humans are responsible for their lives and their world, that things will not just happen. It is, secondly, the conviction that both nature and the place of humans therein are rationally intelligible. It is not a matter of reading tea-leaves or of a blind will: humans can grasp the order of reality and make an informed decision, for good and sufficient reason. However, it should be noted that the Enlightenment conception of reason is emphatically not synonymous with the cartesian calculating ability or the 19th century technical reason, excluding all considerations of meaning and value as a-rational. Descartes and Galileo, with their extentional, mathematico-causal conception of reality, were men of the Enlightenment, but so were Comenius, Shaftsbury and the pansophists, who considered the basic rationality of the cosmos and of human society to be qualitative, meaningful and value laden, rather than mathematico-causal.

The project of domination and exploitation which Foucault attributes to the Enlightenment is really far more typical of the age of "progress" and requires more than the Enlightenment conception of humans as responsible and reality as intelligible. It requires one other factor – the Renaissance conception of self-indulgence as the true form of self-realization. It would be utterly foreign to Kant, who shared the values of the Enlightenment but strictly distinguished between self-respect and self-indulgence – and saw service as life's fulfilment. If we must have a villain in the piece, it should not be the Enlightenment conception of rationality but rather the project of domination and exploitation which follows from the combination of that rationality with the Renaissance conception of self-gratification as the *summum bonum*. The basic opposition is not between reason and the emotions, rationalism and romanticism. Emotions can be narcissistic, reason altruistic. Altruism depends both on emotion – empathy – and reason, being able to grasp the way the world appears to the other. Egotism is likewise both an emotion and a mode of reasoning. The basic difference is one between two possibilities both of which are a part of the Enlightenment – that of the theoretical reason which asks "what is true" and consigns the question of goodness to irrationality, and practical reason which starts with the question, "what is good?" and bases intelligibility – or "rationality" – upon the goodness of being.

With that, we have come a full circle, and though I would not wish to claim to have provided an answer, perhaps we have shed some light on the question. In light of all we have said, it would seem that the problem of modernity is not in the first place that it is technological, but that it is deeply, profoundly egocentric, having elevated greed to the status of highest virtue.

Certainly, technology reinforces that egocentrism. It is in contact with living beings who have an agenda of their own that humans learn the autonomy and value of the other. Humans whose life-experience is typically that of constructs and artefacts whose sole purpose is to serve as a tool for human gratification, might well conclude that they are both the seat and source of all value and meaning, the conviction basic to egocentrism. It is interaction with the Other who both deserves and demands respect that teaches humans both respect and self-respect. The technology user is more likely to learn domination and conceit. In that respect, the Heideggerean critique is accurate and important.

Yet egocentrism – greed – is far older than technology and would be most likely to survive technology's demise, which we in all likelihood would not. If life has a future on this earth, it lies not in abandoning technology but in placing it in the service of a more adequate vision of the Good than domination, exploitation and every rising consumption. It means placing technology in the service of practical reason and of a rational vision of what is good. Not, we should stress, in the service of what I like, or what we, our community like: that is an emotional judgement and individual subject relative. Practical reason – or perhaps we should specify, pure practical reason – must start from what is universalizable, the recognition that Being is a value unto itself: whatever is, basically wishes to remain in being and, all other things being equal, values its being. Being bears value intrinsically within itself: *esse qua esse bonum est*. Because it does, its being in turn constitutes value. But because claims to being do conflict – as the owl and the mouse – the *summum bonum* is not simply being but the harmony of being which assures the optimal, sustainable realization to all beings. For humans, that means not "expanding economy" or even "sustainable development," but a quest for a sustainable harmony – and a voluntary simplicity in conformity with the slogan, "Live simply that others may simply live." Not self-gratification but self-transcendence in the name of sustainable harmony of all being is the project that might yet allow life to survive on this earth.

Here, though, we should note that the vision of sustainable harmony of all being is very much an Enlightenment one. Not, perhaps, of those Enlightenment thinkers who, following Descartes, dismiss judgements of meaning and value as a-rational. But, as we have already noted, not only Gassendi and Mersenne, but also Comenius and Shaftsbury were men of the Enlightenment, and to them the theme of cosmic harmony was very familiar. Similarly, the theme of meaningful, value-laden ordering of reality is no less familiar from Husserl. The project of sustainable harmony is not a project of emotional sponataneity: it is a project of practical reason. Nor is rationality coextensive with modernity's project of domination and exploitation. There is also practical reason which critically poses the question, "what is the Good?"

Finally, I do believe that the answer to the crisis of modernity cannot be emotive irrationality, any more than in Husserl's time. Without the constraints of critical reason, the spontaneity of passion is more likely to lead to a Bosnian hell than to a post-modern panacea. Nor can it be an answer of theoretical reason which concerns itself with the technology of nuclear bombs but dismisses the question of their purpose and use as "unscientific." The answer, I think, must be one of practical reason. Given a global commitment to "growth" and a thinning ozone layer, we shall find out, in not too many years, whether that is indeed the answer – and whether it can still come in time.

Boston University

GYORGY MARKUS

THE END OF A METAPHOR: THE BASE AND THE SUPERSTRUCTURE

There are some theoretical terms so closely associated with – assumed to be central and idiosyncratic to – the thought of some thinker that their simple mentioning immediately invokes a name. *Cogito* 'means' Descartes, the 'monad' – Leibniz, and similarly 'base and superstructure' immediately recalls Marx.

There is something rather strange in the strength of this association – both philologically and substantively. As a philological fact, the pair of terms: *Basis* (or sometimes: *Grundlage*) and *Überbau* occurs in the great many thousand pages of Marx's writings (including his manuscripts unpublished during his lifetime) maybe a few dozen times, though some of these cases are undoubtedly emphatic, the most famous being his summary characterisation of the general conclusion and guiding principle of his studies in the Preface to *A Contribution to the Critique of Political Economy*. Substantively, one can hardly imagine a metaphor less suited to illuminate the basic thrust of the Marxian conception of history and society than this old simile borrowed from architecture. Marx's theory was primarily that of *historical process*. It precisely aimed at dissolving the appearance of the thing-like fixity of social institutions, arrangements and relations which for the isolated individual *de facto* are pre-given realities to which it can only adapt him/herself. He attempted to demonstrate how these are day by day produced and reproduced in the on-going collective activity of the socially situated individuals who in this process constantly, and usually unknowingly, recreate and change the social conditions of their own activity. The metaphor of a static relationship between some persisting, substantive elements-entities is singularly misleading in respect of this underlying, central idea of the whole theory.

There can be, however, no doubt that, independently of all reservations concerning the imagery associated with it, the metaphor, 'base and superstructure', articulates an idea equally fundamental to Marx's historical materialism and often formulated by him without the use of these terms: that of a necessary dependence of all phenomena of political, legal and cultural life (the corresponding practices, their insti-

C.C. Gould and R.S. Cohen (eds.), *Artifacts, Representations and Social Practice*, 419–439.

tutional frameworks and their objectivations) upon the "mode of production of the material life", more particularly upon the economic structure of society.[1] As to the exact nature of this dynamic relation of dependency, Marx's writings do not provide an unambiguous answer. There are places where he interprets it rather unmistakably in a *causal* sense[2] (and it is always so understood by Engels whose later qualifications and corrections concerning the "relative independence" of the various constituents of superstructure are consistently formulated in terms of a causal interaction between two unequal fields of forces). On the whole, however, such a view cannot be consistently ascribed to Marx. He specifically underlines[3] that the relationship between the development of the economic base and that of the various realms of superstructure is 'uneven', meaning by this not only some significant time-lag between the two, but also the fact that superstructural changes can in their content *anticipate*, and thereby foster, the development of the 'base' (to this he draws attention not only in respect of the cultural, but also that of the legal components of the superstructure, his own example being the relation between Roman private law and capitalist mode of production). And he equally admits that some elements of the superstructure may retain their social significance and effectivity even after the complete disappearance of the economic base which 'conditioned' them. So this relation of dependence and conditioning (*bedingen*) has to be understood in a weaker sense. In fact it encompasses and sums up in one (rather vague) formula a number of more specific and heterogeneous relations: Economic development constrains, sets limits to, superstructural changes, primarily by determining not only the extent, but also the character of social resources that are mobilizable for other types of social practice. At the same time it provides important impulses and motives (in the form of definite social needs and interests) for, and in this sense exerts directive pressures upon, the latter. In respect of the 'ideological' elements of the superstructure this first of all means that it is social conflicts ultimately determined in the terrain of economy that "men become aware of and fight out" in these forms.[4] Farther, the way superstructural activities are interconnected and integrated with other types of social practice is determined by the general character of division of labour in society being one of the fundamental elements of its economic structure (e.g., the prevalent commodity form of cultural products in capitalist society as opposed to the client-patron relation through which such activities are often organized in pre-capitalist civilisations). Production in this sense constitutes the 'overarching' element in the total process of social reproduction. Lastly there is also the idea of an elusive structural correspondence or homology between some of the characteristics of the two domains of social activity, e.g., between the mutual relations of the agents of a simple commodity exchange to each other, on the one hand, and the ideas of "equality, freedom and Bentham [i.e., utility – G.M.]" as fundamental values articulated by the dominant ideologies of bourgeois society, on the other.[5]

Though the so-conceived idea of economic 'conditioning' is rather unsharp, the 'base/superstructure' metaphor articulating it in general outlines a well circumscribed research program: to *explain* processes of political, legal and cultural *change* by relating them (in various ways and through various mediations) to what are seen as the fundamental trends of economic development, and thereby also to *explicate* the social *significance* (in the sense of historical efficacy) of superstructural phenomena primarily in terms of social conflicts whose ultimate origin is to be sought in the character of the social organisation of material-economic activities. Not to be completely ahistorical, one has to appreciate the strongly *polemical* character of this program. Its formulation by Marx was directly opposed to the then prevalent understanding of the political and cultural spheres as embodiments and expressions of (variously conceived) general interests, sharply differentiated and counterposed to the whole sphere of economy as that of private interests. In particular, 'high culture' was seen as the repository of ultimate and universal human values the accumulation of which is the result of the labour of those geniuses of spirit whose activity is unapproachable by means of mundane social analysis. Marx rejected both sides of this dichotomy: 'private interests' are always socially and differentially articulated interests, and even the most 'elevated' realms of culture are but social practices organized within an encompassing division of labour and through it interconnected with, and dependent upon, mundane-material activities and the interests to which the organisation of these latter gives rise. In this sense the concepts of 'base' and 'superstructure' were constituents of an overarching project of integral historicism which denied the existence of separate and independent histories of politics and spirit[6] – a point which (in this negative generality) is rather commonplace today. And we also ought to remember that even those early Marxist writings about culture that were couched completely in terms of the base/superstructure distinction and which are now quite legitimately criticized for their oversimplifying and reductionist approach, have played in their own time a pioneering role contributing to the formation of new scientific disciplines of culture (the works of Mehring, Hausenstein, Plekhanov in respect of sociology and anthropology of art, that of Hessen in respect of a historical sociology of science, etc.)

At the same time we should note that from the very beginning of the formation of 'Marxism' the relevant ideas of Marx have been transformed in the direction of an extremely simple-minded positivism and economic reductionism, a transformation which subsequently became codified and institutionalized in Soviet Marxism as official ideology. The most fatal step in this reinterpretation was the identification of the distinction between base and superstructure with that between the 'material' and the 'ideal', where this latter was understood as everything the existence of which depends, in varying degrees, on 'ideas', i.e., on consciousness. In this way – in the name of a 'materialism' – the 'economic base' has been transformed into sheer facticity and objectivity, which is simply there in complete independence from any conceptualisation and interpretation, and, as the ontological sphere of Being, is globally counterposed to all phenomena of consciousness which *in toto* belong to (though do not exhaust) the domain of superstructure. Marx himself is hardly guiltless in this misunderstanding: beside his predilection for overemphatic, polemically sharp, but in their meaning highly ambiguous formulations,[7] his whole terminology was implicated in it. In his attempt to secularize the Hegelian distinction between subjective and absolute 'spirit' (*Geist*), he used the term 'consciousness' (*Bewusstsein*) in two different senses. On the one hand, he designated by it the states of awareness and thoughts of the individuals (which themselves are – primarily through the constitutive role of language – 'social products'); on the other, he meant by it – often using the expression 'social consciousness' or 'forms of social consciousness' - the whole complex of cultural objectivations (philosophical and scientific theories, works of art, etc.) as repositories of intersubjective meanings and interpretations characteristic of an epoch or a class. In this *second* sense, 'consciousness' is, of course, a phenomenon of superstructure. But Marx's formulations are (especially in his early writings) sometimes so sweeping that even their context does not make it possible to decide in which of the two senses he actually uses the term at a given place.

Whatever confusions may thus have been caused by some, isolated, statements of Marx, it seems to be not even much disputable that this 'ontological' interpretation of the base/superstructure distinction that transforms it into the social refraction of the dichotomy between matter and mind, completely misrepresents his theory. This latter clearly operates with the presupposition that human activities and their products (of whatever kind) are always *interpreted*, having an intentional sense, resp. meaning, socially codified in language, the basic elements of which are shared by all members of a society making communication between them possible. How elemental and basic this presupposition is for his theory becomes clear from his explication of the concept of labour, one of its central notions. As he underlines,[8] one can speak about labour as specifically human activity only if this activity itself is directed by, and towards, a conscious intention that transcends what is given and perceivably present in the actual situation in which work takes place. Equally it is the intended, consciously envisaged end-use of products of labour which constitutes "their conceptual determination",[9] and production acquires its 'finish' only through that act of consumption which as its posited end internally directed and controlled it from its very beginning.[10] And in the so-called 'Great Manuscript' of 1861–1863 Marx unequivocally states: ". . . social relations among men [are possible] only insofar as they think and possess this capacity of abstraction from the sensuously individual and accidental".[11] At the same time consciousness and thinking exist only in language: ". . . the language *is* the practical, real consciousness, consciousness existing also for other men, and therefore existing for me as well; and language, like consciousness, only arises from the need, the vital necessity of intercourse with other men . . . . Consciousness therefore already from the very beginning is, and remains as long as men exist, a social product."[12] And: "Language itself is just as much the product of a community as, in another respect, it con-stitutes the very existence of the community, its self-expressing (*selbstredende*) existence."[13]

'Base' and 'superstrucure' designate with Marx different domains of *social practices*, i.e., of socially organized and regulated *conscious and intentional* human activities. That economic activities also are in definite, socially prefigured ways interpreted and understood by their agents themselves and that in this way 'consciousness' (definite ideas, beliefs and forms of everyday discourse of the concerned, socialized individuals) belongs to their very constitution as human actions (which are therefore always 'culturally' mediated, in the *broad* sense of this term)

– this hardly world-shattering insight was never denied or doubted by Marx. His theory of fetishism actually depends upon, and elaborates it, under the specific historical conditions of a capitalist society. And all this does not invalidate the distinction between base and superstructure at all, since the latter encompasses 'culture' in quite another sense (in the narrow, value-marked sense of specific social practices like philosophy, science or art). As Adorno rightly remarked: "Also the material basis of society comprising human labour and thinking through which this labour first really becomes social labour, is culture, but this does not lessen its contrast to the superstructure."[14]

Once this transformation of the Marxian distinction in the spirit of a materialist metaphysics has been effected, a number of further consequences followed (even if these latter often also appeared independently of the former – they now became doctrinally fixed). In accordance with the notion of the sole efficacy of the material substance, the idea of the 'conditioning' of the superstructure by the economic base has been reinterpreted in the sense of a *unidirectional causal* dependence. In its result the constituents of superstructure were conceived as some kind of social epiphenomena, and in particular the very meaning of cultural objectivations was reduced to some pre-set economic interest of a class (the 'sociological equivalent') which they – in direct or, more usually, veiled form – express.[15] This was the tendency that Engels in his late correspondence (even if with inadequate conceptual means) so much tried to counteract. Of course, within its framework it becomes impossible to account for the very idea of a critical theory of society as part and parcel of a social practice aimed at its radical transformation – the notion of a political and cultural-ideological *struggle* as such becomes meaningless. To this one needs only to add that while Marx certainly claimed a separate status (that of science as opposed to ideologies) for his own theory, he never made 'truth' (scientific or of any other kind) the precondition of the historical effectivity and real social impact of some cultural formation or activity. So, after he characterized the Monetary System as the alchemy of political economy, he continues: "the illusion about its nature [the nature of money – G.M.], i.e., the fixation of one of its determinations, in the abstract, and the blindness towards the contradictions contained in it, actually gives it – behind the back of individuals – this really magical significance. In fact, it is because of this self-contradictory and therefore illusory determination, because of this abstraction, that it becomes such an enormous instrument in the real development of the social forces of production."[16]

Nevertheless, even if one discards all these accrued interpretations as outright distortions and misunderstandings, there still remain significant and not so easily dispensible difficulties with the employment of the base/superstructure metaphor and with the realisation of a research program articulated by it. The first concerns its scope of historical validity.[17] The thesis about the conditioning of the superstructure through the economic base is among the most general assertions of historical materialism. In whatever way one now understands this relation of 'conditioning', it certainly presupposes that the two concerned domains of social practices can be (at least theoretically) clearly *distinguished* from each other. This does not seem to be, however, the case except in 'classical' (essentially XIXth century) capitalist societies. For in all pre-modern societies, economy does not constitute an institutionally separated sphere of activities which would be organized by its own, *sui generis* relations and mechanisms. 'Economic' activities (i.e., activities ensuring the socially regulated production, exchange and distribution of material goods) are in all these societies, as Karl Polanyi[18] formulated it, 'embedded' in a much broader network of social transactions so that their organisation depends upon the institutions and relations of kinship, religion and/or politics. In this way what are normally regarded as elements of superstructure here are constitutive to the economic base itself. What sense does it make – to give a rough example – to speak about the determination or conditioning of the political institutions and practices by economy in a society where some of the fundamental characteristics of property relations are defined solely by the difference in the political status (e.g., between citizens and metics) among its members? Or: how to characterize the labour (the extent of which can be socially quite significant) spent by a community on the cultivation of a crop which is not consumed at all but used only for ritual-ceremonial purposes?

Questions of such type were, however, not unknown to Marx. In fact, his famous discussion of 'pre-capitalist modes of production' (more accurately: of a series of property forms characterizing them) in the *Grundrisse* is partly based upon the idea of the progressing emancipation of productive activities from their direct 'subordination' to, and restraint by, the requirements of the maintenance of a given form of social organisation which can have under definite circumstances primarily religious or political, etc., character.[19] And he draws from these facts also some methodological consequences: even labour as such, this "simplest abstraction which modern economics places at the head of its discussion and which expresses an immemorially old relation valid for all forms of society, nevertheless appears as practically true in this abstractness only as a category of the most modern society. . . . This example of labour strikingly demonstrates how even the most abstract categories, despite their validity – precisely because of their abstractness – for all epochs, are themselves nevertheless, in this abstract determination, likewise the products of historical relations, and possess their full validity only for and within these relations. Bourgeois society is the most developed and the most complex historical organisation of production. The categories which express its relations, the understanding of its structure, thereby also allow insight into the structure and relations of production of all the vanished social formations out of whose ruins and elements it built itself up, whose partly still unsurmounted remnants it continues to drag along, whose mere portents have achieved within it explicit signification, etc."[20]

Marx here clearly indicates that his own categorial framework is applicable to pre-modern societies only insofar (*cum grano salis*, he says) as they are seen from the perspective of the capitalist present, as steps in a historical evolution leading to it (and undoubtedly it is the teleological understanding of this development which allows him to claim an unrelativized superiority and 'truth' for this perspective itself). And once this is acknowledged, he is *in a sense* quite right to treat the aforementioned difficulties as 'trivial': "The fact that pre-bourgeois history, and each of its phases, also has its own *economy* and an *economic foundation* for its movement, is *au fond* the mere tautology that human life at all times rested on production, *d'une manière ou d'une autre* on *social* production, whose relations are precisely what we call economic relations."[21] I.e., the fact that in pre-capitalist society 'economy' is not *institutionally* differentiated as a separate sphere of activities does not imply that we (from our own standpoint) cannot demarcate in a reasonably sharp manner the domain of such practices. For in all forms of historical life the activities that directly bear upon the continuous reproduction of the material conditions of social existence, i.e., are immediately related to the structured provision of goods and services satisfying the socially determined (and appropriately stratified) aggregate needs within an ecologically stabile environment [22] – these activities are under rather strong system-constraints. That is, they must satisfy definite requirements of 'proportionality' (and disproportionality[23]), both in respect of the employment of the relevant natural and social resources (input-output relations) and in that of their mutual interconnection. Therefore in every society there must be social mechanisms that maintain these required relations and thereby ensure the reproduction of the 'cultural' environment, on the one hand, and the satisfaction of the socially codified needs of groups of individuals who are the bearers of appropriate practical abilities, on the other. It is just these mechanisms (socially enforced and reproduced relations between practices and their agents) that constitute the economic structure of society in the Marxian sense. And since it is possible to demarcate in such a way a domain of economy in all societies (and practitioners of an economic anthropology and economic history are also able to achieve this in relation to pre-capitalist societies in a relatively straightforward manner), the thesis of historical materialism about the 'functional primacy' of economy remains (in some weak sense) at least meaningful: there are good arguments supporting the view about both the 'limiting' and 'impacting' role of economy in respect of *all other* types of social practices. In fact this continues to be a quite fruitful idea, especially in comparative socio-historical research.

All this is, however, insufficient for the validation of the idea about the dependence of the 'superstructure' upon an economic base in respect of pre-modern societies. For in them the so-conceived economic structure will include social relations and mechanisms which are possible only because of the existence of definite forms of kinship, religious, political, etc., organisations that are in this way directly incorporated into the 'base' itself. And just these 'superstructural' factors (as Marx repeatedly underlined it, especially in respect of the so-called 'Asiatic mode of production') often exercised a considerable restraint, acted as 'barriers', upon the possible expansion and development of economy and technology. So while it is possible, due to their specific 'systematicity', to demarcate a domain of economic (or economically relevant) practices, their delimitation does not allow us in these circumstances to clearly distinguish and counterpose them to what the whole Marxist tradition (including Marx) understood by the phenomena collected under the name of 'superstructure'[24] – some of these latter are constitutive to economy and in this role co-determine the character of its whole development. I shall return to some attempts to solve this difficulty in the following. At this point I merely want to indicate that Marx himself, seemingly, never realized it and therefore left unclarified how one can maintain the general validity of the base/superstructure dichotomy in the face of it.

In respect of modern societies of a capitalist type (especially in regard to the 'classical', liberal capitalism of the XIXth century) this difficulty disappears, since the corresponding domains became institutionally sharply separated, and even some contemporary tendencies towards a growing interpenetration of economic and political institutions did not dissolve this differentiation. The *meaningfulness* of the base/superstructure metaphor is here not in question. It is, however, questionable whether the research program outlined by it retains the *critical significance* usually attributed to it in the Marxist tradition.

In the first place one should note the extremely wide and therefore abstract character of the very concept of a superstructure. It minimally encompasses so diverse phenomena as the state and the legal system, religion and the whole multiform sphere of autonomous cultural activities. This abstraction is certainly not unmotivated in Marx. From his viewpoint all these practices and institutions share the common characteristic that their very separation has its historical roots in the internal, antagonistic division of society, in the face of which they appear as compensatory representations of a universality and communality, while in actual fact they merely transpose and misrepresent the actuality of an economically based domination of some particular interests. I have underlined the unmasking, disillusioning function of these ideas in their own time. This function is, however, largely lost today. Whatever illusions continue to accrue in common consciousness to the understanding of the character and functions of the political or the cultural, they are unlikely to run along the line of their Hegelian or Romantic exaltation. What is rationally retainable from the general idea of the dependence of the superstructure upon an economic base, the weak conception of 'conditioning', as outlined earlier, no more represents a genuinely *critical* insight. That political and cultural activities are interconnected with, and in some sense and to some degree dependent upon, economic ones, that practices in the former spheres are often influenced and motivated by variously articulated group interests – in this generality it takes on the appearance of an *empirical fact* today. All depends how this fact – its character, weight and consequences – are understood and evaluated. For the articulation of this, however, the notion of superstructure does not provide an adequate theoretical vehicle – because one thing is clear: in respect of its different elements and constituents this question has to be answered in *basically different* ways.

To retain a critical edge, therefore, the metaphor of base and superstructure has to be employed in a much stronger and specific fashion than the one indicated. In respect of the sphere of culture this inevitably means a tendency towards the interpretation of the meaning-content of its objectivations as the 'reflection' or 'expression' of veiled particular (class or group) interests. This does not correspond, however, to Marx's own practice of analysis of what he considered to be works of genuine cultural significance. For while it is certainly true that he himself made use of such a method of explanation of content through its reduction to a specific configuration of particular 'external' interests, he did so only when dealing with works which he regarded as ephemeral products of a mere apologetics. It is rather conspicuous that when such cultural creations were concerned that in his own opinion represented a genuine and lasting contribution to their field (e.g., Hegel's philosophy or English classical economy) his unflinching critique was no more based on such a method (at most he used it when he explained some accomoda-tive inconsistencies in the views analyzed). He treated such 'epochal' ideologies as the theoretical expressions of a historical perspective connected with a *definite type of society*, hypostatizing by their very method, and thereby transforming into genuine constraints of thought its essential conditions of existence, and not as apologetic manifesta-tions of the concrete, momentary interests of some class or stratum in this society.[25]

Independently, however, of the question about Marx's own views and theoretical practice, such as extreme sociological reductionism is simply untenable. For by trying to establish a direct connection between *particular* works of culture and an economic base, it necessarily isolates the former from the whole context of those institutionalized practices and traditions within which they alone have meaning, and consequently it reduces the question about their meaning (through which they exercise whatever social function they have) to some pre-set and crude formula of their 'social significance'. Nobody today is theoretically defending such a position in an explicit way, though it is far from being extinct in practice. Hence the constant talk about 'relative autonomy' of superstructure, about the 'mediations' between its various constituents and the economic base, etc. These are, of course, completely legitimate qualifications. The problem is only that usually they remain defensive generalities. For to give meaning to the notion of autonomy (however relative it be), one must be capable to indicate not only *from what*, but also *to do what* is the given form of practice (or its agents) 'autonomous'. The dichotomy of base/superstructure, however, lacks precisely this ability: to specify what are the principal characteristics which constitute the various superstructural practices (and institutions) as *sui generis* practices. It not merely does not contain such specifications, it offers no conceptual means to search and to provide for them. The conception is reductionist not because it does not allow the acknowledgment of a 'retroaction' of the superstructure upon the base – it patently does – but because it can conceive this active role only in terms of the single dichotomy between *expressing versus suppressing* definite interests, promoting or hindering definite tendencies of economic development, and this is all what it can say about superstructural institutions and practices.

It seems to me therefore that the critical significance and the theoretical fruitfulness of the base/superstructure metaphor has been exhausted. Its employment today uneasily hovers between the acceptance of an open-ended research program which poses some legitimate, but today rather trivial (in the sense of being uncontested) questions and which, as one of the meaningful directions of investigation, would be underwritten by any sociologist interested in phenomena of culture, on the one hand, and a practical application operating with much stronger presuppositions that, if they were explicitly stated, no more would be regarded as defensible, on the other hand. In this respect it is rather symptomatic that serious theoretical attempts to spell out what remains valid from this conceptualisation today not only have an unmistakable defensive character, but ultimately end up with standpoints that seem to be precisely at cross-purpose with the ideas and intentions that motivated the Marxian introduction of this metaphor. To illustrate this I shall refer schematically to two such attempts which were embedded in quite different theoretical traditions and were striving to revitalize this conceptualisation primarily for the purposes of an analysis of culture, but 'culture' in its opposed (anthropological *versus* narrow, value-marked) senses: to some writings of Maurice Godelier and Raymond Williams.

Godelier departed from the reformulation of the idea of the base/superstructure dichotomy in French 'structuralist' Marxism. To meet some of the earlier indicated difficulties, Althusser proposed a strict distinction between the notions of the 'determinant role of economy in the last instance' and the 'dominant or principal role' of a definite type of social contradiction articulated primarily within some specific level or 'instance' of society (meaning by this latter a relatively autonomous institutionalized domain of social relations and practices like economy, politics, ideology, etc.). Economy constitutes the 'basis' of all social formations only in the sense that it determines which of the various instances can and does function as the dominant one in a given social formation – to presuppose that this latter role is invariably fulfilled by the instance of economy itself is an untenable 'economism'.[26]

This solution, however, is clearly unsatisfactory for an anthropologist like Godelier who is concerned with pre-capitalist social formations. For leaving aside the problem that *the way* economy allegedly 'selects' and 'places' into a dominant position this or that societal instance remains utterly mysterious in Althusser (at best it has the character of an *a priori* presupposition), this reconceptualisation presupposes that one can unambiguously distinguish economy from the other social instances of politics, religion, ideology, etc. What to do, however, in cases (which constitute the rule in pre-capitalist societies) where the domain of economy *internally incorporates* relations and institutions belonging to the 'superstructural instances' of kinship, religion or politics? To answer this question Godelier reinterprets the very meaning of the base/superstructure distinction: it establishes a relation not between various societal instances, levels or institutions, but distinguishes between different *functions* that can be fulfilled by a single institution. So, e.g., kinship relations in a society like the traditional one of the Australian Aborigines, establish not only the rules of marriage, residence, etc., but simultaneously function as direct relations of production, since "consanguineal and affinal relations form the cooperative framework for hunting and gathering and for the distribution of produce".[27] "It is only in certain societies, and particularly in capitalist society, that this distinction between functions happens to coincide with a distinction between institutions."[28]

This interpretation, in my view, correctly brings into focus the fact that under pre-capitalist conditions one can speak about a 'sphere' or 'instance' of economy only in the sense of a functional (sub)system. It also allows Godelier to reconcile formally the Marxian idea of an economic base with the apparent facts of a 'superstructural domination' in pre-modern societies: some institutions or relations of superstructure can acquire a dominant role if and when they, beside their explicit function, also play the role of direct relations of production.[29]

> For a social activity – and with it its corresponding and organizing ideas and institutions – to play a dominant role in the functioning and evolution of society, and hence in the thought and action of the groups and individuals composing it, *it is not enough* for this activity to fulfill several functions; it must necessarily, in addition to its own ostensible purpose and its explicit functions, function *directly and internally as a relation of production*.[30]

There is, however, a rather heavy price to pay for the *verbal* retention of the idea of an 'economic determination in the last instance'. This latter means for Godelier the existence of a universal hierarchy of functions (as to their weight in the organisation and development of the society) and the hypothesis that the dominant role in this hierarchy is always occupied by those institutions which simultaneously play the role of relations of production.[31] The implication is that the actual development of pre-modern societies is primarily determined by the contradictions in those 'superstructural' institutions and relations which occupy the dominant role in it. So Godelier is seemingly quite happy to attribute the dominant role in Indian development to the religious ideas about purity and impurity – a view which Marx specifically makes fun of in *The German Ideology*.[32] In general, it seems that in his view the originating causes of social change, insofar as pre-modern societies are concerned, are primarily to be sought in the development of the *ideatory-interpretative* aspect of the dominant relations and institutions.[33] Thus the basic factor in the transition from classless to class society is the evolution of religious representations concerning invisible realities and forces controlling (in the thought of these societies) the reproduction of the universe and life:

> the monopoly of the means (to us imaginary) of reproduction of the universe and of life must have preceded the monopoly of the visible material means of production, i.e., of those means which everyone could and had to produce in order to reproduce, given their relative simplicity. . . . [T]he religious sphere developed in such a way as to bring about the establishment of stabilized social hierarchies, of aristocracies, thus creating the conditions for the extraction of supplementary labour from the common people.[34] . . . [T]he origins of the State lie in the world of the sacred . . . [35]

So the very sophisticated defense of the base/superstructure dichotomy in Marx ends up with a standpoint that hardly can be characterized otherwise as that of a 'culturological idealism' – a view which certainly has nothing to do with Marx's own fundamental intentions and which does not seem to me – as a *general* hypothesis concerning the mechanisms and course of social evolution – to have more merits than the (equally untenable) theories of an 'economic determinism'.

Questions concerning 'primitive' societies and non-Western civilisations are rather far removed from the concerns and interests of Raymond Williams, the doyen of critical cultural studies in Britain. His work primarily dealt with the development of the modern European drama and novel (and was partly informed by the rich empirical traditions of English Marxist historiography). So when in the early seventies he raised the question about the meaning and significance of the theory of determining base and determined superstructure for a Marxist cultural theory,[36] he understandably meant culture in its narrow, value-marked sense.

At the very point of critical departure his views, however, coincide with those of Godelier: he equally opposes the understanding of the so-designated relationship as any sort of fixed 'correspondence' between two statically conceived domains of 'objects'. His emphasis is, however, on the *processual* historical character of the interrelation and interaction between differently constituted social *practices* as significant responses to an always particular objective situation:

> We have to revalue 'superstructure' towards a related range of cultural practices, and away from a reflected, reproduced or specifically dependent content. And, crucially, we have to revalue the 'base' away from the notion of a fixed economic or technological abstraction, and towards the specific activities of men in real social and economic relationships, containing fundamental contradictions and therefore always in a state of dynamic process.[37]

'Determination' then means the general idea of the setting of limits and the exertion of pressures by a range of practices, with their characteristic contradictions, upon other ones (by being constituents of the objective situation to which these latter react) – a broad methodological guideline whose meaning and validity can only be found out in each concrete case by the investigation of the appropriate, singular historical process.

Such an interpretation is, however (so it seems), also for Williams too broad and vague, opening the way to crass historical empiricism. To counterbalance this danger he introduces – with a reference to Lukacs – the notion of social totality as restriction upon the particularity, and therefore accidentality, of historical processes of change. But this concept again proves to be in itself inadequate, or at least insufficient, since it now creates an opposed danger, that of "withdrawing from the claim that there is any process of determination".[38] Ultimately Williams finds the solution in the Gramscian concept of 'hegemony'. Through processes of education and wider social training, through the organisation of work and leisure, through the selection of tradition, a central system of meanings, values and related cultural practices is created and reproduced which 'ratifies' the existing relations of domination not as some imposed set of abstract notions, but by saturating the consciousness, constituting the framework of the ordinary understanding of lived experience and of practical expectations which are then confirmed by practice. This hegemony is never singular or static, it is active and adjusting, continually renewed, recreated and defended. In its dynamic complexity it allows new experiences to emerge, by which potentially it can be challenged, but as a rule it succeeds in incorporating and thereby neutralizing them.

The *sui generis* cultural practices in large part accept, and thereby reinforce, this dominant meaning- and value-system: "whatever the degree of internal controversy and variation, they do not in the end exceed the limits of the central corporate definitions".[39] But cultural life is never reducible to dominant culture alone, it always contains alternative and oppositional, both residual and emergent, forms and elements. How far these can be incorporated by the dominant one, what is their social significance and effectivity, etc. – all this depends on the concrete conjunction of the actual social and political forces. Theoretically, however, for a *critical* theory of culture, the decisive question concerns the *sources* of emergent practices and meanings. "[O]ur hardest task, theoretically, is to find a non-metaphysical and non-subjectivist explanation of emergent cultural practice".[40]

Williams's answer to this question consists of two parts. On the one hand, very early on he makes a certainly emphatic and rather curious remark:

> [T]he key question to ask about any notion of totality in cultural theory is this: whether the notion of totality includes the notion of intention. . . . For while it is true that any society is a complex whole of . . . practices, it is also true that any society has a specific organization, a specific structure, and that the principles of this organization and structure can be seen as directly related to certain social intentions, intentions by which we define the society, intentions which in all our experience have been the rule of a particular class.[41]

Since no more is forthcoming, it is rather difficult to decide how to understand the notion of this class-bound social intention. But the idea must be taken seriously, since it offers the only way to interconnect with the main body of the argumentation his (otherwise completely unelaborated) reference to the emergence of a new class and its coming to consciousness as a source of new practices and meanings. Without it this seems to be just a polite gesture to Marxist orthodoxy. In any case it is the second point made in this context that receives more attention: since the dominant order of society always selects a definite range of practices and meanings, it by necessity also neglects and excludes other ones emerging outside this dominant mode. These latter emerge, so it seems, partly spontaneously, in immediate personal relationships, partly in particular fields, like science and art, which have a specifically innovative character.[42] So if the first answer is rather enigmatic, the second is tautological: it transforms the question itself into the solution – new meanings and practices emerge, because they always do so.

Whatever be the intrinsic value of this rudimentary analysis, it is clear that it again has nothing to do with the original conception of base and superstructure. Neither social intentions somehow ascribed to classes, nor the inexhaustibility of human interactions, nor the specifically innovatory character of some forms of cultural practice fits into this scheme, however liberally interpreted. So Williams's paper also ends in an impasse: from the viewpoint of the fundamental question concerning the explanation of cultural change, the Marxian dichotomy, whose contemporary meaning it set out to explicate, practically proves to be of no use. Actually one of the most interesting points about the paper consists in the fact that in the very course of its argumentation Williams moves away from the paradigm he set out to explicate and defend. For when he concludes it with the postulate: "we should look not for the components of a product but for the conditions of a practice",[43] he *de facto* signals the transition to another way of conceptualizing the phenomena of culture equally originating with Marx: their understanding as *specific forms of social production.* In his late writings[44] Williams attempts – through the notions of *sui generis* cultural forces and relations of production – to extend the Marxian paradigm of production to the sphere of culture itself and takes a decisive farewell from the metaphor of basis and superstructure: "We should not assume in advance that the basic structural relations between different kinds of activity are themselves ahistorical, yielding regular uniformities and laws which can then be applied to any specific social and historical situation."[45]

The University of Sydney

NOTES

[1] Being part of a larger work dealing with Marxist theories of culture, this paper is concerned primarily with the adequacy of the conceptualisation offered by the basis/superstructure metaphor in its application to cultural phenomena. The points raised by it, however, seem to me to be valid in respect of its broader, i.e., political and legal, application as well.

[2] E.g., "With the change of the economic foundation the entire immense superstructure is less or more rapidly transformed" (Marx-Engels: *Werke*, Berlin, Dietz [in the following: *MEW*] Vol. 13, p. 9).

[3] See especially Marx: *Grundrisse der Kritik der politischen Ökonomie*. Berlin, Dietz, 1953, pp. 29–31.

[4] *MEW*, Vol. 13, p. 9. But see also Marx's remarks about the natural sciences, e.g., "Even this 'pure' natural science receives its aim as well as its material primarily through commerce and industry, through the sensuous activity of men" (*MEW*, Vol. 3, p. 44).

[5] See Marx: *Grundrisse*, pp. 152–160 and 910–915. Some of Marx's related formulations seem to come close to the Weberian idea of 'elective affinities', see, e.g., his remarks about the relationship between Christianity and capitalism in *MEW*, Vol. 26/3, pp. 441–442.

[6] "We know only a single science, the science of history . . . Ideology itself is only one of the aspects of this history". And: "There is no history of politics, law, science etc., of art, religion etc." (*MEW*, Vol. 3, pp. 19 and 539).

[7] It also should be pointed out that in the first comprehensive exposition of a materialist understanding of history, in *The German Ideology*, Marx identifies the premisses of his approach: the concrete individuals producing their life under definite material conditions and in a definite form of intercourse with each other, with premisses that can be established strictly by way of *empirical observation* (*MEW*, Vol. 3, pp. 20, 25, etc.). This naive empiricist positivism, however, is completely absent from Marx's mature reflections on the method of his economical analysis (Cf. Marx: *Grundrisse*, pp. 21–29).

[8] See *MEW*, vol 23, p. 193. This idea (in fact of Hegelian origin) is, however, already present in the earliest analysis of labour by Marx, in the Paris manuscripts of 1844 (cf. Marx-Engels: *Gesamtausgabe* [in the following: *MEGA*]. Berlin, Dietz. Part I, vol. 2, pp. 239–241), and is consistently repeated, in one or another formulation, in all his relevant writings.

[9] *MEGA*, Part II, vol. 3/1, p. 63.

[10] Cf. Marx: *Grundrisse*, pp. 12–13. See also *MEGA*. Part II, vol. 3/1, pp. 54–55.

[11] *MEGA*. Part II, vol. 3/1, p. 210.

[12] *MEW*, vol. 3, pp. 30–31. See also: "To compare money with language is no less erroneous. Ideas are not transformed into language in a way which would dissolve their specificity and give their social character an existence alongside themselves, like prices exist alongside commodities. Ideas do not exist separately from language." (*Grundrisse*, p. 80)

[13] *Ibid.*, p. 390.

[14] Adorno, T. W: *Jargon der Eigentlichkeit*. Frankfurt, Suhrkamp, 1964, p. 83.

[15] This whole 'theory' is undoubtedly so crude and so elementarily contradicted by the simplest reflections upon history and social life that it never – not even in the period

of Stalinism – has been consistently maintained. But this merely meant the acceptance of *ad hoc* qualifications when the absurdity of its consequences became too evident.

[16] Marx: *Grundrisse*, pp. 136–137.

[17] This problem has been recognized already by Lukacs, in the early twenties. See Lukacs, G: *Geschichte und Klassenbewusstsein.* (*Werke*, Vol. 2) Neuwied, Luchterhand, 1974, esp. pp. 405–408.

[18] See Polanyi, K: *Primitive, Archaic and Modern Economies*. Garden City, Doubleday. Part I.

[19] Marx: *Grundrisse*, pp. 375–388, 393–395, etc.

[20] *Ibid.*, p. 25–26.

[21] *Ibid.*, p. 388. Cf. also *MEW*, vol. 23, p. 96 fn.

[22] This characterisation is applicable only to pre-modern societies, but we are concerned here only with them. One of the most fundamental features of capitalist modernity consists in the fact that its economic mechanisms make both possible and necessary a constant technical progress that allows the satisfaction of a dynamic system of needs under (on the long run) *unstable* ecological conditions.

[23] On this latter, important point see *MEW*, Vol. 24, p. 475. The general idea of these 'material', i.e., independent from the particular character of their social organisation, system-constraints of economic activities was broadly formulated by Marx in his letter to Kugelmann, from 11. 7. 1868 – see *Ibid.*, Vol. 32, p. 552).

[24] Of course, the characterisation of these institutions, practices, etc., as religious or political, again, belongs to us – in most cases 'native' taxonomies do not operate with such categories, they classify their practices otherwise. But all the time we are dealing here with our understanding of these societies – the notion of economy as outlined above is again a theoretical term in our vocabulary.

[25] In greater detail see my paper 'Concepts of ideology in Marx'. *Canadian Journal of Politics and Social Theory* 7 (1983) No. 1–2.

[26] Althusser, L: *For Marx*. Harmondsworth, Penguin, 1969, pp. 99–116, 200–216. – This construction relied primarily on a single formulation of Marx in the first volume of *Capital*. There he criticized the view restricting the 'foundational' role of economy to modern capitalist societies alone "where material interests are preponderant", in counter-distinction to the Middle Ages, allegedly dominated by Catholicism, or to Athens and Rome, dominated by politics. In his rejoinder Marx stated: "At least one thing is clear: the Middle Ages could not live on Catholicism, nor could the ancient world on politics. On the contrary, it is the manner in which they gained their livelihood which explains why in one case politics, in the other Catholicism, *played the main role*." (*MEW*, vol. 23, p. 96 fn. – emphasis mine, G.M.) – It is worthwhile to compare Althusser's solution with the way J.-P. Vernant explicates, in respect of the classical *polis*, the same formulation of Marx: "The conflicts which towards the fourth century involved the different social groups within the framework of the city-state, were neither baseless nor purely ideological: they were rooted in the economy of these societies. Human groups came into conflict because of their material interests. But these material interests did not derive either directly or exclusively from the position the individuals occupied in the process of production. They always stemmed from the position which these individuals occupied in political life, which played the dominant role in the system of the *polis*. In other words, the economic function of different individuals – which determined their material

interests, fashioned their social needs and oriented their social and political behaviour in alliance with or in opposition to other groups – was mediated via political status." (Vernant, J.-P: *Myth and Society in Ancient Greece*. London, Methuen, 1980, p. 10)

[27] Godelier, M: 'Infrastructures, Societies and History'. *Current Anthropology*, 19 (1978), No. 4, p. 764.

[28] *Ibid.*, p. 765.

[29] There are isolated formulations in Marx which suggest that such a solution was not alien to him, either. Cf. "If man has made his relation to his own nature, to the external nature and to other men independent in *religious form*, so that he is dominated by these representations, then he needs the *priests* and their *labour*. However, with the disappearence of the religious form of consciousness and its relations, this labour of the priests also ceases to be incorporated into the social process of production". (*MEW*, Vol. 26/3, pp. 486–487)

[30] Godelier, M.: *Op. cit.*, p. 765. This formulation is then almost verbatim repeated in Godelier's later book. Cf. *The Material and the Mental*. London, Verso, 1986, p. 147.

[31] Even in itself this hypothesis, as it is stated, seems to me rather useless for concrete research, in view of the *segmented and institutionally heterogeneous* nature of most pre-modern economies. So according to Godelier in ancient Mesopotamia religious relations played the dominant role, since they functioned directly as relations of production. Now there can be no doubt that religious institutions, practices and representations were internally embedded in the functioning of a Sumerian temple-economy. But this economy existed, and could subsist only, in a close symbiosis with the 'private economy' of the city-dwellers which largely functioned within the framework of *sui generis economic* institutions ('banking houses', civil guilds, etc.), in a constant interaction with the pastoralist nomads of the countryside whose economic activities were organized along *kinship* lines, and in dependence on a long-distance foreign trade which was primarily influenced by *political* factors. (On all this see: Oppenheim, A. L: *Ancient Mesopotamia*. Chicago U.P., 1964, chapt. 2) Given the fact that all these relations and institutions were internally necessary for the functioning of this particular system of production as a whole – what determines then the hierarchic relations *between them*?

[32] *MEW*, Vol. 3, p. 39.

[33] Godelier – and on this point I have no dispute with him – underlines: "[A]ll social relations *arise and exist simultaneously both in thought and outside of it* – . . . all social relations contain, from the outset, an *ideal* element which is not an *a posteriori* reflection of it, but a condition for its emergence and ultimately an essential component." ('Infrastructures . . .', p. 766) This point, however, he couples with the following hypothesis: ". . . for relations of domination and exploitation to be formed and reproduced in a lasting fashion, they must be presented as an exchange, and as an exchange of services. . . .[A]mong the factors which, in the course of history, have effected the internal differentiation of social statuses and the formation of hierarchies founded upon divisions into estates, castes and classes, the fact that the services rendered by the dominant have been predominantly concerned with the invisible forces controlling the reproduction of the universe has always been crucial." (*The Mental and the Material*, p. 160)

[34] 'Infrastructures . . .', p. 467–468. See also his paper 'L'Appropriation de la nature' *La Pensée* (1976), No. 198, pp. 38–45.

[35] *The Mental and the Material*, p. 16.

[36] 'Base and superstructure in Marxist cultural theory' in Williams, R: *Problems in Materialism and Culture*. London, Verso (1980), originally published in 1973.
[37] *Op. cit.*, p. 34.
[38] *Ibid.*, p. 36.
[39] *Ibid.*, p. 40.
[40] *Ibid.*, p. 42.
[41] *Ibid.*, p. 36.
[42] *Ibid.*, pp. 43–44.
[43] *Ibid.*, p. 48.
[44] See especially *Culture*. London, Fontana, 1981.
[45] R. Williams: 'Culture', in the volume *Marx: The First 100 Years* (ed. by D. McLellan). London, Fontana, 1983, p. 41.

WILLIAM McBRIDE

THE MARXIAN VISION OF A (BETTER) POSSIBLE FUTURE: END OF A GRAND ILLUSION?

> I have changed. . . . The retrospective illusion is in smithereens; martyrdom, salvation, immortality, everything is collapsing, the building is falling into ruins. . . . I see clearly, I am disabused, I know my real duties, I certainly deserve a civics prize; since about ten years ago I have been a man who is waking up, cured of a long, bitter-sweet insanity which he can't get over, and who can't recall his former missteps without laughing and who no longer knows what to do with his life. I have once again become the [train] traveller without a ticket that I was [play-acting] at the age of seven. . . .
>
> – Sartre, *Les Mots* (1964), pp. 210–211, my translation

What do we mean by 'illusion'? Complementarily, what is disillusionment? Like Augustine concerning time, many of us may believe that we know the answers to these questions – until we start examining them more carefully. Then we become aware that the belief itself was illusory.

Sartre's generation had lived through at least three disillusionments of world-historical proportions at the time at which he wrote these words: Stalinism and its continuation or rebirth under Khrushchev,[1] the rise of Fascism/Naziism and its culmination in World War II, and, before all that, the previous shattering of European civilization during their childhoods that was World War I. It was this last-mentioned experience that Jean Renoir attempted to capture in his great film of 1937, *La Grande Illusion*. Although it must seem obvious to any viewer of the film that its title was very well chosen, the question may still be asked, in precisely *what* did the principal illusion consist? That some aristocratic military officers witnessed a final end to the dominance of their class when the war broke out was no doubt sad for them but not for others, and it hardly constitutes an illusion/disillusionment of world-historical proportions. Did any informed person living just prior to World War I *really* anticipate an era of perpetual peace, despite all the constant

C.C. Gould and R.S. Cohen (eds.), Artifacts, Representations and Social Practice, 441–461.

tensions and intermittent slaughters that kept erupting in the Balkans and elsewhere during those years – was *that* the great illusion referred to in the film's title? Perhaps, but it seems highly implausible on the face of it. To be sure, there were those who really believed, once the war itself had broken out, that it would somehow, perhaps by its sheer magnitude and ferocity, put an end to all wars.[2] (Of the two French officers in *La Grande Illusion* who successfully escape, one, Maréchal, hints at such an outcome near the end of the film, but his comrade, Rosenthal, immediately dismisses this suggestion out of hand, as an illusion.) To call such a belief an 'illusion' is to give much credit to pure stupidity:[3] an illusion must require *some* factual basis, *some* evidence in lived experience in order to be certified as a 'genuine' illusion – must it not?

The gleeful, almost manic carving up and rearrangement of Austria-Hungary, Germany, the Ottoman Empire, etc. by the victorious Allied leaders in Versailles that marked the end of that war and that is now being re-examined for clues to understanding some of the horrors of the present decade of the 1990's was followed by a truly manic decade, the 1920's, and then of course the ensuing deep depression. It was the sense of real, existential absurdity – of political, economic, artistic, etc., contradictions out of control – pervading the European between-wars world of Sartre's youth that motivated his early, absurdist 'illusions,' so brilliantly depicted in his novel, *Nausea*, and then in his first philosophical *magnum opus, Being and Nothingness* ("Later, I gaily demonstrated that man is impossible"),[4] and repudiated in the citation above. As World War II waned, he became greatly interested, in part by force of circumstances and in part through the influence of his colleague, Maurice Merleau-Ponty, in Marxist thought. The result was what Marx Wartofsky has referred to, in an important essay of his, as "the problematic Marxism of the French phenomenologists, such as Sartre and Merleau-Ponty."[5]

In this essay, which focuses on Husserlian rather than Sartrean or Merleau-Pontyan phenomenology, Wartofsky strongly suggests that the belief that it is possible successfully to conjoin Marxism and phenomenology is itself an illusion. The principal reason for this, he concludes, is phenomenology's "passivity and ahistoricity," which "are in effect functions of its divorce from historical praxis."[6] By contrast, Wartofsky insists, Karl Marx stressed the connection between theory and praxis, which undermines philosophy's traditional self-conception (still retained by Husserl). Paradoxically, however,

> the very [Marxian] critique of philosophy which argues for its transformation into a theorized praxis is still philosophy. But it is philosophy disabused – 'disillusioned,' as Marx liked to say, no longer caught in its own self-mystification.[7]

Feelings of illusion, disabusing, and disillusionment, then (whatever the exact meanings of these terms may be), seem to abound, and one has good reason to ask oneself whether there will ever be an end to it all. In any event, the shattered *grande illusion* being featured at the present time has to do, *à son tour*, with the dramatic, seemingly final 'failure' and collapse of the political regimes that called themselves Communist and that claimed the philosophy of Marx as their own. Of course, there is a relative aspect even to the charge of 'failure.' Yugoslavia under Titoism was arguably better off on the whole than the devastated region pockmarked with killing fields that it (at least large parts of it) has since become. There are many grounds for viewing Russia during the Commonwealth period as worse off than the old USSR. And so on. But enough! It remains the case that the aforementioned disillusionment is vast.

These regimes and these philosophical ideas have, over the years, been objects of great interest both to Marx Wartofsky and to me, as they were to Sartre. It is possible to retrace, through Sartre's writings, his own gradual abandonment of the view, to which he still adhered in the late 1950's when he was writing his *Critique of Dialectical Reason* and working on his autobiography, that the Soviet Union and the nations within its orbit constituted a 'deviation' or 'detour,' albeit a truly massive one, on the road to a future socialist society. (Indeed, this is a central theme of the posthumously-published second volume of the *Critique*, which focuses above all on Stalinist Russia during the mid-1930's.[8]) It is also significant to note that, in addition, he eventually renounced for himself the 'Marxist' label, which he had at one time affixed not only to himself but indeed to an entire epoch.[9] Should we say, as it would certainly be facile to say, that he finally concluded that the related expectations and beliefs had constituted just one more illusion?

To come closer to home, should we say this about our own, respective (and no doubt somewhat divergent) expectations, mine and Marx Wartofsky's? I entitled the penultimate chapter of my 1977 book on

Marx's philosophy 'Vision of a Possible Future'[10]; it is this formula that I have invoked in the title of the present essay. I also, to take one especially salient example among many possible ones, contributed an essay entitled 'Marković's Philosophy and the Spirit of Community' to a *Festschrift* for the outstanding Yugoslav '*Praxis* philosopher,' Mihailo Marković, that was published in Belgrade only a few years ago, in 1987; despite all the government persecution that Marković and a number of his colleagues had endured, and despite the serious political difficulties that the country was already experiencing at the time, this essay exudes a certain sense of optimism.[11] Marx Wartofsky's paper, 'Matter, Action and Interaction,'[12] later reprinted, which portrays conceptual scientific activity as "a (relatively isolated) subsystem of the material world" and hence as an instantiation of "self-active matter," was originally given at the only World Congress of Philosophy to be held in a Soviet-bloc country, the XV World Congress in Varna, Bulgaria, in 1973. At the time, there still appeared to be good reasons to believe that a freeing up of intellectual life might soon begin to take place or might already have begun to take place in that part of the world, and that this Congress might herald or even facilitate such a development. A few years later, in 1979, Wartofsky and Robert S. Cohen prefaced the collection of essays by Yugoslav 'Praxis' writers, edited by Marković and Gajo Petrović, in their *Boston Studies in the Philosophy of Science* series with expressions of highest praise, admiration, and, in the face of the political obstacles that I have already mentioned, encouragement for this entire philosophical movement.[13]

What Sartre called, in the citation with which I began, "the *retrospective* illusion" refers to the self-image that he had once (and for many years) had of an heroic bearer of destiny, who would be seen as such from the standpoint of some future time. (In his childish play-acting about the train trip, for example, he began by admitting to the conductor that he had indeed boarded without a ticket, but then asserted that he was on a secret journey to Dijon that was of the highest importance for the future of France and perhaps humanity itself, and thus in fact more fully entitled to be there than any other passenger.) When I reflect upon his admission that this illusion is "in smithereens" (*en miettes*), I cannot help but draw some analogy with the many writings and events of the past quarter century connected with Marxism and the various attempts to build socialism, such as those to which I alluded in the previous paragraph, that seemed to hold out the possibility of overcoming enormous obstacles for the sake of an heroic destiny to which all humanity was potentially heir. Much of that, at least on the side of *praxis* if not that of theory, seems to be *en miettes* both figuratively and, often enough, literally – both in the present and retrospectively. Research has shown, for example, that younger Bulgarian philosophers were virtually excluded from the Varna World Congress,[14] an ominous anticipation of the enormous generation gap to come. Dubrovnik, where Sartre himself vacationed, and at the Inter-University Centre at which so many of the most important encounters in more recent years between Yugoslav philosophers and those from Western Europe took place, has been severely damaged by bombardment, some of the loveliest historic edifices smashed and *en miettes*, the Centre itself burned to the ground. The very country, as it formerly existed, has been splintered and fragmented. The same can be said of many of the relationships, once so cordial, that obtained among intellectuals there as well as elsewhere in Eastern Europe.

But was it all in fact an illusion from the start? Were the visions, largely but by no means exclusively nourished by the Marxian tradition, of those who, *apparently* in good faith, wrote and acted to try to build a better social alternative to a world driven and determined by impersonal 'free market' economic forces really without substance from the start? Are they, in any event, without substance now? I would like to make a partial contribution towards a negative answer to these questions by touching on five areas about which Marxian thought has been believed to have important things to say: scientific truth, *praxis*, religion, history, and community. I shall draw upon a few of the writings of Marx Wartofsky, as well as some Marxian and Sartrean ideas and my own and others' insights, in formulating these replies.

(1) *Scientific truth.* The so-called 'orthodox' Marxism of the countries of the Soviet Bloc placed enormous emphasis on the idea that there was one Monolithic Truth, embodied in Marxism, that was completely in accord with the findings of modern natural science. A siege mentality was encouraged, from the standpoint of which enemy schools of philosophy were attacked. While it was considered permissible, by the guardians of this orthodoxy, for philosophers to articulate ways in which philosophy and science supposedly interacted at a theoretical level, and to attempt to resolve apparent anomalies and aporias (e.g., the exact nature and status of the *social* sciences) within this worldview, and while

this situation actually stimulated some interest in the philosophy of science in many of the countries in question, theoretical exploration was always constrained by the threat of incurring anathemata from these same guardians. Indeed, this actually occurred often enough.[15] It is this state of affairs to which I am referring in my shorthand expression, 'scientific truth.'

Long before the collapse of the regimes in question, a general mood of skepticism and eclecticism, which can conveniently be assigned the vague label of 'postmodernism' without entering into the details of differences among different thinkers or theoretical problems concerning the label itself, had begun to sweep intellectual circles in the West, where rigid Marxist orthodoxy had never claimed very many adherents anyway. Versions of this mood have gone so far as to question not only dogmatic systems of that extremely rigid sort, however, but also the very idea of 'totalizing thinking' about reality, which is frequently identified with metaphysics. Marxism, even in its 'unorthodox' versions, has often enough been taken to be a salient example of such thinking; indeed, a number of the most prominent postmodernists, at least during the first wave, had begun their careers having a considerable sympathy for some Marxian ideas. They eventually became, as it is said, 'disillusioned.'

Why? Each individual case is undoubtedly different, but for many, if not most, the stagnant and ultimately – if I may be permitted the indulgence of using this adjective without having to spell out its significance in this context – reactionary character of the political regimes that clutched the 'Marxist' banner must have had some psychological effect. The ideological use to which they put the notion of 'truth,' and their close identification of truth with rigid, scientifically-based determinism, made the entire package extremely distasteful to many. These considerations undoubtedly help explain Sartre's assertion, late in his career, that he no longer considered himself a Marxist, an assertion which I have discussed elsewhere.[16] (Sartre is generally regarded as one of the figures against whom the beginnings of postmodernism were directed, but I have also attempted, elsewhere, to show that this is at best only partly and superficially valid, since his later work anticipates some of the main theoretical directions taken by his successors.[17]) More philosophically interesting than the 'why' question is that of whether the wholesale rejection of the entire Marxian tradition that took place and is continuing to take place among some postmodern thinkers is truly justified.

A negative answer to the latter question should include both historical and conceptual components. Historically speaking, there is the whole complex question of the 'spin' that was given to Marx's thought, in the direction of a rigidly deterministic worldview, by certain interpretations of Engels' interpretations of Marxism that became dominant for many decades early in this century and that continued to dominate Soviet Marxism until its demise. Lenin's 'reflection theory of cognition,' understood as photographically representational of reality, epitomizes this.[18] On the other hand, all or nearly all of the principal figures of what has come to be called 'Western Marxism,' as well as (or including, depending on one's lines of demarcation) the Yugoslav *Praxis* thinkers, simply resisted this line of interpretation.

Conceptually, we can derive a great deal of value, in showing the untenability of believing that the 'orthodox' conception of scientific truth is the sole possible conception compatible with Marx's philosophy, from Marx Wartofsky's extensive writings over the years in the philosophy of science. While this is by no means my own area of specialization, I have profited from his extensive discussion of the use of *models* in scientific understanding, models that can and ultimately must be creative rather than rigidly representational in the Leninist or 'mirror of nature' sense upon which some writers have capitalized in order to try to bring the very idea of truth-oriented representation into philosophical disrepute.[19] I have also been impressed anew by a re-reading of his essay on Marx's labor theory of value (which I heard him present originally at an American Philosophical Association symposium commemorating the centenary of Karl Marx's death), in which he expresses agreement with Engels' superficially surprising assessment of that theory as Marx's most significant contribution to science.[20] From the essay on models and from this article on the labor theory as explanatory of what needed most to be explained at the time of its formulation, namely, capitalism's massive *exploitation* of human beings, there emerges a living, vibrant, somewhat (but not infinitely) flexible notion of scientific truth that is light-years distant from that of Marxist 'orthodoxy' in its day and that nevertheless owes much to the Marxian tradition. Wartofsky has always insisted, in these essays and elsewhere, on the impossibility of completely jettisoning metaphysics, properly understood as systematic philosophic thinking deploying categories not necessarily identical with the entities featured in the language of the scientific disciplines themselves, when attempting to explain anything within the realm of material reality –

be it a process studied by the natural sciences or a phenomenon in the domain of psychology.[21] One of the key *explanantes*, in Wartofsky's world of scientific models that explain the past and present and give us clues to the future, is, precisely, *praxis*.

(2) *Praxis*. *Praxis*, or practice, has been central to a number of alternative Marxisms as well as to the thought of Marx Wartofsky, and it therefore has its part to play in the phenomenon of disillusionment that I have been attempting to depict and to comprehend. Because of the central role played by the word 'praxis' in Marx's early writings, particularly his *Theses on Feuerbach*, it became to some extent, as I have put it elsewhere,

> a code-word for an interpretation of Marxism that rejects the allegedly more deterministic variety identified with Soviet 'orthodoxy'; to stress praxis implies viewing Marxism as an ethical doctrine advocating human action for radical social change.[22]

However, even the 'orthodox' Marxists always made much of the idea that Communism was "the union of theory and practice." (When writing in English, they generally preferred the common English word to the Greco-German term that Wartofsky and many others, in order to highlight its somewhat technical meaning and use, have favored.) They, too, saw it as supporting radical social change, in accordance with Marx's famous final dictum in the *Theses* that "the point is, to *change* it" (i.e., to change the world), except in those parts of the world where regimes that they supported held sway.

Disillusionment in this regard has its source, obviously, in just this capacity of *praxis*-talk to become a vehicle for self-adulation, for elaborate rationalizations of regimes and institutional practices, alleged to epitomize the sought-for union of theory and practice, that are caricatures or even outright repudiations of Marx's and others' initial intentions in invoking the notion of *praxis*. I myself have expressed this sense of disillusionment in the past,[23] and could today invoke numerous new examples from the case histories of philosophers, including '*Praxis* philosophers,' who have entered into the past few years into political arenas in Central and Eastern Europe, by way of documenting the pitfalls of trying to turn theory into practice. "*Wenn*," as Hegel wrote wisely and presciently,

> *wir dieses Schauspiel der Leidenschaften betrachten und die Folgen ihrer Gewalttätigkeit, des Unverstandes, der sich nicht nur zu ihnen, sondern selbst auch*

> *und sogar vornehmlich zu dem, was gute Absichten, rechtliche Zwecke sind, gesellt, in der Geschichte uns vor Augen halten, . . . wenn wir auf die Individuen mit tiefstem Mitleid ihres namenlosen Jammers [blicken], so können wir nur mit Trauer über diese Vergänglichkeit überhaupt, und indem dieses Untergehen nicht nur ein Werk der Natur, sondern des Willens der Menschen [ist], noch mehr mit moralischer Trauer, mit der Empörung des guten Geists, wenn ein solcher in uns ist, über solches Schauspiel enden.*[24]

Here Hegel, to whom, more than to anyone else, Marx and we owe the modern understanding of the significance of *praxis*, is certainly not falling prey to the temptation, to which he sometimes succumbs, to gloss over the dark, even murderous side of historico-political reality. Instead, he is confronting it, as the more saccharine versions of the "union of theory and practice" idea fail to do. 'The Dilemma of Philosophy and Politics,' as Marx Wartofsky has so well denominated it in a relatively early essay of his, cannot be conjured away; it is a part of the human condition; and the facile tendency to blame, to 'guilt-trip' politically or socially involved philosophers, from Plato in Syracuse to those caught in the current morass, must be set aside, or at the very least postponed, in favor of trying to understand actual situations in all their ambiguities. Instead of rejecting outright, then, the analysis of *praxis* to which the Marxian tradition has made such important contributions, we ought to recognize the *enormous* value of the concept, which illuminates and helps us to live the at times almost unbearable human dilemma. As Wartofsky puts it in concluding his essay, the dilemma

> does not resolve itself by the sweet work of dialectical reason. It can be resolved, if it is resolvable at all, only in the dialectic of *praxis*, in that fusion of thought and action, of political theory and political praxis which guarantees the sanctity of nothing but human striving.[25]

If, however, it is the case that nothing is guaranteed *sacred* except human striving, what place, if any, is left for religion?

(3) *Religion.* Karl Marx's dismissal of religion as the 'opium of the people' is famous, as is the anti-religious stance adopted, with greater or lesser intensity depending on time and place, by Communist regimes over the decades. Officially and on the surface, the societies dominated by these regimes were thoroughly secular, at best sanctioning continued religious practice on the part of those generally older and less well informed segments of the population who might be thought, realisti-

cally speaking, incapable of changing their ways. It is equally plain and well-known at this point that religious symbols and beliefs have come to the fore over the past few years in most if not all of these countries: intensifying the divisions among Serbs, Croats, and Muslims in former Yugoslavia (where the three-fingered Orthodox fashion of symbolically representing a cross has become a very important Serbian nationalist gesture); serving, through traditional feasts and rituals, to counter the pain and humiliation experienced by so many Russians as their once-powerful nation has collapsed and virtually dissolved; dictating fiercely repressive anti-abortion legislation in Poland; and so on.

In fact, the relationship between the Marxian tradition and religion was never quite so thoroughly antagonistic as the 'opium of the people' remark might have led one to believe. (Indeed, the remark itself, when analyzed carefully, is not thoroughly antagonistic. Opium relieves the pain of those who, unlike the ordinary people of Marx's time, can afford it and thus plays some positive role in a hostile world. It may be said that it does so in an illusory way, and that this was the point about religion that Marx intended with his analogy. But here again, I would like to suggest, the precise meaning of 'illusion' is itself elusive.) There have been many efforts at 'dialogue' between Marxists and religious people, especially Christians, and no one seriously doubts the importance of Marxian influences in the evolution of liberation theologies and philosophies particularly in Latin America. Moreover, even many Marxian thinkers without formal religious ties or, at least as far as their public utterances have indicated, beliefs have placed a great deal of emphasis on what some have not hesitated to call the 'spiritual' aspect of human life; *praxis*-oriented philosophers have been especially insistent on this.[26] Marxian criticism of the alienation of human values under capitalism has frequently identified this latter phenomenon as 'materialism,' this word being understood in its ordinary-language sense; on the level of the explicit language used, such Marxian denunciations have sometimes been indistinguishable from similar passages in Papal Encyclicals.

There is, obviously, the possibility of enormous confusion, to say nothing of sheer hypocrisy and/or deception, in this area. Nevertheless, it should be possible to draw some lines – should it not? When Marx Wartofsky, for instance, in the introductory section of his magisterial study of Ludwig Feuerbach, criticizes most philosophical humanism as "bland, uncritical, and too often sentimental"[27] and identifies his own position as one of (nonreductionist) "critical materialism,"[28] he establishes a clear, self-consistent position that it would be impossible to identify, at least without introducing a language reform of revolutionary proportions, as a religion *cachée*. But then we must recall the strange case of Feuerbach himself, Wartofsky's subject-matter, who in his most influential work treated his own humanism as a distillation of 'the essence of Christianity.' True, Feuerbach's thinking evolved in a more thoroughly secular direction in his later work, but a good case can be made for contending, as Wartofsky does, that that work was less creative. What Feuerbach should have introduced but failed to, as Wartofsky points out toward the end of his study, was a new model of creative praxis[29] – a lacuna that was presumably filled, in large measure, by Karl Marx, as the *Theses on Feuerbach* portend. The difficulty – or the opportunity, depending on one's point of view – is that the source of the ur-idea of creative *praxis*, as Wartofsky makes clear in this context, and as so many other writers and thinkers within the vast and sometimes bewilderingly diverse literature of Marx-interpretation have also pointed out,[30] is, precisely, religious thought.

If it is nothing else, Marxism is dialectical. The dialectician – Marxian, Hegelian, Sartrean, Wartofskyan, or whatever – scorns all versions of positivism, which tries to limit thinking to the single dimension of the here-and-now. By contrast, dialectics points to the *transcendence* of the here-and-now, in various forms. But these forms in their variety – from Transcendence with a capital 't' to transcendence as the Spirit of History to transcendence as creative *praxis* or 'human striving' towards a different and better possible future – have a disconcerting tendency to engage in a sort of transvestism of the spirit: perpetual transmutations from one to the other that contain, logically speaking, no internal self-stopping principle until the mystical or mythical or actual End of History has been reached. That is why, despite the flourishing of anti-religious attitudes in the old Marxist-Leninist regimes, and despite one rather arcane text about a world beyond religion and atheism in Marx's *1844 Manuscripts*,[31] Marxist philosophy will continue to have a great deal to say to, about, and even – in opposition to anti-humanist currents of all kinds – *with* religious thought for the foreseeable future.[32] That is also why we need next to consider the illusions and disillusionments of History.

(4) *History*. "The vengeance of history is more powerful than the most powerful General Secretary."[33] How true! It is perhaps above all with

respect to their alleged perspective on history that the current disillusionment with Marxist theory and Communist practice/propaganda has achieved its apogee. The slogans about 'Communism – The Bright Future of Mankind' now sound hollow and mocking, as does the assertion, in the *Communist Manifesto*, that the "fall [of the bourgeoisie] and the victory of the proletariat are equally inevitable."[34] There are few indeed now who would dare to predict an imminent 'victory' for socialism worldwide, and in fact there was even a brief and exceedingly odd flurry of enthusiasm, a few years ago, over the claims of a Reaganite writer that free market capitalism should be understood as the correct placeholder for Hegel's talk about 'the end of history.'[35]

"I think I may truly say," to paraphrase and parody Hobbes in his angry denunciation of the historically subversive effects of studying Greek and Latin authors' treatises favorable to democracy, "that there was never any thing so dearly bought, as these western parts have bought"[36] the standard interpretation of Marxism as a theory of historical inevitability. But Hobbes was demonstrably wrong in much of his interpretation of what Aristotle, Cicero, and the other objects of his wrath actually said on these topics, and arguably perverse in espousing the highly authoritarian sociopolitical values that undergird this denunciation. Just so, it is my contention that, while the *values* – call them, if you will, 'utopian,' but without intending the pejorative edge that Engels and many following him gave to that term – of a future socialist (or small "c" communist) society that have been endorsed by the Marxian tradition and that were, in the past, frequently 'reinforced' emotionally by claims about the alleged inevitability of their eventual realization *are* truly positive values, the Marx-interpretation that takes such inevitability to be entailed by Marx's own thought is untenable. I have consistently maintained this, particularly as the outcome of an analysis of Marx's account, in *Capital*, of the structural reasons behind the apparent tendency of the rate of profit to fall under capitalism, and of the 'counteracting influences' to this 'tendency.' Since a tendency can in principle be contravened indefinitely, there can be no guarantee of its ever definitely coming to prevail within any given time-period.[37] Marx Wartofsky's discussion of the labor theory of value, which treats it as a debatable but scientific theory and ends by analyzing some of the same material in Marx, seems to me compatible with my own.[38] It is clear to me, then, that to find much that is of value in Marxist theory does not require, and never has required, the sort of predictive confidence about the path that future history will follow which one used to find exuded in the writings of the Marxist-Leninists.

But I want to go much further, in reconsidering the treatment of history within the Marxian tradition, than the simple contention that its implications were misunderstood both by many who claimed to be his followers and perhaps, often enough, by Marx and especially Engels themselves in their congenital optimism. It seems to me that Marx made an invaluable contribution to the history of thought by reconfiguring philosophy, formerly regarded as the interplay of timeless, ghostly abstractions, as a complex kind of (historical) narrative. This process of reconfiguration had been begun by Hegel, though subject to the limitations of his idealism; it has continued under many guises, including some under which the Hegel-Marx tradition has been denounced and ostensibly repudiated, right up to the present time. A good, though very incomplete, illustration of this approach to thought and society is Sartre's own later work, culminating in the *Critique of Dialectical Reason* and the mammoth study of Flaubert and his times, *The Family Idiot*. This body of work is heavily indebted to the techniques of dialectical, 'critical investigation' of history and society, treating them as what Sartre calls 'totalizations' as opposed to positivism's view of history as "one damn thing after another," that were first developed by Marx.

By 'totalization' Sartre means an ongoing *praxis*-process (i.e., the joint product of contemporary human actors, but within an environment that is only partly of their own making), viewed in its incompleteness and from the standpoint of those participating in it, as distinguished from a completed whole (a 'totality'). History could rightly be viewed in the latter fashion only from the standpoint of a Totalizer, a God. Thus, there can be no guarantee of a 'happy outcome,' or in fact of any particular outcome, and thus no meaningful conception of 'progress' towards a *telos*. Nevertheless, progress *can* be made in *understanding* both a particular strand of historical development, such as the Stalinist 'deviation' of socialism (Sartre's expression), or the collapse of capital 'c' Communism, and, along with that, in understanding at least some aspect of its place in the larger unfolding of History conceived as a complex and indefinitely large set of such strands. As Sartre puts it at one point, summarizing the results of his (partial) analysis of sovereignty in a dictatorial regime such as the Soviet Union's: "In its very nature of praxis-process, we have established – it is our only optimism – that it was intelligible as a *constituted dialectic*."[39]

That there *is* an intelligibility to history, accessible by means of tools derived above all from the Marxian tradition, is, today, a more optimistic claim than many thinkers are willing to make, even though it is a far cry indeed from the 'bright future of mankind' about which so many certitudes were once expressed. To say, however, that this is 'our *only* optimism' may sound like a call to retreat to a life of private intellectual self-gratification, a move that Hegel, identifying one possible outcome to the feelings of revulsion at history's tragedies and catastrophes that he anticipates in the passage cited earlier in this essay, characterizes as "*in die Selbstsucht zurücktreten, welche am ruhigern Ufer steht und von da aus sicher des fernen Anblicks der verworrenen Trümmermasse genießt.*"[40] This passive contemplation is not what Sartre intended, nor do I. His commitment to social *praxis*, action, continued unabated – indeed, at a higher level in some respects than during the days of his greatest notoriety as the existentialist intellectual *vedette* – until serious illness felled him towards the end of his life. But in what name, given 'disillusionment' with the idea of a happy and (in the long run) sure outcome to history, can such unselfish activity be carried out? Since 'communism' has become tainted by recent history, and even 'socialism' (to which Sartre himself continued to express adherence) has to some extent fallen under this taint, perhaps a more acceptable name today is 'community.'

(5) *Community.* It is with respect to the possibility of human community, conceived as (future) social idea, perhaps even more than with respect to expectations of inevitable historical progress, that disillusionment with the Marxian and Marxist-Leninist traditions alike is most thoroughgoing. Of course, the word can be understood in multiple senses, but all of these senses imply some notion of commonality and togetherness. In territories traversed by bitter civil war, such as portions of former Yugoslavia and the former Soviet Union, many communities – professional, political, and personal – that once seemed to flourish across republican borders have totally dissolved, and even many once close-knit communities within individual regions or individual towns have been destroyed, often amid extreme brutality. As I was completing this paper, the case of a philosopher of Muslim nationality from Sarajevo, who had once worked in apparent harmony with philosophers of other ethnic backgrounds in that multi-cultural city, was brought to my attention: his apartment deliberately ransacked, his book collection of a lifetime

destroyed, he was thought lucky by some to have escaped the city alive. Sarajevo, the spark for the conflagration which was touched off by a decision reached rather casually by political leaders of the so-called 'European Community' (the terrifying effects of which, however, were of course made possible only because the tinder was already in place), has come to epitomize the death and destruction of community in the contemporary world. All the fine talk about communes and soviets and self-managed societies seems to have come to nought, or worse, to have served as a hypocritical mask for dark motives of the starkest self-interest, in light of what has happened.

And yet – and yet – were those past appearances of community and harmony *mere* appearances – that is, 'really' illusions? Is the extensive philosophical literature on community, a topic which has in fact been undergoing a revival in the past few years in the United States and other English-speaking countries, really just so much material for the book pyres? It is obvious that my own answer to both these questions can only be 'no.' The actual professional communities of Yugoslav philosophers, for instance, in which I and Professor Wartofsky and a number of our mutual acquaintances were privileged at least briefly to share at one time or another, really *did* flourish across ethnic and other boundaries – once upon a time. Nor were philosophers a unique breed of human being in this regard – far from it. (In fact, despite all the horrors of the war, it remains inaccurate to speak of such communities in former Yugoslavia *entirely* in the past tense – but that is another story.) The fact of the destruction of so many of these communities does not obliterate their past existence, though it forces us here in the present to reinterpret that past. But then, did anyone ever really forget that nothing human lasts?

As far as the literature on community is concerned, one of the strongest American contributors to it over many years has been Carol Gould. From her unpublished Doctoral dissertation, in which she undermined Heidegger's claims to have delineated the nature of authenticity through a notion of *Mitsein* that she demonstrated to be unsupported by the structure of his thought,[41] through a number of subsequent articles such as 'Action, Creation and the Concept of Community,' published in Poland in 1979,[42] and her book, *Marx's Social Ontology: Individuality and Community in Marx's Theory of Social Reality*,[43] on to later work on democracy in which she applies her prior insights about human interaction, she has made a strong case for the continuing importance and viability of the concept of community. It is highly significant that this concept, as it informs Gould's normative vision of more just and democratic future societies, owes so much to the Marxian tradition and to Marx's own writings, particularly to the hitherto-underestimated *Grundrisse*.

One of the richest, most frequently-mined 'classic' texts for exploring some conceptual connections between insights about human community as a norm and feminist insights is a single page in Marx's early writings, located at the beginning of the brief manuscript, entitled 'Private Property and Communism,' from the ending of which the remarks about socialism and atheism to which I referred earlier are taken. In effect, Marx says there that the status of women in a given society is a good index of the level of community to which it has attained. Virginia Held, another important contributor to contemporary social philosophy and feminism, has written a helpful gloss on this text and some of its implications, entitled 'Marx, Sex, and the Transformation of Society,' which is printed in the ground-breaking anthology of feminist philosophy, *Women and Philosophy*, that Carol Gould and Marx Wartofsky co-edited two decades ago.[44] The contribution immediately preceding hers is written in the same spirit; it is Mihailo Marković's essay, 'Women's Liberation and Human Emancipation,' which among other points rightly decries the degeneration of Marxist theory that occurs whenever it "becomes the official ideology of a given social system."[45] Marx and Marxism, then, can usefully *point to* a possible future community in which, centrally, a more egalitarian relationship between the sexes – as well as, presumably, among diverse races, religions, and ethnic groups – reigns, though Marxism turned into doctrine is poorly equipped to serve as the *foundation* of such a society. Perhaps the best foundation for a philosophy of community, Gould seems to suggest in her later as well as her earlier work, is human freedom.[46] By making this value central, she aligns herself, in important ways, with Sartre, with Marković in the essay just cited, and with a wide spectrum of Western interpreters of the Marxian tradition.

Of course, it is not always easy to *implement* freedom as a value in the effort to generate a 'spirit of community' in human populations that are riven with conflict and infested with ideologies antipathetic to that value. Indeed, it is not *ever* easy to do so. The present time is one in which this truism is abundantly manifesting itself in the frequently tragic historical realities to which I have been alluding throughout this essay. In this respect, the present time is no different from most past times: 'progress' has once again proved 'illusory.' To acknowledge this, however, is not to denigrate the enduring value of freedom itself. Nor does it entail abandoning the Sisyphean commitment to a sociopolitical *praxis* directed towards conceivable future human communities – at all levels, from local to global or 'cosmopolitical'[47] – in which human freedom *would* be much more profoundly respected, in practice and not just in principle, than it is anywhere at present.[48]

From the standpoint of this future-oriented ideal, let me suggest by way of contributing to an answer to the question with which I began this essay, it is the shambles of the collapsing, impoverished, and war-torn regimes of the present time that are the real illusion. If defending the possibility of that unrealized ideal be considered idealism, then let us make the most of it. Whatever it may be called, this type of activity has always been at the heart of philosophy's mission,[49] and the extensive ongoing writings, edited volumes, and other cultural contributions of Marx Wartofsky – which are very real, in no way illusory – are *oysgetseyknt* instantiations of that mission.

Purdue University

NOTES

[1] See his 'Le Fantôme de Staline,' *Les Temps Modernes 129–131* (Nov.–Dec. 1956–Jan. 1957), pp. 577–697; *The Ghost of Stalin*, tr. M. Fletcher (New York: George Braziller, 1968).

[2] This attitude is famously, sardonically exposed in the Irish folk ballad, 'The Green Fields of France,' in which the singer sits down by the grave of a soldier by the name of Willie McBride who died in that war and addresses him.

[3] Of course, the alleged phenomenon of 'stupidity' itself is by no means as unproblematic as this remark might lead one to think – any more than 'illusion' is. Sartre gives us a brief but suggestive analysis of 'stupidity,' indicating the dubious character of claims that it is an innate quality, in his *Cahiers pour une morale*, and he also discusses the important, manipulative use made of the concept by Flaubert, at various stages in his life, in *L'Idiot de la famille*. See my *Sartre's Political Theory* (Bloomington: Indiana University Press, 1991), indexed reference to 'Stupidity' and to its connection with 'oppression,' for further elaboration of this.

[4] Sartre, *The Words*, tr. B. Frechtman (New York: George Braziller, 1964), p. 252.

[5] Wartofsky, 'Consciousness, Praxis, and Reality: Marxism vs. Phenomenology,' in F. Elliston and P. McCormick, eds., *Husserl: Expositions and Appraisals* (Notre Dame: University of Notre Dame Press, 1977), p. 304.

[6] *Ibid.*, p. 313.
[7] *Ibid.*
[8] This volume is the object of Ronald Aronson's important study, *Sartre's Second Critique* (Chicago: University of Chicago Press, 1987).
[9] The claim about the epoch is made in *Search for a Method*, tr. H. Barnes (New York: Alfred A. Knopf, 1963), p. 7. The renunciation is formulated in Sartre's transcribed interview (with M. Rybalka, O. Pucciani, and S. Gruenheck) in Paul Schilpp, ed., *The Philosophy of Jean-Paul Sartre* (La Salle: Open Court, 1981), p. 20.
[10] *The Philosophy of Marx* (London: Hutchinson/New York: St. Martin's, 1977), pp. 127–140.
[11] "Markovićev Jezik i Duh Zajednice," tr. J. Kovačević, in D. Mićunović, ed., *Filozofija i Društvo* (Belgrade: Centar za filozofiju i društvenu teoriju, 1987), pp. 33–44. Marković and his *Festschrift* editor were soon to become leading figures of rival parties within the Serbian political spectrum.
[12] Wartofsky, 'Matter, Action, and Interaction,' *Proceedings of the XV World Congress of Philosophy* (Varna), Vol. 6 (Sofia: *s.n.*, 1973), pp. 655–662.
[13] M. Marković and G. Petrović, eds., *Praxis: Yugoslav Essays in the Philosophy and Methodology of the Social Sciences*, Vol. XXXVI of the *Boston Studies* . . . (Dordrecht, etc.: D. Reidel, 1979), pp. v–vii. Among the many interesting contributions, made even more interesting in the light of later political developments and alliances, to this volume is the particularly bitter and foresighted denunciation of the Yugoslav situation in 1974, 'Culture and Revolution,' by Dobrica Ćosić (pp. 217–225). Originally presented as a speech by this popular writer to the Serbian Philosophical Society, the essay advises fellow intellectuals who shared "the communist and humanistic vision of our generation" to abandon all optimistic belief in historical progress and to confess that "we have been deceived, but we are also the deceivers." (p. 220) In my review of this volume (*East European Quarterly* XV, 2, June 1981, pp. 265–269), I commented that "The effect of this con-tribution, even when read in translation rather than heard in the original, is stunning" (p. 268). Ćosić, of course, became the President of the new, reduced country of Yugoslavia after the dissolution of the former state in 1991–92.
[14] See Ivanka Raynova and W. L. McBride, 'Visions from the Ashes: Philosophical Life in Bulgaria from 1945 to 1992,' in B. Smith, ed., *Philosophy and Political Change in Eastern Europe*, supplementary volume to *The Monist*, in press (Open Court).
[15] See *ibid., passim*, for a detailed documentation of the way in which the system worked in one country, Bulgaria.
[16] See 'Sartre and His Successors: Existential Marxism and Postmodernism at our *Fin de Siècle*,' *Praxis International 11*, 1 (April 1991), p. 86.
[17] *Ibid.* and 'Sartre e il postmodernismo,' tr. (fr. French) Schipa, *Segni e comprensione 16* (May–Aug. 1992), pp. 21–29.
[18] The classical site of this theory is Lenin's *Materialism and Empirico-Criticism*. It was defended, on the whole, by Roger Garaudy and his generation of French Communist Party intellectuals. *Reflection Theory* was the title of a 1945 book by Todor Pavlov, the most prominent definer of philosophical 'orthodoxy' in immediate post-World War II Bulgaria, and the theory became the object of considerable debate in that country from 1961 on. See Raynova and McBride, *op. cit.*
[19] Wartofsky, *Models: Representation and the Scientific Understanding*, vol. XLVIII of *Boston Studies in the Philosophy of Science* (Dordrecht, etc.: D. Reidel, 1979).

20 Wartofsky, 'Karl Marx and the Outcome of Classical Marxism, or: Is Marx's Labor Theory of Value Excess Metaphysical Baggage?,' *The Journal of Philosophy 80* (Nov. 1983), p. 721.

21 See the admiring review of *Models* . . . by Martin Curd, *Isis 72*, 1, 261 (1981), pp. 106–107.

22 'Praxis,' in L. C. and C. B. Becker, eds., *Encyclopedia of Ethics*, vol. II (New York and London: Garland, 1992), p. 1006.

23 'The Practical Relevance of Practical Philosophy: Philosophers' Impact on History,' *Philosophy in Context 13* (1983), pp. 31–44; reprinted in J. C. Nyiri, ed., *Perspectives on Ideas and Reality* (Budapest: Posztgraduális és Információs Központ, 1990), pp. 66–84.

24 G. W. F. Hegel, *Sämtliche Werke*, ed. Johannes Hoffmeister, Bd. 18a, *Die Vernunft in der Geschichte*, 5th ed. (Hamburg: Felix Meiner, 1955), pp. 79–80. There was a time when I could not read this spectacular passage without thinking of the tragic American involvement in Viet Nam, on which discussions with some of my Yugoslav friends eventually helped me to obtain a certain perspective. I cannot now read it without thinking of the tragedy of former Yugoslavia.

One of the standard English translations, none of which I find as powerful as the original, reads: "When we contemplate this display of passions, and consider the historical consequences of their violence and of the irrationality which is associated with them (and even more so with good intentions and worthy aims); . . . when we are moved to profound pity for the untold miseries of individual human beings – we can only end with a feeling of sadness at the transience of everything. And since all this destruction is not the work of mere nature but of the will of man, our sadness takes on a moral quality, for the good spirit in us (if we are at all susceptible to it) eventually revolts at such a spectacle." – Hegel, *Lectures on the Philosophy of World History: Introduction*, tr. H. B. Nisbet (Cambridge: Cambridge University Press, 1975), p. 68.

25 Wartofsky, 'The *Republic* as Myth: The Dilemma of Philosophy and Politics,' in *The Philosophy Forum* (DeKalb, Illinois) *10* (1971), p. 265.

26 Here, I am thinking *especially* of Gajo Petrović. See, among many examples, his article, 'Man and Freedom,' in Erich Fromm, ed., *Socialist Humanism* (Garden City: Doubleday, 1965), pp. 273–279.

27 Wartofsky, *Feuerbach* (Cambridge: Cambridge University Press, 1977), p. 23.

28 *Ibid.*, p. 24.

29 *Ibid.*, p. 325.

30 Among the many that come to mind are the theologian, Arend Th. VanLeeuwen's, two volumes, *Critique of Heaven* and *Critique of Earth* (New York: Charles Scribner's Sons, 1972 and 1974), and, from an entirely different intellectual background, *Marxism and Christianity*, by Denys Turner (Totowa: Barnes and Noble Books, 1983). Turner's actual chapter (pp. 9–23) on 'Praxis,' focusing, as much of it does, on an analysis of the praxis of *kissing*, may seem an unpromising entrée to his central theme that 'Morality is Marxism,' but I have tried to explain what this theme means in a review article dealing with this book and with Louis Dupré's illuminating work, *Marx's Social Critique of Culture* (New Haven: Yale University Press, 1983). See *Faith and Philosophy 4*, 1 (Jan. 1987), pp. 108–115.

31 "Since the *real existence* of man and nature has become [in socialism] practical, sensuous and perceptible . . . the question about an *alien* being, about a being above nature and man – a question which implies the admission of the inessentiality of nature and of

man – has become impossible in practice. *Atheism*, as the denial of this inessentiality, has no longer any meaning, for atheism is a *negation of God*, and postulates the *existence of man* through this negation; but socialism as socialism no longer stands in any need of such a mediation . . . Socialism is man's *positive self-consciousness*, no longer mediated through the annulment of religion, just as *real life* is man's positive reality, no longer mediated through the annulment of private property, through *communism*." – Marx, *Economic and Philosophic Manuscripts of 1844*, tr. M. Milligan (Moscow: Foreign Languages Publishing House, 1961), p. 114.

[32] The same applies to Sartre's philosophy, which, at once despite and because of its avowed atheism, holds a profound interest for many philosophers with religious orientations. It is interesting to note that the context of the Sartrean remarks, from his autobiography, about his loss of illusions with which this paper opens is his admission that, although as an adolescent he had decided one day, without any apparent trauma, that God did not exist, he had for many years thereafter acted like a believer, a 'militant,' with respect to his own existentialist worldview.

[33] This statement, attributed to Trotsky, is the theme epigram for a recent book by Alex Callinicos, *The Revenge of History: Marxism and the East European Revolutions* (State College: Penn State Press, 1991). I have reviewed this book without enthusiasm, noting the self-referential nature of its title, in *Radical Philosophy Review of Books 6* (1992), pp. 42–45.

[34] *Communist Manifesto*, intro. Harold J. Laski (New York: Random House, 1967), p. 149.

[35] Francis Fukuyama, *The End of History and the Last Man* (New York: Free Press, 1992).

[36] *Leviathan* Ch. 21; Oakeshott ed. (Oxford: Basil Blackwell, 1960), p. 141.

[37] *The Philosophy of Marx*, Ch. 6 ('Prediction'), pp. 116–126.

[38] 'Karl Marx and the Outcome of Classical Marxism,' pp. 719–730.

[39] Sartre, *Critique of Dialectical Reason*, vol. 2: *The Intelligibility of History*, tr. Q. Hoare (London: Verso, 1991), p. 183.

[40] Hegel, *loc. cit.* The English translation reads: "We retreat into that selfish complacency which stands on the calmer shore and, from a secure position, smugly looks on at the distant spectacle of confusion and wreckage." – *op. cit.*, p. 69.

[41] Gould, 'Authenticity and Being-with-Others: A Critique of Heidegger's *Sein und Zeit*,' Yale University, fall term 1971.

[42] Gould, 'Action, Creation and the Concept of Community,' *Dialectics and Humanism, the Polish Philosophical Quarterly*, summer 1979, pp. 53–59. This journal was recently retitled *Dialogue and Humanism* and now devotes considerable space to discussions of Catholic thought.

[43] Gould, *Marx's Social Ontology: Individuality and Community in Marx's Theory of Social Reality* (Cambridge, Mass.: MIT Press, 1978).

[44] Gould and Wartofsky, eds., *Women and Philosophy* (New York: Putnam, 1976). The original version of this volume appeared as a special double issue of *The Philosophical Forum* 5, 1–2, in fall/winter 1973–74. Virginia Held's essay is on pp. 168–183 in the book version.

[45] In *Ibid.*, p. 151.

[46] Gould, *Rethinking Democracy, Freedom and Social Cooperation in Politics, Economy, and Society* (Cambridge: Cambridge University Press, 1988).

[47] See *ibid.*, Chapter 12 'Cosmopolitical Democracy: moral principles among nations,' pp. 307–328.
[48] Lest it be thought that this allusion to Sisyphus implies some latent, perhaps even unconscious, endorsement of Albert Camus' later attitude of thoroughgoing disillusionment with political activity over, for example, Sartre's ongoing commitment to it, I should point out that Sartre, in publicly accusing his former friend of having "fait votre Thermidor" by writing and publishing *The Rebel*, asks plaintively, "Où est Sisyphe?" In other words, Sartre mourns the Camus of *The Myth of Sisyphus*. See "Réponse à Albert Camus," in Sartre, *Situations*, IV (Paris: Gallimard, 1964), p. 91.
[49] See my article, 'The Philosophy Job Crisis and the Future of Our Culture,' in *The Philosophical Forum 10*, 2–4, winter/summer 1978–79, pp. 371–384.

THOMAS McCARTHY

ON THE COMMUNICATIVE DIMENSION OF SOCIAL PRACTICE

> . . . the separation of *logos* from *praxis* is impossible . . .

In the book from which my epigraph is taken, Marx Wartofsky focuses his attention on the role of representation in 'cognitive praxis.'[1] He is quite clear, however, that there are other aspects to the indissoluble link between logos and praxis: "[Karl] Marx's striking aphorism, 'Language is practical consciousness,' requires the elaboration that it is also social consciousness, that is, the medium of communication and expression in the contexts of social interaction . . ."[2] In this paper I offer some observations on the social-interactional aspect of the connection between reason and practice. Like Wartofsky, I attempt to capture the contextual and pragmatic features of social practice without renouncing the universal and ideal import of the claims of reason. To capture the former, I exploit some recent contributions to the sociology of everyday life by Harold Garfinkel and other ethnomethodologists;[3] to capture the latter, I draw upon Jürgen Habermas's analysis of communicative reason.[4] The challenge, of course, will be to integrate these two very different and often opposed approaches into one coherent account of reason in practice. The question of how my account of communicative praxis might be integrated with Wartofsky's account of cognitive praxis will have to be left to another occasion.

I

In the Parsonian paradigm that came to dominate sociological thought after World War II, social order was explained chiefly through the institutionalization and internalization of values and norms. Correspondingly, social action was represented as normatively regulated behavior such that deviations from institutionally established patterns would, regularly enough, be sanctioned; and social actors were depicted as

C.C. Gould and R.S. Cohen (eds.), Artifacts, Representations and Social Practice, 463–482.

psychologically committed, in consequence of the 'need dispositions' instilled by socialization, to normatively prescribed patterns of action. This conceptualization was effectively challenged in the 1960s by approaches to social action developed in the social interactionist tradition stemming from George Herbert Mead and in the tradition of phenomenological sociology stemming from Alfred Schutz. Particularly in the sociologies of everyday life elaborated by Erving Goffman and Harold Garfinkel and their followers, the picture of social actors as mere rule followers was rendered implausible beyond repair. Let us begin by distilling a few key ideas from the ethnomethodological critique: (1) accountability, (2) reflexivity, (3) indexicality, (4) the dialectic of general and particular, (5) temporality, and (6) the orientation to mutual understanding.

1. In his *Studies in Ethnomethodology*, Garfinkel objects to the representation of social agents as "judgmental dopes" who simply act "in compliance with preestablished and legitimate alternatives of action that the common culture provides," or whose choices among alternatives are "compelled on the grounds of psychiatric biography, conditioning history, and the variables of mental functioning."[5] Such representations treat as epiphenomenal the agents' own knowledge of social norms and their active use of that knowledge to produce and manage the very scenes in which they are involved. As against this, Garfinkel focuses on practical reasoning in everyday settings and treats the "rational properties of practical activities" as the "ongoing accomplishments" of actors themselves.[6] The key to these accomplishments is the way we use our common-sense mastery of social organization to make situations of interaction *intelligible and accountable*. The socially standardized understandings we bring to action settings comprise both shared background knowledge *and* common normative expectations, and thus they serve simultaneously as cognitive *and* normative presuppositions of interaction. We orient our behavior to these shared schemes of interpretation and expectation, mutually attributing awareness of them to one another and reciprocally holding one another accountable in terms of them.[7]

This altered view of the cognitive-normative basis of social order requires corresponding changes in received views of socialization and the motivation to conform. Becoming a competent social actor cannot be a matter merely of depth-psychological formation or behavioral conditioning, for the core of interactive competence is an extensive and detailed understanding of social organization, of the 'natural-moral' facts of life in a society, of the rules of its different games. To be sure, the taken for granted background assumptions that enter into the production and recognition of 'perceivedly normal' courses of action remain largely implicit, unformulated. We routinely bring to situations of interaction a tacit awareness of the normative expectations relevant to them and an intuitive appreciation of the consequences that might follow from breaking them. We implicitly attribute the same awareness and anticipation to our interaction partners, as they do to us, and we reciprocally hold one another accountable by reference to such shared expectations. If called to account, we may cite the relevant norms as reasons for normatively appropriate behavior, whereas violations usually require special considerations if they are to be excused or justified. Unwarranted departures from shared background expectations may be regarded as willful deviations and sanctioned by everything from negative affective responses and breakdown in cooperation to explicit reprimands and punishments. The possibility of being called to account for what we do and our awareness of the differential accountability for behavior that deviates from, versus that which conforms to, socially shared expectations are central to the normative constraint and motivated compliance that characterize routine interactions; as such, they are essential ingredients in any stable social order.

2. One aspect of the rational basis of social interaction upon which ethnomethodological studies have shed light is the way in which competent agents use their knowledge of what is normatively expected in a situation to shape their own conduct. They perform their actions with a view to how the latter will be perceived and assessed by others. Aware that they will be held accountable for what they do, they make what they do accountable; that is, a concern with accountability figures already in the 'design' of their actions.[8] And the very considerations that enter into that design may, if they are called to account, be cited to explain and justify those actions. *Accountability is thus 'reflexive'*: it tacitly informs the very performance of the actions for which agents may be held accountable. Thus there is an inbuilt relation between the sorts of 'reasonable consideration' that go into shaping our conduct and the sorts of 'reasons' to which we may appeal in giving accounts of it. Rationally accountable agents are, in short, expected to produce account-

ably rational actions, that is, actions which can be described, explained, and justified as reasonable in their contexts. In the case of 'perceivedly normal' courses of conduct, this may involve no more than a mention of elements of the common culture and existing institutions which serve as socially warranted grounds of inference and action, together with an indication of relevant features of the situation at hand. In the case of perceived breaches of standard expectations, something more will be required.

In either case, our accounts of our actions also contribute to the sense of the situations in which they are given. They do not merely provide 'second-order' commentary on scenes of action defined independently of them. They are part and parcel of those very scenes, reflexively reconstituting the situations to which they refer. (Compare: "I didn't mean to ignore you. I just didn't hear you say 'hello'." and "I heard you, but after what you did last night I want nothing to do with you.")[9] The reflexivity of accountability and accounting means that reasons and reason-giving belong to the warp and woof of social interaction from the start. The type of critical-reflective discourse usually connected with the idea of autonomy thus marks a significant change in degree, but not the creation of a wholly new kind of interaction. It is, as it were, the everyday reflexivity of accounting practices become reflective.

3. Competent social actors are supposed to be able not only to understand what is going on and to give an account of themselves in standard situations, but to use their mastery of the rules of the game actively to manage social situations. Like Wittgenstein, Garfinkel rejects the received models of 'rule-governed' behavior. On his account, social action is not simply a matter of following set rules in predefined situations, for social norms are neither exhaustive nor unambiguous, neither fully spelled out nor algorithmically applicable, and, as we have seen, social situations are not defined independently of the participants' own activities. By their very nature, general rules *must* understate the complexity of the concrete circumstance to which they apply. They are *supposed* to cover an indefinitive range of specific instances, and those instances are not marked out as such in advance of any interpretive work. This is a *fortiori* true of the largely implicit, approximate, ambiguous, and open-ended norms that structure interaction in everyday life. Like any general rules, but even more so, they come without a detailed, exhaus-

tive set of instructions for applying them. Thus they cannot determine, and hence cannot fully explain, specific, concrete actions in specific, concrete situations. Their 'application' requires not unthinking conformity but competent practical reasoning to deal with contingencies as they arise, and competent for-all-practical-purposes judgments in the light of concrete circumstances. And this of course means that there is always an element of the *ad hoc* and discretionary about their meaning-in-practice.[10] That meaning is elaborated continually as the agents whose behavior is guided by them negotiate the definitions of the ever-changing situations in which they are applied. As competent agents, they will be expected not only to know what everyone knows, but to supply the practical reasoning and good judgment required to put that knowledge into practice.

Not only decontextualized general rules and norms but also abstract general concepts, standards, criteria, principles, schemes, ideals, and so forth allow for an indefinite number of possible contextualizations, so that their particular, situated meaning has to be practically determined through participants' on-the-spot interpretations. This is at the root of another pervasive feature of social action, which Garfinkel, recalling the semantic properties of deictic terms, refers to as its *indexicality*. It is not only terms like 'here' and 'there,' 'then' and 'now,' 'I' and 'you' whose concrete sense is determined by the circumstances of utterance. This holds for general terms across the board. Thus the specific referent of typical referring terms (e.g. 'the medicine,' 'the market') and the specific sense of typical descriptive terms (e.g. 'is expensive' as predicated of either) will also be dependent on when and where and by whom they are uttered. As a result, communication in natural language always depends on the 'interpretive cooperation' of speakers and hearers. A speaker has no choice but to 'trust' hearers to do the contextualizing 'work' needed to make concrete sense of the terms with whose general meaning they are all familiar. Thus, mutual intelligibility is not given once and for all with socialization into a common language and culture; the latter is merely a precondition for something that has to be 'ongoingly accomplished,' in ever-changing circumstances, and always only 'for all practical purposes.'[11]

As Garfinkel points out, the indexicality of meaning holds not only for linguistic expression but also for utterances as actions, as well as for actions generally. Their concrete sense can be understood only with reference to the who, when, where, what, and why of their performance.[12]

Furthermore, the indexicality of action applies to accounting practices too. ("Why is *he* giving *this* account of *that*, *here* and *now*") That is to say, everyday accounts are not merely representations of action. They may be that, but much more besides. For they are themselves practical actions with a point, designed to organize and manage social settings, and like other such actions they can function and signify in a variety of ways.

4. Because no two situations are exactly alike, general terms, rules, norms, standards, techniques, schemes, and the like *have to abstract* from determinate circumstances of application *if they are to be at all useful*. This is not a defect but a condition of possibility of speech and action.[13] And indexicality is just the other side of this abstract indeterminacy. Languages are spoken and understood, and actions are performed and recognized by particular individuals in particular situations. And that particularity inevitably contributes to the concrete sense that is made by and of speech and action. Contextual considerations are, then, always at play in the production and interpretation of meaning. This is the truth of contextualism – but it is not the whole truth. For it is precisely *the dialectic of the general and the particular* that sustains the play of signifiers. Some light has been thrown on just how this works by ethnomethodological studies of conversation. What seems to be required to make conversation work are formal structures with 'local particularization potential,' that is, abstract, general structures which are, on the one hand, relatively invariant to the parties speaking, but are also, on the other hand, able to accommodate the variations they introduce.[14] The dialectic of the general and the particular here takes the form of an interplay between structure and context carried through by competent actors who can orient to both. Reciprocally imputed, mutually assumed, and socially sanctioned structures of interaction are 'locally' and 'interactionally' managed by participants themselves through on-the-spot interpretations and judgments.[15] The key mechanism here is 'recipient design,' that is, the shaping of utterances and actions by participants so as to display not only a knowledge of the common culture but a sensitivity to the particular context, including the particular others involved and the history of their relationships.[16] Competent, practically rational agents are, then, the lifeblood of the dialectic. It is only through their 'work' that the essential indeterminacy of language and culture

can achieve the practical determinacy of speech and action in concrete contexts.

Sensitivity to context is itself exhibited in organized and accountable ways shaped by the general structures at play in a situation. The use of general referring or descriptive terms, for instance, establishes an orientation for whatever contextualizing work has to be done by a hearer to make (practically) determinate sense of a speaker's utterance of them on a particular occasion. "The description and its context elaborate each other. The description evokes a context to be searched and, in turn, the results of the search elaborate the specific sense of a description . . . The sense that is proposed by the speaker and achieved by the hearer will therefore be particular and locally situated. The [general] term elaborated by its circumstances of use will be *sufficient* [for all practical purposes, TMc]."[17] In short, the abstract indeterminacy of general terms need not be, and usually is not, a problem *in practice*. Combined with the extensive and detailed knowledge and the capacity for practical reasoning of competent language users, it is rather the wellspring of language's inexhaustible resourcefulness.

5. It is also important to note that the achievement of (practically) determinate sense in concrete circumstances happens *over time*. As social action always has a past and a future, its sense always has retrospective and prospective dimensions. What came before and what comes after are routinely understood as relevant to the sense of what is presently said or done; they figure in the determination of what is 'in fact' and 'really' meant by an utterance or action. (Example: "After what I said to him yesterday, his refusal to return my greeting is probably intended as a rebuke." "As it turns out, he was merely distracted and hadn't taken offence after all.") Not only the determination of an utterance's sense, but also the determination of its validity – of its truth, say, or of its sincerity – is a step-by-step affair that involves dealing with potential discrepancies, dismissing or 'normalizing' them, and the like. If social interaction is to remain stable in the face of a vast and unpredictable range of potentially relevant considerations, participants must routinely adopt what Garfinkel calls a 'wait-and-see' attitude toward one another's utterances and actions so as to learn "what will have been said or done." For their meaning is inherently open to clarification, confirmation, alteration, and so forth by the future course of events in general and by

how speakers and hearers notice and deal with considerations in particular. For this reason too, mutual understanding and mutual agreement can only be *ongoingly* achieved.

As noted, this openness of speech and action to reinterpretation and reevaluation places a burden of cooperation on members if mutual understanding is to be maintained in the face of the never-ending contingencies and exigencies of social life. For instance, speakers have to trust hearers to supply relevant contextualizations, and hearers are held accountable for implementing the usual methods of doing so. On the other hand, as speakers' utterances reflexively contribute to the sense of speech situations in which they are implicated, they are held accountable for employing the usual procedures to make their meanings sufficiently determinate for the purposes at hand. The 'usual' methods and procedures include, as we have seen, orienting to the organizational patterns and structural features of a shared way of life, to the details of ever shifting contexts (e.g. to particular individuals, circumstances, and purposes), and to the temporal structures of sequences of interaction. This last aspect of interactional competence has been examined in some detail and to great effect by conversational analysts.[18] In designing what to say and in recognizing what is said, speakers and hearers routinely attend, tacitly but minutely, to the ways in which current speech acts form the context for subsequent speech acts. The temporal orderliness of conversation is continuously produced, displayed, recognized, and used by them as grounds for inference and action. Thus, for instance, the 'proposals' contained in speech acts like questioning, inviting, requesting, and the like define situations to which subsequent speech acts will normally be, and be heard as, oriented – for instance, answering a question, accepting or declining an invitation, granting or refusing a request. The projection of relevant next activities and the orientation thereto are not merely 'in the heads' of participants but 'in the culture': they function as shared schemes of intelligibility and accountability. By such means, intersubjective understanding can be continually displayed and updated in and through the organizational details of conversation without having constantly to be thematized. That is to say, hearers can exhibit their understanding of prior speech acts indirectly by what they do with their 'turns'; and speakers can by the same means monitor and correct how they have been understood – so that, for example, not using one's turn to correct a previously displayed understanding is normally taken to mean that one has been understood correctly.

6. Another feature of the cognitive/normative structures of ordinary conversation illustrates very strikingly how an *orientation to mutual understanding* is built into the finest details of linguistic interaction. "There is a bias intrinsic to many aspects of the organization of talk which is generally favorable to the maintenance of bonds of social solidarity."[19] For example, the types of proposal we considered above may be accepted or rejected, but the two responses are normally accomplished in quite different ways. While acceptances (e.g. of invitations, offers, requests, and the like) are normally simple and straightforward, refusals are normally delayed, qualified or mitigated, and treated as requiring an explanation. This differential patterning of 'preferred' and 'dispreferred' responses is relatively invariant across speakers, and even when it is not implemented it is oriented to. Delayed acceptances, for example, are usually heard as 'reluctant,' and early, straightforward refusals as 'hostile' or 'rude.' It is significant that the "preferred format responses" are "uniformly affiliative actions which are supportive of social solidarity, while dispreferred format responses are largely [disaffiliative and] destructive of social solidarity."[20] Moreover, the preferred explanations for refusals are 'no fault' accounts (e.g. an inability – rather than an unwillingness – to accept) that 'save face' and preserve relationships. And prefatory delays before dispreferred responses are routinely used by speakers to modify or revise their offers before rejection occurs so as to mitigate the 'face-threatening' potential of the latter. Structures of this sort, through which conflict is avoided and solidarity promoted, are pervasive in conversation. And participants are not only expected to implement them, they are held accountable for doing so. Ordinary talk thus exhibits a decided normative-structural bias in favor of maintaining the intersubjectivity of mutual understanding.[21]

This analysis of the structure of social action cuts not only against explanations of social order that ignore practical reasoning altogether while emphasizing unconscious processes, conditioning, habit, custom, or the like, but also against explanations that do focus on practical reasoning but construe it exclusively in instrumental terms. To *reduce* practical reasoning to calculations of rational choice is to ignore an equally fundamental dimension of social interaction: normality is normatively provided for, stability is underwritten by accountability. In a slogan: doing the right thing is as pervasive an action orientation as doing the smart thing. Garfinkel attempts to capture this side of things in his characterization of social order as a moral order. "For Kant, the moral order 'within' was an awesome mystery; for sociologists, the moral order 'without' is a technical mystery. A society's members encounter and know the moral order as perceivedly normal courses of action . . ."[22] The shared understandings that *constitute* the common-sense meaning of social situations are at the same time reciprocal expectations that *regulate* behavior in them. The very framework of intelligibility which is *presupposed* by any concrete representation of social interests, social conflicts, or social force, and which cannot therefore be reduced to them, is a framework of normative *accountability*. Mutual understandings and normative expectations are as basic to the structure of social action as teleological orientations to means and ends.

This last observation suggests the deep affinities between Garfinkel's account of the routine grounds of everyday activities and Habermas's account of the structure of communicative interaction. But there are important differences as well, and they turn on precisely the elements that Habermas regards as central to his conception of communicative *reason*: context-transcending validity claims and idealizing presuppositions.[23] Can the transcendence and idealization stressed by Habermas somehow be reconciled with the indexicality and practicality emphasized by Garfinkel? In part II I shall argue that they can, that communicative rationality may be understood temporally (it is an ongoing accomplishment), pragmatically (that is, never absolute but always only for all practical purposes), and contextually (in ever changing circumstances), *without surrendering* transcendence (it turns on validity claims that go beyond the particular contexts in which they are raised) or idealization (and rests on pragmatic presuppositions that function as regulative ideas).

II

Context-transcendence through the idealizing projection of a horizon of unlimited validity has always been a function of reason in our philosophical tradition. From the time of Socrates, reason has been viewed as a capacity to go beyond the immediacies of sense, place, time, upbringing, authority, convention, tradition, and the like and to appeal to considerations that any reasonable person could accept. To renounce strong notions of reason is to forfeit this potential for self- and situation-transcendence. On the other hand, our philosophical tradition is also replete with sceptical critiques of such notions. In current practice, criticism often takes the form of countering idealizing presuppositions with an array of ontological presuppositions that seem to give them the lie: the former are seen to be contrary to fact when we attend to the latter. Claims to unconditional validity are everywhere found to be conditioned by factors beyond the immediate ken of the claimants and hence, it seems, in vain. Under the rubrics of 'language,' 'power,' 'Being,' '*différance*,' and the like, deconstructionist critics have heightened our awareness of the vast and heterogeneous 'unconscious' that underlies all conscious practices, including such paradigmatically rational practices as theorizing, and criticizing. Critical social theorists have certainly not been unaware of such conditionality. It was, after all, Marx who taught us that consciousness is not pure but 'afflicted' from the start with the 'burdens' of matter, that it is "directly interwoven with the material activity and the material intercourse of human beings."[24] And the members of the original Frankfurt School did not hesitate to add cultural and psychological 'burdens' to the socioeconomic ones. Yet, as I have argued elsewhere, the desublimated notions of reason and rational autonomy that the early Horkheimer and his colleagues in the Institute for Social Research worked with were still too strong to withstand the latest round of attacks on Western rationalism.[25]

I now want to argue that Habermas's account of communicative reason, if read through the sort of pragmatizing lens provided by ethnomethodology, is able to accommodate ontological as well as idealizing assumptions. Even if the former always render the latter counterfactual, that is, even if the ideal conditions supposed in the anticipation of universal agreement never in fact obtain, ideas of reason are not thereby rendered useless or pernicious. One of their salutary functions is precisely to serve as ongoing correctives to actually accepted validity claims. This perspective underscores the importance of critical traditions and the institutionalization of critical discourses. As we have seen, practical reasoning and public accounting are already pervasive in everyday interaction. Competent actors are sanctionably expected to produce accountable actions, that is, to act in ways that are publicly explicable and publicly defensible. The accounts called for typically involve nothing more than an appeal to conventional, customary, taken-for-granted warrants of various kinds – shared beliefs, standards, norms, values, and the like – and an explication of the action situation in terms of them. But conventional warrants can lose their unquestioned character and, under certain circumstances, this may result in validity claims being thematized in critical-reflective discourse.

In the context of theoretical discourse, the regulative force of the idea of truth is supposed to be given full scope and the traditional bounds of the communication community to be expanded without limit. In practice, of course, rational acceptance by all competent judges is not something we can hope to accomplish all at once or once and for all. Rather, we try to establish the warranted assertability of a truth claim *in a particular forum* and implicitly assume responsibility – Habermas might say, issue a warranty – for demonstrating its rational acceptability in other relevant forums as well.[26] The unlimited inclusiveness of the claim to truth is, thus, never more than potential. It is a promissory note issued across the full expanse of social space and historical time.[27] Whether or not this promise can be made good always remains to be seen. As Garfinkel might put it, warranted assertability is always 'until further notice'; we always have to 'wait and see' whether the reasons we found compelling in discursive conditions we assumed to be ideal (or close enough) were actually so. It is because validity claims are redeemed by the grounds or reasons offered in support of them, and not by agreement as such, that any existing consensus is open to reconsideration. In the end, there is no way of determining which is 'the better argument' apart from how competing arguments fare over time, that is, how they stand up to the ongoing give-and-take of argumentive discourse. The redemption of truth claims, the establishment of their warranted assertability or rational acceptability, is thus an intrinsically temporal, open-ended process. Because 'for us' all the evidence is never in and all the conditions are never ideal, it is a potentially misleading hypostatization to speak, as Habermas sometimes does, of "rationally motivated consensus" in anything but processual terms – all the more so once we have abandoned traditional notions of demonstrative certainty and realized that in most areas we have to make do with a 'principle of insufficient reason.'[28] Intersubjective recognition of truth claims has to be ongoingly accomplished through rationally persuading one audience after another that it is 'reasonable' to accept them, that is, that there are good reasons for doing so, better reasons than for accepting any of the available alternatives.

Claims to scientific truth, for instance, are raised and discussed in culturally and institutionally established forums that promote their critical examination over time. Scientists have reflexively to anticipate the scrutiny of their work by others who share the cognitive, normative, and evaluative presuppositions marking them as members of the same scientific subcommunity. If they can convince that particular audience, they have reason to expect they could convince other relevant audiences as well. That is to say, in virtue of its presumed competence, that audience can plausibly stand in for the 'universal audience' of all rational beings able to judge in the matter.[29] It is this temporally extended, in principle never-ending, to-and-fro of claim and criticism in institutionalized forums, vis-a-vis particular audiences authorized to represent the universal audience, that constitutes scientific discourse. And this ongoing, dialectical give-and-take lives from the unconditionality of the truth claims raised – the implicit promise they contain to convince a universal audience. Without that, argumentive discourse would degenerate into the 'I think'/'you think,' 'we think'/'they think' of argument in its colloquial sense. With it, the 'conversation' continues, as Rorty might say – in significant part, through the ever-renewed efforts of critics to discover the hidden conditions of what is allegedly unconditional and to take them explicitly into account. Scientific truth, as the continually revised outcome of this open-ended dialectic, can only be an ongoing accomplishment. But the process itself makes no sense apart from the supposition of an objective world and the repeated claims to have captured some truth about it. They are what prompt critique and fuel the dialectical movement.

The unavoidability of ontological presuppositions of various sorts does entail that there will always be a gap between the ideal and the real. But it does not follow from this that we ought to keep ourselves pure by forsaking participation for the standpoint of the uninvolved spectator. That is not an option for us: we cannot but be participants and thus cannot avoid undertaking the presupposition this entails.[30] Nor is ironic participation, distancing ourselves from suppositions we regard as illusory but cannot avoid making, more than an occasional possibility. In fact, the very notion that the idealizing assumptions underlying communicative interaction are 'illusions' smacks of a nostalgia for presuppositionless starting points. The postmetaphysical question is, rather, what does the gap between the ideal and the real mean *in practice*. I have suggested that it is, among other things, an invitation to *critical participation*, which has to be understood in process terms. Becoming conscious of unconscious determinants of our thought and action is not a fully achievable state of affairs but an orientation for ongoing efforts to deal reflectively with real problems, actually experienced, in concrete situations: critical self-awareness is always only 'for all practical purposes.' This is as true of the sorts of 'ontological unconscious' stressed by deconstructionists as it is of the 'psychological unconscious' that interested Freud, the 'cultural unconscious' attended to by hermeneutics, and the 'social-structural unconscious' unearthed by Marx. In all these dimensions, even if we suppose the possibility in principle of bringing any *particular* unconscious factor to consciousness, that by no means amounts to supposing that they might all be made conscious all at once. The hope for such self-transparency disappeared with German Idealism. And this means that at every moment and in every situation unconscious factors – in this very broad sense – will inevitably play a role in shaping our interpretive and evaluative schemes. The moral of this story, however, is not a deconstructionist renunciation of any non-ironic employment of ideas of reason but a critical-pragmatic warning against treating any attempt to specify them as finished, final, complete. It reminds us that universal validity claims can be only provisionally redeemed, that the – less than logically compelling – warrants which suffice to convince one audience in one set of circumstances will not suffice to convince all other audiences in all other circumstances, and thus that 'convincing a universal audience' can never be anything more than an orientation to essentially open-ended discursive processes.[31]

The dialectic of the ideal and the real, like that of the general and the particular discussed in Part I, is borne along by the on-the-spot actions and utterances of accountable agents. It is, as we saw, only through their situated exercise of practical reasoning and judgment that the inherent – and functionally necessary – indeterminacy of linguistic and cultural patterns can achieve the practical determinacy of everyday speech and action. But competent agents can just as well render the determinate indeterminate, the unproblematic problematic, the accepted unacceptable. That is to say, since mutual intelligibility and mutual understanding have to be ongoingly accomplished by the parties to interaction themselves, the outcomes of interaction are not determined in advanced by preexisting structures of whatever sorts. Represented, as they typically are, apart from the agents that bear them, social and cultural structures may seem to be devoid of any inherent dynamic of change. It is to agency that we must look if we hope to comprehend that dynamic, and particularly if we hope to understand the indeterminacy, innovation, proliferation, and the like to which poststructuralists so often appeal in opposing the hegemony of logocentrism, while at the same time espousing holistic conceptions of language, culture, and society that leave little or no room for agency.[32] Practically reasoning agents are a condition of possibility of determinacy *and* indeterminacy, stability *and* innovation, repetition *and* proliferation. And this means, in the case of all such apparent oppositions, that you can't have one without the other. Of course, the agents in question are always socialized agents and their agency is always informed by the sociocultural lifeworlds they take for granted and, in taking for granted, renew. The circle connecting presuppositions to accomplishments is at work here as well; but it is a non-vicious circle and one that is continually being redrawn.

The critical potential of the role that the theory of communicative action reserves to autonomous agents sometimes gets obscured by the stress Habermas places on agreement and consensus. But there are sufficiently unambiguous indications of the negative freedoms built into the discourse model as well, as when he underscores "the individual's inalienable right" to agree or disagree, his or her "uninfringeable freedom to respond with a 'yes' or 'no' to critizable validity claims."[33] It was in this latter spirit that Foucault, asked to comment on the consensus model, insisted on regarding consensus as a "critical" rather than a "regulating" principle, "because starting from the point where you say regulating principle, you grant that it is indeed under its governance that the phenomenon has to be organized . . . I would say, rather, that it is perhaps a critical idea to maintain at all times: to ask oneself what proportion of nonconsensuality is implied in [a] power relation and whether that degree is necessary or not . . . The farthest I would go is to say that perhaps one must not be for consensuality, but one must be against nonconsensuality"[34] Habermas, of course, wants to argue that these are two sides of the same coin. Rational consensus as a regulative principle already makes agreement depend on the uncoerced 'yes' or 'no' of each participant, already confers upon everyone involved the right to withhold assent unless and until convinced. The threat that Foucault, with good historical reason, to be sure, sees looming behind positive versions of any such principle can be met only if we place equal stress on the negative version. Kant, it will be recalled, already balanced the universal-law formulation of his categorical imperative with a formulation interdicting the treatment of others as less than responsible subjects: "Act in such a way that you always treat humanity, whether in your own person or in the person of any other, never simply as a means, but always at the same time as an end!"[35] In explaining this formulation, he noted that it entailed respecting the competence of others to give or withhold their assent, that is, it enjoined treating them as *mündige*, accountable and autonomous agents capable of conducting their own lives on the basis of their own insights, deliberations, and judgments.[36]

As explained in Part I, the concrete, situated meaning of *any* general norm, value, ideal, standard, or the like has to be practically determined by participants' on-the-spot interpretations and judgments. This belongs to the very nature of general normative schemes; they are supposed to cover an indefinite range and variety of specific instances. And while previously accomplished applications may serve as a guide to and, in varying degrees, as a constraint upon future interpretations, they are never sufficient to determine them exhaustively and thus to remove the need for competent practical reasoning and good judgment. To repeat a point made earlier, this is not a defect of general normative structures, but a condition of their possibility. The omnipresent dialectic of the general and the particular becomes all the more central to the institutions of post-traditional, pluralistic societies. General norms have to be accepted as binding by socially and culturally heterogeneous groups. Their legitimacy can no longer rely upon a shared comprehensive view of the meaning and value of life. In binding all alike, they must at the same time allow for diversity and difference. These apparently conflicting requirements can be met only by abstract, general norms that are applied with situation-specific inflections. In this respect, abstract generality is not a threat to difference but a license for it. The currently widespread notion that concrete, specific norms and values are somehow more respectful of difference is a 'logical howler' pure and simple. There is no *need* to impose general conceptions rigidly on variegated situations. Abstraction *need* not mean homogenization, nor universality uniformity. Here, as elsewhere, everything depends on how they are put into practice. As Garfinkel has reminded us, *all* "the rational properties" of our activities have to be ongoingly accomplished by the actors themselves. If we want, for instance, to think sensibly about what 'impartiality' does or should mean in the judicial system, we have to examine how it is or should be put into practice in all the different settings and by all the different actors in that domain. The same holds for 'objectivity' in science, 'effectiveness' in technology, 'efficiency' in business, 'self-determination' in politics, 'self-fulfillment' in personal life, and so on. The point, in short, is to reject the either/or opposition between decontextualized-because-generalized norms and values, on the one hand, and contextualized-because-particularized judgments, on the other. Especially in modern pluralistic societies, we cannot help but have it both ways, that is, agree upon decontextualized – abstract, general, formal – norms, principles, rights, procedures, and the like, which must then be ongoingly contextualized – interpreted, elaborated, applied – in particular situations. As there is no way of predicting the variety and multiplicity of actors, settings, purposes, and so forth that will thereby come into play, there is no possibility of fully predetermining their meaning-in-practice. It is precisely because abstract, formal structures and procedures leave a much broader scope for 'local particularization potential' than do prescriptions of concrete patterns of behavior that they are so central to modern societies. And it is for the same reason that their concrete meaning-in-action is inherently open to discussion, that is to say, essentially contestable.

Northwestern University

NOTES

[1] Marx Wartofsky, *Models: Representation and Scientific Understanding* (Dordrecht: Reidel, 1979). The line quoted appears on p. 173.

[2] *Ibid.*, p. xviii.

[3] I will be relying primarily on Garfinkel's earlier work, especially the essays collected in *Studies in Ethnomethodology* (Englewood Cliffs: Prentice Hall, 1967), republished by Polity Press in 1984. My reading of his work is indebted to the excellent study by John Heritage, *Garfinkel and Ethnomethodology* (Oxford: Polity Press, 1984). It should be mentioned that this line of interpretation would not be accepted by ethnomethodologists who adopt a more radically reflexive and anti-theoretical stance – including, perhaps Garfinkel himself. On the other hand, it is akin to the line followed by the conversational analysts to whom I shall also be referring. As my concern here is solely to indicate another way of thinking about practical reason, I shall ignore exegetical disputes as such.

[4] On the notion of communicative reason, see Jürgen Habermas, *The Theory of Communicative Action*, Vol. I, trans. by T. McCarthy (Boston: Beacon Press, 1984), chapter I, pp. 1–141.

[5] *Studies*, pp. 67–68.

[6] *Studies*, p. 33.

[7] In this paper I shall leave to one side questions of cultural difference within and between cultures and focus on the basic ingredients of mutual understanding in a common culture. This may slant the discussion in a 'conformist' direction, and that slant can indeed be detected in Garfinkel's account of the "routine grounds of everyday activities." However, in my view there is nothing inherently conservative about this approach. Analyzing socially

shared and sanctioned schemes of interpretation and evaluation can be put to effective critical use, as ethnomethodological studies of social science research technique illustrate.

[8] "The activities whereby members produce and manage settings of organized everyday affairs are identical with members' procedures for making those settings 'accountable'. The 'reflexive' or 'incarnate' character of accounting practices and accounts makes up the crux of [my program]." *Studies*, p. 1.

[9] Heritage elaborates on the reflexivity of social action generally and of accounting practices particularly by analyzing the case of greetings, in *Garfinkel and Ethnomethodology*, pp. 106–120. For instance, an initial greeting may transform a situation of disengagement into one of engagement. The greeted party is now faced with the choice of returning or not returning the greeting, and in this way or that. A return greeting ratifies the proposed engagement; failure to return the greeting is open to different interpretations, which reconstitute the situation in different ways: the recipient was a preoccupied friend or a misidentified stranger, s/he did not recognize the greeter or thought s/he was making a pass, s/he deliberately snubbed the greeter to display irritation or to start a quarrel. And so forth and so on.

[10] Which is not to say that it is arbitrary. Judgments of applicability are themselves accountable; they can be contested and criticized, discussed and defended.

[11] Thus Garfinkel stresses that "common understanding" has an "operational structure" consisting of a "course of interpretation" (*Studies*, p. 25) and that "shared agreement" has less to do with "the intersection of overlapping sets" than with "social methods for accomplishing members' recognition" (*Studies*, p. 30).

[12] Compare: "He's always so distracted, he never returns greetings. Think nothing of it." "He's usually so friendly. I wonder if what I said yesterday offended him." "He's always been so formal. Could that warm greeting have been meant flirtatiously?" What is done (not noticing, showing irritation, flirting), like what is said, depends on the context of action, including the identities of those involved and the prehistory of their relationships. The all-purpose term 'context' gets used to cover everything from the very local to the very large, from the sequencing of speech acts to formal-institutional settings. Garfinkel insists that whatever its scope, 'context' not be treated simply as exogenous to interaction; it is endogenously produced and reproduced in and through the details of interaction. (Cf. Heritage, p. 283.) This is one aspect of the dialectic of agency and structure that Anthony Giddens, among others, has elaborated upon. See his *The Constitution of Society* (Berkeley: University of California Press, 1984).

[13] In their fixation on the homogenizing effects of logocentrism, poststructuralists sometimes miss this evident truth.

[14] Cf. Harvey Sacks, Emanuel Schegloff, and Gail Jefferson, 'A Simplest Semantics for the Organization of Turn-Taking for Conversation,' in J. N. Schenckhein (ed.), *Studies in the Organization of Conversational Interaction* (New York: Academic Press, 1978), p. 10. Garfinkel himself would likely reject this formulation in favor of a more radically reflexive approach. See, for example, the article he coauthored with Harvey Sacks, 'On the Formal Structures of Practical Actions,' in J. C. McKinney and E. A. Tiryakian (eds.), *Theoretical Sociology* (New York: Appleton Century Crofts, 1970), pp. 338–66, esp. pp. 345–46.

[15] 'A Simplest Semantics,' pp. 40–42. That is to say, 'local management' means that social actors orient *both* to general – linguistic, cultural, social, etc. – structures and to

the particulars of their situations, and that the capacity to do so is reciprocally imputed to and sanctionably expected of competent agents.

[16] *Ibid.*, p. 43.

[17] Heritage [Note 3, above] pp. 147–48.

[18] On what follows, see Heritage, pp. 233–292.

[19] Heritage, p. 265.

[20] *Ibid.*, p. 268.

[21] Thus, for example, Sacks, Schegloff, and Jefferson point out that in the organization of conversational turn-taking, talk addressed to problems encountered in understanding previous utterances is given priority, so that the normal order may be superseded for that purpose (p. 33). And they note that turn-taking organization, with its reliance on cues, displays, and the like, "builds an intrinsic motivation for listening to all utterances in a conversation" and "translates a willingness or potential desire to speak into a corollary obligation to listen" (pp. 43–44).

[22] *Studies*, p. 35. He is obviously using 'moral' in a much wider sense than is usual in philosophy or sociology, that is, as more or less coextensive with 'normative.' But this is a useful rhetorical device for stressing the omnipresence of the 'ought.' By contrast, even when rational choice theorists do acknowledge the pervasiveness of action guided by social norms, they typically account for this phenomenon in reductionistic terms. Thus, for example, Jon Elster conceptualizes norm-guided behavior as behavior that is *not* outcome oriented but 'pushed' by forces resulting from internalization processes. As such, it is not rationally motivated behavior. (See *The Cement of Society* (Cambridge University Press, 1989) and *Nuts and Bolts for the Social Sciences* (Cambridge: Cambridge University Press, 1989).) Recently, Elster has gone so far as to concede to considerations of justice an independent weight in political deliberations; but at that point he is diverging from a strictly rational-choice approach. (See 'The Possibility of Rational Politics,' in D. Held (ed.), *Political Theory Today* (Cambridge: Cambridge University Press, 1991), pp. 115–142.)

[23] Garfinkel does, to be sure, attend to the "rational properties of practical activities," among which he numbers efficiency, efficacy, effectiveness, intelligibility, clarity, consistency, coherency, planfulness, typicality, uniformity, reproducibility, adequate demonstration, adequate reporting, sufficient evidence, necessary inference, and "every topic of 'logic' and 'methodology'." (*Studies*, pp. 33–34). But he insists that these "contingent ongoing accomplishments of organized artful practices" (p. 11) be treated solely as "*phenomena* for ethnomethodological study but not otherwise" (p. 33), in particular that they not be described or assessed by "using a rule or standard obtained outside [their] actual settings" (p. 33). Like its direct ancestor, phenomenological bracketing, "ethnomethodological indifference" enjoins us simply to describe members' activities and accounts "while abstaining from all judgments of their adequacy." ('On Formal Structures of Practical Actions', p. 345). This is the point at which ethnomethodology parts ways with critical theory. Habermas does not think such indifference is a live option. (See *The Theory of Communicative Action*, vol. I, pp. 128–130.)

[24] Karl Marx and Friedrich Engels, *The German Ideology* (New York: International Publishers, 1947), pp. 19, 14.

[25] T. McCarthy, 'The Idea of a Critical Theory and Its Relation to Philosophy,' in *On Max Horkheimer: A Retrospective*, ed. by S. Benhabib, W. Bonss, and J. McCole (Cambridge, MA: The MIT Press, 1993).

[26] For an elaboration of the notion of a 'forum of argumentation,' see S. Toulmin, R. Rieke, and A. Janik, *An Introduction to Reasoning* (New York: Macmillan, 1979). See also Habermas's discussion of it in *The Theory of Communicative Action*, vol. I, pp. 22–42.

[27] In 'Individuation through Socialization' Habermas remarks that Kant's intelligible world has to be rendered both "socially concrete" and "temporally dynamic" (p. 184). He also notes that temporal dynamics play an increasingly important role in modern life. For one thing, the present is "ever more plainly interpreted" in light of the future: "[T]he supposition that present action will be placed under premises that anticipate future presents . . . applies to systemic processes (such as long-term political commitments, debt-financing, and so forth) as well as to simple interactions" (p. 188). Such modes of anticipation, i.e. of reflexive accountability, more and more come to be "socially expected," i.e. culturally imputed and institutionally demanded.

[28] This phrase is used by Hans Blumenberg on p. 447 of 'An Anthropological Approach to the Contemporary Significance of Rhetoric,' in K. Baynes, J. Bohman, and T. McCarthy (eds.), *After Philosophy* (Cambridge, MA: The MIT Press, 1987), pp. 429–458. He is concerned there with "a theory of man outside the realm of Ideas, forsaken by evidentness," (432) and with rhetoric as the theory and practice of "the production of mutual understanding [and] agreement" in that state (433). Thus, rhetoric is presented as a "rational way of coming to terms with the provisionality of reason" (452).

[29] On the notion of a 'universal audience,' see Ch. Perelman and L. Olbrechts-Tyteca, *The New Rhetoric* (Notre Dame: University of Notre Dame Press, 1969) and Habermas's discussion of it in *The Theory of Communicative Action*, vol. I, pp. 22–42.

[30] Once we make the turn to a critique of 'impure' reason, the empirical description of general structures and inescapable presuppositions can no longer claim for itself the same status 'outside the world' as transcendental and phenomenological inquiry once did. It is and must remain a form of 'mundane' – historical, sociological, cultural, linguistic, psychological – inquiry into structures of speech and action. The idea of getting 'always already there' structures descriptively right carries with it all the basic presuppositions of 'being in the world.' Compared Foucault's discussion of the 'empirico-transcendental doublet' in *The Order of Things* (New York: Random House, 1970), pp. 318–322.

[31] Working out the relation between actually persuading particular audiences and ideally addressing a universal audience, is, I think, the key to grasping the dialectic of immanence and transcendence in argumentation theory.

[32] See my critiques of Rorty, Foucault, and Derrida in this regard in T. McCarthy, *Ideals and Illusions* (Cambridge, MA: The MIT Press, 1991), Part I.

[33] J. Habermas, *Moral Consciousness and Communicative Actions*, trans by C. Lenhardt and S. Weber (Cambridge, MA: The MIT Press, 1990), p. 202.

[34] 'Politics and Ethics: An Interview,' in P. Rabinow (ed.), *The Foucault Reader* (New York: Pantheon, 1984), p. 379.

[35] I. Kant, *Groundwork of the Metaphysics of Morals*, trans. by H. J. Paton (New York: Harper and Row, 1964), p. 96.

[36] Maria Herrera develops this side of Kantian moral theory in seeking to balance Habermas's stress on consensus with an equal stress on difference, in 'Justification and Application in Habermas's Discourse Ethics,' unpublished ms, and in 'Equal Respect among Unequal Partners: Gender Difference and the Constitution of Moral Subjects,' *Philosophy East and West 42* (1992): 263–275.

CHEYNEY RYAN

THE BREAD OF FAITHFUL SPEECH

> And the secondary senses of the ear
> Swarm, not with secondary sounds, but choirs,
> Not evocations but last choirs, last sounds
> With nothing else compounded, carried full,
> Pure rhetoric of language without words.
> – Wallace Stevens
> "*Credences of Summer*"[1]

For me, no better evidence exists for the beneficent smilings of fate than my landing at Boston University in the fall of 1970. For the philosophy department there was truly a remarkable group, and became more so as the years progressed. They were at once wise and generous, tolerant yet committed – such a rare combination that it now all seems like Brigadoon, something that appears all too briefly, and then recedes into the mist. At the center of it all, their chief assembler, was Marx Wartofsky, to whom I owe pretty much everything. When I landed there – and 'landed' is the right word, for I had been projected out of my previous institution – it was largely due to him, as it was largely through his machinations that I managed to finish my education without further interruption. With Alasdair MacIntyre, Marx supervised my thesis on philosophy of economics. What follows are further reflections on that topic, inspired by the developments of subsequent years. I hope that they will be received as the expressions of gratitude and love that they are.

I

If economics is the dreary science, much recent *philosophy* of economics has been dreariness twice over. There have been important exceptions, of course. Daniel Hausman's *Capital, Profits, and Prices*[2] is a model of how philosophical analysis can illuminate a particular economic dispute – in this case, the Cambridge Controversy of the 1960's

C.C. Gould and R.S. Cohen (eds.), Artifacts, Representations and Social Practice, 483–500.

over the status of the neo-classical production function. But on the whole (and admittedly my view is a jaded one), reflections on the nature and status of economic 'science' have provided little more illumination than that already found in, say, Mill's *Principles of Political Economy*. Moreover, and this bears on their dreariness, they are unduly prone to that philosophical vice of confusing seriousness with tedium. But the tedium shows signs of lifting with the meta-economic writings of Donald McCloskey and Arjo Klamer, which, since their appearance in the mid-1980's, have introduced a certain post-modern moment into discussions of these matters.[3] Their work, as one might expect, champions rhetoric over logic, 'free-wheeling' conversation over 'constraining' methodology, tentative agreement over final truth. It is post-modernist but decidedly not post-structuralist, in that it has little time for the latter's sense of contradiction or apocalypse. Its heroes are William James and Kenneth Burke, not Derrida or Deleuze.

McCloskey is by far the most important figure in this discussion, largely because he occupies a position very much like Richard Rorty's in philosophy. He is, like Rorty, something of an apostate: an accomplished theorist of the Chicago School, a prominent figure in the 'positivist armies' (his phrase) which transformed economic history in the 1960's and 1970's, but one who experienced increasing doubts about the hegemonic positivist methodology – and the grim academic culture it sometimes fosters.[4] Like Rorty's, then, McCloskey's past endows his voice with a certain authority. He speaks as one who has been through it all and has come out the other side, as one who can now place his earlier concerns in a certain weary perspective. The voice can be dismissive, but it is far from unsettling. For despite his jibes at the 'bourgeois' nature of methodology, McCloskey's tone (again, like Rorty's) is essentially a complacent one. His irony conveys the sense that economists, when they aren't berating each other with their 'methodology', are basically doing what they should be doing. Stanley Fish likens the 'rhetoric of economics movement' to the Critical Legal Studies movement, but in many ways the two are poles apart[5]. Critical Legal Studies' professed purpose is (as Chairman Mao used to say) to 'bombard the headquarters', its method is self-consciously one of 'trashing', whereas McCloskey's attitude is more 'let a hundred flowers bloom'.[6] He would have economists pretty much go about their business, only he would have them do it with more gentility and wit. All of which, he believes, will be brought about by more attention to the rhetorical dimension of their discipline.

Perhaps because of its tone, McCloskey's work has thus far occasioned none of the rancor that has been generated by Critical Legal Studies. Response to it from within the profession, like that of A.W. Coats, has in fact been rather muted.[7] One wonders though whether this peace can long survive: whether economic theory can continue conducting business as usual with its positivist self-conceptions under attack, or whether (as it has in other fields) this post-modernist moment will compel a revisioning of the economic project itself.[8] In this essay I shall consider some of the more unsettling implications of this rhetorical turn for economic theory, both neo-classical and Marxist. To highlight the political stakes here, in this age of capitalist 'reconstruction' of former state socialist societies, I shall be particularly concerned with how the rhetorical turn may effect changes in our fundamental conceptions of 'the market'.

II

The association of rhetoric and economics is an old one, as old perhaps as money itself. One ecounters it in Zeno, who likens 'well ordered discourses' to the 'coins of Alexander', and of course one encounters it in the Sophists. The criticism of the Sophists by Plato and Aristotle, their insistence that we keep distinct economics proper and 'chrematistics', introduces the fear that both rhetorical and monetary exchanges can be sites of substantial duplicity and subterfuge.[9] Contemporary interest in the interface of rhetoric and economics can probably be traced to Derrida's essay 'White Mythology'. There, Derrida excavates an old piece of Nietzsche's, 'On Truth and Lie in an Extra-Moral Sense', to focus attention on metaphor as a kind of conceptual coin.[10] But Derrida is concerned with economics as economic practice, specifically monetary practice. In his work and the work it has inspired, economic *theory* goes largely unaddressed.

Yet the concern that economic theory might be (or be perceived as) little else than rhetorical gesture is also an old one. It was to *counter* this concern, in fact, that Dudley North first suggested that economic theory be cast in a deductive structure.[11] In this century, prior to

McCloskey at least, the theorist most sensitive to the rhetorical dimension of economic theory is Thorstein Veblen, who is particularly attentive to metaphor's capacity to sustain while obscuring bias. The following is typical:

> It is precisely in this use of figurative terms for the formulation of theory that the classical normality still lives its attenuated life in economics; and it is this facile recourse to inscrutable figures of speech as the ultimate terms of theory that has saved the economist from being dragooned into the ranks of modern science.[12]

But in my view, Marx remains that theorist most sensitive to the interplay of rhetoric and economics in all its dimensions. He speaks of 'the language of objects' – meaning both economic theory, as a discourse *about* commodities, and economic practice, as the discourse *of* commodities. Talk about money is talk about money talking, talk about persons who put their money where their mouths should be. The 'absurdity' in all this is the concern of his 'critique of political economy', which consistently contrasts 'the language of objects' with a language 'truly human' – what Wallace Stevens terms 'the bread of faithful speech'.[13]

McCloskey's concern with rhetoric arises from, and stands at the heart of, his critique of 'economic modernism' (his term for positivist methodology and its various vices). What does he mean by 'rhetoric'? 'Rhetoric is the art of speaking", he writes. "More broadly, it is "the study of how people persuade."[14] But rhetoric is not just the 'study' of how people persuade, it is the *practice* of persuading. 'Rhetorical' thus describes both the analysis of our means of persuasion and the means of persuasion themselves. McCloskey acknowledges from the start how rhetoric and economics double back on themselves. He seeks a rhetoric of economics, mindful that "rhetoric is an economics language" – by which he means that it concerns "the proportioning of means to desires in speech." Oddly, McCloskey construes this 'proportioning' in terms of efficiency. Rhetoric concerns how "scarce means are allocated to the insatiable desires of people to be heard."[15] Construing it instead as seeking the most *effective* means of persuading others would draw attention to some of the central problems with this approach.

It is the prominence that McCloskey and others in the 'rhetoric of economics movement' ascribe to *persuasion* that raises the greatest anxieties among their critics. Those anxieties are of two sorts. One pertains to the *epistemic* status of economic discourse, thus conceived.

It wonders how, or whether, economic discourse's concern with persuading others connects with a concern for presenting the truth. A second anxiety pertains to what might be termed the *ethics* of economic discourse, thus conceived. In its sharpest form it asks: If the aim of the economic discourse is just to persuade others, mightn't the most effective means be the most devious and insidious? Mightn't the 'art' of rhetoric be just the art of the conjurer?

McCloskey recognizes the danger of rhetorical artifice, indeed he presents and portrays his own enterprise as in part a prophylactic against it. "Metaphors", he writes, "evoke attitudes that are better kept in the open and under the control of reasoning."[16] And as one might expect, the victory over darkness promises freedom: the championing of 'honest rhetoric' provides an alternative to the 'coercion' of 'philosophy, politics, or method.'[17] Still, numerous questions abound. Some of them concern the intelligibility of rhetoric's policing itself in this way. If an entire field *rests* on a 'master' metaphor, as economics is said to rest by McCloskey and Klamer on the 'rational choice' metaphor (or, as I would say, it rests on the 'market' metaphor), can we really master *that* metaphor – keep its limits constantly in mind, say – without undermining the enterprise it grounds? Other questions concern the impulses behind rhetoric's self-policing. Talk of 'honest' rhetoric sounds nice, but what motive is there for a rhetoric bent on persuasion to *be* honest? It seems a short step in fact from regarding economics as a project of persuasion to regarding it as an inherently political project of power, which is exactly how Critical Legal Studies views legal discourse. Should we take the same view of economic discourse?

(1)

I want to begin with some words on the issue of persuasion and power. My starting point will be the views of an earlier student of rhetoric and economics – Adam Smith.

Smith, it should be remembered, comes to economics from the study of rhetoric, but from the start his conception of rhetoric is infused with economic notions. Most important is the notion of *transaction*. Smith's rhetoric focuses on 'communication', but 'communication' is at bottom a kind of 'exchange' – the exchange of 'ideas', mainly of ideas embodying 'sentiments'. Just as Smith's economics studies the logic of market exchange, in its proper or 'natural' form, his rhetoric studies the logic of rhetorical exchange, in its proper or 'natural' form. And, just as Smith's economics inveighs against the 'artificial' barriers to market relations posed by Mercantilism, his rhetoric inveighs against the barriers to communicative relations posed by the 'artifices' of excessive metaphor. On this latter score, Shaftesbury is a particular object of scorn. Smith chastises his championing of ornate language over plain speech for producing a discourse of 'dark and perplexed' expressions, 'absurdities' that wallow in a 'dungeon of metaphorical obscurity'.[18] But here, as in his economics, Smith's criticism of 'unnatural' policies presumes a vision of the proper or 'natural' order of things. How then does he conceive of healthy 'communication'?

A logical way to approach this question is by pursuing the analogy between communication and exchange, and the 'natural' forms of each. The results are surprising, for Smith suggests that if communication can be understood as kind of exchange, exchange can be understood as a kind of persuasion! This is the lesson, for example, of these remarks from the *Lectures on Jurisprudence*:

> If we should inquire into the principle in the human mind on which this disposition of trucking (Smith's term for buying and selling – CR) is founded, it is clearly the naturall (*sic*) inclination every one has to persuade. The offering of a shilling, which to us appears to have so plain and simple a meaning, is in reality offering an argument to persuade one to do so and so as it is in his interest. Men always endeavor to persuade others to be of their opinion even when the matter is of no consequence to them ... And in this manner, everyone is practicing oratory on others thro (*sic*) the whole of his life.[19]

This picture of exchange is remarkable, most notably for the amount of dissemblance it sees in that relation. What appears 'plain and simple' is not: it seems that in exchange I seek only to satisfy my material interest, whereas in reality I seek mainly to get my way; it seems I satisfy my material interest by appealing to your material interest, whereas in reality I get my way by convincing you that you have such an interest. 'Material interest' emerges here as more an artifact of social interaction than the occasion of it. This might seem surprising in someone like Smith, but it is in keeping with much that Smith says about labor – particularly in his *Theory of Moral Sentiments*, where he suggests that the propensity to labor is more a construct of performative interactions than the reflection of any material need.

I shall consider the implications of this for our understanding of rhetoric and economic theory. But first, let me remark on its more immediate implications for our conception of economic practice.

(2)

Smith presents us with a rather disturbing vision of the rhetoric of economic practice, of market exchange as persuasive act. For to see such exchange as primarily an assertion of the social self is to see it as a strategic gesture, even an assertion of power. Now it is hardly novel to regard market exchange as occurring within a *context* of differential power relations. In economics, this is most identified with the Institutionalist approach, though it is most developed in the writings of the Legal Realists.[20] I shall return to their views in a moment. On the whole, it seems to me that Smith accords a greater role to power than any of these in suggesting that even the *presuppositions* of commodity exchange – the notion of 'material interests', for example – may be constituted through the powers of persuasion. This implies a rather deep sense in which our status as economic subjects, hence the status of the 'economic' *per se*, are socially constituted. I shall say more about how they are so constituted in section (3).

As I have said, the writings of Legal Realism provide the most subtle analyses thus far of the interconnections between markets and power, an analysis that it employed with great success in criticizing traditional doctrines of 'freedom of contract'. A classic in this regard, but one largely forgotten, is Robert Hale's *Freedom Through Law*, a work that extends with great energy Marx's claim that the very necessity to exchange goods *as* commodities is inherently coercive.[21] (This is a thought that G.A. Cohen has recently explored with respect to labor power.) Critical Legal Studies has done the most in recent years to explore that interplay of power and markets, in particular how power animates not just the act of exchange but the structure of property rights that underlies it. But theorists like Roberto Unger press this point to insist on the *plasticity* of the market, its capacity to assume radically different institutional forms – hence, more equitable and liberatory forms.[22] In contemporary Leftist thinking, this view fits in with a certain return to Proudhon that one observes among theorists of work and technology as well.[23] Of course, the alternative to seeing the market as something that can take a potentially infinite number of forms is to regard 'the market' as a concept that is almost without meaning. Again, I shall say more about this in section (3).

Let us return now from the rhetoric of economic practice to the economics of rhetorical practice.

First encountered, Smith's focus on 'communication' rather than 'persuasion' seems to assuage some of the worries about rhetoric's truthfulness and ethics. But upon examination, rhetoric (as Smith conceives it) proves to be animated by power – not just in its ends, but in its means. What I have in mind here is best illustrated by considering the workings of metaphor. I spoke above of Nietzsche's view of metaphor as kind of conceptual coin, whose progressive circulation serves to obscure its metaphorical origin much as in circulation a coin's inscription is rubbed off. But prior to its entrance into circulation a metaphor is constituted by an initial act of exchange, the replacement of one word by another. This exchange seems 'plain and simple', but Smith's economics seems to suggest that it too may be an instance of persuasive self-assertion. But if rhetoric, like the market, is constituted by exchange thus conceived, can we hope to keep its machinations (in McCloskey's words) "in the open and under the control of reasoning"?

III

Any such hopes will hinge in part on rhetoric's relation to the articulation of truth. Let us consider how McCloskey addresses this issue.

The epistemic anxiety concerns how, or whether, economic discourse's aim of persuading others connects with a concern for presenting the truth. McCloskey has several responses to this, but the most substantial lies in his reconception of what economic theory is all about. Such theory, for him, does not aspire to represent the facts nor does it aspire to "the prediction and control assigned to it by modernist social engineering." Rather, its achievement lies in "the making sense out of economic experience."[24] Its epistemic claim rests not in being true to the facts but in being truthful to our experience. And its rhetorical dimension is presumably more compatible with this aim insofar as theory's capacity to persuade us seems connected to its capacity to engage our experience. I noted at the start McCloskey's rather benign view of received economic theory. It is evidenced here in his conviction that such theory unquestionably does "make sense out of economic experience". And any who doubt "that economics *is* science, a successful sort at that", he writes, "can become persuaded by reading Mancur Olsen's *Logic of Collective Action* or Thomas Schelling's *Micromotives and Micro-behavior* or another of the accessible jewels of the discipline."[25]

But what precisely does McCloskey have in mind in speaking of '*experience*', or '*economic* experience'? McCloskey addresses these questions only by implication, and on the whole he adopts these notions basically without criticism. He fails to consider the extent to which 'experience' may itself be constructed, in part by the discourses we employ; hence he fails to consider the extent to which our sense of an experience *as* an 'economic' experience may also be constructed – in part by the theoretical discourses we employ. I have observed that McCloskey regards rhetoric as a kind of economics, but his view of economics, hence rhetoric, is preoccupied with questions of distribution and allocation. Had he adopted a more classical view, he might have considered rhetoric's more *productive* possibilities – how, in the words of A.L. Macfie, "Economics in the full sense does not just examine the facts, it rationalizes, indeed creates the experience with which it deals."[26]

(1)

The notion that experience and subjectivity are socially constituted is one that is now associated with various post-modernist thinkers, but it has a long history. Some time ago, Lukacs argued in *History and Class Consciousness* that one can read Marx's writings as providing an account of how our experiences in and of capitalist society are constructed by that society's internal logic. But Lukacs's account is distinguished from some post-modernist ones on several points. First, he insists on the extent to which the very *quality* of our experiences is shaped by the discourses of capitalist society. For example, he maintains that there is something about commodity society which constitutes the relation of subjects to their experience as a basically *passive* one: they experience their experience as something given to them, fixed and immutable, rather than experiencing their experience as something that is itself a social product. Lukacs's term for this is '*reification*', and much of *History and Class Consciousness* is a reflection on its import for social theory: how bourgeois social theory sustains this basically passive relation to experience, and why Marxist theory is capable of piercing it. A second point pertains to the basis of this fact. Recent theory has stressed the extent to which experience is a construct of our discourses and their rhetorical resources. Lukacs agrees with this, but he insists that our conception of discourse be expanded to include forms of practice as well as reflection – what Marx termed 'the language of objects'.

How can we escape from the prisonhouse of this language? Lukacs himself accords particular importance to the fact that the working class engages not only in exchange but in production. He maintains that a critical perspective on the effects of commodification can be found in the experience of labor. But there are other places where it can be found. One of them is *history*: we can come to appreciate the extent to which experience is a social construct by exploring the extent to which it is a historical construct. I shall return to this below. Another place to look for critical perspective is *anthropology*. Lukacs's book is partly a dialogue with Rosa Luxemburg, who is remembered for her belief that capitalism's contradictions may be most apparent in places where that system has yet to achieve full hegemony – say, in 'underdeveloped' countries or in less 'developed' countries (like her native Poland). Luxemburg applied this insight to the economic realm, but it may also be applied to the experiential realm, to commodification's construction of everyday life. A notable case in point is the anthropologist Michael Taussig's *The Devil and Commodity Fetishism in South America*.[27]

But perhaps we needn't look beyond the boundaries of the capitalist world. If commodification is always an ongoing process, if, even within a capitalist society, there are always dimensions of the social order being 'colonized' by commodification and its mind-set, then one should find the basis for a more critical perspective wherever 'the language of objects' is extending its domain. Let us consider how one might find such a basis within economic theory itself.

The works by Olsen and Schelling that McCloskey cites above as 'jewels' of the discipline are certainly classics of orthodoxy. They are noteworthy in one further respect, that both are concerned with extending the economic viewpoint to (apparently) non-economic realms. Indeed, this makes for their 'accessibility' for the non-economist, which McCloskey praises. I want to consider an even more accessible work in this genre, one that owes much to the work of Downs and Schelling – Richard McKenzie and Gordon Tullock's *The New World of Economics*.

Published in 1975, the methodology of this work is professedly what

McCloskey would term 'modernist', for in proper Milton Friedman-esque fashion it rejects any concern with 'realism', staking its claim instead on its capacity to make 'correct predictions'.[28] But few predictions are made in the work, and none are assessed for their correctness. (This evokes one of McCloskey's better jibes at this methodology, that "One is not doing science merely because one has promised ultimately to do it."[29]) All of this suggests that what the authors themselves find compelling in their work lies elsewhere. And where it lies is suggested by the subtitle of their work, 'Explorations into the Human Experience'. The talk of 'human' here should not be taken too seriously (the authors' first appeal to the notion of 'human experience' in fact rejects it as a concern of science).[30] But the talk of 'experience' should be taken seriously, for their concern is with how matters apparently non-economic can be experienced *as* economic.

The nature of that concern is well reflected in their title. To speak of economics' 'New World' is explicitly to liken the authors' 'exploration' to that of previous voyagers. Implicitly, and more critically, it raises questions about their venture, not only about its goals (their quest for the realms of gold?), but about its mind-set (for Columbus was notoriously capable of mistaking imposed meanings for given ones). Are these associations too uncharitable? McCloskey himself is impressed with some of the parallels. He speaks of Gary Becker, to whom McKenzie and Tullock are much indebted, as the 'Kipling' of the 'economic empire' whose 'imperialism' involves extending its reach to "history, law, politics, crime, and the rest".[31] But McCloskey finds little troubling in this parallel, in part because, by likening Becker to Kipling rather than, say Cecil Rhodes or William Walker, he portrays the economist's 'white man's burden' as one that only proceeds through poetry, through the turns of metaphor rather than the twisting of arms – through the arm of criticism rather than the criticism of arms, as someone once said.

My favorite part of *The New World of Economics* is its chapter on 'sexual behavior'. Here the authors reconceptualize the world of erotic encounter in the terms of marginal analysis, and with mind-bending results. They suggest that we regard sex as a 'service', whose 'production' and 'consumption' can then be analyzed in terms of 'benefits' and 'costs'. Because this analysis only works if sex's 'benefits' and 'costs' are regarded as commensurable with other goods, the authors are compelled to advance these claims: (1) "The utility that one receives from a satisfying sexual experience may in the psychic realm be similar, if not identical, to the satisfaction a person receives from eating a good peanut butter sandwich" (though they note that "for most persons", sexual intercourse "probably" provides more satisfaction than eating a sandwich).[32] (2) The sexual experience has 'costs': it has *opportunity* costs, in the fact that "participants must generally forgo some opportunity which has value to them . . . The actual experience requires at least a few minutes, and this of course implies that one cannot normally do anything else of consequence at the same time." To which they add "One can, perhaps, imagine eating an apple or reading a book while producing sex for someone else," but not "playing a successful game of pool."[33] (3) Finally, let us not forget that the sexual experience has *direct* costs, most notably (in the authors' view) the "expenditure of effort." "The male orgasm alone", they note, "requires approximately 200 calories)"![34]

What are we to make of all this?

We are confronted here with the frontiers of commodification, or its rhetoric. But it is difficult to say what precisely is wrong with it. It strikes me, for example, as importantly inadequate to say that the account here rendered is simply '*false*' – if only because the analysis is in fact one that engages our experience: it engages our sense that the erotic *is* like this in certain realms, singles bars for example, and it engages our sense that the erotic in general *could* be like this, were we to conceive of and conduct our lives differently. In that event, love would become 'love' – which is how McKenzie and Tullock in fact refer to it as something in quotes, a mere figure of speech. That the erotic could be like this marks the fact that our practices are partly constituted by the concepts we apply to them. That we find this possibility troubling, that we experience this rendering of the erotic not as false so much as *weird* – this marks a further fact: that our experiences have an integrity of their own which leads them to *resist* the transformations which practice or reflection would inflict upon them.

McCloskey, it should be noted, is mindful of the weirdness here. He speaks of Gary Becker's views – which liken children to 'durable goods', specifically refrigerators (why not lampshades?) – as 'bizarre'.[35] But because he perceives the rhetorical dimension of economics as only a literary affair, he does not consider the extent to which that rhetoric may be productive of its object, hence he does not consider the reasons for *resisting* that production. Much post-modernist thinking fails to explore this resistance as well, but for opposite reasons. Because it is *too* impressed with the productive powers of discourse, because it regards that production as a kind of *creatio ex nihilo*, it can find no place for what I have termed the integrity of experience. What both should appreciate is the extent to which the relation between discourse and experience is, to coin a term, dialectical.

I read Marx's analysis of commodity fetishism as having much to say about what I have been terming the experience of weirdness, but Marx is as much concerned with our *failure* to experience the commodity order as weird – as marked by the fact that we may only experience it as such when it ventures into 'New Worlds'. This harkens back to my earlier point about the quality of our experiences, and it has much to do with how we interpret 'the language of objects', or how that language interprets itself to us – its oscillation back and forth between the figurative and the literal, much like the oscillation Derrida places (in 'White Mythology') at the heart of all metaphysical discourse. Thus Marx suggests that our resistance to looking beyond the 'appearances' of commodity society involves a penchant for mistaking the figurative for the literal. We accept money at face value, not as a metaphor for something else. But this resistance in turn expresses a penchant for mistaking the literal for the figurative: we live in a society in which money truly talks, but we do not experience the full absurdity here because we construe the fact solely as metaphor. Marx's critique of political economy provides a hermeneutic that helps reveal the full absurdity of the text.

(2)

The other path to a critical perspective on our experience is through history. While we have just considered the weirdness of conceiving the erotic as an 'economic experience', specifically a 'market' experience, our resistance seems isolated to the personal realm. That was not always so, however; for history reveals a complex process by which members of our own society came to terms with the 'market' experience – as something they found deeply troubling, but also deeply attractive.

One of the more compelling accounts of this history is Jean-Christoph Agnew's *Worlds Apart*.[36] Agnew's title refers to important facts about the *market* (that institution which I regard as constitutive of the 'economic' as we know it). For the market began as literally a 'world apart', a self-contained and much constrained place in which certain practices of buying and selling were conducted. But as it expanded under capitalism to subsume other social dimensions within it, the market continued to be experienced as a 'world apart' – as a world mysterious and disturbing in its effects. We can look to Marx, and his discussion of the Enclosure Movement, for a catalogue of the material disruptions brought on by the market's expansion. But coincident with these were the symbolic disruptions: the crisis of representation brought on by the expansion of the monetary form. The world of the market was *experienced* as one in which the relation of appearance and reality was probematic in unprecedented respects.

Only by reclaiming this fact can we reclaim the importance of that other 'world apart' of concern to Agnew, the world of the *theater*. The theater of that time gave voice to this transformation of experience, both in its structure (its players were ones who professionally marketed illusion) and in its themes (its plays were ones which made the distance between appearance and reality its central concern). But implicit throughout was the transformation of theater as a metaphor for reality as a whole. For if the notion of *theatricum mundi* once evoked the inessentiality of worldly actions, as mere appearances, it came to evoke the essentiality of our actions' appearances, in a world constituted by the market. Smith is a central figure here, not just for his theatricalization of moral sentiments but for his privileging of rhetoric – as that medium through which we impress ourselves on the eyes of others.[37]

To see experience as partly constituted by metaphor, though, is to see the expansion of the market experience is partly constituted by the expansion of a certain metaphor. 'Central place theory', for example, may study the geographical processes by which the market organism extends itself outward.[38] But the process was also a symbolic one, which we are hard pressed to describe because, as Bernard Barber has pointed out, there has been so remarkably little attention to the history of the 'market' concept.[39] Agnew relates some of its history, of how the 'market' was initially conceived of as a distinct set of actions occurring in a particular time and place, and it later was conceived as a rather indistinct and all pervasive process. But another way to put this is that over time, a time and place bound institution became the metaphor for a timeless and spaceless process. And as the 'market' has expanded, the metaphor has become continuously more attenuated: under the rubric of the 'market', the abstract process becomes an even more abstract perspective or attitude: witness the views of McKenzie and Tullock.

I have spoken of resistance to the market experience, but perhaps that resistance should involve rethinking the extent to which the 'market' is itself a rhetorical construct. There is currently work being done on the genealogy of our notions of the 'economic', some of it inspired by Foucault.[40] But on the whole, I am amazed how little critics otherwise so sensitive to the power of discourse (I am thinking of Frederic Jameson) have raised questions about the very notion of the 'market'.[41] Yet the resources for such questioning are already at hand in Karl Polanyi's writings on the deep transformations of the 'market' through time, and Fernand Braudel's attempts at distinguishing 'market' relations from capitalist relations.[42] To pursue a genealogy of the 'market' would mean questioning economic theory's current focus on the 'market' realm in much the same way that Marx questioned classical political economy's focus on the 'material' realm – questioning not only whether the concept had become too vacuous to sustain meaningful analysis, but whether, in its very vacuity, it had become hopelessly ideological.

I began by suggesting that we need a better appreciation of power's constitutive role in market exchange. I am now suggesting that a full appreciation of rhetoric's constitutive role in the economic compels us to question economics' constitutive concept, the 'market', and with it the 'economic' itself. Such a theoretical endeavor would certainly engage some of my experiences with how the world's current changes present themselves – my sense that neither Marxist orthodoxy nor market apologetics provide a speech faithful to these transformations. But the details of this endeavor must be left for another time.

University of Oregon

NOTES

[1] *The Collected Poems of Wallace Stevens* (New York: Alfred A. Knopf, 1956), p. 274.
[2] New York: Columbia University Press, 1981.
[3] Donald N. McCloskey, *The Rhetoric of Economics* (Madison: University of Wisconsin Press, 1985), hereafter cited as McCloskey; Arjo Klamer, *Conversations with Economists* (Totowa, N.J.: Rowman and Allanheld, 1983).

[4] McCloskey speaks of how "(t)he unreasonable dogmatism of both sides of any debate involving Chicago ... suggested to me that all was not well with the way economists ran their arguments." (McCloskey, p. xi.)

[5] Stanley Fish, 'Comments from Outside Economics', in *The Consequences of Economic Rhetoric*, ed. by Arjo Klamer, Donald N. McCloskey, and Robert Solow (Cambridge: Cambridge University Press, 1988, hereafter cited as Klamer, McCloskey, and Solow, p. 22.

[6] Mark Kelman, 'Trashing', *Stanford Law Review* Jan 1984 pp. 293–348.

[7] See Bruce Caldwell and A.W. Coats, 'The Rhetoric of Economists: A Comment on McCloskey', *Journal of Economic Literature* 22 (June, 1984), pp. 575-578, and A.W. Coats, 'Economic Rhetoric: The Social and Historical Context', in Klamer, McCloskey, and Solow, pp. 64–84. For a response by Marxist economists, see Stephen Resnick and Richard Wolff, 'Marxian Theory and the Rhetorics of Economics', in Klamer, McCloskey, and Solow, pp. 47–63. By far the most alarmist reaction I have seen is from philosophers, like Alex Rosenberg. See Alex Rosenberg, 'Economics is Too Important to Be Left to the Rhetoricians', in *Economics and Philosophy 5*. No. 1 (1988), pp. 129–149.

[8] Two fields where post-modern ideas have been particularly influential are history and anthropology. See Hayden White, *Metahistory* (Baltimore: Johns Hopkins University Press, 1973), and James Clifford, *The Production of Culture* (Cambridge: Harvard University Press, 1988).

[9] For discussions of Zeno, the Sophists, Plato and Aristotle on these matters see Marc Shell, *The Economy of Literature* (Baltimore: Johns Hopkins University Press, 1978), p. 38, pp. 36–39, p. 24. Shell's writings are a key reference point for anyone interested in the topics of this essay.

[10] Jacques Derrida, 'White Mythology', in Derrida, *Margins of Philosophy* (Chicago: University of Chicago Press, 1982). Nietzsche's essay can be found in his *Early Greek Philosophy* (New York: Russell and Russell, 1964).

[11] On Dudley North, see William Letwin, *The Origin of Scientific Economics* (London: Methuen and Co. Ltd., 1968), pp. 182–204.

[12] 'Economics and Evolution', in Thorstein Veblen, *The Place of Science in Modern Civilization* (New York: Viking Press, 1942), p. 66. I discussed what Veblen terms the classical 'normality' in my doctoral dissertation, particularly as it animates classical political economy's doctrines of economic crisis (see Cheyney Ryan, *Value, Capital, and Crisis: A Study in Ideology* (Boston University, 1974)). But I did not appreciate then the role that metaphor might play in all this.

[13] See Marx, 'James Mill, *Elements d'economie politique*', in *Marx/Engels Gesamtausgabe* (Berlin, 1927–1935), V. 3, pt. 1, pp. 545–546. The line is from Wallace Stevens, "Notes Toward a Supreme Fiction" in *op. cit.* p. 380.

[14] McCloskey, p. 29.

[15] *Ibid.*, p. xviii.

[16] *Ibid.*, p. 82. See also p. 175 and p. 183.

[17] *Ibid.*, p. 29.

[18] Adam Smith, *Lectures on Rhetoric and Belles Lettres* (Oxford: Oxford University Press/Clarendon Press, 1983), p. 8. For a general discussion of Smith's views on rhetoric, see Wilbur Howell, 'Adam Smith's Lectures on Rhetoric: An Historical Assessment', in *Essays on Adam Smith*, ed. by Andrew Skinner and Thomas Wilson (Oxford: Oxford

University Press/Clarendon Press, 1975). Smith regards the triumph of the market economy as bringing with it the triumph of prosaic over more poetic forms of speech (this is a view which seems to have influenced Hegel). See Kurt Heinzelman, *The Economics of the Imagination* (Amherst: University of Massachussets Press, 1980), p. 75.

[19] Adam Smith, *Lectures on Jurisprudence* (Oxford: Oxford University Press/Clarendon Press, 1978), pp. 352–353. These views are echoed in the early drafts of *The Wealth of Nations*. See *Ibid.*, pp. 571–574.

[20] On contemporary Institutionalism, see *The Economy and a System of Power*, ed. by Marc Tool and Warren Samuels (New Brunswick: Transaction Press, 1989).

[21] Robert Hale, *Freedom Through Law* (New York: Columbia University Press, 1952). Hale's ideas on 'freedom of contract' are developed, among other places, in his 'Coercion and Distribution in a Supposedly Non-coercive State', *Political Science Quarterly 38* (1923), p. 470ff. For an excellent overview of Hale's work, see Warren J. Samuels. 'The Economy as a System of Power and Its Legal Bases: The Legal Economics of Robert Lee Hale', *University of Miami Law Review* 27 No. 3–4 (1973), pp. 262–371.

[22] See Roberto Unger, *The Critical Legal Studies Movement* (Cambridge: Harvard University Press, 1986), and his more recent three volume work, *Politics*.

[23] See Charles Sabel, *Work and Politics* (Cambridge: Cambridge University Press, 1982).

[24] McCloskey, p. 175.

[25] *Ibid*, p. 56.

[26] A.L. Macfie, *An Essay on Economy and Value: Being an Inquiry into the Real Nature of the Economy* (London: MacMillan, 1936), p. 35. I found this quote in Heinzelman, *op. cit.*, p. x.

[27] Michael Taussig, *The Devil and Commodity Fetishism in South America* (Chapel Hill: North Carolina University Press, 1980).

[28] Richard McKenzie and Gordon Tullock, *The New World of Economics* (Homewood, Ill.: Richard Irwin, Inc., 1975), p. 6.

[29] McCloskey, p. 43.

[30] McKenzie and Tullock, *op. cit.*, p. 6.

[31] McCloskey, p. 76.

[32] McKenzie and Tullock, *op. cit.*, p. 50.

[33] *Ibid.*, p. 53.

[34] *Ibid.*, p. 53.

[35] McCloskey p. 76. The bizarreness is obviously not a problem in his profession's eyes. Becker is the most recent Chicago economist to receive the Nobel Prize in his field.

[36] Jean-Christophe Agnew, *Worlds Apart* (Cambridge: Cambridge University Press).

[37] See the article by Howell, cited in note 17.

[38] Central place theory was introduced in the 1930's by German economic geographers Walter Christaller and August Losch. See, for example, E.A.J. Johnson, *The Organization of Space in Developing Countries* (Cambridge: Harvard University Press, 1970).

[39] Bernard Barber, 'Absolutization of the Market: Some Notes on How We Got from There to Here', in *Markets and Morals*, Gerald Dworkin ed. (New York, 1977).

[40] See Nicholos Xenos, *Scarcity and Modernity* (New York: Routledge, 1989).

[41] See the various discussions of the market in Frederic James *Post-Modernism: The Cultural Logic of Late Capitalism.*

[42] See, for example, Karl Polanyi, 'The Economy as Instituted Process', in *Trade and Market in Early Empires*, Karl Polanyi, Conrad Erensberg, and Harry Pearson, eds. (New York, 1957). A good introduction to Braudel's ideas on the market is found in Immanuel Wallerstein, 'Braudel on Capitalism, or Everything Upside Down', *Journal of Modern History, 5*. 63 No. 2 (June 1991), pp. 354–362.

KRISTIN SHRADER-FRECHETTE

UNSAFE AT ANY DEPTH: GEOLOGICAL METHODS, SUBJECTIVE JUDGMENTS, AND NUCLEAR WASTE DISPOSAL

In one of the most ambitious technological programs ever undertaken, the US Department of Energy (DOE) is now developing plans to construct the world's first permanent facility for high-level commercial radioactive waste. Planned for a site in Yucca Mountain, Nevada, the facility is supposed to keep the waste isolated from the biosphere for more than 10,000 years. As of 1991, the DOE has spent more than $3 billion on assessments of site suitability. Apart from the scientific and technical questions raised by the proposal to build such a facility, the Yucca Mountain repository raises a whole host of ethical and policy questions, for example: Is it fair for present persons to impose the burden of nuclear waste on future generations? Is it equitable for most Americans to receive the benefits of fission-generated electricity while Nevadans bear the risks and costs of nuclear waste? Does the permanent disposal of radioactive waste present unacceptable burdens in terms of potential infringements on civil liberties and on rights to due process? Is it ethical to impose nuclear waste on those who have not consented to the risk and who will have received only minimal benefits from it?

1. REPOSITORY RISK ESTIMATES RELY ON PARTIALLY SUBJECTIVE JUDGMENTS

Uncertainty is perhaps the single most important technological and ethical issue raised by the proposed Yucca Mountain facility. How certain is certain enough when one is dealing with a potentially catastrophic technology? How much certainty ought citizens to demand when they face government-imposed risks? Is the best science and the best technology certain enough to deal with permanent disposal of radioactive waste?

Claude Bernard, the foremost French physiologist of his day, wisely warned that "true science teaches us to doubt."[1] It teaches us that all science and technology rely on partially subjective judgments about the merit of factors such as experimental designs, theories, standards of proof,

C.C. Gould and R.S. Cohen (eds.), Artifacts, Representations and Social Practice, 501–524.

observations, explanations, curve-fits, and estimates. However, one measure of the degree to which we should doubt any particular scientific or technological conclusion is the extent to which it relies on hidden, unjustified, or controversial judgments about methods. If a conclusion is extremely sensitive to such a judgment, then it is important to know how plausible the judgment is. Otherwise one could place misguided confidence in an unreliable conclusion.

In arguing for permanent geological disposal of high-level radioactive waste, proponents often rely on the scientific conclusion that repository risk estimates are low. Such estimates, of course, depend on partially subjective judgments. Department of Energy (DOE) researchers, for example, have admitted that their estimates regarding geohydrology, geochemistry, and waste-container lifetimes are 'nonconservative' and 'realistic' regarding the proposed Yucca Mountain waste facility.[2] The DOE scientists justify their 'nonconservative' judgments by claiming that limited data force them to "rely heavily on professional judgment attended by expressions of the level of subjective confidences in findings."[3] Whether we call them 'professional judgments' or 'subjective judgments,' nonconservative risk estimates present a problem for all technological risk assessments. Proponents of permanent repositories for nuclear waste, however, are likely to believe that such subjective judgments do not threaten the objectivity of their claims about the safety and feasibility of disposal. Recognizing the presence of the judgments, they may believe that they have only an insignificant effect on their conclusions. Opponents of geological disposal, on the other hand, are likely to believe that estimates of repository risks are highly sensitive to questionable judgments, and that such judgments could play havoc with allegedly low estimates for the risks of disposal. Who is correct, the proponents or the opponents? How justified are the partially subjective judgments associated with risk estimates for high-level radwaste disposal?

One of the best ways to determine both the defensibility of disposal-related judgments – and the degree to which technological risk conclusions are sensitive to them – would be to examine some state-of-the-art technology assessments or quantitative risk assessments (QRAs) of proposed repositories. In the US, there is only one site, Yucca Mountain, Nevada, that is currently being considered for a high-level repository. Because the Yucca Mountain studies represent some of the latest, most expensive, and best available risk assessments, they provide a number of clues about the status of methodological judgments used in risk estimation. Indeed the US National Academy of Sciences has confirmed that the Yucca-Mountain assessments are 'state-of-the-art.'[4] In this essay, we shall analyze four of the partially subjective methodological judgments that are central to Yucca-Mountain risk estimates. Most of these judgments are paradigmatic instances of the types of subjective assessments that underlie all assurances of repository feasibility and safety. Moreover, we shall argue that at least some of these subjective judgments threaten the validity of claims about the suitability of Yucca Mountain for high-level radioactive waste. We also shall argue that similar difficulties are likely to plague any optimistic risk estimates associated with permanent geological disposal of nuclear wastes.

Probabilistic risk-assessment conclusions about Yucca Mountain, while typically optimistic about site suitability, are also highly sensitive to a number of assumptions or partially subjective judgments. For example, most site assessors have concluded that, as US regulations require, no radionuclides are likely to reach the water table at Yucca Mountain in less than 1,000 years. Other assessors, particularly those from the state of Nevada, have concluded that groundwater travel time from the disturbed environment to the accessible environment could be as short as 978 years, and that, as a result, Yucca Mountain would not meet government standards for a repository.[5] Conclusions about groundwater travel time, however, are highly sensitive to a number of assumptions about the rate of fracture flow and infiltration, e.g., 0.01 mm/year.[6] Conclusions about radionuclide migration are also highly sensitive to a number of partially subjective judgments about the reliability of the simulations used, the validity of a variety of extrapolations and interpolations, the lack of need for more site-specific data, and so on. If different methodological judgments and assumptions were used, then quite different conclusions about radionuclide migration and repository containment could very likely be drawn. As one group of assessors put it, for example: "the use of a low net infiltration rate for this problem made it not unexpected that there was no predicted radionuclide release to the water table. Other rates would probably result in considerably different transport results."[7]

Similarly, in a review of the Yucca Mountain Site Characterization Plan, evaluators criticized the fact that much of the report is comprised of theoretical designs based on large assumptions. As a result, claimed the reviewers, the likely success of the designs proposed within the report is highly questionable from a technical point of view. Further, the evaluators charged that the problematic designs are based on technical errors and on oversimplifications. They concluded that many of the concepts within the Yucca Mountain plan have not been developed to the degree necessary to insure scientific integrity.[8] The purpose of this essay is to examine several examples of the ways in which the Yucca Mountain risk assessments – despite the areas in which they are technically successful – may have gone wrong, either through logically questionable inferences or problematic, partially successful, methodological judgments, and to learn what, if anything, these difficulties tell us about the general question of permanent geological disposal.

The forthcoming analysis of problematic methodological judgments will uncover a number of specific difficulties which, although they are neither mutually exclusive nor exhaustive, are typical of quantitative risk assessments (QRAs) of proposed repositories. Subjective judgments about model reliability, for example, often intersect with other judgments about the plausibility of certain simplifications in the treatment of site phenomena. Our point in assessing some of these partially subjective, methodological judgments is neither that our survey is exhaustive nor that methodological judgments ought to be avoided. Such judgments are unavoidable in both science and in technological risk assessment. Hence, the problem is not the methodological judgments, as such, but (1) the frequent absence of any ideas about the limits of error or uncertainty associated with such judgments and (2) the tendency of technological risk assessors to be overconfident about the effects of their methodological judgments.

2. SUBJECTIVE JUDGMENTS ABOUT LABORATORY PREDICTIONS

One of the most problematic methodological judgments used in estimating the risk at Yucca Mountain is that laboratory analyses of samples provide an adequate basis for conclusions about field behavior.[9] Rather than doing a sufficient number of experiments in the field, for example, hydrologists and geologists at Yucca Mountain have done laboratory experiments on cores of tuff taken from the field.[10] Using laboratory data about site geochemistry, about behavior of crushed (not solid) tuff, about sorption, and about retardation of radionuclide transport, for example, technological risk assessors extrapolate from laboratory results in order to draw conclusions about all these factors on site.[11] One of the things assessors wish to determine in their laboratory studies is whether 'fracture healing' (in which fractured rock samples become sealed when water flows through them at elevated temperatures) can prevent leakage of radionuclides.[12] Because the assessors' conclusions about field behavior are *extrapolations* from premises about laboratory behavior, rather than deductions from premises about field behavior, they are obviously questionable. Moreover, because of the many disanalogies between laboratory and field behavior at Yucca Mountain, the conclusions are questionable even on inductive grounds.

Obviously, there are many disanalogies between laboratory and field behavior of a core of tuff, any one of which could account for erroneous estimates of future radionuclide transport at sites like Yucca Mountain. Admittedly, some assessors have been sensitive to these disanalogies and have emphasized the importance of field studies to check models and samples.[13] Nevertheless the very use of samples investigated in a laboratory setting requires one to make untested methodological judgments about the degree of fit between the laboratory conclusions and those of the field. Like judgments about sampling, judgments about the accuracy of using laboratory data to describe field conditions occur in many areas of science. The problem is not with the methodological judgment that one can use the laboratory data in this way, but with technological risk assessments of sites like Yucca Mountain, studies that have failed to test this judgment as completely as possible by experiments in the field. In other words, the problem is not with laboratory data and associated methodological judgments, but with applications that appear lacking in scientific conservatism and adequate field confirmation.

3. SUBJECTIVE JUDGMENTS ABOUT FRACTURED, UNSATURATED MEDIA

Another methodological judgment central to technological risk assessment of many proposed high-level repositories, like Yucca Mountain, is that despite the discontinuous, nonhomogeneous nature of the fractures in the geological substrate, nevertheless it is possible to predict, with reasonable accuracy, the path and rate of potential radioactive migration, once containment of the canisters has been breached. This value judgment is a potential problem for virtually any repository because there might be fracture flow, and because it is 'generally the case' that all rock sites are fractured.[14] Yet, most fractures are not uniform, and hence they cannot easily be modelled. Such heterogeneities prevent a 'comprehensive flow analysis,' as at the Stripa Mine in Sweden.[15] Modelling the fractured substrate is especially problematic for the unsaturated zone at Yucca Mountain, because very little is known about it. Also, the volcanic tuff itself is quite heterogeneous, with eight orders of magnitude of variation sometimes occurring in the saturated hydraulic conductivities.[16] Nevertheless, risk assessors claim to have demonstrated that "methodologies are available for characterizing fracture flow" through saturated fractures.[17] They say that such methods are 'suitable' for determining fracture flow, or 'adequate' and 'valid.'[18] An alternative methodological judgment, one to which representatives of the state of Nevada appear to subscribe, is that given the discontinuities and heterogeneities in the fractures, such characterizations are impossible because the flow is likely to be random and unpredictable. Indeed, recognizing that travel time through fractures is faster than through any rock matrix,[19] representatives of the state of Nevada have warned that the great uncertainly regarding flow through heterogeneous, fractured material is one of their seven major hydrogeological 'concerns' about Yucca Mountain.[20]

At Yucca Mountain, technological risk assessors are attempting to model the site and to predict possible releases of radionuclides through the fractures.[21] Because they have inadequate empirical data,[22] and because behavior of heterogeneous, partially saturated, or saturated sites is not at present mathematically tractable, they use models that assume a more homogeneous hydrogeological situation. The models assume, for example, uniformly spaced or parallel fractures.[23] Often they ignore the fact that drying is more or less rapid along different fractures.[24] Yet, all these heterogeneities could be enough to alter substantially the predicted rates of radionuclide migration at Yucca Mountain and at other proposed sites. Such predictions are scientifically suspect, at least in part, because they are not derived from any general scientific laws about fracture flow in a heterogeneous environment. Indeed there are no precisely predictive laws for such an environment. The methodological problem with partially subjective judgments about fracture flow is not, however, that they rely on no predictive general laws, since much of science involves no such laws. Rather, the real difficulty here is that US government regulations for repositories require precise predictive power over thousands of years, whereas the science available for QRAs of repository sites is not as precise and predictive as regulations appear to require.

One of the major methodological judgments about hydrogeologically modelling a heterogeneous site has to do with rate of infiltration. Infiltration of water into a site is largely a function of precipitation which, as we mentioned already, is highly variable. Infiltration, in turn, is highly variable from one spot to another at Yucca Mountain, for example, because of differences in the underlying hydrogeology, especially the presence of fractures. In some Yucca Mountain studies, researchers showed that rainwater on site evaporated and had no influence at depth; yet they cautioned, for example, that five feet of water entered boreholes, via fractures, at one of their sampling spots.[25] This sort of heterogeneity, with no infiltration at depth anywhere, except for one place where five feet of water accumulated, is exactly the sort of nonuniformity that could play havoc with methodological judgments about our ability to predict radwaste migration.

Infiltration is particularly important at sites like Yucca Mountain because although less than three percent of precipitation onsite likely infiltrates deeply enough to recharge the saturated zone, nevertheless concentrations of infiltration in time or space, according to some assessors, could supply enough flux to cause fracture flow on site.[26] Assessors have warned that the rock at Yucca Mountain could act as a sink for infiltrating water, and that "significant winter recharge could occur . . . due to snow melt."[27] Climate change to wetter conditions could also cause more infiltration and enhanced fracture flow.[28] Given the possibility of fracturing because of the heat of the waste,[29] and given the possibility of intense rainfall at a particular location, the assessors' claim that fracture flow is 'unlikely' or 'not credible'[30] is a questionable methodological judgment. It is questionable in part, because fracture flow is so sensitive to a relatively small change in the percolation rate. If the percolation rate increases by only one order of magnitude above conservative values, this might be enough to initiate fracture flow. Fracture flow, in turn, could significantly reduce the travel time of water between the repository and the water table.[31]

A few assessors have warned about the reliability of the models for fractured media,[32] cautioning that the hydraulic parameters are 'poorly known,' and basic issues are 'unresolved.'[33] Such worries are likely part of the reason why representatives for the state of Nevada, responding to the conceptual design of the Yucca Mountain Site Characterization Plan, claimed that "one of the most disturbing aspects is the lack of attention given to the geohydrologic setting and its performance at the Yucca Mountain site."[34] Likewise, when the 14 DOE peer reviewers assessed the 1992 *Early Site Suitability Evaluation* (ESSE) for Yucca Mountain, they said that the biggest conceptual issue in the ESSE was whether there was likely to be fracture flow at Yucca Mountain and therefore groundwater travel time from the disturbed zone to the accessible environment in less than 1,000 years.[35] The DOE peer reviewers also criticized the ESSE for lacking a detailed discussion of fracture flow and for relying on simplified models for a heterogeneous site.[36] Hence even the outside evaluators appear disturbed by the partially subjective judgments about precise prediction of flow in fractured media at Yucca Mountain.

Some researchers, however, appear to believe that the site models can be 'verified' or are 'valid.'[37] One difficulty with methodological judgments about using models to predict fracture flow, however, is that past judgments about flow through fractures (at other radioactive radwaste sites) have been overly optimistic about radioactive containment, and they have resulted in serious error. Indeed the methodological judgment at the Maxey Flats radwaste facility, that no flow through fractures was likely, was one of the key mistakes that led to offsite release of radionuclides, including transuranics, less than a decade after the facility was opened.[38] This methodological judgment about Maxey Flats is uncomfortably similar to the claim that fracture flow is 'unlikely' or 'not credible' at Yucca Mountain.[39] On the surface, the Maxey Flats, Kentucky facility appears to have little to teach us about sites like Yucca Mountain. In most respects, the Nevada and Kentucky cases are quite dissimilar. The Nevada site, in an area of low rainfall, is for deep burial of high-level, heavily shielded radwastes, while the Kentucky facility, in a region of high rainfall, was for unshielded, shallow land burial of low-level and transuranic wastes. Nevertheless, the two sites have at least two apparent similarities. Both are hydrogeologically heterogeneous sites, and both are plagued by fractures in the underlying rock.

The Kentucky predictions went wrong, in part, because the scientists focused on the alleged impermeability of the shale underlying the site but underestimated the potential for migration through its hairline fractures. Similarly, at least some risk assessors have argued that Yucca Mountain is full of fractured tuff through which contaminated water could migrate.[40] Another similarity between the two sites is that, once the high-level radwaste canisters and the fuel cladding at Yucca Mountain have been breached, allegedly between 300 and 1000 years after emplacement, then the underlying geology and hydrology will determine the extent of the waste migration because of precipitation, infiltration, and leaching. Ultimately, therefore, both high-level and low-level sites face the same problems of geological containment. Hence, some of what happens regarding low-level radwaste migration, decades after the waste is stored, could provide a preview of part of what happens regarding high-level radwaste migration centuries after it is deposited. Thus, low-level sites, like Maxey Flats, and proposed high-level sites, like Yucca Mountain, may have some pregnant similarities, despite their obvious dissimilarities. These similarities suggest that our mistakes at Maxey Flats may have something to teach us about potential problems at sites like Yucca Mountain.

At Maxey Flats, the radioactive releases occurred within nine years, although they were predicted to occur after centuries or even after thousands of years.[41] Scientists employed by promoters of the facility presented simple, low-permeability models for the underlying shale, and policymakers concluded that the possibility of offsite migration was "essentially nonexistent."[42] The US Geological Survey (USGS) Project Director, however, later explicitly said that any simple, quantitative model for the site was impossible because Maxey Flats is a poorly permeable, fractured, geological system. Any prediction of flow paths was impossible, he said, because of the highly irregular hairline fractures and the fracture intensities.[43] Because prediction of fracture flow at Maxey Flats would require detailed information about the spatial distribution and hydraulic properties of each of many hairline fractures in a variety of successive strata (e.g., Nancy Shale, Henley Shale, Ohio Shale, and so on), the USGS Project Director concluded that groundwater flow at Maxey Flats could not be predicted.[44] Any model, he said, would presuppose conditions for its accuracy that were not met at Maxey Flats, conditions such as "uniform transmission of water through the rocks."[45] Although his warnings were ignored by state promoters of the site, he concluded that "hydraulic conditions do not meet the requirements for the method of analysis."[46] His argument years earlier, about the Maxey

Flats site, is quite similar to current claims, by Yucca Mountain assessment reviewers, that the untested Yucca Mountain models are not appropriate for the complex site hydrogeology.[47] In both cases, a major threat was/is fracture flow, and in both cases, risk assessors were/are using untested models that may not account adequately for fracture flow.

Given the fractured tuff and the heterogeneity of the hydrogeological site, caused in part by incomplete information, by irregular fractures, and by no information about the long-term future, one might well be justified in concluding at Yucca Mountain, just as one concluded at Maxey Flats: the hydraulic and geological conditions do not meet the requirements for the idealized methods of analysis that presuppose uniformities. Indeed, at any proposed geological repository facing similar heterogeneities, the same charge would hold. If site conditions such as fractured tuff do not meet the conditions for analysis, then it could be quite dangerous to make the methodological judgment that one can accurately predict radwaste migration in the heterogeneous, fractured media at sites like Yucca Mountain. Just as the DOE peer reviewers warned in their 1992 evaluation, methodological judgments about fracture flow could represent a serious deficiency in Yucca Mountain QRAs and in repository studies generally.

4. SUBJECTIVE JUDGMENTS THAT INTERPOLATIONS ARE ACCEPTABLE

Still another major methodological judgment in many repository risk assessments, including some of those at Yucca Mountain, is that accurate interpolation of sparse and irregularly spaced data is possible. A number of technological risk assessors make this assumption,[48] even though the data are not uniform, even though infiltration, for example, occurs at some locations but not at others,[49] and even though low groundwater-flow rates are guaranteed only for unsaturated conditions where infiltration has not compromised the repository site.[50] Heterogeneities in the hydrology and geology onsite indicate that curve-fitting techniques could fail to provide an accurate picture of hydrogeology at sites like Yucca Mountain. This may be why, for example, the state of Nevada listed uncertainties regarding infiltration and groundwater travel time as among its seven major 'concerns' regarding site hydrogeology.[51] Of course, the problem here is not with partially subjective judgments about interpolation, as such, because all scientists must make judgments about interpolation. The problem is that the use of such judgments about interpolation at Yucca Mountain appears insufficiently careful and conservative, given the heterogeneity of the site.

5. SUBJECTIVE JUDGMENTS THAT HUMAN ERROR IS NOT SIGNIFICANT

Another problematic judgment that is implicit in a number of repository risk assessments, including many of those at Yucca Mountain, is the presupposition that human error is not a significant contributor to risk at the site. In one of the official US Department of Energy (DOE) documents, this value judgment is made quite clear: "Because of the plans to mark the site and to use other passive means to prevent human intrusion into the repository, and because of the intention to ensure that the site will have little value in unique, large-scale human activities that would inadvertently affect the repository are not likely to take place at the site. Accordingly, the only scenarios currently considered for this case are those involving occasional random, exploratory drilling at the site."[52] Indeed, US DOE risk assessors seem almost completely to ignore the effects of human error in thwarting repository containment. The DOE environmental assessment for Yucca Mountain concludes: "No impediments to eventual complete ownership and control [of Yucca Mountain] by the DOE have been identified. Therefore, on the basis of the above evaluation, the evidence does not support a finding that the site is not likely to meet the qualifying condition for post-closure site ownership and control."[53]

As the policy position of the DOE makes clear, the DOE discounts not only large-scale human error and inadvertent activities causing intrusion into Yucca Mountain, but also organized sabotage and terrorism. Indeed, virtually 99 percent of the DOE research on Yucca Mountain, as revealed in its own Program Assessment, is on physical-science topics related to the risk. There is almost nothing on possible socioeconomic problems caused by the facility or on human aspects of risk.[54] Indeed, in the 1992 ESSE for Yucca Mountain, the DOE researchers bluntly stated: "unmitigatable social and/or economic impacts are not expected to occur" at Yucca Mountain.[55]

Typical Yucca Mountain risk assessments, apparently following the position of the DOE, also make the same controversial methodological judgment that human error at the site will be insignificant. For example, assessors assume that risk managers will have to deal only with radioactive "releases that would be expected from the repository under undisturbed conditions," that is, conditions where human accident, error, or sabotage does not disturb the site.[56] After explicitly ignoring human errors, accidents, and other disruptive features, the same technological risk assessors conclude that "even for the highest credible flux . . . releases of radioactivity to the accessible environment in 10,000 years after closure are significantly less than limits imposed in the draft standards (40 CFR 191) for environmental radiation protection."[57] After ignoring such human disruptions of the site, risk assessors typically conclude that Yucca Mountain will meet government environmental requirements; indeed, after ignoring these human factors, they usually put a quite precise figure on the risk involved. For example, they say that the site will cause "less than one health effect every 1,400 years."[58] Such confident and optimistic predictions seem questionable in the face of assessors' tendency to ignore an important class of risk factors, namely, human causes of repository failures.

Because the vast majority of the DOE-funded Yucca Mountain studies deals with purely physical, non-human processes, it appears that DOE policymakers and assessors presuppose either that human error is unimportant at Yucca Mountain, or that it is poorly understood and therefore can be ignored in favor of what is better understood (physical-transport mechanisms for radwaste). Both judgments about Yucca Mountain fly in the face of previous inductive evidence about estimates of human error at other facilities and at Yucca Mountain. For example, several risk assessors calculated the likelihood of human disruption onsite after the first century. They said that the probability of drilling at Yucca Mountain could be as high as 5×10^{-2} to 2×10^{-3} per year.[59] If the estimates of some scientists about human disruption at Yucca Mountain are so high, it is questionable whether many assessors ought simply to ignore this human factor, and then conclude that the risk of disruption at Yucca Mountain is quite low. Given that there is an active gold mine within 15 miles of Yucca Mountain and several inactive mines within 10 miles,[60] it does seem reasonable to believe that future drilling might occur nearby. Moreover, it is conceivable that future generations might try to use the buried waste, because of its "potentially valuable energy."[61]

The questionable judgment by the DOE – that one can ignore the human-caused risks at Yucca Mountain – is all the more problematic, both because the US Environmental Protection Agency (EPA) does take account of human error in its assessment of the impacts associated with repositories,[62] and because the EPA claims that "the most important cause of radionuclide release appears to be human intrusion."[63] Also, it is obvious that many precise aspects of human behavior far into the future are not predictable.[64] Moreover, even in a probabilistic sense, it is not clear that good reasons support the judgment that human error at the Yucca Mountain site will be insignificant. After all, if the EPA claim is correct, and if the human contribution to technological accidents is typically as high as 90 percent (in the case of the airlines and the chemical industry),[65] then human error/intrusion could be a significant repository risk. After all, US Office of Technology Assessment authors conclude that more than 60 percent of accidents involving hazardous materials are the result of human error.[66] Hence, it is not surprising that some reviewers have explicitly criticized the DOE failure to give sufficient attention to human error at Yucca Mountain. They have charged that "the most striking weakness of risk assessments for the proposed Yucca Mountain repository is their failure to address the social amplification of risk."[67]

Human error is one of the major contributors to risk in the cases where government radioactive-waste facilities have caused large-scale contamination and threats to welfare. At the Maxey Flats site, for example, the facility was closed only a decade after it was opened in 1962. Offsite migration of radionuclides caused the closure, and the facility received much government attention and adverse publicity. Yet, more than 15 years after its well documented problems of poor containment, the government has failed to oversee and clean up Maxey Flats effectively. Indeed, social and human factors amplified the risk even after is was brought to public attention. As late as 1987, wells to monitor contamination on site revealed that tritium levels in the Maxey Flats groundwater exceeded the Environmental Protection Agency limit by 6 orders of magnitude.[68] The Maxey Flats situation suggests not only that humans cause a significant amount of the risk from radioactive waste management, but that once their failures are brought to light, both individuals and institutions often amplify, rather than correct, the risk.[69] One important question also raised by the Maxey Flats case is this: Why should one believe that humans at Yucca Mountain and at other proposed repositories will not amplify the radwaste risk, given that they have done so at Maxey Flats, at Fernald

(Ohio), at Hanford, at Savannah River, at Oak Ridge, at Rocky Flats, at Idaho Falls, and at Lawrence Livermore? Indeed, many corrections for safety and environmental problems were not made adequately at these and other facilities. The risk was amplified, in large part, because the government did not wish to bear the cost.[70] The DOE has admitted that it caused soil and water contamination at all these places, that it withheld information from the public, that it violated its own regulations, that it failed to tell workers when they were handling plutonium, and that it used retaliation against employees who were whistleblowers.[71]

The US General Accounting Office (US GAO) pointed out, for example, that Savannah River, Rocky Flats, Hanford, Idaho Falls, Maxey Flats, and West Valley all have extensive onsite contamination. Indeed, the US GAO identified 1,061 DOE sites that currently require a massive cleanup; the government estimates that this cleanup will require $130 to 300 billion.[72] Presumably the existence and continuation of contamination at these major DOE facilities indicate that no hazardous site is free from problems caused by human error.[73] Indeed, the US GAO said that contamination at DOE facilities is so extreme that the government knows precisely neither the extent of it nor what is needed to clean it up. Government representatives have also testified that the DOE has lost its credibility,[74] and yet that the complacency at the DOE is similar to that existing at Chernobyl before the accident.[75]

At the nuclear feed materials plant at Fernald, Ohio, for example, the company (under DOE supervision and instruction) repeatedly broke environmental regulations, contaminated the site, and illegally stored radioactive materials. Fernald workers even processed plutonium for 22 years without wearing respirators.[76] DOE managers gave the facility repeated permission to break its own regulations and to make regular discharges of uranium into nearby rivers, permission that was given from the very beginning of the operation of the plant.[77] In-house recommendations for correcting safety problems and reducing risks were ignored, even when workers died.[78] At the Fernald facility, DOE authorities made "no attempt" to monitor releases or to "take corrective measures" once problems had occurred.[79] Indeed, the DOE shipped 2,500 tons of thorium sludge in corroding steel containers to Fernald, to be stored there, even though Fernald had no use for the waste.[80]

At the Maxey Flats facility, the government-supervised company (Nuclear Engineering Company, NECO, later called "US Ecology") illegally dumped radioactive materials offsite and refused to admit investigators to the facility once offsite radioactivity had become a problem.[81] For government administrators even to allow Maxey Flats to dump 950 pounds of special nuclear materials (plutonium, uranium-233, and enriched uranium 235), in an area with average annual rainfall of 48 inches, in shallow soil trenches 25 feet deep,[82] suggests human error, ignorance, mismanagement, or corruption. Moreover, given DOE's previous violations of environmental regulations and its secrecy, as illustrated in such problems as the Fernald facility, it is clear that assessors may be wrong to discount the problem of human error at any repository, including Yucca Mountain. Discounting human error at the proposed repository is extremely questionable because external reviewers have already criticized the DOE for lack of coordination at the Yucca Mountain site,[83] for ignoring the social contributors to risk,[84] and for allegedly using Yucca Mountain as a justification for its own self- and pre-determined policies.[85]

Representatives from the Hanford nuclear facility and from Yakima nation, near Hanford, likewise have criticized "DOE's reckless environmental and safety practices at Hanford."[86] Texans have testified before Congressional committees that DOE's handling of the proposed radwaste site in their state caused them to have "no confidence" in the DOE's technical ability and responsiveness to legitimate public concerns.[87] Residents of Mississippi accused the DOE of not sharing data about their state's possible radwaste site with them,[88] and numerous state geologists have accused the DOE of technical deficiencies caused by its own political agenda.[89] Nevadans have charged DOE with secrecy in the face of Yucca Mountain studies, with not consulting with them, and with failure to fund Nevada requests for research about the site.[90] Indeed, Nevada had to sue the DOE to get funding for independent technical studies of the Yucca Mountain site.[91] As a result, the US General Accounting Office (GAO) concluded that the DOE has not allowed the states and Indian tribes to participate in the high-level waste siting program to the extent intended by the Nuclear Waste Policy Act of 1982.[92] All these problems suggest not only that human error is likely to occur at Yucca Mountain and other proposed geological repositories, contrary to what many government risk assessors claim, but also that such error may arise in part through the bias, mismanagement, and lack of concern of the very government agencies that ought to be protecting the public interest.

One common denominator among many of the problems of human

error may be the institutional shortcomings of agencies such as the DOE. To the degree that the difficulties are institutional, then the human risk-amplification problem may be even greater than we have suggested so far. Hence, making the judgment that human error will not be a significant contributor to repository risks – at places like Yucca Mountain – seems highly questionable, in part because the judgment appears capable of compromising QRA conclusions about repository safety.

6. CONCLUSION

Many of the four methodological judgments (e.g., about the reliability of conclusions based on interpolations) surveyed in this essay are widely used throughout science and technological risk assessment. Hence, the difficulty is not with using these partially subjective judgments, as such, but with drawing stronger conclusions from them than are warranted by the existing scientific theories and inductive data. At Yucca Mountain, for example, we argued that some of the DOE site-suitability conclusions appeared to be stronger than were warranted by the limited quantity and the questionable representativeness of the evidence. In general, however, there are no basic problems with employing many methodological judgments (such as those related to sampling, extrapolation, and use of laboratory data) in science and technological risk assessment. Indeed, in many areas of science, there simply are no viable alternatives to making a variety of such judgments.

However, a number of methodological judgments used in the Yucca Mountain case do present difficulties. They appear to exemplify 'bad science.' They suggest that, were certain judgments – regarding risk estimation, for example – handled differently at Yucca Mountain, the scientific conclusions about the site might be somewhat different. Some of these examples of 'bad science' include the judgments that laboratory data provide an accurate account of field conditions; that one can precisely predict site hydrology, despite fractured, inhomogeneous conditions; and that, despite the site heterogeneities, one can interpolate to obtain precise and representative hydrological and geological data.

Obviously, if 'bad science' were responsible for all of the questionable methodological judgments about technological risk estimation that we have criticized at Yucca Mountain, then these problems would argue only against repository sites chosen on the basis of flawed studies. 'Bad science' would not argue generally against permanent, geological disposal of radwaste. But this raises the question whether the problems we have exposed in this essay represent difficulties that might be avoidable at other proposed repository sites or in other technological risk studies. Do these problems with partially subjective judgments about risk estimation argue only against 'bad science' and 'bad risk assessment' or also against any permanent radwaste repositories?

At least one of the methodological judgments criticized in this essay appears to present obstacles to any permanent, geological disposal of radwastes anywhere. This is the judgment that human error and risk amplification do not threaten the security of a permanent repository. If the arguments in this essay have been correct, then both short-term scientific data and the potential for human error militate against any precise, optimistic predictions regarding long-term (10,000 years) repository suitability and safety. Indeed, as we argued, and as the Yucca Mountain peer reviewers and various social scientists admitted, beyond 50 or 100 years, precise predictions about human and institutional errors and behavior are not reliable. This being so, short-term geological data and the potential for human intrusion into the repository exacerbate the more general problem of accurate, long-term prediction.

US regulations stipulate that precise, thousand-year predictions – regarding either geological factors like seismicity or social factors like repository intrusions aimed at securing natural resources – are required to guarantee the safety of a permanent repository. If such accurate, long-term predictions are at present impossible, in part because of the potential for human error, then a scientifically defensible siting of a permanent repository, at least at present in the US, likewise appears impossible.

University of South Florida

NOTES

[1] C. Bernard, *Bulletin of the New York Academy of Medicine 4* (1928), 997.

[2] US Department of Energy, *Nuclear Waste Policy Act, Environmental Assessment, Yucca Mountain Site, Nevada Research and Development Area, Nevada*, DOE/RW-0073, 3 vols. (Washington, DC: US DOE, 1986), vol. 3, p. C.5–55; hereafter cited as: DOE, NWPA-Yucca.

[3] DOE, NWPA-Yucca, p. C.5–56. The same admission is made in US DOE, *Nuclear*

Waste Policy Act, Environmental Assessment, Reference Repository Location, Hanford Site, Washington, DOE/RW-0070, 3 vols. (Washington, DC: US Department of Energy, 1986), vol. 2, p. 6–294; hereafter cited as: DOE, NWPA-Hanford.

[4] See I. Amato, 'Dangerous Dirt: An Eye on the DOE' *Science News 130*, no. 14 (4 October 1986), 221.

[5] See, for example, R. Bryan, *State of Nevada Comments on the US Department of Energy Draft Environmental Assessment for the Proposed High-Level Nuclear Waste Site at Yucca Mountain*, 2 vol. (Carson City, NV: Nuclear Waste Project Office, Office of the Governor, 1985), vol. 1, pp. I-42 and I-43; R. Peters, *The Effect of Percolation Rate on Water Travel Time in Deep, Partially Saturated Zones*, SAND85-0854 (Albuquerque, NM: Sandia National Labs., 1986), p. 32; see G. Sawyer, 'Statement,' in US Congress, *Nuclear Waste Program*, Hearings Before the Committee on Energy and Natural Resources, US Senate, 100th Congress, First Session on the Current Status of the Department of Energy's Civilian Nuclear Waste Activities, January 29, February 4 and 5, 1987, Part 1 (Washington, DC: US Government Printing Office, 1987), pp. 709–712.

[6] R. Bernard and H. Dockery, *Technical Summary of the Performance Assessment Calculational Exercises for 1990 (PACE-90), Volume 1: 'Nominal Configuration' Hydrogeologic Parameters and Calculation Results*, SAND90-2726 (Albuquerque, NM: Nuclear Waste Repository Technology Department, Sandia National Labs., 1991).

[7] Barnard and Dockery, *Technical Summary*, p. 5–6.

[8] Thompson Engineering Company, *Review and Comment on the US Department of Energy Site Characterization Plan Conceptual Design Report*, NWPO-TR-009-88 (Carson City, NV: State of Nevada, Agency for Projects/Nuclear Waste Project Office, October 1988) (Item 329 in US DOE, DE90006793).

[9] H. Fuentes *et al.*, 'Solute Leaching from Resin/Tuff Media in Unsaturated Flow,' *Radioactive Waste Management 10*, no. 4 (June 1988): 285–320 (Item 10 in US DOE, DE 89005394); see Triay 1989; Rundberg *et al.*, 'Observation of Time Dependent Dispersion'; Voegele 1985; V. Oversby and R. McCright, 'Laboratory Experiments Designed to Provide Limits on the Radionuclide Source Term for the NNWSI Project,' in *Workshop on the Source Term for Radionuclide Migration from HLW or Spent Nuclear Fuel* (Albuquerque, NM: Lawrence Livermore National Lab., 1985) (Item 122 in US DOE, DE88004834).

[10] See, for example, E. Weeks and W. Wilson, *Preliminary Evaluation of Hydrologic Properties of Cores of Unsaturated Tuff, Test Well USW H-1, Yucca Mountain, Nevada*, Report 84-4193 (Denver, CO: US Geological Survey, 1984) (Item E111 in US DOE, NVO-96-24 (REV. 5).).

[11] See R. Loux, *State of Nevada Comments on the U.S. Department of Energy Consultation Draft Site Characterization Plan, Yucca Mountain Site, Nevada Research and Development Area, Nevada*, 2 vols. (Carson City, NV: State of Nevada, Agency for Nuclear Projects/Nuclear Waste Project Office, 1988), vol. 1, pp. I-9, I-10, II-2, II-3.

[12] US DOE, 'Project History,' in *Yucca Mountain Project Bibliography, 1988–1989.*

[13] See, for example, F. Heuze, 'Geomechanics in Hard Rock Mining-Lessons from Two Case Histories,' *Preprint, Society of Mining Engineers of AIME* 82–364 (September 1982), 14 (Item 134 in US DOE, DE89005394).

[14] I. Roxburgh, *Geology of High-Level Nuclear Waste Disposal* (London: Chapman and Hall, 1987), p. 183; hereafter cited as: Geology.

[15] Roxburgh, Geology, p. 181.

[16] R. Peters *et al.*, *Fracture and Matrix Hydrologic Characteristics of Tuffaceous Materials from Yucca Mountain, Nye County, Nevada*, SAND84-1471 (Albuquerque, NM: Sandia National Labs., 1984) (Item D170 in US DOE, NVO-96-24 (REV. 5).), p. i.

[17] D. Evans *et al.*, 'Fracture System Characterization for Unsaturated Rock,' in *Waste Management '87: Waste Isolation in the US, Technical Programs, and Public Education*, ed. R. Post (Tucson, AZ: University of Arizona Nuclear Engineering Dept., 1987) (Item 196 in US DOE, DE89005394).

[18] W. Daily and A. Ramirez, 'Geophysical Tomography for Imaging Water Movement in Welded Tuff,' in *Second International Congress on Nuclear Waste Management* (CA: Lawrence Livermore National Lab., 1986) (Item 142 in US DOE, DE90006793); K. Pruess and T. Narasimhan, 'Numerical Modeling of Multiphase and Nonisothermal Flow in Fractured Media,' in *International Conference on Fluid Flow in Fractured Rocks* (Atlanta, GA: Lawrence Berkeley Lab., 1988) (Item 43 in US DOE, DE89005394); see E. Majer *et al.*, 'VSP [Vertical Seismic Profiling] and Cross Hole Tomographic Imaging for Fracture Characterization,' in *Nuclear Waste Isolation in the Unsaturated Zone: FOCUS '89* (CA: Lawrence Berkeley Lab., 1989) (Item 108 in US DOE, DE90006793).

[19] E. Klavetter *et al.*, *Experimental Plan for Investigating Water Movement Through Fractures: Yucca Mountain Project* (Albuquerque, NM: Sandia National Labs., 1989) (Item 172 in US DOE, DE90006793).

[20] Nevada NWPO, *State of Nevada Comments on the U.S. Department of Energy Consultation Draft Site Characterization Plan, Yucca Mountain Site, Nevada: Volume 1* (Carson City, NV: US DOE, 1989) (Item 335 in US DOE, DE90006793); C. Malone, 'The Yucca Mountain Project,' *Environ. Sci. Technol. 23*, no. 12 (1989), 1453; Loux, *State of Nevada Comments*, vol. 3, p. 6.

[21] See Karasaki *et al.*, 'Building of a Conceptual Model'; J. Yow, 'Block Analysis for Preliminary Design of Underground Excavations,' in *US Symposium on Rock Mechanics* (Golden, CO: US DOE, 1990) (Item 55 in US DOE, DE91000566); Costin, 'Application of Models for Jointed Rock'; J. Tillerson *et al.*, 'Uncertainties in Sealing a Nuclear Waste Repository in Partially Saturated Tuff,' in *Proceedings of an NEA/CEC Workshop* (Paris: OECD, 1989) (Item 88 in US DOE, DE91000566); R. Zimmerman and G. Bodvarsson, 'Combined Analytical/Numerical Approaches to Solving Fluid Flow Problems in the Unsaturated Zone at Yucca Mountain,' in *International Conference for High-Level Radioactive Waste Management* (Las Vegas: US DOE, 1990) (Item 90 in US DOE, DE91000566); K. Lee *et al.*, 'Application of Geophysical Methods for Fracture Characterization,' in *International Conference for High-Level Radioactive Waste Management* (Las Vegas, NV: US DOE, 1990) (Item 114 in US DOE, DE91000566); L. Pyrak-Nolte and N. Cook, 'A Stratified Percolation Model for Saturated and Unsaturated Flow through Natural Fractures,' in *International Conference for High-Level Radioactive Waste Management* (Las Vegas, NV: US DOE, 1990) (Item 115 in US DOE, DE91000566).

[22] L. Greenwade and G. Cederberg, 'Preliminary Geochemical/Geophysical Model of Yucca Mountain,' in *Symposium on the Scientific Basis for Nuclear Waste Management* (Los Alamos, NM: Los Alamos National Lab., 1987) (Item 74 in US DOE, DE90006793).

[23] See, for example, T. Buscheck and J. Nitao, *Preliminary Scoping Calculations of Hydrothermal Flow in Variably Saturated Sign Test at the Yucca Mountain Exploratory*

Shaft Test Site (CA: Lawrence Livermore National Lab., 1988) (Item 130 in US DOE, DE90006793); K. Erickson *et al.*, 'Approximate Methods to Calculate Radionuclide Discharges for Performance Assessment of HLW Repositories in Fractured Rock,' in *Waste Management '86. Volume 2: High-Level Waste*, ed. R. Post (Albuquerque, NM: Sandia National Labs., 1986) (Item 175 in US DOE, DE89005394); see J. Cuderman, *Design and Modeling of Small Scale Multiple Fracturing Experiments* (Albuquerque, NM: Sandia National Labs., 1981) (Item 121 in US DOE, DE89005394); for use of single flow equations see, for example, R. Peters and E. Klavetter, 'Continuum Model for Water Movement in an Unsaturated Fractured Rock Mass,' *Water Resources Research 24*, no. 3 (March 1988): 416–430 (Item 38 in US DOE, DE89005394).

[24] See, for example, A. Ramirez and W. Daily, 'Electromagnetic Experiment to Map *in Situ* Water in Heated Welded Tuff: Preliminary Results.' In *Rock Mechanics: Proceedings of the 28th U.S. Symposium on Rock Mechanics* (Tucson, AZ: A.A. Balkema Publishers, 1987) (Item 154 in US DOE, DE90006793).

[25] US DOE, 'Project History,' in *Yucca Mountain Project Bibliography, 1988–1989*, p. xiii; see Norris *et al.*, 'Infiltration at Yucca Mountain.'

[26] See, for example, S. Sinnock and T. Lin, *Preliminary Bounds on the Expected Postclosure Performance of the Yucca Mountain Repository Site, Southern Nevada*, SAND84-1492 (Albuquerque, NM: Sandia National Labs., 1984), pp. 14–15.

[27] Linderfelt, *Characterization of Infiltration.*

[28] Sinnock and Lin, *Preliminary Bounds*, p. 17.

[29] Sinnock and Lin, *Preliminary Bounds*, p. 24.

[30] Sinnock and Lin, *Preliminary Bounds*, pp. 41, 16.

[31] Peters, *The Effect of Percolation Rate*, p. i; Travis *et al.*, *Preliminary Estimates of Water Flow*, p. 4.

[32] K. Pruess *et al.*, *Effective Continuum Approximation for Modeling Fluid and Heat Flow in Fractured Porous Tuff: Nevada Nuclear Waste Storage Investigations Project* (CA: Lawrence Berkeley Lab., 1988) (Item 42 in US DOE, DE89005394).

[33] Evans *et al.*, 'Fracture System Characterization for Unsaturated Rock'; J. Rulon *et al.*, *Preliminary Numerical Simulations of Groundwater Flow in the Unsaturated Zone, Yucca Mountain, Nevada* (CA: Lawrence Berkeley Lab., 1986) (Item 181 in US DOE, DE89005394); see T. Rasmussen and D. Evans, *Unsaturated Flow and Transport Through Fractured Rock Related to High-Level Waste Repositories: Final Report, Phase 2* (Tucson, AZ: Arizona University Department of Hydrology and Water Resources, 1987) (Item 216 in US DOE, DE89005394); Thompson, *Laboratory and Field Studies Related to the Radionuclide Migration Project: Project Report, October 1, 1985–September 30, 1986.*

[34] Thompson Engineering Company, *Review and Comment*, p. 13.

[35] See J. L. Younker, W. B. Andrews, G. A. Fasano, C. C. Herrington, S. R. Mattson, R. C. Murray, L. B. Ballou, M. A. Revelli, A. R. Ducharme, L. E. Shepard, W. W. Dudley, D. T. Hoxie, R. J. Herbst, E. A. Patera, B. R. Judd, J. A. Docka, and L. R. Rickertsen, *Report of the Early Site Suitability Evaluation of the Potential Repository Site at Yucca Mountain, Nevada*, SAIC-91/8000 (Washington, DC: US DOE, 1992), pp. 214, 2–13; hereafter cited as: Younker, Andrews, *et al.*

[36] Younker, Albrecht, *et al.*, pp. 181, 490, 240, 427–430, 472, 506.

[37] Pruess and Narasimhan, 'Numerical Modeling.'

[38] R. Blanchard *et al.*, *Supplementary Radiological Measurements at the Maxey Flats*

Radioactive Waste Burial Site – 1976–1977, EPA-520/5-78-011 (Montgomery, AL: US EPA, 1978), pp. 1, 29; see K. Shrader-Frechette, 'Models, Scientific Method, and Environmental Ethics', in *Upstream/Downstream*, ed. D. Scherer (Philadelphia: Temple University Press, 1990).

39 Sinnock and Lin, *Preliminary Bounds*, pp. 41, 16.

40 C. Malone, 'Geologic and Hydrologic Issues Related to Siting a Repository for High-Level Nuclear Waste at Yucca Mountain, Nevada, USA,' *Journal of Environmental Management 30* (1990), 381–396. J. Raloff, 'Fallout Over Nevada's Nuclear Destiny,' *Science News 137*, no. 1 (16 January 1990), 11–12.

41 J. Neel, 'Statement,' in US Congress, *Low-Level Radioactive* Waste Disposal, Hearings Before a Subcommittee of the Committee on Government Operations, House of Representative, 94th Congress, Second Session, February 23, March 12, and April 6, 1976 (Washington, DC: US Government Printing Office, 1976), p. 258; see also A. Weiss and P. Columbo *Evaluation of Isotope Migration – Land Burial*, NUREG/CR-1289 BNL-NUREG-51143 (Washington, DC: US NRC, 1980), p. 5; Meyer, 'Maxey Flats Radioactive Waste Burial Site: Status Report,' p. 9.

42 Neel, 'Statement,' p. 258; EMCON, *Geotechnical Investigation and Waste Management Studies.*

43 Zehner, *Hydrologic Investigation*, pp. 35, 40; Werner, *Joint Intensity Survey*, p. 45; see Wilson and Lyons, *Ground-Water Levels and Tritium Concentrations*; Shrader-Frechette 'Values and Hydrogeological Method'; Shrader-Frechette, 'Idealized Laws, Antirealism, and Applied Science'; Shrader-Frechette, 'Models, Scientific Method, and Environmental Ethics.'

44 Zehner, *Hydrologic Investigation*, p. 3.

45 Zehner, *Hydrologic Investigation*, p. 132.

46 Zehner, *Hydrologic Investigation*, p. 134.

47 Loux, *State of Nevada Comments*, vol. 2, p. 2.

48 See R. Williams, *A Technique for the Geothermic Modeling of Underground Surfaces: Nevada Nuclear Waste Storage Investigations Project* (Albuquerque, NM: Sandia National Labs., 1988) (Item 171 in US DOE, DE90006793); K. Campbell, *Kriging for Interpolation of Sparse and Irregularly Distributed Geologic Data* (Los Alamos, NM: Los Alamos National Lab., 1986) (Item 73 in US DOE, DE90006793).

49 Norris *et al.*, 'Infiltration at Yucca Mountain.'

50 W. Glassley, 'Evaluation of the Postimplacement Environment of High Level Radioactive Waste Packages at Yucca Mountain, Nevada,' in *Waste Management '89* (CA: Lawrence Livermore National Lab., 1989) (Item 150 in US DOE, DE90006793); see R. Aines, 'Estimates of Radionuclide Release from Glass Waste Forms in a Tuff Repository and the Effects on Regulatory Compliance,' in *Nuclear Waste Management II*, ed. W. Passchier and B. Bosnjakovik (Westerville, OK: American Ceramic Society Inc., 1986) (Item 155 in US DOE, DE90006793).

51 Nevada NWPO, *State of Nevada Comments*, vol. 1.

52 US DOE, *Yucca Mountain Project Bibliography, 1988–1989*, pp. 3–18.

53 DOE, NWPA-Yucca, vol. 1, pp. 1–5.

54 US DOE, *Yucca Mountain Project Bibliography, 1988–1989.*

55 Younkers, Andrews, *et al.*, pp. 3–46.

56 Thompson *et al.*, *Preliminary Upper-Bound Consequence Analysis*, p. i.

[57] Thompson *et al.*, *Preliminary Upper-Bound Consequence Analysis*, p. i.
[58] Thompson *et al.*, *Preliminary Upper-Bound Consequence Analysis*, pp. vi–vi.
[59] Smith *et al.*, *Population Risks*, p. 222.
[60] US Congress, *Nuclear Waste Program*, p. 195.
[61] D. J. Evans, US Senator, State of Washington, 'Statement,' in US Congress, *Nuclear Waste Program*, Hearings before the Committee on Energy and National Resources, US Senate, 100th Congress, April 29 and May 7, 1987, part 3 (Washington: US Government Printing Office, 1987), p. 17.
[62] US EPA, *Draft Environmental Impact Statement, 40 CFR, Part 191, Environmental Standards for Management and Disposal of Spent Nuclear Fuel, High Level, and Transuranic Radioactive Wastes* (Washington, DC: US EPA, 1982), pp. 54, 57; hereafter cited as: EPA, EIS.
[63] EPA, EIS, p. 109.
[64] R. Goble *et al.*, *Potential Retrieval of Radioactive Wastes at the Proposed Yucca Mountain Repository: A Preliminary Review of Risk Issues*, NWPO-SE-010-88 (Carson City, NV: State of Nevada, Agency for Nuclear Projects/Nuclear Waste Project Office, June 1988), p. 40.
[65] J. Emel *et al.*, *Yucca Mountain Socioeconomic Project*, #RA001–RA005 (Las Vegas, NV: Coopers and Lybrand, 1987), p. 44; D. Golding and A. White, *Guidelines on the Scope, Content, and Use of Comprehensive Risk Assessment in the Management of High-Level Nuclear Waste Transportation*, NWPO-TN-007-90 (Carson City, NV: State of Nevada, Agency for Projects/Nuclear Waste Project Office, December 1990), p. 5.
[66] Golding and White, *Guidelines*, pp. 5–8; see also S. Tuler *et al.*, *The Effects of Human Reliability in the Transportation of Spent Nuclear Fuel*, NWPO-SE-007-88 (Carson City, NV: State of Nevada, Agency for Nuclear Projects/Nuclear Waste Project Office, June 1988) (Item 305 in US DOE, DE90006793), p. 91.
[67] Emel *et al.*, *Yucca Mountain Socioeconomic Project*, pp. 108, 110; J. Emel *et al.*, *Risk Movement and Organizational Systems for High-Level Radioactive Waste Disposal: Issues and Priorities*, NWPO-SE-008-88 (Carson City, NV: State of Nevada, Agency for Projects/Nuclear Waste Project Office, September 1988), p. 9; Golding and White, *Guidelines*, pp. 2, 28; W. Burns *et al.*, *Social Amplification of Risk: An Empirical Study*, NWPO-SE-027-90 (Carson City, NV: State of Nevada, Agency for Projects/Nuclear Waste Project Office, September 1990), p. ii; Tuler *et al.*, *The Effects of Human Reliability*; see J. Petterson, *Goiania Incident Case Study*, NWPO-SE-015-88 (Carson City, NV: State of Nevada, Agency for Nuclear Projects/Nuclear Waste Project Office, June 1988); H. Peters and L. Hennen, *The Accident at Gorleben: A Case Study of Risk Communication and Risk Amplification in the Federal Republic of Germany*, NWPO-SE-012-88 (Carson City, NV: State of Nevada, Agency for Nuclear Projects/Nuclear Waste Project Office, July 1988). For discussion of the social amplification of risk see, for example, J. F. Short, 'On Defining, and Explaining Elephants (and Reactions to Them): Hazards, Disasters, and Risk Analysis,' *International Journal of Mass Emergencies and Diasters 7*, no. 3 (November 1989): 397–418; J. F. Short, 'Hazards, Risks, and Enterprise,' *Law and Society Review 24*, no. 1 (1990): 179–198; W. Freudenburg, 'Nothing Recedes Like Success? Risk Analysis and the Organizational Amplification of Risks,' *Risk: Issues in Health and Safety 3* (1992): 1–35.
[68] M. Lyverse, *Records of Wells for the Period June 13, 1984 to December 4, 1986 at*

the Maxey Flats Radioactive Waste Disposal Site, Report 87-214 (Louisville, KY: US Geological Survey, 1987), pp. 15–16.

[69] See Smith *et al.*, *Population Risks*, p. 222 and R. Lipschutz, *Radioactive Waste* (Cambridge: Ballinger, 1980), pp. 157ff.

[70] US Congress, *Safety of DOE Nuclear Facilities*, Hearing Before the Subcommittee on Energy and Power of the Committee on Energy and Commerce, House of Representatives, 101st Congress, First Session, February 22, 1989, Serial no. 101-1 (Washington, DC: US Government Printing Office, 1989); US Congress, *DOE: Pollution at Fernald, Ohio*, Hearing Before the Subcommittee on Transportation, Tourism, and Hazardous Materials of the Committee on Energy and Commerce, House of Representatives, 100th Congress, Second Session on H.R. 3783, H.R., 3784, and H.R. 3785, October 14, 1988, Serial no. 100-236 (Washington, DC: US Government Printing Office, 1989).

[71] US Congress, *DOE Nuclear Facility at Fernald, OH*, Hearing Before the Subcommittee on Energy Conservation and Power of the Committee on Energy and Commerce, House of Representatives, 99th Congress, Second Session, August 13, 1986, Serial no. 99-163 (Washington, DC: US Government Printing Office, 1987), pp. 2ff, 9ff. See note 12, chapter one, for further information on DOE violations. See also DOE, NWPA-Yucca, vol. 3, pp. C.2-6 through C.3-23 and DOE, NWPA-Hanford, pp. C.2-9 through C.8-21.

[72] H. Reid, US Senator from Nevada, 'Statement,' in *Nuclear Waste Policy Act*, Hearing Before the Subcommittee on Energy and the Environment of the Committee on Interior and Insular Affairs, House of Representatives, 100th Congress, September 18, 1987 (Washington, DC: US Government Printing Office, 1988), p. 46; hereafter cited as: NWPA-88. See the discussion and notes in chapter one for additional information and references.

[73] US Congress, *Safety of DOE Nuclear Facilities*, p. 36.

[74] Rep. P. Sharp, 'Statement,' NWPA-88, p. 3; A. Bringloe, 'Statement,' in US Congress, *High-Level Nuclear Waste Issues*, Hearings Before the Subcommittee on Nuclear Regulation of the Committee on Environment an Public Works, US Senate, 100th Congress, April 23; June 2, 3, 18, 1987 (Washington, DC: US Government Printing Office, 1987), p. 237; hereafter cited as HLNWI.

[75] C. Fultz, 'Statement,' in US Congress, *Safety of DOE Nuclear Facilities*, pp. 36, 41–45; US Congress, *DOE: Pollution at Fernald, Ohio*; Emel *et al.*, *Risk Management and Organizational Systems*, pp. 68–74.

[76] US Congress, *DOE Nuclear Facility at Fernald, OH*, pp. 143–146.

[77] US Congress, *Pollution Facility at Fernald, Ohio*, pp. 52, 27–28. See also US Congress, NWPA-88, pp. 211ff., 393ff.

[78] US Congress, *Pollution Facility at Fernald, Ohio*, p. 54.

[79] US Congress, *Pollution Facility at Fernald, Ohio*, p. 67.

[80] US Congress, *Pollution Facility at Fernald, Ohio*, p. 134.

[81] Shrader-Frechette, 'Values and Hydrogeological Method,' p. 127.

[82] W. Carey *et al.*, *Hillslope Erosion at the Maxey Flats, Radioactive Waste Disposal Site*, Report 89-4199 (Louisville, KY: US Geological Survey, 1990), p. 1.

[83] GeoTrans, *Review of Modeling Efforts*, p. 1.

[84] See, for example, C. Malone, 'Environmental Performance Assessment: A Case Study of an Emerging Methodology,' *J. Environmental Systems 19*, no. 2 (1990): 171.

[85] J. Lemons *et al.*, 'America's High-Level Nuclear Waste Repository: A Case Study of Environmental Science and Public Policy,' *Intern. J. Environmental Studies 34* (1989): 31; M. Winsor and C. Malone, 'State of Nevada Perspective on Environmental Program Planning for the Yucca Mountain Project,' *The Environmental Professional 12*: 197, 205; Golding and White, *Guidelines*.

[86] US Congress, *Nuclear Waste Program*, p. 271.

[87] US Congress, *Mission Plan for the Civilian Radioactive Waste Management Program*, Hearing Before the Subcommittee on Energy Research and Development of the Committee on Energy and Natural Resources, United States Senate, 99th Congress, First Session on the Department on Energy's Mission Plan for The Civilian Radioactive Waste Management Program, September 12, 1985 (Washington, DC: US Government Printing Office, 1986), p. 325; US Congress, *Nuclear Waste Program*, p. 245.

[88] US Congress, *Mission Plan*, p. 625.

[89] US Congress, *Mission Plan*, p. 736. See notes 174, 177.

[90] US Congress, *Nuclear Waste Program*, pp. 216, 70ff.

[91] US Congress, *Nuclear Waste Program*, p. 726.

[92] US Congress, *Nuclear Waste Program*, p. 923.

LORENZO C. SIMPSON

COMMUNITY AND DIFFERENCE: REFLECTIONS IN THE WAKE OF RODNEY KING

Sharing Marx Wartofsky's concern for matters of social justice, I offer these exploratory reflections on the occasion of his 65th birthday.

In what today gets called postmodern times, there is among what Dan Quayle called the 'cultural elite' a keen awareness of the salience of matters of difference. For example, the idea of American society as a 'melting pot', where the social ideal is for differences to be smoothed out and homogenized, is now widely regarded as an outmoded metaphor. And this sensibility is clearly salutary in that it stands as a constant challenge to those who would implicitly privilege one of the ingredients in that mixture as the one to which the others should assimilate themselves, and allows us to call such a privileging by its proper name, namely, ethnocentrism, provincialism and so on. The other side of this, however, is that the rejection of a common grid wherein the points of difference can even be articulated, let alone reconciled, exposes us to the threat of relativism and irreconcilable conflict. The question then becomes, What are the prospects for community, and not only a community of liberal tolerance but also one of genuine recognition of and respect for difference, given this emphasis upon difference and diversity? How can we negotiate differences in our public life? Unlike some, I do not believe that doing justice to matters of difference need necessarily compromise the notion of community.[1] As a gesture in the direction of making good on this claim, in this essay I want first to motivate the idea of community as a negotiated, unfinished project. Secondly, I want to consider this idea in relation to the case of Rodney King, the first trial of the involved officers and the subsequent uprising, perhaps the events in our recent national life that have done the most to fuel attitudes of cynicism and despair regarding the possibility of real community here in the U.S.

C.C. Gould and R.S. Cohen (eds.), Artifacts, Representations and Social Practice, 525–542.

I

How we respond to difference depends upon what we make of differences. First, it is important to attend to the analytic distinctions among the various proximal sources of group difference. For example, there are differences that arise from a group's position in the political economy, differences arising from class and power differentiation. These differences have their origin in contingent and often objectionable social arrangements. Some of these differences we may want to celebrate; some we may want to tolerate; some we may want to eliminate. On the other hand, we can think of group differences as products of the working, over time, of cultural and historical forces, where the features of the group in question can be thought to characterize it with some degree of independence from the political economy. These two descriptions of the provenance of difference ought perhaps best be thought of as descriptions of ideal types, for in many if not most cases both sorts of factor will be operative. This is particularly true in cases of cultural confrontation when the descriptions that serve to pick out groups for social oppression also serve to pick out distinctive cultural patterns. And of course such social targeting often occurs as a consequence of the perception of cultural difference. Debates concerning the relative roles of the experience of slavery, on the one hand, and of west African cultural retentions, on the other, in the constitution of African-American culture testify to the overdetermination of cultural distinctiveness.

Along with matters of provenance there are in addition various positions on the significance of difference no matter what its derivation may be. Two obvious possibilities are, one, in the manner of Enlightenment universalism, to make of differences between groups of people something that is accidental or inessential to what really counts about them or, two, to make of those differences something that is essential, something that essentially distinguishes one group from the other. I think that in the public realm where our concern is with the democratic regulation of social interaction, and particularly in a society that is as diverse as ours, that neither of these possibilities forms an adequate basis for a response to difference. If we think of the groups so distinguished as composed of racially and/or culturally classified members, these two possibilities have formed the basis for competing strategies for dealing with matters of racial difference. In the first case, where the universality of human characteristics is emphasized, where emphasis is placed upon similarities among persons, and matters of difference are relegated to the accidental, we tend to find ourselves having to do with abstract, deracinated, bloodless creatures somehow hovering over history and culture. Such a view clearly does not do justice to *difference*. When, on the other hand, the differences are emphasized to the point at which those features that constitute the differences are taken to determine an 'essence' for the members of a group, then we move dangerously close to racism, where the other is seen as essentially or absolutely different. "Look, a Negro!," the title of one of the chapters of Frantz Fanon's *Black Skin, White Masks*, was the expression of a white child in whose eyes Fanon's blackness *defined* Fanon, making him utterly different, alien and therefore frightening. Postmodernists, particularly feminist postmodernists, have taken the lead in indicating how such essentializing conceptions and the practices based upon such conceptions are oppressive, are sexist as well as racist. For them, differences are not matters of rigid hierarchical distinctions to be oppressively deployed, most often to the detriment of people of color, the poor and women. They are rather marks of a diversity to be celebrated. Nonetheless, I find the postmodernists in general to be unable to give a satisfying account of community. Not surprisingly, I view the accidental vs. essential dichotomy that has informed much discussion of difference as a false one to be overcome, and in the course of this essay about the possibilities of community, I hope to point to a way beyond it.

So far, speaking quite generally, I have invoked two ways of responding to difference in our public world. One is to ignore the differences, for what counts about humans is what we all share, for example, the capacity to reason, which is the basis of the dignity we accord to all persons, a dignity recognized by the attribution of fundamental rights to all citizens in a liberal democratic society. The other is to take the differences to mark a crucially important distinction among persons, one so fundamental that the search for commonalities or universal descriptions that bind persons together is seen as wrongheaded. This latter view has both its 'benign' postmodern variants and its malignant racist, sexist and xenophobic versions. I want to suggest a third response to matters of difference in our public life. This is the idea of taking seriously the possibility of an always provisional consensus effected through dialogical interaction and conversation on the part of participants who bring their differences with them to the negotiation. This would hold out the possibility of a moment of community that would not be assimilationist.

My view of the possibilities of community is inspired by my reading of the work of Jürgen Habermas, who has perhaps done the most to theorize the notion of a non-relativist, democratic practical discourse, and that of Hans-Georg Gadamer, who has provided the most compelling account of the nature of hermeneutic interpretation. Though often taken to be antagonists, together their work helps to inform a model for thinking about how the conflict that arises from difference can be attenuated if not resolved, and in a way in which the parties to the conflict need not give up what makes them special.

From Habermas, I find it helpful to keep in mind the ideas of practical discourse, of the ideal speech situation, and of a generalizable interest. From Gadamer, I find useful the ideas of dialogue and of what he calls the fusion of horizons. Let me first briefly characterize these ideas and some of their implications and then relate that discussion to the idea of community.

Habermas proposes a version of what is called a consensus theory of truth. What is perhaps most novel about his theory is the suggestion that in the very process of communicating through meaningful speech, we necessarily anticipate or presuppose a dialogue situation which is ideal and in which the truth claims of our speech could be confirmed in an unforced consensus. Such a dialogue situation is what Habermas calls an ideal speech situation and is structured by conditions which insure that the partners in dialogue are swayed only by the force of the stronger argument. The conditions of the ideal speech situation are meant to insure that no participant in dialogue be coerced to give her consent to an assertion against her will. The idea is that everyone be free to make assertions, to challenge others' assertions, to question, to reveal their inner feelings, to suggest rules for the discussion, and so on. These conditions further allow the freedom to move from a given way of framing the discussion to increasingly abstract levels of discussion and to alternative ways of framing a discussion, that is, to call into question and modify the originally accepted conceptual framework. For example, in a discussion structured by a particular normative ethical framework, participants are free to move to a metaethical discourse that calls that framework into question. The importance of this freedom becomes apparent once we recognize that the cogency of an argument will depend partly upon the conceptual or linguistic framework in which it is formulated. For example, theologically based arguments increasingly lose their cogency after the 17th century in the West. This feature of the ideal speech situation, what Habermas calls its reflexive aspect, is also important because it allows for the negotiation or forging of a framework that might accommodate seemingly opposed points of view. In general, the point of the ideal speech situation is to protect participation in discourse and acceptance of truth claims from the coercion of power.

Now, if empirical or scientific speech aims at truth, then practical discourse or social, political and ethical discussion aims at what Habermas calls a generalizable interest. Basically, interests are needs that we become aware of when they are not being met, needs, for example, like that in the black community for respectful treatment by police. Furthermore, Habermas denies that all needs or interests are merely subjective and, therefore, that they would necessarily lead to irreconcilable conflict when groups with differing interests must share the same social space. Those interests that *are* irreconcilable in this way, he refers to as particular interests. If *all* interests were sheerly particular then we *would* have to resign ourselves to what we would call "an impenetrable pluralism of apparently ultimate [and incommensurable] value orientations."[2] If this were so, we would seem to have no recourse other than to depend upon the liberal state to referee those fundamentally conflicting values. (Though, as I shall suggest, reflection upon the hermeneutic dimension from the standpoint of a possible fusion of horizons may help to soften conflict even at this more purely cultural level.) Opposed to these purely particular interests, generalizable interests for Habermas are those that can be communicatively shared, those that reflect an unforced consensus about what *all* could want. The question to keep in mind when trying to determine whether an interest is generalizable or merely particular is, Would the guiding principle or law that reflects a given interest be one such that *every* person affected by the principle can with good reason accept the consequences and side-effects that are expected to result from the *general* observance of that principle?[3] A particular interest is one that fails to meet the conditions of this test.

Generalizable interests are not simply found but are rather shaped and discovered through communication in discussions about ethical, social and political life among all persons concerned.[4] Both the social norms regulating and reflecting interests or needs and the interpretation of the needs themselves form the subject matter of practical discussions. That is, among the items discussed are the interpretations of needs. The requirements of the ideal speech situation, and in particular the required freedom to choose levels of discussion or alternative ways of framing a discussion, mean that a participant need not accept any need interpretation in which she cannot recognize what she truly wants. This points to the importance of there being a common language in which the needs of everyone concerned can be given a voice.

For example, think of women whose needs were (are?) interpreted within a linguistic or conceptual framework structured in the following way: either you restrict yourself to home and family or consider yourself selfishly ambitious. Could the need for self-respecting self-fulfillment find an appropriate interpretation in such a framework? Must the need to be responsible be interpreted as a need to relinquish autonomy and self-fulfillment? The interpretation of needs must take place within some linguistic or conceptual framework or other. This example highlights the acceptance of social arrangements (and the correspondingly prevailing conceptual resources for understanding social life) that are structured in such a way that needs are perceived as being met only in certain arbitrarily (that is, unreasonably) limited ways. Needs are interpreted as needs for x in linguistic or conceptual framework y. It is therefore important that the chosen framework not be inappropriate to express what we really want. As a final example, think of the public discourse of a society structured purely along classically liberal lines, in particular, a society where grievances can be discussed only in terms of the violation of individual rights. The advocacy of group-based needs and remedies, remedies such as affirmative action programs, is notoriously difficult to square with such a framework.

The point of social and political discussion is to search for ways to frame the needs of everyone concerned in such a way that common or generalizable interests can be found. The idea of generalizable interests then suggests one possible basis for the discursive construction of community, albeit perhaps of a rather thin form of community founded in overlapping interests. Further, the particular interests that find themselves incorporated into such a general interest do not find themselves having been erased thereby, but rather redescribed. But no matter how much it is altered, no need interpretation can be valid in which an individual cannot recognize what she wants. Such a community would respect diversity in two senses: one, the title of 'generalizable interest' would be associated only with need interpretations in which *everyone* can recognize what they can want, and two, interests not so incorporable are adjudicated according to principles of fair compromise.

Nevertheless, on this view, an interest is *presumed* generalizable or shareable until proven particular. The situation that we are faced with is, I believe, one where it is always a matter of mutual adjustment between *supposedly* valid principles or social and political norms, on the one hand, and concrete interests, on the other, so that sometimes a norm will have to be revised in order to take a newly proposed concrete interest into account, and sometimes a concrete interest will have to be rejected as ungeneralizable in light of its conflict with an achieved, supposedly rational, consensus. There probably are no *a priori* criteria, no standards that can be applied before such a discussion takes place, to tell us under what conditions to do which.

II

Communicative discourse, for Habermas, is informed by an ideal of mutual understanding. In an ideal, mutually respectful dialogue, the expression of forms of life different from our own is an invitation perhaps to expand our vision of what all humans could want and perhaps of what it is to be human. The communicative goal of mutual understanding requires what can be called a 'reversibility of perspectives,' not in the sense of my collapsing myself into you or you into me, but in the sense that I try to understand – but not necessarily agree with – what you take your life to be about and you do the same with respect to me. Such reversibility does not threaten to extinguish difference though it may well *transform* difference. The result of the conversation that effects the incorporation of the novel understanding of the world would be what Hans-Georg Gadamer would call a 'fusion of horizons' in which something new gets acknowledged. By the idea 'fusion of horizons,' Gadamer refers to the product of an interaction between a text and a reader or between members of two cultures such that a new meaning gets produced, a new meaning that cannot be reduced to the horizons of either of, say, the two cultures but which incorporates some of both. Our horizon, for example, composed of background assumptions and values that shape our interpretations of the world, can be broadened in such a way that our background assumptions and values can be situated as just one possibility alongside the different assumptions and values of a formerly unfamiliar culture; they can be situated as shaping just one way to be human. In a fusion of horizons, our vocabulary for discussing social life, and now I am suggesting cultural life as well, gets expanded.

Respectful understanding of another is *non-strategic critical* understanding. We want to put ourselves in a position to take seriously the reasons and the reasoning of the other. Though this will require an appreciation for how things look from the inside of the other's hermeneutic situation, it does not require nor does it encourage identification with the other or the *adoption* of her point of view. I think that what is involved is a *mutual* dialectic of recognition where each side has an understanding of what the other *takes* herself to be doing and can raise critical questions about it. Each side will be able to issue the other reciprocal rejoinders. There are at least two sorts of critical questions that can be raised, corresponding to two senses in which we can say the other's view of what she is doing is a subjective view. There is always the question of whether what we think we want is what we really want, and, secondly, whether what we want as individual subjects, or as members of a given group, is just or can be accommodated within a just society. We of course can also raise questions related to an immanent critique, and these of course will require an *understanding* of the cultural context from which the other's claims emerge.

Through mutually respectful interaction, parties engaged in practical discourse attempt to forge a language which is neither just the self's nor just the other's, neither just mine nor just yours.[5] We attempt to forge a common language, a language that will have as categories dimensions of experience that reflect the space of human possibilities. For a given dimension of experience, say that having to do with our understanding of the moral relationship of the individual to the community, we then seek to indicate the *alternative* positions on a common scale that you and I, ego and alter, occupy. We can engage in a mutual location or triangulation of each other. This will require a language whose constituting categories *beg questions from neither side*. In such a negotiated mutual understanding you may come to alter the way in which you understand yourself and I undoubtedly may find that listening to you leads me to alter my self-understanding. But again, we need to *understand* the other in order to know where to place her, i.e., to know *that* she is different in some respect and *how* she is different. This hermeneutic encounter with the other enables us to become more explicitly aware of who we are – and we may not like what is revealed – and enables our awareness that we occupy only one of the positions in the space of human possibilities, that is, it enables an awareness of difference. Furthermore, the common language within which this encounter comes to expression need not be seen as hegemonically usurping the original languages. Rather, think of it as a dialect in the making, or perhaps as a specialized vocabulary of each. Such a common language can serve as the basis for a community structured upon an ever expanding shared vocabulary for moral and cultural identity.

Now, it is true that we always understand from within our hermeneutic horizon and that we are therefore vulnerable to all sorts of category error when seeking to understand others. But it cannot be settled *a priori* just what the limits to our understanding might be in a given situation, and it behooves us to try to push our understanding as far as we can. We cannot know what meanings are unshared until we have made the hermeneutic attempt to understand the situation of the other, how she experiences or interprets social life, the descriptions under which she understands her actions and those of others. And indeed, there may already be more community than we think. Thinking about matters of race, white middle class Americans might find, *pace* the culture of poverty theorists, for example, that they share a great deal more with poor blacks in regard to values and aspirations than they would otherwise think.[6] Once one has made the real attempt to understand, *then* one can speak of genuine differences. How are we even to *know* that others have different values from ours absent a hermeneutic investigation, an attempt to understand from the inside?

Accordingly, I think that it is unhelpful abstractly to oppose, as do some postmodernist writers, a recognition of the fundamental opacity of the other, on the one hand, and an ideal of complete transparency or fusion on the other.[7] We can acknowledge the ways in which the other will always transcend any given understanding that we may achieve by saying that the fusion of horizons is a potentially infinite task. The recognition that full transparency may be an unrealizable ideal does not make it unreasonable for us to hope and strive to get some light to shine through a glass darkly. To foreclose the importance of such a project's success at the outset and to not act *as if* it *could* succeed in a significant measure is to fail to take seriously, is to fail to respect, the rational agency of those who share our moral community. For such respectful recognition would require our making the same attempt to maximize the reasonableness of what they say and do that we would expect them to make in our case. I think that it is important both to preserve a *sense* of difference, in order to guard against premature interpretive closure, and to maintain an ideal of understanding that urges us to recognize and take seriously our common humanity.

In line with this, we can recognize that if we think of the purposes, goals and projects of a group as what provide its members with a sense of their identity, those purposes can themselves be redescribed, relativised and expanded as a result of encounters with other groups. Think of the 'we' of a group or of a community as capturing the sense of the community as it is held together by shared purposes. The 'we' of a group is capable of redescription, redescriptions suggested by other 'we's, other 'we's with whom it comes into contact who offer not only alternative means to its ends but also redescriptions of those ends themselves, and maybe even different ends, different things to want. The recognition of the fact that there are no *a priori* limits to the redescription of ends, that there is no *a priori* established line demarcating what is internal to a self-description *vs* what is external to a description of self, suggests that mutually recognized descriptions or self-images resulting from a 'fusion of horizons' attest to the possibility of a contingently expandable sense of the 'we.'

Furthermore, there is no reason to think that we cannot actually *change* in certain ways without losing our identity in the process of such a fusion. It will be a matter of judgment what counts as change and what counts as having lost identity. However, it makes sense to conceive of identity as a sort of cluster concept in that few if any basic beliefs or professions of value, taken separately, are essential to an identity, although a large portion of such beliefs or professions of value must be maintained if a community is to retain its identity. So our identities need not be construed as being identical to our prevailing purposes.

What these remarks suggest is the possibility of a notion of humanity as an unfinished project, as the result of historical negotiation. Here it is inviting to invoke the notion of a space, a virtual space wherein a group can be unsettled about its perspectives and purposes. This is the space of a virtual or ideal 'we,' of a regulative ideal that challenges and relativises a group's purposes, a space wherein each party to an encounter can learn from the others. This allows us to appeal in a non-provincial way, in a way not limited to the 'we-intentions' or purposes of a particular group, to the human as a standard, to a tentatively achieved and revisable consensus, to a standard that is not handed down from the heavens and that is historically contingent. Postmodernism's distrust of such an appeal to the human arises from its understandable scepticism towards what can be called *essentialist* notions of humanity, towards concepts of humanity that claim to reflect an independently existing reality. But postmodernists err in assuming that such essentialist notions are the only ones.

Let me briefly indicate one of the consequences of rigidly valorizing identity preservation above all else. Drawing upon Anthony Giddens's theory of subjectivity and Julia Kristeva's theorization of the 'abject,' Iris Young formulates a notion of unconscious phobia that results from such an investment in identity.[8] She argues that even if one is free of conscious racism, sexism or homophobia, for example, one may still be subject to unconscious aversions that are fueled by meanings and reactions that are occasioned by perceived threats to basic identity security. As a result, aversive judgments are likely to be made in interactive contexts, judgments that stereotype, devalue and degrade some groups.[9] Arising from the primal repression in which the infant struggles to separate from the mother's body, from the infant's reluctant struggle to establish a separate corporeal schema and to become a self, what Kristeva calls 'abjection' produces the moment of separation, the border between the 'I' and the other. This border can be established only by 'expelling' the mother. The expelled becomes menacing because it threatens to re-enter and obliterate the border. Feelings of aversion and repulsion are the defense of the separated self, are the way of keeping the border firm and of staving off disintegration. For Kristeva, according to Young, abjection is prior to the emergence of a subject in opposition to an object, and ultimately makes that distinction possible. Abjection is thus typically unavailable for, because presupposed by, conscious control. The abject provokes fear and loathing because it exposes the border between the self and the other as fragile, and it threatens to dissolve that border. The abject arises potentially in whatever disturbs identity, and it occasions the exclusion of the other, the different, in order to maintain identity.

Although perhaps controversial in some of its details, this approach offers an intriguing framework for understanding certain aversive reactions that are seemingly recalcitrant to reason. Young suggests that racism, sexism, homophobia and so on are partly structured by such a process.[10] In the case of *groups*, social and historical factors lead to the targeting of a particular group as the abject in that members of that group come to be perceived as marked by an identity threatening difference. Young suggests as an antidote a self-understanding that is plural and heterogeneous.[11] She argues that if phobias and aversions toward others have a source in identity loss, then the urge for a homogeneous self may be part of the problem. The more expansive is one's self-conception, the less is one compelled to reject others as threateningly different. While I would not support Young's call for a fractured, decentered self, the pluralization of identity and of norms of evaluation and a notion of the 'subject in process' are what I try to capture in my more Hegelianized project of humanity as an unfinished project. As both Young and, in a somewhat different context, Yael Tamir argue, "[a]lthough cultural choices are neither easy nor limitless, cultural memberships and moral identity are not beyond choice," and they can be made the subject matter for a politicized discussion oriented towards bringing these emotional processes to discursive consciousness.[12]

I would like now to pursue, albeit in only a provisional way, the notion of humanity as an unfinished project. A crucially important issue concerns the *limits* to the redescription and change mentioned in my discussion of an historicized notion of humanity. As features of a framework that would accommodate this concern, I propose the following notions: the virtual or ideal 'we' of a humanity that is a negotiated, unfinished project functioning as an *ideal* community, a notion that makes a virtue of being open to the other, and a notion of *self-recognition* functioning as a boundary condition for the conversation wherein the negotiation takes place.

To take an example, consider gendered and cultured notions of the 'we', in particular those aspects of it having to do with identity as moral agents or with identity insofar as it is informed by moral development. As is by now well known, Carol Gilligan has argued that female moral development seems to be most adequately construed in terms of contextual notions such as those of equity, care and responsibility. On the other hand, male development, at least in Western cultures, is often articulated in terms of non-contextual principles of abstract equality and notions of rights.

Furthermore, many scholars have referred to the curious coincidence, with regard to ethics, between African world views and what in feminist literature is labeled a distinctively feminine world view. Though it is of course rather dangerous to speak of *the* African or of *the* feminine world-view, many observers have noted the similarities between, on the one hand, African and African-American communal self-understanding and, on the other, what is called here a contextual self-understanding, with both contrasting with the individualistic, atomistic self-understanding of Euro-American and western European males. So we can think of a 'we' defined in terms of a concrete ethics of care, where social life is understood to be comprised of fields of concern that are internally related by care and an ethics of responsibility and where the self is defined through relations with others, on the one hand, and, on the other, a 'we' that owes its identity to a vocabulary centered around the notion of rights, where social reality is understood to consist of isolated atoms that are only externally related through contracts based upon rights. The notion of humanity as an unfinished, negotiated project would stand in tension with the 'ghettoization' of moral identity, implying instead that, for example, notions of care, assuming that they are not already, would be appropriate for men as well as for women, and those of rights, for women as well as men.

That a genuine fusion of horizons is conceivable here, issuing in something new, as opposed to the assimilation of one 'we' to another or to an unmediated moral schizophrenia, is suggested by the philosopher Ronald Dworkin's reading of 'the right to equality'. His reading *might* be a key to how we might accomplish the preservation and combination of the horizons of care and rights, of equity and equality, in such a way that we might see how a general agreement on this matter might look. I have in mind here his understanding of the right to equality, not as the right to equal treatment, but as the right to treatment as an equal, which is the right to be treated with the same concern and respect as anyone else.[13] Treatment of others as equals requires, on his terms, not an abstraction from, but a sensitivity to, the contexts and needs of others. The achievement of such a fusion of horizons would be a step towards the production of the sort of redescription of moral identity that I have in mind.

But the question remains, What sorts of limits or controls prevent a 'we', willing to take its part in the drama of humanity as an unfinished project, from being swept along and swallowed up in the rush to fuse horizons in such a way that it loses its identity or relinquishes what matters to it most? As I pointed out earlier, though participation in practical discourse may result in the reconceptualization or redescription of what matters most to the parties in question, no one need accept a formulation in which she cannot *recognize* or find herself. So, taking the risk genuinely to raise the questions, Am I playing the right game? Are my purposes the right ones? need not entail the risk of losing oneself.

III

But what is there to be said about community here, barely a year since the infamous trial of the police officers whose brutal behavior occasioned last year's uprising in Los Angeles and almost 30 years after the Watts upheaval, where we still find ourselves as a nation groping in the dark when it comes to knowing how to think and talk about matters of race? Despite seemingly ubiquitous calls for healing and a second, federal trial whose outcome was more just, most would cede that, if anything, race relations have deteriorated since last year's events in Los Angeles.

The first Rodney King verdict, the ensuing events and the reactions to those events can be read as metaphors for what Gunnar Myrdal in 1944 aptly described as the American dilemma, by which he meant the so-called 'Negro problem' in a modern democracy. It seems clear that unless we can negotiate ways of speaking and acting with regard to matters of race, the color line that W. E. B. DuBois referred to as the problem of the 20th c. threatens to hold us all hostage well into the 21st.

Among the contributions to be made to this negotiation will be a critical exploration of the texture of the mainstream white reception of so-called black related events and issues. We must ask, What is the family of perceptual and conceptual resources and constraints that many whites tend to bring to bear when perceiving and thinking about blacks, situations involving blacks and black related issues?

What are we to make of the original jurors' failure to discern or to perceive police brutality in Rodney King's case? The expression 'police brutality' failed to refer, for them, to the record of social reality captured on the video tape. *Even* if we resist the all too understandable temptation to describe this as an explicit and conscious *racist* failure, it, I would argue, was at the very least a *race-linked* failure.

To begin to flesh out this account, let me say a word or two about the emotional basis of perception. It can be argued that there is a significant social component of our emotional lives.[14] Mature human emotions are socially constructed on several levels. Children, for example, are taught both deliberately and implicitly what their group defines as appropriate responses to certain situations and certain people. Remember Fanon's experience with the young white child. Furthermore, emotional responses involve judgments; they have what is called an intentional structure. They involve concepts, socially constructed ways of organizing and making sense of the world.[15] All of this is to say that the emotions are, partly, learned responses to situations or objects, real or imagined, and are responses to situations and objects *under a description*. Consequently, emotions are both made possible by and *limited* by the conceptual and linguistic resources of a group. The emotions that we experience reflect prevailing social constructions of reality.

So the social emotional responses of members of a group will be determined by what such persons take themselves to have observed. These responses will depend upon the way things and events get perceived. If there is a socially common perception of black bodies within a group, particularly of black male bodies, a perception informed by that group's resources for imagining the lives of blacks, then that group's socialized response to what they see will in all likelihood follow suit.

There is a circle here. Perception partly determines emotional response, and emotions partly shape perceptions. What members of a group observe will in some measure be determined, directed, shaped or partially defined by emotions.[16] Observation is not only, as Thomas Kuhn and Norwood Hanson pointed out, theory-laden, or, as Marx Wartofsky has suggested, *praxis*-laden, or even as I suggested above, determined by social context; it is also to some extent emotion-laden. What might have been the social, cultural and emotional dynamic that helped shape the jurors' perception of King? This is of course complexly determined, but as I suggested above, cultural issues of identity may well have social implications, and I think that the Rodney King case illustrates one aspect of this rather well. Issues of identity preservation may well have contributed to King's being projected as a menacing, bestial presence.

What is particularly troubling is that the jurors from Simi Valley did not necessarily constitute an idiosyncratic collection. That group of Simians was more broadly representative than we might like to think. For them, 'police brutality' could not describe the behavior they witnessed because they had learned to perceive black male bodies, particularly urban black male bodies, as those of wild animals, of beasts. So it was easy for them to see Rodney King in particular as a threatening 'bear' and indeed even sexually threatening to a white police woman. For them this could not be gratuitous brutality they were witnessing but just the force necessary to bring this brute under control, to domesticate it. For them, the police were the 'thin blue line' separating civilization from the abyss.[17] For them – and the defense attorneys were quite skillful in reinforcing this – a claim that a behavior in which a white policeman beat a black man constituted 'police brutality' is the sort of thing that would be said only by a black apologist, or by a bleeding-heart white liberal. To them it would seem perverse, idiosyncratic. In short, for those jurors the expression lacked what can be called semantic authority; it lacked descriptive power. It did not describe social reality.

But, why should *this* description of this bit of social life be taken seriously by others? After all, the defense attorneys and the jurors could say, it is just special pleading, is just another group's description of social life. Well, simply put, because it would be more just to do so than not to. As many in the tradition of what has come to be known as deliberative or discursive democracy have maintained, just social decisions are those that result from deliberations where all affected by a decision attempt to appreciate the perspectives of all others. The idea of the 'reversibility of roles' should be thought of as a device for ensuring that a claim made by a particular social group has a claim on all of us, that it be recognized as a general claim. If the results of our deliberations are to be *as if* they were produced by a genuine reversibility of perspectives, then we should do all we can to listen sympathetically to others.

We want to respect the claim of the other for the *general, public* semantic authority of the other's descriptions of social life. We do not want to treat such descriptions as reflecting merely, say, the woman's perspective, or the black perspective. This is part of what it means to accord everyone equal respect, it seems to me. And indeed, we should treat those claims as criticizable; this too is one of the consequences of that respect. Failure to seek to understand, and to take such claims seriously *as claims*, is to fail to give the other her due respect. I worry about treating sexual harassment and police brutality as merely descriptions of social interaction from the points of view of women and blacks, respectively, with no presumption that these descriptions will have *general semantic authority*. Such a restriction would allow these issues to be understood as simply idiosyncratic matters of 'their perception', where their perception has become our collective problem, a problem to be handled perhaps strategically, rather than to be understood as matters of what their perception *reveals* about our common social reality.

In the idiom of the philosophy of science, these descriptions are to be contested as observation terms, not as idiosyncratic theoretical terms.

Again, what is needed is an expanded common vocabulary, in this case, a vocabulary for describing social reality, upon which we can begin to wrest community from the ashes. What is particularly noxious is that those in power have a language that suits their purposes, but *they also claim, despite evidence to the contrary, that that language will suit everyone's purposes*, in particular those of the oppressed. The recent Republican administration's talk about 'family values' comes to mind here.[18] In Habermas's terms, what is objectionable is the claim that social arrangements that actually serve particular interests serve a generalizable interest, that what all could want is represented in the purposes served by those arrangements.

Necessary to having the relevant perceptions here may well be critical challenges to rigid identity formations. My call – implicit in the project of humanity as an unfinished, negotiated project – to not be so unyieldingly invested in a particular identity structure, if heeded, might serve to undermine the basis for the degrading images and perceptions of black bodies. It would allow those of the dominant culture not only to *know* but also to *feel* that blacks, like they, are human too.

As not only the Rodney King verdict but also the Clarence Thomas hearings indicate, neither 'sexual harassment' nor 'police brutality' have yet achieved semantic authority in a very wide sense. It is not just a matter of imagining new vocabularies, or of having them accepted within particular social groups; it is also a matter of having the power – the social authority to have one's speech taken seriously – to make those vocabularies stick and make a difference in the wider society. If that authority is not already institutionally secured, as, almost by definition, it is not for members of marginalized groups, it must be struggled for. Needless to say, it would be an understatement to say that there is much left to struggle for. Our struggle must be to forge forms of democratic community that would reflect what all could want in ways that would be inclusive of matters of difference, be they cultural, gendered, or ethnic, while being mindful of the ways that being in touch with the humanity of each contributes to the humanity of all.

University of Richmond

NOTES

[1] For example, Iris Marion Young's recent book, *Justice and the Politics of Difference*, presents a case for the incompatibility of the ideal of community and a genuine respect for difference (see especially pp. 226–236).
[2] J. Habermas, *Legitimation Crisis*, trans. Thomas McCarthy (Boston: Beacon Press, 1975), p. 108.
[3] J. Habermas, 'A Reply to My Critics,' in *Habermas: Critical Debates*, p. 257.
[4] Habermas, 'A Postscript to *Knowledge and Human Interests*,' *Philosophy of the Social Sciences* 3 (1973): 177.
[5] See Charles Taylor, 'Understanding and Ethnocentricity,' in *Philosophy and the Human Sciences* (Cambridge: Cambridge University Press, 1985), p. 125.
[6] One of the deepest flaws of neoconservatism is what might be called the fallacy of inferring values from behavior. Contrary to what many of these analysts seem to believe, values are not observable in any straightforward way, and certainly not in the way in which behavior is. Values are held within some conceptual scheme or other and emerge in behavior in a mediated way.

Rather than viewing behavior as an immediate reflection of values, it can be more adequately understood as a manifestation of valuation within the context of a particular conception of social life. In order, therefore, to understand what sort of valuing is going on, one must gain access to the view of the social world held by the agents in question. In other words, one must seek an answer to the hermeneutic question, What do they take themselves to be doing?
[7] See, for example, Young, pp. 231, 233–234.
[8] Young, pp. 131, 133–135, 141–148.
[9] *Ibid.*, p. 133.
[10] *Ibid.*, p. 145.
[11] *Ibid.*, pp. 148, 153.
[12] Yael Tamir, 'Liberal Nationalism,' *Philosophy and Public Policy 13* (Winter/Spring 1993): 4.
[13] Ronald Dworkin, *Taking Rights Seriously* (Cambridge: Harvard University Press, 1978), pp. 226–227. See also his more recent attempt to mediate the abortion controversy in *Life's Dominion* (New York: Knopf, 1993).
[14] Allison Jagger, 'Love and Knowledge,' in *Women, Knowledge and Reality*, ed. by Ann Garry and Marilyn Pearsall (Boston: Unwin Hyman, 1989), pp. 129–155.
[15] *Ibid.*, p. 135.
[16] *Ibid.*, p. 138.
[17] Robert Gooding-Williams, 'Look a Negro!,' in *Reading Rodney King/Reading Urban Uprising*, ed. Robert Gooding-Williams (New York: Routledge, 1993), p. 166.
[18] For a discussion of the 'values issue' as it concerns the plight of the American 'underclass,' see my 'Values, Respect and Recognition: On Race and Culture in the Neoconservative Debate,' *Praxis International 7* (1987): 164–173.

WILLIS H. TRUITT

PARTISANSHIP, UNIVERSALISM, AND THE DIALECTICS OF MORAL CONSCIOUSNESS

Upon the collapse of the Socialist governments of Eastern Europe and the Soviet Union, there was very quickly manifested a rash of nationalisms, parochialisms, racisms, and other forms of fiendishness. However, there was also a somewhat gentler manifestation that took the form of abandoning Marxist philosophy and bringing East European and Russian philosophy into a closer alignment with what might be called the more standardized philosophical traditions of the capitalist world. One example of this change was the formation of the International Society for Universalism in Poland, which grew out of the group of Marxist humanists who published the journal *Dialectics and Humanism*, the name of which was altered to *Dialogue and Humanism*. The purpose of this group was that of identifying and codifying a set of universal values. It is against this background that we have written this paper.

Is there a universal moral standpoint? There is indeed. It is the standpoint or the position that moral reasons and moral values, in order to be legitimate, must exhibit a character of universal disinterestedness with regard to social class interests or special group interests. This is sometimes called the principle of impartiality of the principle of fairness. This standpoint has a distinguished philosophical pedigree which I shall consider later. And, I might add, it is one of the ideological cornerstones of bourgeois ethical theory.

On the other hand, some Marxists have argued that there is no such thing as an impartial moral position and that partisanship toward the interests of the working class is moral in itself. For example, Kai Nielsen writes:

> If one is serious about defending socialist revolution and socialism generally, one must appeal, not disinterestedly to the interests of everyone alike, but to the class interests of the proletariat and their allies. Since on Marx's reckoning the proletariat is the vast majority, we are appealing to what is in fact the interests of the majority . . .[1]

Now others have argued against the position of working class partisanship that to assume such a posture is non-moral. Richard Miller,

C.C. Gould and R.S. Cohen (eds.), Artifacts, Representations and Social Practice, 543–550.

for instance, believes that for Marx, morality is not the basis for political choice and that Marx had no ethical theory. Rather, according to Miller, the most important aspects of Marx's thinking, in regard to ethics, is that he attacks pervasive philosophical assumptions about morality, and this distances him from typical moral philosophers much more than they are separated from one another.[2] Miller's argument would seem to support a kind of disinterested impartiality with regard to ethical questions, if he believes that ethical questions are of any importance at all.

It is true, of course, that Marx condemned 'bourgeois' morality as 'ideological rubbish'.[2] But this does not mean that Marx had no moral convictions. What he condemned as rubbish was the position that putative moral judgments based on the furthering of the interests of a particular social class were intrinsically non-moral and that only a universal, disinterested standpoint is truly moral. If we are thinking about the interests of a minority class (slaveholders, aristocrats, industrialists) the realization of which can be achieved only at the expense of the majority of people, then there may be merit in the claim that arguments based on class interests are non-moral or even immoral. But such is not the thrust of Marx's moral perspectives and arguments. In other words, for Marx, what is moral for a ruling class may not, probably is not, moral for a subjected class. Disinterested universalism is an illusion. As Nielsen has pointed out:

> On a Marxist account – and this is part of its canonical core – proletarian emancipation, which is the key to the creation of a classless society, will provide the conditions for a general human emancipation. The defender of the class interest thesis does not have to choose between pursuing class interest and pursuing what is disinterestedly good, for by pursuing [working] class interests, she *thereby* in fact in most circumstances also pursues what is disinterestedly good.[3]

Does pursuing working class interests always work for a disinterested good? The answer is, not if that pursuit leads to the perpetuation of dogmatism and dictatorship. We are historical witnesses to such a deformation of Marxist ethics. But because bourgeois universalism is spurious, and because it was (is) the co-inheritor of the tradition of ethics, it becomes very difficult to establish a framework for moral judgment. Let me try to show why this is so by discussing the total absence of reciprocity between Marxist ethics and the Anglo-American tradition with which I am most familiar.

It could be argued, with some validity, that the failure of Anglo-

American philosophy to come to grips with Marxist ethics is a matter of pure anti-communism. Although such a view cannot be discounted, it would be an oversimplification. I think there are deeper philosophical, or one might charge ideological, reasons. To show what more may be involved I would like to start by talking briefly about morality prior to modernity, i.e., before the development of capitalism.

The most widely accepted interpretation of Aristotle's ethics is that he created a philosophical science or discipline called 'ethics' which is still carried on today. Foreshadowings both of this discipline and of Aristotle's ethical doctrines appear in many of Plato's dialogues which Aristotle thought represented a turning away, by Socrates and Plato, from physics to ethics, from natural philosophy to moral philosophy. It is also generally held that Aristotle created another science called 'politics' and that his ethics and politics were more closely related than ours are today. In fact, for him, politics was applied ethics and the study of ethics was an introduction to the science of politics. And since Aristotle's politics does not confine itself to institutions we call political, but deals with all social institutions, a more appropriate name for it might be 'social science', a sub-division of which he called 'ethics'.

Another interpretation is, however, possible. It is that Aristotle did not create a discipline that he called 'ethics', that this was a post-Aristotelian invention. What is at work throughout the *Nicomachean Ethics* is politics, or political considerations. For him, politics meant the art and science of the *polis* or of an autonomous city. Politics ceases to exist when the *polis* ceases to exist. This means what the modern discipline that comes closest to Aristotle's conception of politics, but still only remotely, would be the moral philosophy of the eighteenth century that included jurisprudence and political economy. Aristotle's politics was both more general and more specific that ours today. Its more general aspect was that it treated all social institutions as well as ethical issues. It was more specific in that it confined itself exclusively to the *polis*. When Aristotle said that man is a political animal (*zoon politokon*), he meant that he is an animal that becomes humanized in the *polis*. For he felt that nations and empires were too large to enhance humanization; they diminish it. And finally, the transition at the end of the *Nicomachean Ethics* is not from ethics to politics, but within politics itself. It is simply a shift from considering political principles to a discussion of political constitutions, laws, educational and other institutions. A conclusion that is derivable from either of these two interpretations is that with Aristotle moral choices are not personal. They are the result of careful political training and are an essential aspect of social cooperation and cohesion. Ethics, then, for Aristotle, is everybody's business. A personal or private morality is unthinkable. Indeed, this was the thrust of Marx's condemnation of the modern (capitalist) separation of social, political, and moral 'forces' in his *Critique of Hegel's Philosophy of Right*. For Marx, like Aristotle, they were all of a piece.

The disintegration of the classical *polis* was followed by the destruction of the medieval syncretic culture in the transition from feudalism to capitalism. There followed a cultural fragmentation and privatization with the ascendancy of capitalism and industrialization. Privatization took place in all imaginable shares. First in property through politics and law. Then it spread to religion through protestantism in the right of the individual to privately interpret the scriptures, and then even to epistemology with the empiricist conception that experience is private to each individual, thus giving rise to the problems of solipsism, the existence of other minds, and the many other problems elaborated by David Hume.

This privatization and fragmentation of experience had the effect of transforming questions of why be just – a social question – into the question why should I be moral. Marx and Marxists seldom, if ever, address this last question. It only becomes a question for bourgeois ethics, i.e., the ethics of capitalist society, and notice that the question "Why should I be moral?" requires a universal answer.

Terry Eagleton has offered a parallel analysis.[5] He writes that the notoriously formal and empty nature of moral judgment in Kant is a consequence of social changes or changes in social relations that are taken for granted by ethicists in such a way that moral questions no longer have the same meaning that they had traditionally and even, at times, become unintelligible. In pre-capitalist society, the question of what one ought to do was largely determined by a person's location within the social structure. Normative requirements were self-evident and were linked organically with convention. Thus moral discourse came easily and was generally unproblematic, e.g., Aristotle unhesitatingly stated that all men agree upon the nature of the good. In most of the Classical Period and in the Middle Ages, certain obligations, duties, privileges, and responsibilities were essentially functions of the constitution or structure of society so that there is a unity between purely sociological *descriptions* of what was and *prescriptions* about what a member of society ought to do (one might note an effort to reconstitute a similar unity within the context of Marxism).

But with the development of capitalism, there also develops a reification of fact (the empiricist tradition) which stood as an obstacle to the possibility of ethics and as the main challenge that the Kantian morality attempted to overcome. With empiricism we have a reconstruction of the individual self as transcendentally prior to any particular set of social relations. These 'atomic' individual selves become ontologically prior to the social structures they inhabit (J. S. Mill's notorious decontextualized individuals). In this ideological transition, the politically and cognitively based ethics of pre-capitalist societies is refuted and the discipline of ethics enters into a period of crisis. What one ought to do could no longer be inferred from what or who one was in terms of one's place in society. There was no longer any way of describing what people actually were that might have pointed to what they could become, or what they could not become.

In this situation ethical terms began to float free from cognitive analysis (non-cognitivism)[6] and this led to the various forms of intuitionism, finalism, and moral sense theories. In Kant, for example, this problem was resolved tautologically in the form, if one can give no social answer to the question why be moral, then virtuous behavior is an end in itself: one ought to be moral because it is moral to be moral! (One ought to do one's duty because it is one's duty to do one's duty.) Or, alternatively, as above, the ethical sphere, separated from its political and cognitive roots, falls into the fuzzy and uncertain dimension of the sensibilities. As Hume said in the *Treatise on Human Nature*, morality is more properly felt than judged of. This constitutes a shrinkage of, and a highly specialized relocation of, ethics and ethical discourse. Ethics becomes autonomous but at the same time shapeless and enigmatic. The result was that moral questions became privatized and were translated into issues concerning pleasure, fulfillment, and self-realization. Or, in accordance with a dialectical principle, ethics floated off into a self-grounded, tautologous, formal or abstract sphere bearing no relation to cognitive inquiry – as a function of what Kant called 'practical reason' in contrast to 'pure reason', i.e., cognition.

All of this had its origins in that momentous transition from feudalism to capitalism. One of Kant's main arguments was that to try to derive the moral imperatives prescribing what ought to be done from what actually is done is reprehensible. There was no way any longer of moving from social practices to values, or as Hume pointed out, one could not deduce normative conclusions from factual premisses. If one derives one's ideological values from the brute facts of the market place, then the result is the worst kind of values: egoism, self-interest, aggressiveness, mutual antagonism. This is where ethics becomes duplicitously ideological, i.e., more than a direct reflection of practice. Because to say that moral values cannot follow from the facts is to insist that ideology (in this case moral ideology) no longer reflects prevailing social practices, but has the task of mystifying and legitimizing them. Thus, ethics in fact divides itself along the lines of class interests, whereas, at the same time, it must appear as a disinterested point of view, the very opposite of what it actually is. And this necessarily constrains the moral point of view in nearly all modern and contemporary bourgeois ethical theories and even in some 'Marxist' discussions of ethics. Or it leads to metaethics as a recognition of the rampant confusion of moral discourse.

The robbing of ethics of its cognitive content at first led to open subjectivism. But subjectivism being unsatisfactory gave way to a 'third person' disinterestedness in which the main task of the ethicist was seen as that of analyzing the terms of moral discourse independently of their real world connections, i.e., independent of the subject-object relation. This move was carried forward by invoking the charge of the 'naturalistic fallacy'. This could only happen when it was presupposed that there was an unbridgeable gap between the non-moral and the purely moral. At a subsequent stage in the narrowing of the scope of ethics, the discipline came very close to disappearing altogether. This was when, under the verification canons of logical positivism, it was 'discovered' that moral terms were unverifiable in principle and were, therefore, literal nonsense, or mere emotive ejaculations (emotivism).[7]

This theoretical disintegration of ethics, that took place in Anglo-American philosophical development, was the inevitable consequence of the separation of the moral and non-moral spheres of life in which 'what is' and 'what ought to be' seemed forever divided. It took the crises of the nineteen-sixties and seventies (the civil rights struggle and the Vietnam War) to compel philosophers to attempt to recover an ethical stance that would enable normative judgments to be made with any degree of confidence.

But what was recovered was meagre indeed and in reality could not move bourgeois ethics beyond the conceptual framework produced by the development of capitalism in the eighteenth century. For modern Anglo-American ethics, then, the moral point of view became distinct from self-interest, class interest or the insinuation of any interest, i.e., the moral point of view became a neutralized standpoint. And this is why Marx found the 'universality' embodied in such a moral neutrality to be 'ideological rubbish'. But this does not mean that Marx was a non-moralist. What it does mean is that the narrowed conceptual categories of most modern and contemporary Anglo-American ethics (after Hume and Kant), say, for example, in the writings of Mill, Sidgwick, Baier or Rawls, at no point comes into contact with the main thrust of the critical morality of Marxism. This is because the Marxists are anti-utilitarian (as a systematic approach to judgment), anti-formal, anti-intuitionist, anti-emotivist, anti-contractarian, etc., but they are not anti-moral. Or to put it differently, the Marxist concern with the determinative or conditioning power of economic and social systems and class interests precludes the setting of the question of what an individual ought to do as the paramount problematic of ethical inquiry. The question of what private individuals ought to do is, for Marxism, an ideological smokescreen behind which the interests of the capitalist class prevail. It is for these reasons that there has been little or no common ground on which Marxists and Anglo-American moral philosophers could engage in fruitful discourse.

In a funny twist, faced with the charge that Marxism is non-moral, Marxists might find it tempting to reply that it is the modern and contemporary ethics of capitalist society that is non-moral (as indeed it was for a while), and not Marxism – the irony of the dialectic. And this is why there is so much thrashing around in pursuit of 'applied ethics' because bourgeois ethics is, in a sense, 'unapplicable', i.e., ethics and the marketplace seem to be incommensurable.

The development of Marxist ethical theory is different. This development can be traced historically in terms of moral consciousness and action. It is evident that in historical periods of the expansion of the productive forces and the stabilization of the mode of production (and its economic structure), social consciousness is heavily determined by the prevailing material conditions and legitimating ideologies. But when class antagonisms and structural contradictions begin to accrue, questions of human action come to the forefront and moral issues become vital. What this means is that historical transitions bring with them variations in the ethical perceptions of different classes. Kant's moral admonition in Eighteenth Century Oligarchical Prussia, to treat people as ends in themselves and never as means becomes in socialism, with Lenin, a law which forbids anyone to exploit another's labor – a mockery of which was executed under the brutal pragmatisms of Stalin and Mao Tse-Tung. Or, in the revolutionary struggle of the working class, it was found necessary to subordinate ethical considerations to tactics. This is why attempting to deal with *the* Marxist moral standpoint devoid of historical context and in a purely conceptual way must fail.

And this brings us to the lately much talked about issues of human rights and democracy. In an age of peace and cooperation, we must recognize – both in capitalism and socialism – that there are no abstract, universal human rights. There are only specifically, historically determined human rights that are often in opposition to one another, e.g., the right to health care for each person as opposed to the right of corporations to profit from the sickness of a person, or the right of a person to run a business for personal gain as opposed to overall, centralized state planning, or the right of a person to protest in opposition to the demands of internal state security. In a time of peace, these oppositions and antagonisms are no longer incommensurable for socialism. In capitalism, the resolution still needs to be addressed. And this means that in capitalism, moral partisanship remains the active moral principle.

University of South Florida

NOTES

[1] Kai Nielsen, 'Arguing About Justice: Marxist Immoralism and Marxist Moralism', *Philosophy and Public Affairs, 17*, 1988, p. 216.

[2] Richard Miller, *Analyzing Marx* (Princeton, Princeton University Press, 1984), pp. 20–21.

[3] Nielsen, *op. cit.*, pp. 222–23.

[4] By syncretic culture I mean a culture or community of commonly shared moral convictions.

[5] See his article 'The Ideology of the Aesthetic' in *Art as a Social Phenomenon*, W. Truitt and B. Dziemidok, eds. (New York, Haven Publishers, 1989).

[6] Non-cognitivism I take as the position that moral statements do not state moral facts.

[7] The two crucial formulations are to be found in A. J. Ayer, *Language, Truth, and Logic* (London, 1936), and Ogden and Richards, *The Meaning of Meaning* (London, 1956).

NAME INDEX

Boston Studies in the Philosophy of Science

Editor: Robert S. Cohen, *Boston University*

1. M.W. Wartofsky (ed.): *Proceedings of the Boston Colloquium for the Philosophy of Science, 1961/1962.* [Synthese Library 6] 1963
ISBN 90-277-0021-4
2. R.S. Cohen and M.W. Wartofsky (eds.): *Proceedings of the Boston Colloquium for the Philosophy of Science, 1962/1964.* In Honor of P. Frank. [Synthese Library 10] 1965 ISBN 90-277-9004-0
3. R.S. Cohen and M.W. Wartofsky (eds.): *Proceedings of the Boston Colloquium for the Philosophy of Science, 1964/1966.* In Memory of Norwood Russell Hanson. [Synthese Library 14] 1967 ISBN 90-277-0013-3
4. R.S. Cohen and M.W. Wartofsky (eds.): *Proceedings of the Boston Colloquium for the Philosophy of Science, 1966/1968.* [Synthese Library 18] 1969
ISBN 90-277-0014-1
5. R.S. Cohen and M.W. Wartofsky (eds.): *Proceedings of the Boston Colloquium for the Philosophy of Science, 1966/1968.* [Synthese Library 19] 1969
ISBN 90-277-0015-X
6. R.S. Cohen and R.J. Seeger (eds.): *Ernst Mach, Physicist and Philosopher.* [Synthese Library 27] 1970 ISBN 90-277-0016-8
7. M. Čapek: *Bergson and Modern Physics.* A Reinterpretation and Re-evaluation. [Synthese Library 37] 1971 ISBN 90-277-0186-5
8. R.C. Buck and R.S. Cohen (eds.): *PSA 1970.* Proceedings of the 2nd Biennial Meeting of the Philosophy and Science Association (Boston, Fall 1970). In Memory of Rudolf Carnap. [Synthese Library 39] 1971
ISBN 90-277-0187-3; Pb 90-277-0309-4
9. A.A. Zinov'ev: *Foundations of the Logical Theory of Scientific Knowledge (Complex Logic).* Translated from Russian. Revised and enlarged English Edition, with an Appendix by G.A. Smirnov, E.A. Sidorenko, A.M. Fedina and L.A. Bobrova. [Synthese Library 46] 1973
ISBN 90-277-0193-8; Pb 90-277-0324-8
10. L. Tondl: *Scientific Procedures.* A Contribution Concerning the Methodological Problems of Scientific Concepts and Scientific Explanation.Translated from Czech. [Synthese Library 47] 1973 ISBN 90-277-0147-4; Pb 90-277-0323-X
11. R.J. Seeger and R.S. Cohen (eds.): *Philosophical Foundations of Science.* Proceedings of Section L, 1969, American Association for the Advancement of Science. [Synthese Library 58] 1974 ISBN 90-277-0390-6; Pb 90-277-0376-0
12. A. Grünbaum: *Philosophical Problems of Space and Times.* 2nd enlarged ed. [Synthese Library 55] 1973 ISBN 90-277-0357-4; Pb 90-277-0358-2
13. R.S. Cohen and M.W. Wartofsky (eds.): *Logical and Epistemological Studies in Contemporary Physics.* Proceedings of the Boston Colloquium for the Philosophy of Science, 1969/72, Part I. [Synthese Library 59] 1974
ISBN 90-277-0391-4; Pb 90-277-0377-9

14. R.S. Cohen and M.W. Wartofsky (eds.): *Methodological and Historical Essays in the Natural and Social Sciences.* Proceedings of the Boston Colloquium for the Philosophy of Science, 1969/72, Part II. [Synthese Library 60] 1974 ISBN 90-277-0392-2; Pb 90-277-0378-7
15. R.S. Cohen, J.J. Stachel and M.W. Wartofsky (eds.): *For Dirk Struik.* Scientific, Historical and Political Essays in Honor of Dirk J. Struik. [Synthese Library 61] 1974 ISBN 90-277-0393-0; Pb 90-277-0379-5
16. N. Geschwind: *Selected Papers on Language and the Brains.* [Synthese Library 68] 1974 ISBN 90-277-0262-4; Pb 90-277-0263-2
17. B.G. Kuznetsov: *Reason and Being.* Translated from Russian. Edited by C.R. Fawcett and R.S. Cohen. 1987 ISBN 90-277-2181-5
18. P. Mittelstaedt: *Philosophical Problems of Modern Physics.* Translated from the revised 4th German edition by W. Riemer and edited by R.S. Cohen. [Synthese Library 95] 1976 ISBN 90-277-0285-3; Pb 90-277-0506-2
19. H. Mehlberg: *Time, Causality, and the Quantum Theory.* Studies in the Philosophy of Science. Vol. I: *Essay on the Causal Theory of Time.* Vol. II: *Time in a Quantized Universe.* Translated from French. Edited by R.S. Cohen. 1980 Vol. I: ISBN 90-277-0721-9; Pb 90-277-1074-0 Vol. II: ISBN 90-277-1075-9; Pb 90-277-1076-7
20. K.F. Schaffner and R.S. Cohen (eds.): *PSA 1972.* Proceedings of the 3rd Biennial Meeting of the Philosophy of Science Association (Lansing, Michigan, Fall 1972). [Synthese Library 64] 1974 ISBN 90-277-0408-2; Pb 90-277-0409-0
21. R.S. Cohen and J.J. Stachel (eds.): *Selected Papers of Léon Rosenfeld.* [Synthese Library 100] 1979 ISBN 90-277-0651-4; Pb 90-277-0652-2
22. M. Čapek (ed.): *The Concepts of Space and Time.* Their Structure and Their Development. [Synthese Library 74] 1976 ISBN 90-277-0355-8; Pb 90-277-0375-2
23. M. Grene: *The Understanding of Nature.* Essays in the Philosophy of Biology. [Synthese Library 66] 1974 ISBN 90-277-0462-7; Pb 90-277-0463-5
24. D. Ihde: *Technics and Praxis.* A Philosophy of Technology. [Synthese Library 130] 1979 ISBN 90-277-0953-X; Pb 90-277-0954-8
25. J. Hintikka and U. Remes: *The Method of Analysis.* Its Geometrical Origin and Its General Significance. [Synthese Library 75] 1974 ISBN 90-277-0532-1; Pb 90-277-0543-7
26. J.E. Murdoch and E.D. Sylla (eds.): *The Cultural Context of Medieval Learning.* Proceedings of the First International Colloquium on Philosophy, Science, and Theology in the Middle Ages, 1973. [Synthese Library 76] 1975 ISBN 90-277-0560-7; Pb 90-277-0587-9
27. M. Grene and E. Mendelsohn (eds.): *Topics in the Philosophy of Biology.* [Synthese Library 84] 1976 ISBN 90-277-0595-X; Pb 90-277-0596-8
28. J. Agassi: *Science in Flux.* [Synthese Library 80] 1975 ISBN 90-277-0584-4; Pb 90-277-0612-3

29. J.J. Wiatr (ed.): *Polish Essays in the Methodology of the Social Sciences.* [Synthese Library 131] 1979 ISBN 90-277-0723-5; Pb 90-277-0956-4
30. P. Janich: *Protophysics of Time.* Constructive Foundation and History of Time Measurement. Translated from the 2nd German edition. 1985 ISBN 90-277-0724-3
31. R.S. Cohen and M.W. Wartofsky (eds.): *Language, Logic, and Method.* 1983 ISBN 90-277-0725-1
32. R.S. Cohen, C.A. Hooker, A.C. Michalos and J.W. van Evra (eds.): *PSA 1974.* Proceedings of the 4th Biennial Meeting of the Philosophy of Science Association. [Synthese Library 101] 1976 ISBN 90-277-0647-6; Pb 90-277-0648-4
33. G. Holton and W.A. Blanpied (eds.): *Science and Its Public.* The Changing Relationship. [Synthese Library 96] 1976 ISBN 90-277-0657-3; Pb 90-277-0658-1
34. M.D. Grmek, R.S. Cohen and G. Cimino (eds.): *On Scientific Discovery.* The 1977 Erice Lectures. 1981 ISBN 90-277-1122-4; Pb 90-277-1123-2
35. S. Amsterdamski: *Between Experience and Metaphysics.* Philosophical Problems of the Evolution of Science. Translated from Polish. [Synthese Library 77] 1975 ISBN 90-277-0568-2; Pb 90-277-0580-1
36. M. Marković and G. Petrović (eds.): *Praxis.* Yugoslav Essays in the Philosophy and Methodology of the Social Sciences. [Synthese Library 134] 1979 ISBN 90-277-0727-8; Pb 90-277-0968-8
37. H. von Helmholtz: *Epistemological Writings.* The Paul Hertz / Moritz Schlick Centenary Edition of 1921. Translated from German by M.F. Lowe. Edited with an Introduction and Bibliography by R.S. Cohen and Y. Elkana. [Synthese Library 79] 1977 ISBN 90-277-0290-X; Pb 90-277-0582-8
38. R.M. Martin: *Pragmatics, Truth and Language.* 1979 ISBN 90-277-0992-0; Pb 90-277-0993-9
39. R.S. Cohen, P.K. Feyerabend and M.W. Wartofsky (eds.): *Essays in Memory of Imre Lakatos.* [Synthese Library 99] 1976 ISBN 90-277-0654-9; Pb 90-277-0655-7
40. B.M Kedrov and V. Sadovsky (eds.): Current Soviet Studies in the Philosophy of Science. (In prep.) ISBN 90-277-0729-4
41. M. Raphael: Theorie des geistigen Schaffens aus marxistischer Grundlage. (In prep.) ISBN 90-277-0730-8
42. H.R. Maturana and F.J. Varela: *Autopoiesis and Cognition.* The Realization of the Living. With a Preface to 'Autopoiesis' by S. Beer. 1980 ISBN 90-277-1015-5; Pb 90-277-1016-3
43. A. Kasher (ed.): *Language in Focus: Foundations, Methods and Systems.* Essays in Memory of Yehoshua Bar-Hillel. [Synthese Library 89] 1976 ISBN 90-277-0644-1; Pb 90-277-0645-X
44. T.D. Thao: *Investigations into the Origin of Language and Consciousness.* 1984 ISBN 90-277-0827-4

45. A. Ishimoto (ed.): Japanese Studies in the History and Philosophy of Science. (In prep.) ISBN 90-277-0733-3
46. P.L. Kapitza: *Experiment, Theory, Practice.* Articles and Addresses. Edited by R.S. Cohen. 1980 ISBN 90-277-1061-9; Pb 90-277-1062-7
47. M.L. Dalla Chiara (ed.): *Italian Studies in the Philosophy of Science.* 1981 ISBN 90-277-0735-9; Pb 90-277-1073-2
48. M.W. Wartofsky: *Models.* Representation and the Scientific Understanding. [Synthese Library 129] 1979 ISBN 90-277-0736-7; Pb 90-277-0947-5
49. T.D. Thao: *Phenomenology and Dialectical Materialism.* Edited by R.S. Cohen. 1986 ISBN 90-277-0737-5
50. Y. Fried and J. Agassi: *Paranoia.* A Study in Diagnosis. [Synthese Library 102] 1976 ISBN 90-277-0704-9; Pb 90-277-0705-7
51. K.H. Wolff: *Surrender and Cath.* Experience and Inquiry Today. [Synthese Library 105] 1976 ISBN 90-277-0758-8; Pb 90-277-0765-0
52. K. Kosík: *Dialectics of the Concrete.* A Study on Problems of Man and World. 1976 ISBN 90-277-0761-8; Pb 90-277-0764-2
53. N. Goodman: *The Structure of Appearance.* [Synthese Library 107] 1977 ISBN 90-277-0773-1; Pb 90-277-0774-X
54. H.A. Simon: *Models of Discovery* and Other Topics in the Methods of Science. [Synthese Library 114] 1977 ISBN 90-277-0812-6; Pb 90-277-0858-4
55. M. Lazerowitz: *The Language of Philosophy.* Freud and Wittgenstein. [Synthese Library 117] 1977 ISBN 90-277-0826-6; Pb 90-277-0862-2
56. T. Nickles (ed.): *Scientific Discovery, Logic, and Rationality.* 1980 ISBN 90-277-1069-4; Pb 90-277-1070-8
57. J. Margolis: *Persons and Mind.* The Prospects of Nonreductive Materialism. [Synthese Library 121] 1978 ISBN 90-277-0854-1; Pb 90-277-0863-0
58. G. Radnitzky and G. Andersson (eds.): *Progress and Rationality in Science.* [Synthese Library 125] 1978 ISBN 90-277-0921-1; Pb 90-277-0922-X
59. G. Radnitzky and G. Andersson (eds.): *The Structure and Development of Science.* [Synthese Library 136] 1979 ISBN 90-277-0994-7; Pb 90-277-0995-5
60. T. Nickles (ed.): *Scientific Discovery.* Case Studies. 1980 ISBN 90-277-1092-9; Pb 90-277-1093-7
61. M.A. Finocchiaro: *Galileo and the Art of Reasoning.* Rhetorical Foundation of Logic and Scientific Method. 1980 ISBN 90-277-1094-5; Pb 90-277-1095-3
62. W.A. Wallace: *Prelude to Galileo.* Essays on Medieval and 16th-Century Sources of Galileo's Thought. 1981 ISBN 90-277-1215-8; Pb 90-277-1216-6
63. F. Rapp: *Analytical Philosophy of Technology.* Translated from German. 1981 ISBN 90-277-1221-2; Pb 90-277-1222-0
64. R.S. Cohen and M.W. Wartofsky (eds.): *Hegel and the Sciences.* 1984 ISBN 90-277-0726-X
65. J. Agassi: *Science and Society.* Studies in the Sociology of Science. 1981 ISBN 90-277-1244-1; Pb 90-277-1245-X

66. L. Tondl: *Problems of Semantics.* A Contribution to the Analysis of the Language of Science. Translated from Czech. 1981 ISBN 90-277-0148-2; Pb 90-277-0316-7
67. J. Agassi and R.S. Cohen (eds.): *Scientific Philosophy Today.* Essays in Honor of Mario Bunge. 1982 ISBN 90-277-1262-X; Pb 90-277-1263-8
68. W. Krajewski (ed.): *Polish Essays in the Philosophy of the Natural Sciences.* Translated from Polish and edited by R.S. Cohen and C.R. Fawcett. 1982 ISBN 90-277-1286-7; Pb 90-277-1287-5
69. J.H. Fetzer: *Scientific Knowledge.* Causation, Explanation and Corroboration. 1981 ISBN 90-277-1335-9; Pb 90-277-1336-7
70. S. Grossberg: *Studies of Mind and Brain.* Neural Principles of Learning, Perception, Development, Cognition, and Motor Control. 1982 ISBN 90-277-1359-6; Pb 90-277-1360-X
71. R.S. Cohen and M.W. Wartofsky (eds.): *Epistemology, Methodology, and the Social Sciences.* 1983. ISBN 90-277-1454-1
72. K. Berka: *Measurement.* Its Concepts, Theories and Problems. Translated from Czech. 1983 ISBN 90-277-1416-9
73. G.L. Pandit: *The Structure and Growth of Scientific Knowledge.* A Study in the Methodology of Epistemic Appraisal. 1983 ISBN 90-277-1434-7
74. A.A. Zinov'ev: *Logical Physics.* Translated from Russian. Edited by R.S. Cohen. 1983 ISBN 90-277-0734-0
    See also Volume 9.
75. G G. Granger: *Formal Thought and the Sciences of Man.* Translated from French. With and Introduction by A. Rosenberg. 1983 ISBN 90-277-1524-6
76. R.S. Cohen and L. Laudan (eds.): *Physics, Philosophy and Psychoanalysis.* Essays in Honor of Adolf Grünbaum. 1983 ISBN 90-277-1533-5
77. G. Böhme, W. van den Daele, R. Hohlfeld, W. Krohn and W. Schäfer: *Finalization in Science.* The Social Orientation of Scientific Progress. Translated from German. Edited by W. Schäfer. 1983 ISBN 90-277-1549-1
78. D. Shapere: *Reason and the Search for Knowledge.* Investigations in the Philosophy of Science. 1984 ISBN 90-277-1551-3; Pb 90-277-1641-2
79. G. Andersson (ed.): *Rationality in Science and Politics.* Translated from German. 1984 ISBN 90-277-1575-0; Pb 90-277-1953-5
80. P.T. Durbin and F. Rapp (eds.): *Philosophy and Technology.* [*Also* Philosophy and Technology Series, Vol. 1] 1983 ISBN 90-277-1576-9
81. M. Marković: *Dialectical Theory of Meaning.* Translated from Serbo-Croat. 1984 ISBN 90-277-1596-3
82. R.S. Cohen and M.W. Wartofsky (eds.): *Physical Sciences and History of Physics.* 1984. ISBN 90-277-1615-3
83. É. Meyerson: *The Relativistic Deduction.* Epistemological Implications of the Theory of Relativity. Translated from French. With a Review by Albert Einstein and an Introduction by Milić Čapek. 1985 ISBN 90-277-1699-4

84. R.S. Cohen and M.W. Wartofsky (eds.): *Methodology, Metaphysics and the History of Science.* In Memory of Benjamin Nelson. 1984 ISBN 90-277-1711-7
85. G. Tamás: *The Logic of Categories.* Translated from Hungarian. Edited by R.S. Cohen. 1986 ISBN 90-277-1742-7
86. S.L. de C. Fernandes: *Foundations of Objective Knowledge.* The Relations of Popper's Theory of Knowledge to That of Kant. 1985 ISBN 90-277-1809-1
87. R.S. Cohen and T. Schnelle (eds.): *Cognition and Fact.* Materials on Ludwik Fleck. 1986 ISBN 90-277-1902-0
88. G. Freudenthal: *Atom and Individual in the Age of Newton.* On the Genesis of the Mechanistic World View. Translated from German. 1986 ISBN 90-277-1905-5
89. A. Donagan, A.N. Perovich Jr and M.V. Wedin (eds.): *Human Nature and Natural Knowledge.* Essays presented to Marjorie Grene on the Occasion of Her 75th Birthday. 1986 ISBN 90-277-1974-8
90. C. Mitcham and A. Hunning (eds.): *Philosophy and Technology II.* Information Technology and Computers in Theory and Practice. [*Also* Philosophy and Technology Series, Vol. 2] 1986 ISBN 90-277-1975-6
91. M. Grene and D. Nails (eds.): *Spinoza and the Sciences.* 1986 ISBN 90-277-1976-4
92. S.P. Turner: *The Search for a Methodology of Social Science.* Durkheim, Weber, and the 19th-Century Problem of Cause, Probability, and Action. 1986. ISBN 90-277-2067-3
93. I.C. Jarvie: *Thinking about Society.* Theory and Practice. 1986 ISBN 90-277-2068-1
94. E. Ullmann-Margalit (ed.): *The Kaleidoscope of Science.* The Israel Colloquium: Studies in History, Philosophy, and Sociology of Science, Vol. 1. 1986 ISBN 90-277-2158-0; Pb 90-277-2159-9
95. E. Ullmann-Margalit (ed.): *The Prism of Science.* The Israel Colloquium: Studies in History, Philosophy, and Sociology of Science, Vol. 2. 1986 ISBN 90-277-2160-2; Pb 90-277-2161-0
96. G. Márkus: *Language and Production.* A Critique of the Paradigms. Translated from French. 1986 ISBN 90-277-2169-6
97. F. Amrine, F.J. Zucker and H. Wheeler (eds.): *Goethe and the Sciences: A Reappraisal.* 1987 ISBN 90-277-2265-X; Pb 90-277-2400-8
98. J.C. Pitt and M. Pera (eds.): *Rational Changes in Science.* Essays on Scientific Reasoning. Translated from Italian. 1987 ISBN 90-277-2417-2
99. O. Costa de Beauregard: *Time, the Physical Magnitude.* 1987 ISBN 90-277-2444-X
100. A. Shimony and D. Nails (eds.): *Naturalistic Epistemology.* A Symposium of Two Decades. 1987 ISBN 90-277-2337-0
101. N. Rotenstreich: *Time and Meaning in History.* 1987 ISBN 90-277-2467-9
102. D.B. Zilberman: *The Birth of Meaning in Hindu Thought.* Edited by R.S. Cohen. 1988 ISBN 90-277-2497-0

103. T.F. Glick (ed.): *The Comparative Reception of Relativity.* 1987
ISBN 90-277-2498-9
104. Z. Harris, M. Gottfried, T. Ryckman, P. Mattick Jr, A. Daladier, T.N. Harris and S. Harris: *The Form of Information in Science.* Analysis of an Immunology Sublanguage. With a Preface by Hilary Putnam. 1989 ISBN 90-277-2516-0
105. F. Burwick (ed.): *Approaches to Organic Form.* Permutations in Science and Culture. 1987 ISBN 90-277-2541-1
106. M. Almási: *The Philosophy of Appearances.* Translated from Hungarian. 1989
ISBN 90-277-2150-5
107. S. Hook, W.L. O'Neill and R. O'Toole (eds.): *Philosophy, History and Social Action.* Essays in Honor of Lewis Feuer. With an Autobiographical Essay by L. Feuer. 1988 ISBN 90-277-2644-2
108. I. Hronszky, M. Fehér and B. Dajka: *Scientific Knowledge Socialized.* Selected Proceedings of the 5th Joint International Conference on the History and Philosophy of Science organized by the IUHPS (Veszprém, Hungary, 1984). 1988 ISBN 90-277-2284-6
109. P. Tillers and E.D. Green (eds.): *Probability and Inference in the Law of Evidence.* The Uses and Limits of Bayesianism. 1988 ISBN 90-277-2689-2
110. E. Ullmann-Margalit (ed.): *Science in Reflection.* The Israel Colloquium: Studies in History, Philosophy, and Sociology of Science, Vol. 3. 1988
ISBN 90-277-2712-0; Pb 90-277-2713-9
111. K. Gavroglu, Y. Goudaroulis and P. Nicolacopoulos (eds.): *Imre Lakatos and Theories of Scientific Change.* 1989 ISBN 90-277-2766-X
112. B. Glassner and J.D. Moreno (eds.): *The Qualitative-Quantitative Distinction in the Social Sciences.* 1989 ISBN 90-277-2829-1
113. K. Arens: *Structures of Knowing.* Psychologies of the 19th Century. 1989
ISBN 0-7923-0009-2
114. A. Janik: *Style, Politics and the Future of Philosophy.* 1989
ISBN 0-7923-0056-4
115. F. Amrine (ed.): *Literature and Science as Modes of Expression.* With an Introduction by S. Weininger. 1989 ISBN 0-7923-0133-1
116. J.R. Brown and J. Mittelstrass (eds.): *An Intimate Relation.* Studies in the History and Philosophy of Science. Presented to Robert E. Butts on His 60th Birthday. 1989 ISBN 0-7923-0169-2
117. F. D'Agostino and I.C. Jarvie (eds.): *Freedom and Rationality.* Essays in Honor of John Watkins. 1989 ISBN 0-7923-0264-8
118. D. Zolo: *Reflexive Epistemology.* The Philosophical Legacy of Otto Neurath. 1989 ISBN 0-7923-0320-2
119. M. Kearn, B.S. Philips and R.S. Cohen (eds.): *Georg Simmel and Contemporary Sociology.* 1989 ISBN 0-7923-0407-1
120. T.H. Levere and W.R. Shea (eds.): *Nature, Experiment and the Science.* Essays on Galileo and the Nature of Science. In Honour of Stillman Drake. 1989
ISBN 0-7923-0420-9

121. P. Nicolacopoulos (ed.): *Greek Studies in the Philosophy and History of Science.* 1990 ISBN 0-7923-0717-8
122. R. Cooke and D. Costantini (eds.): *Statistics in Science.* The Foundations of Statistical Methods in Biology, Physics and Economics. 1990 ISBN 0-7923-0797-6
123. P. Duhem: *The Origins of Statics.* Translated from French by G.F. Leneaux, V.N. Vagliente and G.H. Wagner. With an Introduction by S.L. Jaki. 1991 ISBN 0-7923-0898-0
124. H. Kamerlingh Onnes: *Through Measurement to Knowledge.* The Selected Papers, 1853-1926. Edited and with an Introduction by K. Gavroglu and Y. Goudaroulis. 1991 ISBN 0-7923-0825-5
125. M. Čapek: *The New Aspects of Time: Its Continuity and Novelties.* Selected Papers in the Philosophy of Science. 1991 ISBN 0-7923-0911-1
126. S. Unguru (ed.): *Physics, Cosmology and Astronomy, 1300-1700.* Tension and Accommodation. 1991 ISBN 0-7923-1022-5
127. Z. Bechler: *Newton's Physics on the Conceptual Structure of the Scientific Revolution.* 1991 ISBN 0-7923-1054-3
128. É. Meyerson: *Explanation in the Sciences.* Translated from French by M-A. Siple and D.A. Siple. 1991 ISBN 0-7923-1129-9
129. A.I. Tauber (ed.): *Organism and the Origins of Self.* 1991 ISBN 0-7923-1185-X
130. F.J. Varela and J-P. Dupuy (eds.): *Understanding Origins.* Contemporary Views on the Origin of Life, Mind and Society. 1992 ISBN 0-7923-1251-1
131. G.L. Pandit: *Methodological Variance.* Essays in Epistemological Ontology and the Methodology of Science. 1991 ISBN 0-7923-1263-5
132. G. Munévar (ed.): *Beyond Reason.* Essays on the Philosophy of Paul Feyerabend. 1991 ISBN 0-7923-1272-4
133. T.E. Uebel (ed.): *Rediscovering the Forgotten Vienna Circle.* Austrian Studies on Otto Neurath and the Vienna Circle. Partly translated from German. 1991 ISBN 0-7923-1276-7
134. W.R. Woodward and R.S. Cohen (eds.): *World Views and Scientific Discipline Formation.* Science Studies in the [former] German Democratic Republic. Partly translated from German by W.R. Woodward. 1991 ISBN 0-7923-1286-4
135. P. Zambelli: *The Speculum Astronomiae and Its Enigma.* Astrology, Theology and Science in Albertus Magnus and His Contemporaries. 1992 ISBN 0-7923-1380-1
136. P. Petitjean, C. Jami and A.M. Moulin (eds.): *Science and Empires.* Historical Studies about Scientific Development and European Expansion. ISBN 0-7923-1518-9
137. W.A. Wallace: *Galileo's Logic of Discovery and Proof.* The Background, Content, and Use of His Appropriated Treatises on Aristotle's *Posterior Analytics.* 1992 ISBN 0-7923-1577-4

Boston Studies in the Philosophy of Science

138. W.A. Wallace: *Galileo's Logical Treatises.* A Translation, with Notes and Commentary, of His Appropriated Latin Questions on Aristotle's *Posterior Analytics.* 1992 ISBN 0-7923-1578-2
Set (137 + 138) ISBN 0-7923-1579-0
139. M.J. Nye, J.L. Richards and R.H. Stuewer (eds.): *The Invention of Physical Science.* Intersections of Mathematics, Theology and Natural Philosophy since the Seventeenth Century. Essays in Honor of Erwin N. Hiebert. 1992
ISBN 0-7923-1753-X
140. G. Corsi, M.L. dalla Chiara and G.C. Ghirardi (eds.): *Bridging the Gap: Philosophy, Mathematics and Physics.* Lectures on the Foundations of Science. 1992 ISBN 0-7923-1761-0
141. C.-H. Lin and D. Fu (eds.): *Philosophy and Conceptual History of Science in Taiwan.* 1992 ISBN 0-7923-1766-1
142. S. Sarkar (ed.): *The Founders of Evolutionary Genetics.* A Centenary Reappraisal. 1992 ISBN 0-7923-1777-7
143. J. Blackmore (ed.): *Ernst Mach – A Deeper Look.* Documents and New Perspectives. 1992 ISBN 0-7923-1853-6
144. P. Kroes and M. Bakker (eds.): *Technological Development and Science in the Industrial Age.* New Perspectives on the Science–Technology Relationship. 1992 ISBN 0-7923-1898-6
145. S. Amsterdamski: *Between History and Method.* Disputes about the Rationality of Science. 1992 ISBN 0-7923-1941-9
146. E. Ullmann-Margalit (ed.): *The Scientific Enterprise.* The Bar-Hillel Colloquium: Studies in History, Philosophy, and Sociology of Science, Volume 4. 1992 ISBN 0-7923-1992-3
147. L. Embree (ed.): *Metaarchaeology.* Reflections by Archaeologists and Philosophers. 1992 ISBN 0-7923-2023-9
148. S. French and H. Kamminga (eds.): *Correspondence, Invariance and Heuristics.* Essays in Honour of Heinz Post. 1993 ISBN 0-7923-2085-9
149. M. Bunzl: *The Context of Explanation.* 1993 ISBN 0-7923-2153-7
150. I.B. Cohen (ed.): *The Natural Sciences and the Social Sciences.* Some Critical and Historical Perspectives. 1994 ISBN 0-7923-2223-1
151. K. Gavroglu, Y. Christianidis and E. Nicolaidis (eds.): *Trends in the Historiography of Science.* 1994 ISBN 0-7923-2255-X
152. S. Poggi and M. Bossi (eds.): *Romanticism in Science.* Science in Europe, 1790–1840. 1994 ISBN 0-7923-2336-X
153. J. Faye and H.J. Folse (eds.): *Niels Bohr and Contemporary Philosophy.* 1994
ISBN 0-7923-2378-5
154. C.C. Gould and R.S. Cohen (eds.): *Artifacts, Representations, and Social Practice.* Essays for Marx W. Wartofsky. 1994 ISBN 0-7923-2481-1
155. R.E. Butts: *Historical Pragmatics.* Philosophical Essays. 1993
ISBN 0-7923-2498-6

Boston Studies in the Philosophy of Science

156. R. Rashed: *The Development of Arabic Mathematics: Between Arithmetic and Algebra.* 1994 (forthcoming) ISBN 0-7923-2565-6
157. I. Szumilewicz-Lachman (ed.): *Zygmunt Zawirski: His Life and Work.* With Selected Writings on Time, Logic and the Methodology of Science. 1994 (forthcoming) ISBN 0-7923-2566-4
158. N. Haq: *Names, Natures and Things.* The Alchemist Jābir ibn Ḥayyān and His *Kitāb al-Aḥjār* (Book of Stores). 1994 ISBN 0-7923-2587-7

Also of interest:

R.S. Cohen and M.W. Wartofsky (eds.): *A Portrait of Twenty-Five Years Boston Colloquia for the Philosophy of Science, 1960-1985.* 1985 ISBN Pb 90-277-1971-3

Previous volumes are still available.

KLUWER ACADEMIC PUBLISHERS – DORDRECHT / BOSTON / LONDON

MIX
Papier aus verantwortungsvollen Quellen
Paper from responsible sources
FSC® C105338

If you have any concerns about our products,
you can contact us on
ProductSafety@springernature.com

In case Publisher is established outside the EU,
the EU authorized representative is:
Springer Nature Customer Service Center GmbH
Europaplatz 3, 69115 Heidelberg, Germany

Printed by Libri Plureos GmbH
in Hamburg, Germany